Handbook of Experimental Pharmacology

Volume 161

Editor-in-Chief

K. Starke, Freiburg i. Br.

Editorial Board

G.V.R. Born, London
M. Eichelbaum, Stuttgart
D. Ganten, Berlin
F. Hofmann, München
B. Kobilka, Stanford, CA
W. Rosenthal, Berlin
G. Rubanyi, Richmond, CA

Springer
*Berlin
Heidelberg
New York
Hong Kong
London
Milan
Paris
Tokyo*

Pharmacology and Therapeutics of Asthma and COPD

Contributors
P.J. Barnes, W. Busse, L. Dziadzio, E.M. Erin,
L.B. Fernandes, R.G. Goldie, N.J. Gross, I.P. Hall,
T.T. Hansel, P.J. Henry, N.A. Jones, S. Kharitonov,
R. Lever, W. Neaville, C.P. Page, D. Spina, A.J. Tan,
R.C. Tennant, N.C. Thomson

Editors
Clive P. Page and Peter J. Barnes

 Springer

Professor
Clive P. Page
Sackler Institute of Pulmonary Pharmacology
GKT School of Biomedical Sciences
5th Floor, Hodgkin Building
Guy's Campus
London SE1 1UL, UK
e-mail: clive.page@kcl.ac.uk

Professor
Peter J. Barnes
National Heart & Lung Institute
Department of Thoracic Medicine
Imperial College
Dovehouse Street
London SW3 6LY, UK
e-mail: p.j.barnes@ic.ac.uk

With 39 Figures and 13 Tables

ISSN 0171-2004
ISBN 3-540-00464-5 Springer-Verlag Berlin Heidelberg New York

Library of Congress Cataloging-in-Publication Data
Pharmacology and therapeutics of asthma and COPD / contributors, P.J. Barnes....[et al.]. ; editors, Clive P. Page and Peter J. Barnes. p.; cm. – (Handbook of experimental pharmacology; v. 161) Includes bibliographical references and index.
ISBN 3-540-00464-5 (hard : alk. paper)
1. Antiasthmatic agents. 2. Asthma–Chemotherapy. 3. Lungs–Diseases, Obstructive–Chemotherapy. I. Page, C.P. II. Barnes, Peter J., 1946- III. Series. [DNLM: 1. Anti-Asthmatic Agents–pharmacology. 2. Asthma–drug therapy. 3. Pulmonary Disease, Chronic Obstructive–drug therapy. 4. Asthma–physiopathology. 5. Pulmonary Disease, Chronic Obstructive–physiopathology. WF 553 P5358 2004]
QP905.H3 vol. 161   [RC591]   615'. s–dc21   [615'.72]   2003054274

This work is subject to copyright. All rights are reserved, whether the whole or part of the material is concerned, specifically the rights of translation, reprinting, re-use of illustrations, recitation, broadcasting, reproduction on microfilm or in any other way, and storage in data banks. Duplication of this publication or parts thereof is permitted only under the provisions of the German Copyright Law of September 9, 1965, in its current version, and permission for use must always be obtained from Springer-Verlag. Violations are liable for Prosecution under the German Copyright Law.

Springer-Verlag is a part of Springer Science+Business Media
springeronline.com

© Springer-Verlag Berlin Heidelberg 2004
Printed in Germany

The use of general descriptive names, registered names, etc. in this publication does not imply, even in the absence of a specific statement, that such names are exempt from the relevant protective laws and regulations and free for general use.

Product liability: The publishers cannot guarantee the accuracy of any information about dosage and application contained in this book. In every individual case the user must check such information by consulting the relevant literature.

Cover design: design & production GmbH, Heidelberg
Typesetting: Stürtz AG, 97080 Würzburg

Printed on acid-free paper   27/3150 hs – 5 4 3 2 1 0

Preface

In 1991 we published volume 98 entitled *Pharmacology of Asthma* in this prestigious series. Since that time an enormous amount of information has been generated in the field of "pulmonary pharmacology", not least the recent recognition that chronic obstructive pulmonary disease (COPD) is also an important lung disease requiring considerable research effort and is a disease that has perhaps been neglected with respect to the effort put into our understanding of asthma and its treatment. This new information is brought together in the present volume written by internationally recognised authorities.

The aim of the book is to provide an in-depth review of our current understanding of the pathogenesis of both asthma and COPD. This volume also discusses the promising new options for pharmacological intervention of these diseases.

It is hoped this book will be invaluable for research scientists and clinicians involved in research and treatment of asthma and COPD. It is also envisaged that it will be a major reference resource for respiratory physicians and those involved in the development of novel drugs.

Each chapter is extensively referenced, generously illustrated with clear diagrams and photographs, and represents a state-of-the-art review of this important area of lung biology.

C. P. Page · P. J. Barnes, London

List of Contributors

(Their addresses can be found at the beginning of their respective chapters.)

Barnes, P.J. 79, 219, 303, 349
Busse, W. 273

Dziadzio, L. 273

Erin, E.M. 303

Fernandes, L.B. 3

Goldie, R.G. 3
Gross, N.J. 37

Hall, I.P. 287
Hansel, T.T. 303
Henry, P.J. 3

Jones, N.A. 53, 179

Kharitonov, S. 303

Lever, R. 245

Neaville, W. 273

Page, C.P. 53, 179, 245, 349

Spina, D. 153, 179

Tan, A.J. 303
Tennant, R.C. 303
Thomson, N.C. 125

Contents

Part 1. Current Drugs

β-Adrenoceptor Agonists 3
    L. B. Fernandes, P. J. Henry, R. G. Goldie

Anticholinergic Bronchodilators..................................... 37
    N. J. Gross

Theophylline in the Treatment of Respiratory Disease 53
    N. A. Jones, C. P. Page

Corticosteroids... 79
    P. J. Barnes

Mediator Antagonists and Anti-Allergic Drugs 125
    N. C. Thomson

Part 2. Future Drugs

New Bronchodilator Drugs 153
    D. Spina

Selective Phosphodiesterase Inhibitors in the Treatment
of Respiratory Disease 179
    N. A. Jones, D. Spina, C. P. Page

Cytokine Modulators 219
    P. J. Barnes

Inhibitors of Leucocyte–Endothelial Adhesion
as Potential Treatments for Respiratory Disease 245
    R. Lever, C. P. Page

New Antiallergic Drugs 273
    L. Dziadzio, W. Neaville, W. Busse

Pharmacogenetics, Pharmacogenomics and Gene Therapy 287
    I. P. Hall

Evaluation of New Drugs for Asthma and COPD:
Endpoints, Biomarkers and Clinical Trial Designs. 303
    P. J. Barnes, E. M. Erin, T. T. Hansel, S. Kharitonov,
    A. J. Tan, R. C. Tennant

Novel Anti-inflammatory Therapies 349
    P. J. Barnes, C. P. Page

Subject Index 373

Part 1
Current Drugs

β-Adrenoceptor Agonists

L. B. Fernandes · P. J. Henry · R. G. Goldie

Western Australian Institute for Medical Research, Pharmacology Unit,
School of Medicine and Pharmacology, The University of Western Australia,
35 Stirling Highway, 6009, Crawley, WA, Australia
e-mail: rgoldie@receptor.pharm.uwa.edu.au

1	Introduction	4
1.1	Asthma	5
1.2	COPD	5
2	The β-Adrenoceptor and Its Associated Signal Transduction Processes	6
2.1	β-Adrenoceptor Subtypes	6
2.2	Adenylyl Cyclase	7
3	Distribution and Density of β-Adrenoceptors in the Lung	8
4	β-Adrenoceptor Agonists	9
4.1	Adrenaline	9
4.2	Isoprenaline	10
4.3	Selective β_2-Adrenoceptor Agonists	10
4.4	Long-Acting β_2-Adrenoceptor Agonists	10
4.5	Delivery of β_2-Adrenoceptor Agonists	12
5	Major Sites of Therapeutic Action	12
5.1	Airway Smooth Muscle	12
5.2	Tracheobronchial Microvessels	12
6	Other Potential Therapeutic Tissue Targets	14
6.1	Inflammatory Cells	14
6.2	Secretory Cells	15
7	Adverse Reactions to β_2-Adrenoceptor Agonists	16
7.1	Primary Adverse Reactions	16
7.2	Other Significant Direct Adverse Reactions	16
7.3	Stereoisomers of Salbutamol	16
8	Combination Therapy	17
8.1	Long-Acting β_2-Adrenoceptor Agonists and Glucocorticoids	17
8.2	D_2-Receptor Agonists	19
9	Pharmacogenetics of β_2-Adrenoceptor Agonists in Asthma	19
10	Clinical Application	21
10.1	Asthma	21
10.2	COPD	23
11	Concluding Remarks	24
	References	25

Abstract β_2-Adrenoceptor agonist bronchodilators are widely used in the treatment of both asthma and chronic obstructive pulmonary disease (COPD). They provide rapid and effective symptom relief principally by opposing the bronchoconstriction induced by excitatory airway mediators. While asthma is associated with episodic increases in baseline airway tone, it is defined as an inflammatory disease of the airways, and accordingly, therapy generally involves the use of anti-inflammatory as well as bronchodilator therapy. When delivered directly to the lungs by inhalation, β_2-adrenoceptor agonist bronchodilators provide rapid and effective reversal of acute airway obstruction caused by bronchoconstriction, with minimal acute adverse effects on the patient. Importantly, the rapidity of relief provided by inhaled β_2-adrenoceptor agonists is a significant feature of this class of drugs and helps to explain why they are used so widely to reverse the potentially life-threatening effects of bronchoconstriction in asthma. Short-acting β_2-adrenoceptor agonist bronchodilators, such as salbutamol, have durations of action of 4–6 h and provide rapid symptom relief in a large proportion of asthmatics. Long-acting β_2-adrenoceptor agonists, which include salmeterol, have durations of action of up to 12 h and provide effective treatment in asthmatic individuals whose symptoms were not adequately managed with short-acting agents. Given the complementary roles of β_2-adrenoceptor agonists and glucocorticoids in the treatment of asthma, combination therapy using these drugs has been shown to improve disease control and lower exacerbation rates. Accordingly, the consensus is that these β_2-adrenoceptor agonist bronchodilators should not be used as monotherapy in asthma in any but the most mild of cases. Indeed, the powerful bronchodilator actions of short- and long-acting β_2-adrenoceptor agonists may mask the onset and/or deterioration of airway inflammation in asthmatics. COPD is characterized by shortness of breath, cough, sputum production and exercise limitation, with acute exacerbations resulting in worsening of symptoms. While short-acting inhaled β_2-adrenoceptor agonist bronchodilators reduce respiratory symptoms and improve the quality of life in COPD patients, these drugs fail to alter the progression of this disease in the long term. In these individuals however, the recent introduction of long-acting β_2-adrenoceptor agonists has had a positive impact on quality of life. The combined use of bronchodilator/corticosteroid regimes further assists in the management of COPD.

Keywords Asthma · Chronic obstructive pulmonary disease · β_2-Adrenoceptor agonists · Delivery devices · Combination therapy

1
Introduction

β_2-Adrenoceptor agonist bronchodilators are widely used in the treatment of both asthma and chronic obstructive pulmonary disease (COPD) where they can provide rapid and effective symptom relief. The major role of these agents in these diseases is to oppose airway smooth muscle contraction caused by a variety of excitatory airway mediators.

1.1
Asthma

In addition to elevated bronchial tone, a major defining characteristic of asthma is that it is an inflammatory airway disease. The combined effects of these two elements results in a disease involving reversible airway obstruction which may cause persistent systems such as dyspnea, chest tightness, wheezing, cough and sputum production. Variable airflow obstruction and airway hyperresponsiveness to both endogenous and exogenous stimuli are also distinguishing features of asthma. Chronic inflammation of the airways is accompanied by structural changes to the bronchial wall and these phenomena are collectively referred to as airway remodelling. These changes to the normal architecture of airway mucosal and submucosal tissues underlie the development and continued maintenance of this disease. The inflammatory response in the airways is characterized by mucosal and bronchial wall oedema, lymphocyte and eosinophil infiltration, damage to and loss of airway epithelium, and hypersecretion of mucus that may cause plugging and occlusion of the airway lumen. Accordingly, asthma therapy in the modern era has tended to emphasize anti-inflammatory drug approaches since these are predicted to have a positive impact on processes driving airway remodelling.

However, it must be remembered that asthma also involves episodic increases in baseline airway tone resulting from active shortening of airway smooth muscle, causing reduced bronchial airflow and thus impaired lung ventilation. Contraction of airway smooth muscle, like airway wall remodelling, oedema and hypersecretion of mucus, contributes significantly to bronchial obstruction. As a result, the use of bronchodilators remains at the forefront of modern approaches to asthma therapy. This is despite the continuing research and therapeutic emphasis on airway inflammation as a driver of asthma progression and maintenance.

β_2-Adrenoceptor agonist bronchodilators in particular, delivered directly to the airways by inhalation, provide rapid and effective reversal of acute airway obstruction caused by bronchoconstriction, with minimal acute adverse effects on the patient. Importantly, the rapidity of relief provided by inhaled β_2-adrenoceptor agonists is a significant feature of this class of drugs and helps to explain why they are used so widely to reverse the potentially life-threatening effects of bronchoconstriction in asthma.

1.2
COPD

COPD is defined as "a disease state characterized by airflow limitation that is not fully reversible. The airflow limitation is usually both progressive and associated with an abnormal inflammatory response of the lungs to noxious particles or gases" [Global Initiative for Chronic Obstructive Lung Disease (GOLD) Workshop Report 2001]. COPD is characterized by inflammation throughout the respiratory system, including bronchial and bronchiolar airways, parenchyma and pulmonary vasculature. There are increased numbers of macrophages, T lymphocytes (predominantly CD8$^+$) and neutrophils in the airways (Jeffery 1998; Pesci et al. 1998). These activated inflammatory cells release a variety of mediators, including leukotriene B$_4$ (LTB$_4$), interleukin (IL)-8 and tumour necrosis factor (TNF)-α that contribute to widespread degenerative structural changes to the respiratory tract and promote neutrophilic inflammation (Keatings et al. 1996; Mueller et al. 1996; Yamamoto et al. 1997; Pesci et al. 1998; Hill et al. 1999). There is also evidence in COPD for an imbalance in proteases that digest elastin and other structural proteins and antiproteases that protect against this damage (Chapman and Shi 2000). Oxidative stress may also contribute to the pathogenesis of this disease. It is likely that cigarette smoke and other COPD risk factors initiate an inflammatory response in the airways that can lead to this disease. As in asthma, the airflow obstruction seen in COPD patients is often accompanied by airway hyperresponsiveness. The chronic airflow obstruction particularly affects small airways, and lung elasticity is also lost due to enzymatic destruction of the lung parenchyma, resulting in progressively worsening emphysema. Thus, COPD is characterized by shortness of breath, cough, sputum production and exercise limitation, with acute exacerbations resulting in worsening of symptoms. Inhaled bronchodilators, including β_2-adrenoceptor agonists, have been shown to reduce respiratory symptoms and improve the quality of life for COPD patients and are recommended in the management of acute exacerbations of this disease. However, in the long term, β_2-adrenoceptor agonists fail to alter the progression of this disease.

Much of what is known of the basic and clinical pharmacology of β_2-adrenoceptor agonist bronchodilators, such as salbutamol, was established in studies from the 1970s and 1980s. The actions of β_2-adrenoceptor agonists in the lung, particularly in relation to asthma, have previously been extensively reviewed by us (Goldie et al. 1991) and it is appropriate to revisit some of the issues raised at that time. However, significant advances in β_2-adrenoceptor agonist development and therapy in asthma and COPD have been made in recent years and these will also be highlighted in this review.

2
The β-Adrenoceptor and Its Associated Signal Transduction Processes

2.1
β-Adrenoceptor Subtypes

The β-adrenoceptor is a single polypeptide glycoprotein moiety (Gilman 1987), embedded in the plasma membrane of the cell (Stiles et al. 1984). At least three functionally distinct subtypes (β_1, β_2, and β_3) are known to exist and have been cloned. β-Adrenoceptors are found throughout the respiratory tract and, in human bronchial smooth muscle, are entirely of the β_2-subtype (Harms 1976; Goldie et al. 1986b). While β_1-adrenoceptors predominate in human cardiac tis-

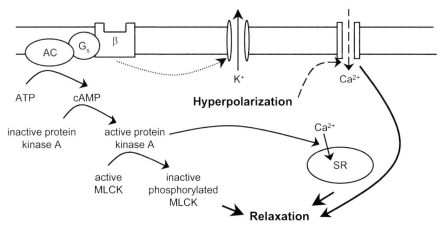

Fig. 1 Main pathways promoting airway smooth muscle relaxation associated with the β-adrenoceptor-effector system. The *dashed lines* indicate inhibition of Ca^{2+} entry into the cell via voltage-gated Ca^{2+} channels. The *dotted line* indicates that the exact mechanism is yet to be defined. (Adapted from Thirstrup 2000)

sue, a small functional population of $β_2$-adrenoceptors is also present. More recently, $β_3$-adrenoceptors have been found in cardiac muscle, intestinal smooth muscle as well as in white and brown fat. However, $β_3$-adrenoceptor mRNA has not been detected in human lung (Mak et al. 1996) and $β_3$-adrenoceptor agonists failed to induce relaxation in human isolated bronchi (Martin et al. 1994).

2.2
Adenylyl Cyclase

Agonist-binding to all β-adrenoceptors subtypes activates the membrane-bound enzyme adenylyl cyclase via a guanine nucleotide regulatory protein (G_s) to convert adenosine 5′-triphosphate (ATP) to cyclic adenosine 3′,5′-monophosphate (cAMP; Benovic et al. 1985) (Fig. 1). Cyclic AMP is produced continuously following β-adrenoceptor activation and is inactivated by hydrolysis to 5′-AMP, through the action of phosphodiesterases. Cyclic AMP acts as an intracellular messenger to regulate many aspects of cellular function including the contraction of smooth muscle. Thus, cAMP activates cAMP-dependent protein kinases to modify cellular function by phosphorylation. For example, relaxation of airway smooth muscle results from phosphorylation, and thus inactivation of myosin light chain kinase, which precludes its interaction with the contractile protein myosin (Thirstrup 2000). In addition, β-adrenoceptor agonists may also decrease airway smooth muscle tone via an interaction with plasma membrane potassium channels (Fig. 1). This results in hyperpolarization of the cell membrane and inhibition of calcium influx via voltage-dependent calcium channels.

Rho, a small monomeric G protein of the Ras superfamily of guanosine triphosphate (GTP)ases, has been shown to control airway smooth muscle tone following activation of G protein-coupled receptors (Seasholtz et al. 1999; Amano et al. 2000; Schmitz et al. 2000; Somlyo and Somlyo 2000; Pfitzer 2001). Rho activates Rho-kinase (Kimura et al. 1996) which phosphorylates and thus inhibits myosin light chain phosphatase. The latter enzyme acts to de-phosphorylate myosin light chain and promote smooth muscle relaxation. However, the result of Rho-kinase activity is blockade of this process and thus maintenance of smooth muscle tone (Seasholtz et al. 1999; Amano et al. 2000; Somlyo and Somlyo 2000). While Y27632, an inhibitor of Rho-kinase, has been shown to potentiate the relaxant effects of β-adrenoceptor agonists in airway smooth muscle (Iizuka et al. 2000; Nakahara et al. 2000), a linkage between Rho-kinase and the β-adrenoceptor-effector system has yet to be clearly defined.

3
Distribution and Density of β-Adrenoceptors in the Lung

Radioligand binding and autoradiographic studies have been critical to evaluations of the distribution of β-adrenoceptors in both animal and human lung tissue. Over 20 years ago, Rugg and coworkers (1978) using rabbit and rat lung membranes and Szentivanyi (1979) using human lung membrane preparations demonstrated the presence of high densities of β-adrenoceptors (Rugg et al. 1978; Szentivanyi 1979). Subsequent studies in guinea-pig (Barnes et al. 1980; Engel et al. 1981) and hamster lung (Benovic et al. 1983) confirmed that the lung was densely populated with β-adrenoceptors. Furthermore, lung parenchyma contained heterogeneous populations of both β_1- and β_2-adrenoceptors (Dickinson et al. 1981; Engel et al. 1981; Carswell and Nahorski 1983). However, the most important information, i.e. the location of these receptors within the normal structure of the lung, was not revealed in detail until autoradiographic assessments were completed.

Light-microscopic autoradiography (Young and Kuhar 1979) has enabled the detection and localization of β-adrenoceptor subtypes in mammalian airways and lung parenchyma from many animal species including ferret (Barnes et al. 1982), guinea-pig (Goldie et al. 1986a), rabbit (Barnes et al. 1984), rat (Finkel et al. 1984), mouse (Henry et al. 1990) and pig (Goldie et al. 1986a), as well as from the human (Carstairs et al. 1985; Spina et al. 1989).

In human lung, the greatest density of β-adrenoceptors was found in alveolar septae (Carstairs et al. 1984). Approximately 78% of the total human lung tissue volume consists of alveolar tissue, with only 8% being vascular smooth muscle, 3% as airway smooth muscle and the remaining 11% is connective tissue and cartilage (Bertram et al. 1983). Furthermore, β-adrenoceptor numbers were 3 times higher in the alveolar wall than over bronchial smooth muscle and 1.4 times higher than over bronchiolar smooth muscle. Thus, approximately 96% of the β-adrenoceptor population was located in alveolar tissue. However, as might be expected, significant numbers of β-adrenoceptors were found in bronchial and bronchiolar airway smooth muscle, as well as in airway epithelium and in vascular endothelium and smooth muscle.

4
β-Adrenoceptor Agonists

4.1
Adrenaline

The β_2-adrenoceptor agonists used in the therapy of both asthma and COPD are all structurally related to the endogenous catecholamine adrenaline (Fig. 2). It is the powerful β_2 effects of adrenaline that are most important in asthma treatment, although α- and β_1-adrenoceptors are also activated. Adrenaline is not effective when taken orally because it is rapidly metabolized by gastrointestinal and hepatic monoamine oxidase (MAO). Accordingly, in emergency conditions, when the use of adrenaline in asthma is necessary, it is given by parenteral injection.

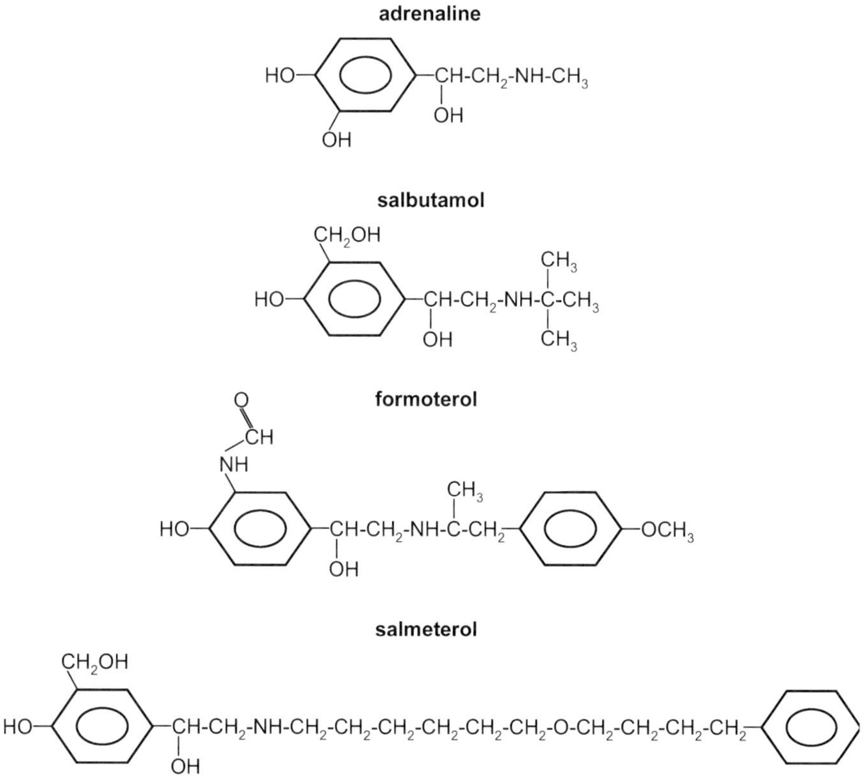

Fig. 2 β-Adrenoceptor agonists

4.2
Isoprenaline

Isoprenaline is an *N*-isopropyl derivative of adrenaline that has no significant agonist effect at α-adrenoceptors, and was the first β-adrenoceptor-selective agonist introduced into asthma therapy. Isoprenaline, like adrenaline, is a catecholamine and so is not useful orally since it is metabolized rapidly by catechol-*O*-methyltransferase (COMT). However, significant cardiac stimulation induced via activation of β_1-adrenoceptors, even after inhalational administration, reduces its acceptability as a bronchodilator.

4.3
Selective β_2-Adrenoceptor Agonists

A major advance occurred with the development and introduction of β_2-adrenoceptor-selective agonists that could be given orally or by inhalation and had extended durations of action compared with adrenaline or isoprenaline. Orciprenaline was the first of this new generation of bronchodilator amines and is a resorcinol derivative rather than a catecholamine and thus is not inactivated by extraneuronal COMT. Furthermore, this tertiary amine is not metabolized by MAO. Although orciprenaline is metabolized by sulpho-conjugation, enough free drug is absorbed to make it an effective oral bronchodilator. However, selectivity for the β_2-adrenoceptor is only slightly improved over that of isoprenaline (Mcevoy et al. 1973).

Salbutamol (Brittain et al. 1968) and terbutaline (Bergman et al. 1969) possess much greater selectivity for β_2-adrenoceptors than orciprenaline. Both compounds are active orally, as well as by inhalation and intravenous injection. All of the newer orally active β_2-adrenoceptor agonists are used in various inhalation formulations. The great virtue of being able to administer these new β_2-adrenoceptor agonists by inhalation is that small but highly effective doses can be delivered to the lung, giving the desired therapeutic effect rapidly (onset 5–10 min) and with an extended duration of action (up to 6 h). Negligible plasma concentrations of the active drug result from these inhaled doses.

4.4
Long-Acting β_2-Adrenoceptor Agonists

Members of the first generation of β_2-adrenoceptor-selective agonist bronchodilators, such as salbutamol, are considered to be short-acting since they have a duration of action of 4–6 h. These agents are effective in a large proportion of asthmatics where they provide rapid symptom relief. Short-acting β_2-adrenoceptor agonists are also used to prevent asthma exacerbations that may, for example, be triggered by exposure to cold air or exercise. However, in some asthmatic patients, these short-acting drugs need to be administered several times a day for adequate symptom relief. In addition, treatment of nocturnal symptoms in susceptible patients may be problematic given the relatively short duration of action of these drugs. In order to provide effective treatment in asthmatic individuals whose symptoms were not adequately managed with short-acting agents, β_2-adrenoceptor agonist bronchodilators were developed that had durations of action of up to 12 h. Their use in asthma has recently been reviewed (Kips and Pauwels 2001). Long-acting β_2-adrenoceptor agonists have been shown to be more effective than salbutamol in reducing asthma symptoms and improving lung function in mild-to-moderate asthmatics (Pearlman et al. 1992; Leblanc et al. 1996; Taylor et al. 1998). Since long-acting β_2-adrenoceptor agonists are relatively new drugs, their safety and efficacy has been compared to both theophylline and the cysteinyl leukotriene receptor antagonist zafirlukast. Here, salmeterol has been found to provide significantly greater improvement in the management of asthma symptoms than either theophylline or zafirlukast (Davies et al. 1998; Busse et al. 1999).

Salmeterol (Ullman and Svedmyr 1988) (Fig. 2) was specifically designed to prolong the duration of action of the short-acting β_2-adrenoceptor agonist salbutamol. While formoterol (Fig. 2) was not deliberately designed to have this property, it was found to have a 12-h duration of action when administered by inhalation (Hekking et al. 1990). The agonist activity profiles of formoterol and salmeterol are distinct, suggesting that the extended duration of action of these agents is achieved via different mechanisms. Furthermore, salmeterol is a partial β_2-adrenoceptor agonist, whereas formoterol has higher intrinsic activity and is a full agonist (Linden et al. 1993; Naline et al. 1994). Unlike salbutamol, which is hydrophilic, both salmeterol and formoterol possess lipophilic properties which allow them to remain in airway tissues in close proximity to the β_2-adrenoceptor. This partly explains why the duration of action of these long-acting β_2-adrenoceptor agonist bronchodilators is at least 12 h. In addition to being lipophilic, formoterol is also water soluble, ensuring rapid access to the β_2-adrenoceptor and thus rapid bronchodilator activity. In contrast, salmeterol, being highly lipophilic, probably diffuses more slowly to the β_2-adrenoceptor laterally through the cell membrane and has a slower onset of action (Lotvall 2001; Kottakis et al. 2002). Salmeterol contains the saligenin head of salbutamol that binds to the active site of the β_2-adrenoceptor. This saligenin head is coupled to a long aliphatic side chain that significantly increases the lipophilicity of salmeterol. The side chain then binds to a discrete "exosite" that anchors it to the receptor and enables repetitive receptor activation (Green et al. 1996). The exact mechanism by which formoterol exerts its prolonged effects is unclear but may result from its lipophilicity, allowing formoterol to enter the plasma membrane where it is held for a prolonged period. From this site, formoterol diffuses over time to activate the β_2-adrenoceptor. Inhaled formoterol has also been shown to have a longer duration of action than orally administered formoterol, probably as a result of high concentrations building in the bronchial periciliary fluid (Anderson et al. 1994).

4.5
Delivery of β_2-Adrenoceptor Agonists

The metered dose inhaler (MDI) is the most commonly prescribed patient-operated device for the delivery of asthma therapies, including β_2-adrenoceptor agonists, with approximately 340 million units used every year world-wide (Partridge 1994; Woodcock 1995). The popularity of the MDI is by virtue of its effectiveness and ability to deliver a wide range of drugs (Woodcock 1995). Indeed, salbutamol, when administered via MDI and spacer is as effective and more cost-effective when compared with delivery via a nebulizer (Newman et al. 2002). MDIs were formulated with a combination of the chlorofluorocarbon (CFC) propellants 11 and 12. However, because of the ozone-depleting potential of CFCs, the ozone-friendly propellant hydrofluoroalkane (HFA) 134a has now replaced CFCs. Studies have demonstrated that the effectiveness and safety of salbutamol/HFA 134a is comparable to that of salbutamol/CFC (Hawksworth et al. 2002; Langley et al. 2002).

5
Major Sites of Therapeutic Action

5.1
Airway Smooth Muscle

Both contraction studies in vitro (Goldie et al. 1982) and autoradiographic studies have confirmed that only β_2-adrenoceptors are expressed and mediate relaxation to β-adrenoceptor agonists in human airway smooth muscle (Spina et al. 1989). This explains why β_1-adrenoceptor-selective agents such as prenalterol given intravenously, elevate heart rate without inducing bronchodilatation (Lofdahl and Svedmyr 1982). Agonist stimulation of β_2-adrenoceptors reverses airway obstruction in asthmatics primarily by causing relaxation of central and peripheral airway smooth muscle. However, given that β_2-adrenoceptors are widely distributed throughout the lung, the beneficial actions of β_2-adrenoceptor agonists may in part be the result of actions at other sites. For example, reversal or blunting of the actions of inflammatory mediators causing airway wall oedema would be expected to relieve that component of bronchial obstruction.

5.2
Tracheobronchial Microvessels

It has long been established that airway wall oedema is an obligatory accompaniment to airway inflammation. This phenomenon involves the exudation of plasma from tracheobronchial microvessels into the extravascular space in these airways and thereby contributes significantly to airway narrowing in asthma and possibly to epithelial shedding and bronchial hyperresponsiveness (Persson et al. 1986). The infiltration of inflammatory cells from the vascular space into the submucosa and thence into the mucosa itself, is a natural consequence of this increased microvascular permeability in response to neuropeptides, histamine, endothelin-1 and other mediators of asthma. The tracheobronchial circulation consists of a subepithelial capillary network, with postcapillary venules as the main site of plasma extravasation. While the mechanisms that provoke plasma protein extravasation are incompletely understood, a variety of stimuli such as antigen, histamine, platelet-activating factor (PAF) and substance P induce direct plasma extravasation from bronchial microvessels. The targets of such mediators and thus the major sites of microvascular plasma leakage leading to a generalized airway wall oedema, are postcapillary venular endothelial cells which contract, leaving intercellular gaps which act as pores facilitating plasma leakage (Persson 1987). Airway oedema contributes to airway narrowing as well as to bronchial hyperresponsiveness. Thus, inhibition of microvascular permeability could improve airway calibre and also reduce airway inflammation, thereby providing both therapeutic and prophylactic benefit.

β_2-Adrenoceptor agonists have the potential to inhibit mediator-induced microvascular plasma extravasation by relaxing post-capillary endothelial cells and thus opposing the spasmogenic actions of various mediators that induce intercellular gap formation (Persson 1986). Indeed, such activity has been demonstrated for β_2-adrenoceptor agonists in vitro (Langeler and Van Hinsbergh 1991) and in vivo (Rippe and Grega 1978; Baluk and Mcdonald 1994). With respect to its effects on endothelial barrier function, cAMP stabilizes endothelial tight junctions, inhibits myosin light chain kinase, reduces actin-non-muscle interaction and the formation of stress fibres and prevents agonist-induced endothelial gap formation (Moy et al. 1993; Siflinger-Birnboim et al. 1993; Adamson et al. 1998).

The long acting β_2-adrenoceptor agonist formoterol reduced histamine-induced microvascular leakage in guinea-pig airways (Erjefalt and Persson 1991; Advenier et al. 1992) and salmeterol reduced both early- and late-phase microvascular plasma leakage in rat. Furthermore, inhaled procaterol inhibited histamine-induced microvascular leakage in not only non-sensitized control guinea-pigs but also in animals sensitized and challenged with ovalbumin (Mirza et al. 1998). This suggests that β_2-adrenoceptor agonists may be effective in reversing oedema in the airway wall associated with allergic inflammation. Furthermore, β_2-adrenoceptor agonists can potentiate the inhibitory effects of both non-selective and selective phosphodiesterase IV inhibitors against antigen-induced microvascular leakage (Planquois et al. 1998). The long acting β_2-adrenoceptor agonist salmeterol may also reduce angiogenesis and vascular remodelling in the airways (Orsida et al. 2001). Another controversial action of β_2-adrenoceptor agonists is their potential to inhibit the release of inflammatory mediators from sensory nerves (Advenier et al. 1992; Verleden et al. 1993). This raises the possibility that β_2-adrenoceptor agonists have an anti-inflammatory impact and might attenuate oedema via this indirect mechanism.

These positive findings are to some extent countered by the observations that formoterol was less effective in the presence of ozone-induced airway inflammation (Inoue et al. 1997). Indeed, it has previously been shown in cases of established airway microvascular leakage that pretreatment with a β_2-adrenoceptor agonist does not always reduce the leakage of molecules induced by a further inflammatory stimulus (Erjefalt et al. 1985; Persson 1987). Furthermore, established oedema in the tracheobronchial model associated with airway inflammation does not resolve rapidly in the presence of a conventional (short-acting) β_2-adrenoceptor agonist. Hence, the therapeutic importance of the relaxant effect of β_2-adrenoceptor agonists on post-capillary endothelial cells is controversial. Despite this misgiving, formoterol has been shown to reduce plasma exudation in induced sputum in normal subjects (Greiff et al. 1998).

6
Other Potential Therapeutic Tissue Targets

6.1
Inflammatory Cells

It has long been known that β_2-adrenoceptors are expressed on inflammatory cells including mast cells (Butchers et al. 1980; Hughes et al. 1983), peripheral blood lymphocytes (Williams et al. 1976; Koeter et al. 1982; Sano et al. 1983), polymorphonuclear leukocytes (PMNL) (Galant et al. 1980; Davis et al. 1986; Nielson 1987), peritoneal macrophages (Schenkelaars and Bonta 1984), alveolar macrophages; (Fuller et al. 1988), platelets (Cook et al. 1987) and eosinophils (Koeter et al. 1982; Kraan et al. 1985). The established effects of β_2-adrenoceptor stimulation in some of these cells may be relevant to the therapeutic benefits of these agents in asthma. For example, it is well established that the response of both normal volunteers and of asthmatics to intramuscular β-adrenoceptor agonists such as adrenaline is for blood eosinophil numbers to fall dramatically (Koch-Weser 1968; Reed et al. 1970), an apparent anti-inflammatory reaction. However, in the case of the inhaled β_2-selective bronchodilator terbutaline, no such decrease in circulating eosinophils was observed.

Arguably, the most important potential anti-inflammatory action of the relatively short-acting β_2-adrenoceptor agonist bronchodilators such as salbutamol and terbutaline, is their capacity to suppress pro-inflammatory mediator release from inflammatory cells. For example, in the case of lymphocytes, inhibition of lymphokine secretion (and of proliferation) is well established (Bourne et al. 1974; Reed 1985). In PMNL, inhibition of superoxide radical generation and leukotriene release has been reported (Busse and Sosman 1984; Mack et al. 1986). In the case of human lung mast cells, salbutamol is a potent inhibitor of antigen-induced release of histamine and leukotrienes (Peters et al. 1982; Church and Young 1983). Indeed, salbutamol is 10–100 times more potent that disodium cromoglycate in this regard (Church and Hiroi 1987).

However, while the early asthmatic response is inhibited by β_2-adrenoceptor agonists, their impact on the late response to allergen is much less impressive (Cockcroft and Murdock 1987). Thus, the anti-inflammatory effects of monotherapy with inhaled, short-acting β_2-adrenoceptor agonist bronchodilators is minimal (Juniper et al. 1990; Haahtela et al. 1991; Van Essen-Zandvliet et al. 1992). Accordingly, it is generally accepted that the airway smooth muscle relaxant activity of β_2-adrenoceptor agonists is the action of primary importance in asthma. Paradoxically, it is this very powerful bronchodilator action that can harbour dangers for the asthmatic, since the sense of relative well-being and control over symptoms that accompanies the use of β_2-adrenoceptor agonists can mask the underlying progression and deterioration of this disease. This potential problem has been recognized and is a driver of recommendations for the combined use of such bronchodilators with an anti-inflammatory glucocorticoid (Kips and Pauwels 2001).

The advent of long-acting β_2-adrenoceptor agonist bronchodilators such as formoterol and salmeterol has re-ignited the question of whether or not a real therapeutic benefit is obtained in terms of the suppression of mediator release from inflammatory cells, even though these agents are also delivered by inhalation. It could be argued that the longer duration of action of these agents increases the likelihood of such an effect. Predictably, long-acting β_2-adrenoceptor agonists have been shown in animal studies both in vivo and in vitro, to effectively suppress pro-inflammatory mediator release and cytokine production and/or release from inflammatory cells. These actions have been demonstrated in human and/or animal T lymphocytes (Sekut et al. 1995; Holen and Elsayed 1998), macrophages (Linden 1992; Baker et al. 1994; Oddera et al. 1998), mast cells (Butchers et al. 1991; Gentilini et al. 1994; Lau et al. 1994; Nials et al. 1994; Bissonnette and Befus 1997; Chong et al. 1998; Drury et al. 1998), eosinophils (Eda et al. 1993; Rabe et al. 1993; Munoz et al. 1995) and neutrophils (Anderson et al. 1996). Furthermore, these agonists are also known to inhibit chemotaxis and recruitment of eosinophils (Whelan and Johnson 1992; Eda et al. 1993; Whelan et al. 1993; Teixeira et al. 1995; Teixeira and Hellewell 1997) and to delay apoptosis in these cells (Kankaanranta et al. 2000). However, it is now clear that monotherapy with long-acting agents such as salmeterol does not provide significant anti-inflammatory effect in asthma (Simons 1997; Verberne et al. 1997).

6.2
Secretory Cells

The deleterious impact of mucous hypersecretion and impaired mucociliary clearance on the effective bronchial lumen diameter and thus on bronchial airflow, can be life-threatening in the poorly controlled, severe asthmatic. Submucosal glands in human airways contain β_2-adrenoceptors (Carstairs et al. 1985), the stimulation of which increases mucus output. Importantly, β_2-adrenoceptor agonists also stimulate increases in ciliary beat frequency (Verdugo et al. 1980; Lopez-Vidriero et al. 1985) and in the movement of water towards the mucosal surface where it can hydrate mucus (Phipps et al. 1980). The net effect of these actions appears to be to improve mucociliary transport in asthmatics (Mossberg et al. 1976). However, in patients with significantly damaged bronchial epithelium, it seems likely that cilia function will be impaired, raising the possibility that in some patients, β_2-adrenoceptor agonist-stimulated mucous secretion could be detrimental.

7
Adverse Reactions to β_2-Adrenoceptor Agonists

7.1
Primary Adverse Reactions

The most widely reported adverse effects of therapeutic doses of β_2-adrenoceptor agonists mediated via β_2-adrenoceptors are skeletal muscle tremor (Larsson and Svedmyr 1977), cardiac effects (Paterson et al. 1979), metabolic changes including hyperglycaemia, hypokalaemia and decreased partial pressure of arterial oxygen (PaO$_2$) (Tai and Read 1967; Smith and Kendall 1984). These effects are seen in both healthy volunteers and in asthmatics. However, tolerance usually develops to the tremorogenic effects of β_2-adrenoceptor agonists in patients receiving long-term treatment (Svedmyr et al. 1976; Paterson et al. 1979). Furthermore, while there is little evidence that recommended aerosolized doses exacerbate pre-existing cardiac arrhythmias, caution should be taken in such cases.

7.2
Other Significant Direct Adverse Reactions

β_2-Adrenoceptor agonist bronchodilators can induce the mobilization of triglycerides resulting in elevated blood levels of fatty acids and glycerol (Smith and Kendall 1984), although it is the β_1-adrenoceptor that is responsible for mediating this effect. Salbutamol, terbutaline and fenoterol can induce mild appetite suppression, headache, nausea and sleep disturbances (Miller and Rice 1980; Pratt 1982). This is consistent with their ability to cross the blood–brain barrier, leading to CNS levels approximately 5% of those seen in plasma (Caccia and Fong 1984).

7.3
Stereoisomers of Salbutamol

Salbutamol, the most widely used β_2-adrenoceptor agonist bronchodilator, is a racemic mixture of equal parts of R-salbutamol and S-salbutamol. β_2-Adrenoceptor-mediated bronchodilatation is stereoselective, with R-salbutamol being wholly responsible for β_2 adrenoceptor-mediated bronchodilation and S-salbutamol being inactive in humans (Prior et al. 1998; Zhang et al. 1998). Since S-salbutamol has previously been shown to cause a small increase in airway reactivity in vitro (Mazzoni et al. 1994; Yamaguchi and Mccullough 1996), it was suggested that the S-enantiomer of racemic β_2-adrenoceptor agonists may cause airway hyperreactivity and even contribute to increased mortality

(Perrin-Fayolle et al. 1996; Handley et al. 1998). This potential safety concern, coupled with the finding that repeated administration of R, S-salbutamol resulted in S-salbutamol accumulation (Gumbhir-Shah et al. 1998; Dhand et al. 1999; Schmekel et al. 1999), resulted in the development of the optically pure R-salbutamol, levalbuterol, recently introduced into the U.S. market. Importantly, studies have now demonstrated that S-salbutamol has no deleterious effect on airway responsiveness to methacholine in asthmatic patients (Cockcroft and Swystun 1997; Cockcroft et al. 1999). Thus, R-salbutamol cannot claim to be safer than R,S-salbutamol based on the argument that S-salbutamol increases airway reactivity. Indeed, R,S-salbutamol has been found to be as safe as R-salbutamol in patients with asthma (Gumbhir-Shah et al. 1998; Nelson et al. 1998; Gawchik et al. 1999). Furthermore, evidence in both adults and children with stable asthma indicates that R-salbutamol is as effective a bronchodilator as equimolar doses of R,S-salbutamol (R-salbutamol 1.25 mg=R,S-salbutamol 2.5 mg) (Nelson et al. 1998; Gawchik et al. 1999). As an added disadvantage, R-salbutamol is likely to be more expensive than a comparable generic racemic salbutamol preparation. Taken together, the evidence indicates that R-salbutamol offers no genuine advantage with respect to safety or clinical efficacy over racemic salbutamol (Ahrens and Weinberger 2001; Boulton and Fawcett 2001).

8
Combination Therapy

8.1
Long-Acting β_2-Adrenoceptor Agonists and Glucocorticoids

The scientific rationale for the use of long-acting β_2-adrenoceptor agonists in combination with a corticosteroid has recently been summarized (Barnes 2002). The use of long-acting β_2-adrenoceptor agonists has been examined in asthmatic patients whose symptoms persisted despite treatment with low-dose glucocorticoids. In a randomized, double-blind, parallel-group trial, 429 adult asthmatics receiving 200 µg twice daily of inhaled beclomethasone dipropionate were selected. These mild-to-moderate asthmatics were symptomatic despite treatment with inhaled glucocorticoids. Subjects were assigned to receive either 50 µg salmeterol plus 200 µg beclomethasone or 500 µg beclomethasone alone twice daily for 6 months (Greening et al. 1994). There were significant advantages in favour of salmeterol plus beclomethasone compared with the higher dose of beclomethasone alone with respect to lung function and symptom control. Woolcock et al. (1996) recruited 738 moderate-to-severe asthmatics, whose symptoms were not controlled by twice daily 500 µg beclomethasone dipropionate. In this study, the administration of either 50 µg or 100 µg salmeterol twice daily with 500 µg beclomethasone had a more rapid and pronounced beneficial effect on control of asthma symptoms and lung function than doubling the dose of beclomethasone (twice daily 1000 µg) (Woolcock et al. 1996). Importantly, the addition of salmeterol was found to not increase bronchial hyperresponsiveness or asthma exacerbation rates (Greening et al. 1994; Woolcock et al. 1996). Furthermore, meta analysis of nine parallel group trials revealed that addition of salmeterol to low to moderate doses of inhaled glucocorticoid in symptomatic patients was superior to doubling the dose of inhaled glucocorticoid (Shrewsbury et al. 2000).

These studies demonstrate that interactions between β_2-adrenoceptor agonists and glucocorticoids are predominantly positive, with combinations of the two drugs improving asthma control and exacerbation rates. While this is particularly true for long-acting β_2-adrenoceptor agonists, the exact mechanism remains unclear. For example, the effects of long-acting β_2-adrenoceptor agonists and glucocorticoids may be merely additive; with the former causing prolonged bronchodilation and the latter reducing or reversing airway inflammation. Alternately, there may be true synergy between these agents with long-acting β_2-adrenoceptor agonists enhancing the effects of glucocorticoids (Kips and Pauwels 2001; Barnes 2002). It has been suggested that long-acting β_2-adrenoceptor agonists may have "steroid-enhancing" or "steroid-sparing" effects. However, it is important to note that monotherapy with long-acting β_2-adrenoceptor agonists is less effective than inhaled glucocorticoids alone, suggesting that these terms need to be used cautiously (Lazarus et al. 2001).

Based on the complementary roles of β_2-adrenoceptor agonists and glucocorticoids, the long-acting β_2-adrenoceptor agonist salmeterol and the glucocorticoid fluticasone have been combined in a single inhaler with the potential to treat both the airway smooth muscle dysfunction and inflammatory components of asthma. Such combination products have the potential to limit overuse of β_2-adrenoceptor agonist bronchodilators in the absence of anti-inflammatory therapy, thus ensuring that β_2-adrenoceptor agonists are not used as monotherapy. However, the use of "fixed" combination inhalers may be associated with the overuse of both drugs in the management of asthma, as control over individual drug dosages is lost.

In spite of these shortcomings, combination inhalers are effective in the treatment of many asthmatics and this format for combination therapy may become the method of choice in the near future in patients with persistent asthma (Barnes 2002). Studies in adults and adolescents have demonstrated improvements in forced expiratory volume in 1 s (FEV$_1$), peak expiratory flow (PEF), and asthma symptoms with a combination product containing salmeterol (50 µg) and fluticasone propionate (100, 250 or 500 µg) delivered via the dry powder Diskus inhaler (Seretide) (Aubier et al. 1999; Chapman et al. 1999; Bateman et al. 2001). Additionally, children aged 4–11 years who were symptomatic while receiving inhaled glucocorticoids, had similar improvements in FEV$_1$, PEF and asthma symptoms with salmeterol/fluticasone propionate (50/100 µg) (Van den Berg et al. 2000). The combination of fluticasone propionate and salmeterol via the Diskus device has also been found to improve lung function and reduce the severity of dyspnea in patients with COPD (Mahler et al. 2002). More recently, a salmeterol/fluticasone propionate MDI has been developed to provide an alternative choice of delivery system. Three strengths of the salmeterol/fluticasone propionate MDI are available each containing a constant dose of salmeterol (25 µg) combined with fluticasone (50, 125 or 250 µg) per actuation. Since each dose is given as two actuations, these preparations are equivalent to the three strengths of the salmeterol/fluticasone propionate Diskus indicated above. The efficacy and safety of salmeterol/fluticasone propionate (50/100 µg) was found to be comparable whether administered via MDI or dry powder Diskus inhaler, allowing a choice of delivery systems (Bateman et al. 2001).

8.2
D$_2$-Receptor Agonists

A different approach to combination therapy is to incorporate multiple pharmacological actions within the one drug molecule. Airway hyperreactivity, a feature of both asthma and COPD, is associated with neural reflex pathways that include sensory afferent nerves. While the receptors that modulate the activity of these airway nerves have yet to be characterized, reflex nerve activity may be controlled by modulating the activity of afferent nerves. For example, dopamine, via stimulation of D$_2$-receptors, may play a role in the control of lung function by reducing the ability of sensory nerves to produce harmful reflex activity. Indeed, D$_2$-receptor mRNA has been detected in rat vagal afferent neurones (Lawrence et al. 1995) and dorsal root ganglia (Xie et al. 1998), nerves associated with reflex pathways. Thus, D$_2$-receptor agonists should reduce reflex bronchoconstriction, dyspnea, cough and mucus production, without any direct bronchodilator activity. A dual dopamine D$_2$-receptor and β_2-adrenoceptor agonist would combine the modulating effects of a dopamine D$_2$-receptor agonist on sensory afferent nerves with the bronchodilator action of a β_2-adrenoceptor agonist in the one molecule. An example of such a compound is AR-C68397AA (Viozan) (Bonnert et al. 1998). Combination therapy of this sort may provide effective symptomatic treatment for both asthma and COPD with the added advantage of reducing neurogenic inflammation in the airways. Interestingly, the benzothiazole structure of the synthetic compound AR-C68397AA has since been found to occur in the natural β_2-adrenoceptor agonist S1319 (4-hydroxy-7-[1-(1-hydroxy-2-methylamino)ethyl]-1,3-benzothiazole-2(3H)-one) found in a marine sponge *Dysidea sp.* (Suzuki et al. 1999).

9
Pharmacogenetics of β_2-Adrenoceptor Agonists in Asthma

Pharmacogenetics is the study of the role of genetic determinants in the variable response to therapy. Within the human population, the β_2-adrenoceptor is polymorphic, with some of these polymorphic receptors having different pharmacological properties. Recent studies have suggested that genetic factors may underlie some of the variability in treatment responses to β-adrenoceptor agonists seen in asthmatics. Both single-nucleotide polymorphisms (SNPs) and variable nucleotide tandem repeats (VNTRs) are genetic polymorphisms that have been shown to have pharmacogenetic effects in asthma. A total of 13 polymorphisms in the β_2-adrenoceptor gene and its transcriptional regulator β-upstream peptide have been identified (Liggett 2000a,b).

Within the β_2-adrenoceptor gene, coding variants at positions 16 and 27, in the extracellular N-terminal domain, have been shown to be functionally important in vitro (Green et al. 1994; Mcgraw et al. 1999). While the Gly-16 receptor exhibits enhanced downregulation in vitro following exposure to an agonist (Green et al. 1994), Arg-16 receptors are more resistant to desensitization. However, N-terminal polymorphisms at position 16 failed to alter either the rates of new receptor synthesis following irreversible alkylation or the rate of agonist-promoted internalization of the receptor to the intracellular pool (Green et al. 1994). Due to linkage disequilibrium, individuals who are Arg/Arg-16 are much more likely to be Glu/Glu-27 and individuals who are Gly/Gly-16 are much more likely to be Gln/Gln-27. Furthermore, the position 27 genotypes influence but do not abolish the effect of position 16 polymorphisms with respect to down-regulation of phenotypes in vitro (Green et al. 1994; Mcgraw et al. 1999). The potentially protective Glu-27 polymorphism has been reported to be associated with decreased airway reactivity in asthma (Hall et al. 1995) but it did not seem to influence nocturnal asthma (Turki et al. 1995) or bronchodilator responsiveness (Martinez et al. 1997). In contrast, the Gln-27 polymorphism has been associated with elevated IgE levels and an increase in self-reported asthma in children (Dewar et al. 1997). Israel and co-workers (2001) noted a decrease in morning peak expiratory flow in patients who were Arg/Arg-16 and who regularly used salbutamol (Israel et al. 2001).

In an attempt to explain the apparent disparity between in vitro and patient data, Liggett has proposed that Gly/Gly-16 individuals are already downregulated as a result of exposure to endogenous catecholamines (Liggett 2000b). As such, desensitization caused by recurrent exogenous β-adrenoceptor agonist exposure would be more apparent in Arg/Arg patients with functional β-adrenoceptors. In this scenario, the initial response to salbutamol in β-adrenoceptor agonist-naïve patients would be depressed in Gly/Gly individuals, since their receptors would have been downregulated to a greater extent due to endogenous catecholamines. The bronchodilator response obtained after administration of a single dose of salbutamol has also been examined (Martinez et al. 1997). Here, β-adrenoceptor agonist-naïve asthmatic and non-asthmatic children in the Arg/Arg-16 group showed a greater bronchodilator response, with Arg/Arg-16 children being 5.3-fold more likely to exhibit a positive bronchodilator response to salbutamol compared with Gly/Gly-16 children.

It is important to note that pharmacogenetic studies of treatment response are often negative (Hancox et al. 1998) or involve small subject numbers (Tan et al. 1997; Lipworth et al. 1999). Larger scale pharmacogenetic studies will need to be conducted in order to detect large effects associated with a SNP. The data obtained so far suggest that β_2-adrenoceptor polymorphisms may alter the response to β-adrenoceptor agonists. However, it is still unclear whether β_2-adrenoceptor polymorphisms will have any great clinical relevance for most patients.

10
Clinical Application

10.1
Asthma

In general, asthma medication can be divided into two groups; reliever and preventer medications. The major group of asthma reliever medications are β_2-adrenoceptor agonist bronchodilators which act quickly and effectively to relieve bronchoconstriction and the associated asthma symptoms of chest tightness, wheezing and cough. The main asthma preventer medications are the glucocorticoids which are used prophylactically and as maintenance therapy to reduce, reverse and prevent airway inflammation. It is vital that all asthmatic patients learn to manage their own asthma and that they have a good understanding of the role of reliever and preventer medications in treating their disease. The goal of asthma management is to achieve and maintain best lung function and an ideal starting point is the institution of an asthma management plan (National Asthma Campaign—Asthma Management Handbook 2002). Typically, the first step in such a plan is the assessment of the patient's asthma severity. The patient may then be treated intensively, with reliever and/or preventer medication, until best lung function is achieved. The types and quantity of drug used can then be back-titrated to the least number of medications and lowest dose required for good control of asthma symptoms and maintenance of best lung function. Since prevention is the key to successful asthma management, an important component of any asthma management plan is the identification and avoidance or control of asthma triggers such as allergen, exercise and cold air. In addition, an individualized action plan needs to be developed to manage any ongoing asthma symptoms and exacerbations. An effective asthma management plan necessitates regular review and ongoing patient education.

The severity of asthma may be classified based on an assessment of asthma symptoms and lung function in combination with the types and quantity of drug required to reduce or avoid symptoms. In this way, patients with asthma may be classified as having mild intermittent, mild persistent, moderate or severe disease (NIH: NHLBI 1997; 1998; NHLBI/WHO Workshop report 1995). The clinical classification of asthma severity forms the basis of the stepwise approach to asthma pharmacotherapy, with the number and frequency of medications increasing (step up) as the severity of asthma increases and decreasing (step down) when asthma is under control (Table 1). However, classifying asthma severity is not intended to restrict the type of drug therapy received by an individual patient, but is intended as a guide to the level of therapy that may be required to achieve symptom control. Furthermore, patients diagnosed with any

Table 1 Classification of asthma severity and the therapeutic use of β_2-adrenoceptor agonists

	Mild intermittent	Mild persistent	Moderate persistent	Severe
Clinical Features:				
Symptom frequency	<1 a week	>1 a week but <1 a day	Daily	Continuous
Nocturnal symptoms	≤2 a month	>2 a month	>1 a week	Frequent
Exacerbations	Brief, asymptomatic and normal lung function between exacerbations	Exacerbations may affect activity and sleep	>2 a week, exacerbations affect activity and sleep	Frequent
Lung function	PEF or FEV_1 ≥80% predicted, variability <20%	PEF or FEV_1 ≥80% predicted, variability 20%–30%	PEF or FEV_1 60%–79% predicted, variability >30%	PEF or FEV_1 < 60% predicted, variability >30%
Drug treatment:				
Reliever: β_2-adrenoceptor agonist	Short-acting inhaled β_2-adrenoceptor agonist should be taken as needed for symptom relief	Regular short-acting inhaled β_2-adrenoceptor agonist	Regular long-acting inhaled β_2-adrenoceptor agonist and short-acting β_2-adrenoceptor agonist as needed	Regular long-acting inhaled β_2-adrenoceptor agonist
Preventer: Glucocorticoid		Low-dose inhaled glucocorticoid	Low-dose inhaled glucocorticoid	High-dose inhaled and oral glucocorticoid

level of asthma may have mild, moderate or severe exacerbations and these exacerbations also require appropriate management.

In patients with mild intermittent asthma, short-acting inhaled β_2-adrenoceptor agonists, including salbutamol and terbutaline, are the treatment of choice and should be used as required to relieve symptoms and prevent those induced by exercise or exposure to allergen. If this regimen fails to control asthma symptoms, an increase in β_2-adrenoceptor agonist use needs to be considered. Usually, the infrequent nature of symptoms in this group of patients does not warrant continuous β_2-adrenoceptor agonist therapy.

Patients with mild persistent asthma should be treated with low-dose inhaled glucocorticoids to treat airway inflammation. In addition, the regular use of short-acting inhaled β_2-adrenoceptor agonists is required for the relief of acute asthma symptoms. If best lung function is not maintained under this treatment regimen, the dose of inhaled glucocorticoid can be increased and/or a long-acting β_2-adrenoceptor agonist used, particularly when breakthrough and/or night-time symptoms persist.

The treatment of moderate persistent asthma involves inhaled glucocorticoids and the regular use of long-acting β_2-adrenoceptor agonists, such as salmeterol and formoterol. The latter are particularly useful for the control of night-time symptoms. The addition of a long-acting β-adrenoceptor agonist to the treatment regimen may also have a steroid-sparing effect in these patients. Short-acting β_2-adrenoceptor agonists may be used in these patients for the rapid treatment of acute symptoms.

In most cases, patients with severe asthma should receive high doses of inhaled glucocorticoids and the regular use of long-acting β_2-adrenoceptor agonists. Short-acting β-adrenoceptor agonist medications should be used for acute symptom relief. Asthma exacerbations in this group of patients may also require a course of oral glucocorticoid therapy.

10.2
COPD

The severity of COPD may be classified into four stages (GOLD 2001; Table 2). However, the management of COPD is driven largely by symptomology and there is often no direct relationship between the degree of airflow limitation and the presence of symptoms. Thus, disease classification provides only a very general indication of the approach to be given to management of COPD.

The goals of effective COPD management are to prevent disease progression, relieve symptoms, improve exercise tolerance, improve health status, prevent and treat complications, prevent and treat exacerbations and reduce mortality. Pharmacotherapy is used to prevent and control symptoms, reduce the frequency and severity of exacerbations, improve health status and improve exercise tolerance. Importantly, existing medications used for the treatment of COPD have not been shown to modify the long-term decline in lung function associated with this disease. Bronchodilator medications including β_2-adrenoceptor ag-

Table 2 Classification of COPD by severity (GOLD 2001)

Stage	Characteristics
0: At Risk	Normal spirometry Chronic symptoms (cough, sputum production)
I: Mild COPD	$FEV_1/FVC<70\%$ $FEV_1 \geq 80\%$ predicted With or without chronic symptoms (cough, sputum production)
II: Moderate COPD	$FEV_1/FVC<70\%$ $30\% \geq FEV_1 <80\%$ predicted With or without chronic symptoms (cough, sputum production)
III: Severe COPD	$FEV_1/FVC<70\%$ $FEV_1<30\%$ predicted or $FEV_1<50\%$ predicted plus respiratory failure or clinical signs of right heart failure

FEV_1 values refer to post-bronchodilator values; FVC, forced vital capacity.

onists, anticholinergics and theophylline, given alone or in combination, have a role to play in relieving symptoms as well as preventing and treating exacerbations (Chrystyn et al. 1988; Vathenen et al. 1988; Gross et al. 1989; Higgins et al. 1991; Anthonisen et al. 1994). These bronchodilators may be used either on an as needed basis for the relief of persistent or worsening symptoms, or on a regular basis to prevent or reduce symptoms. The choice between β_2-adrenoceptor agonist, anticholinergic, theophylline (or related compound) or some combination of these drug therapies depends on the response obtained by the individual in terms of symptom relief and side effects. A combination of bronchodilators may produce additional improvements in lung function and health status while decreasing the risk of side effects compared with increasing the dose of a single bronchodilator (Taylor et al. 1985; Guyatt et al. 1987; Gross et al. 1998; Van Noord et al. 2000).

A key diagnostic feature of COPD is poor reversibility of airflow limitation following inhalation of a short-acting β_2-adrenoceptor agonist. Importantly, β_2-adrenoceptor agonist bronchodilators have been shown to improve hyperinflation, exercise capacity and quality of life in COPD patients, without necessarily producing significant changes in FEV_1 (Guyatt et al. 1987; Jenkins et al. 1987; Cazzola et al. 1995; Boyd et al. 1997). Recent studies have shown that long-acting inhaled β_2-adrenoceptor agonists significantly improve symptoms and increase health-related quality of life in COPD patients (Ulrik 1995; Jones and Bosh 1997; Mahler et al. 1999).

11
Concluding Remarks

The airway smooth muscle relaxant effect of β_2-adrenoceptor agonists is their primary beneficial action in asthma and COPD, although positive therapeutic influences on mucus production and clearance and bronchial oedema may also occur. β_2-Adrenoceptor agonists appear to be largely ineffective in suppressing or controlling airway inflammation in asthmatics and are likely to be equally ineffective in COPD patients. Accordingly, in asthma, despite their relative lack of significant, direct detrimental side effects, there is consensus that β_2-adrenoceptor agonist bronchodilators, whether or not they are long acting, should not be used as monotherapy in any but the most mild of cases. Indeed, the powerful bronchodilator (reliever) actions of both long- and short-acting β_2-adrenoceptor agonists may mask the onset and/or deterioration of on-going airway inflammation in asthmatics. The increased emphasis on anti-inflammatory therapies in recent years is now complemented by the use of β_2-adrenoceptor agonists in therapeutic regimes centred on the combined use of corticosteroids and β_2-adrenoceptor agonist bronchodilators. Indeed, the introduction of single administration formulations of inhaled steroid with a bronchodilator is finding increasing acceptance in the treatment of persistent asthma. Unfortunately, in COPD, bronchodilator therapies do not alter the long-term decline in lung function. However, β_2-adrenoceptor agonist bronchodilators and anticholinergics and theophylline, given alone or in combination, can relieve symptoms and help to reverse exacerbations. The introduction of long-acting β_2-adrenoceptor agonists has produced significant improvements in symptoms in COPD patients and thus has had a positive impact on quality of life in these patients. The use of combination bronchodilator/corticosteroid regimes further assists in the management of this disease.

References

Adamson RH, Liu B, Fry GN, Rubin LL, Curry FE (1998) Microvascular permeability and number of tight junctions are modulated by cAMP. Am J Physiol 274:H1885–94

Advenier C, Qian Y, Koune JD, Molimard M, Candenas ML, Naline E (1992) Formoterol and salbutamol inhibit bradykinin- and histamine-induced airway microvascular leakage in guinea-pig. Br J Pharmacol 105:792–8

Ahrens R, Weinberger M (2001) Levalbuterol and racemic albuterol: are there therapeutic differences? J Allergy Clin Immunol 108:681–4

Amano M, Fukata Y, Kaibuchi K (2000) Regulation and functions of Rho-associated kinase. Exp Cell Res 261:44–51

Anderson GP, Linden A, Rabe KF (1994) Why are long-acting beta-adrenoceptor agonists long-acting? Eur Respir J 7:569–78

Anderson R, Feldman C, Theron AJ, Ramafi G, Cole PJ, Wilson R (1996) Anti-inflammatory, membrane-stabilizing interactions of salmeterol with human neutrophils in vitro. Br J Pharmacol 117:1387–94

Anthonisen NR, Connett JE, Kiley JP, Altose MD, Bailey WC, Buist AS, Conway WA, Jr., Enright PL, Kanner RE, O'Hara P (1994) Effects of smoking intervention and the use of an inhaled anticholinergic bronchodilator on the rate of decline of FEV1. The Lung Health Study. JAMA 272:1497–505

Aubier M, Pieters WR, Schlosser NJ, Steinmetz KO (1999) Salmeterol/fluticasone propionate (50/500 microg) in combination in a Diskus inhaler (Seretide) is effective and safe in the treatment of steroid-dependent asthma. Respir Med 93:876–84

Baker AJ, Palmer J, Johnson M, Fuller RW (1994) Inhibitory actions of salmeterol on human airway macrophages and blood monocytes. Eur J Pharmacol 264:301–6

Baluk P, McDonald DM (1994) The beta 2-adrenergic receptor agonist formoterol reduces microvascular leakage by inhibiting endothelial gap formation. Am J Physiol 266:L461–8

Barnes P, Jacobs M, Roberts JM (1984) Glucocorticoids preferentially increase fetal alveolar beta-adrenoreceptors: autoradiographic evidence. Pediatr Res 18:1191–4

Barnes PJ (2002) Scientific rationale for inhaled combination therapy with long-acting beta2-agonists and corticosteroids. Eur Respir J 19:182–91

Barnes PJ, Basbaum CB, Nadel JA, Roberts JM (1982) Localization of beta-adrenoreceptors in mammalian lung by light microscopic autoradiography. Nature 299:444–7

Barnes PJ, Karliner JS, Dollery CT (1980) Human lung adrenoreceptors studied by radioligand binding. Clin Sci 58:457–61

Bateman ED, Silins V, Bogolubov M (2001) Clinical equivalence of salmeterol/fluticasone propionate in combination (50/100 microg twice daily) when administered via a chlorofluorocarbon-free metered dose inhaler or dry powder inhaler to patients with mild-to-moderate asthma. Respir Med 95:136–46

Benovic JL, Pike LJ, Cerione RA, Staniszewski C, Yoshimasa T, Codina J, Caron MG, Lefkowitz RJ (1985) Phosphorylation of the mammalian beta-adrenergic receptor by cyclic AMP-dependent protein kinase. Regulation of the rate of receptor phosphorylation and dephosphorylation by agonist occupancy and effects on coupling of the receptor to the stimulatory guanine nucleotide regulatory protein. J Biol Chem 260:7094–101

Benovic JL, Stiles GL, Lefkowitz RJ, Caron MG (1983) Photoaffinity labelling of mammalian beta-adrenergic receptors: metal-dependent proteolysis explains apparent heterogeneity. Biochem Biophys Res Commun 110:504–11

Bergman J, Persson H, Wetterlin K (1969) 2 new groups of selective stimulants of adrenergic beta-receptors. Experientia 25:899–901

Bertram JF, Goldie RG, Papadimitriou JM, Paterson JW (1983) Correlations between pharmacological responses and structure of human lung parenchyma strips. Br J Pharmacol 80:107–14

Bissonnette EY, Befus AD (1997) Anti-inflammatory effect of beta 2-agonists: inhibition of TNF-alpha release from human mast cells. J Allergy Clin Immunol 100:825–31

Bonnert RV, Brown RC, Chapman D, Cheshire DR, Dixon J, Ince F, Kinchin EC, Lyons AJ, Davis AM, Hallam C, Harper ST, Unitt JF, Dougall IG, Jackson DM, McKechnie K, Young A, Simpson WT (1998) Dual D2-receptor and beta2-adrenoceptor agonists for the treatment of airway diseases. 1. Discovery and biological evaluation of some 7-(2-aminoethyl)-4-hydroxybenzothiazol-2(3H)-one analogues. J Med Chem 41:4915–7

Boulton DW, Fawcett JP (2001) The pharmacokinetics of levosalbutamol: what are the clinical implications? Clin Pharmacokinet 40:23–40

Bourne HR, Lichtenstein LM, Melmon KL, Henney CS, Weinstein Y, Shearer GM (1974) Modulation of inflammation and immunity by cyclic AMP. Science 184:19–28

Boyd G, Morice AH, Pounsford JC, Siebert M, Peslis N, Crawford C (1997) An evaluation of salmeterol in the treatment of chronic obstructive pulmonary disease (COPD). Eur Respir J 10:815–21

Brittain RT, Farmer JB, Jack D, Martin LE, Simpson WT (1968) α-[-t-butylamino)methyl]-4-hydroxy-m-xylene-α^1,α^3-diol (AH.3365) a seleective β-adrenergic stimulant. Nature 219:862–3

Busse W, Nelson H, Wolfe J, Kalberg C, Yancey SW, Rickard KA (1999) Comparison of inhaled salmeterol and oral zafirlukast in patients with asthma. J Allergy Clin Immunol 103:1075–80

Busse WW, Sosman JM (1984) Isoproterenol inhibition of isolated human neutrophil function. J Allergy Clin Immunol 73:404–10

Butchers PR, Skidmore IF, Vardey CJ, Wheeldon A (1980) Characterization of the receptor mediating the antianaphylactic effects of beta-adrenoceptor agonists in human lung tissue in vitro. Br J Pharmacol 71:663–7

Butchers PR, Vardey CJ, Johnson M (1991) Salmeterol: a potent and long-acting inhibitor of inflammatory mediator release from human lung. Br J Pharmacol 104:672–6

Caccia S, Fong MH (1984) Kinetics and distribution of the beta-adrenergic agonist salbutamol in rat brain. J Pharm Pharmacol 36:200–2

Carstairs JR, Nimmo AJ, Barnes PJ (1984) Autoradiographic localisation of beta-adrenoceptors in human lung. Eur J Pharmacol 103:189–90

Carstairs JR, Nimmo AJ, Barnes PJ (1985) Autoradiographic visualization of beta-adrenoceptor subtypes in human lung. Am Rev Respir Dis 132:541–7

Carswell H, Nahorski SR (1983) Beta-adrenoceptor heterogeneity in guinea-pig airways: comparison of functional and receptor labelling studies. Br J Pharmacol 79:965–71

Cazzola M, Matera MG, Santangelo G, Vinciguerra A, Rossi F, D'Amato G (1995) Salmeterol and formoterol in partially reversible severe chronic obstructive pulmonary disease: a dose-response study. Respir Med 89:357–62

Chapman HA, Jr., Shi GP (2000) Protease injury in the development of COPD: Thomas A. Neff Lecture. Chest 117:295S-9S

Chapman KR, Ringdal N, Backer V, Palmqvist M, Saarelainen S, Briggs M (1999) Salmeterol and fluticasone propionate (50/250 microg) administered via combination Diskus inhaler: as effective as when given via separate Diskus inhalers. Can Respir J 6:45–51

Chong LK, Cooper E, Vardey CJ, Peachell PT (1998) Salmeterol inhibition of mediator release from human lung mast cells by beta-adrenoceptor-dependent and independent mechanisms. Br J Pharmacol 123:1009–15

Chrystyn H, Mulley BA, Peake MD (1988) Dose response relation to oral theophylline in severe chronic obstructive airways disease. BMJ 297:1506–10

Church MK, Hiroi J (1987) Inhibition of IgE-dependent histamine release from human dispersed lung mast cells by anti-allergic drugs and salbutamol. Br J Pharmacol 90:421–9

Church MK, Young KD (1983) The characteristics of inhibition of histamine release from human lung fragments by sodium cromoglycate, salbutamol and chlorpromazine. Br J Pharmacol 78:671–9

Cockcroft DW, Davis BE, Swystun VA, Marciniuk DD (1999) Tolerance to the bronchoprotective effect of beta2-agonists: comparison of the enantiomers of salbutamol with racemic salbutamol and placebo. J Allergy Clin Immunol 103:1049–53

Cockcroft DW, Murdock KY (1987) Comparative effects of inhaled salbutamol, sodium cromoglycate, and beclomethasone dipropionate on allergen-induced early asthmatic responses, late asthmatic responses, and increased bronchial responsiveness to histamine. J Allergy Clin Immunol 79:734–40

Cockcroft DW, Swystun VA (1997) Effect of single doses of S-salbutamol, R-salbutamol, racemic salbutamol, and placebo on the airway response to methacholine. Thorax 52:845–8

Cook N, Nahorski SR, Barnett DB (1987) Human platelet beta 2-adrenoceptors: agonist-induced internalisation and down-regulation in intact cells. Br J Pharmacol 92:587–96

Davies B, Brooks G, Devoy M (1998) The efficacy and safety of salmeterol compared to theophylline: meta-analysis of nine controlled studies. Respir Med 92:256–63

Davis PB, Simpson DM, Paget GL, Turi V (1986) Beta-adrenergic responses in drug-free subjects with asthma. J Allergy Clin Immunol 77:871–9

Dewar JC, Wilkinson J, Wheatley A, Thomas NS, Doull I, Morton N, Lio P, Harvey JF, Liggett SB, Holgate ST, Hall IP (1997) The glutamine 27 beta2-adrenoceptor polymorphism is associated with elevated IgE levels in asthmatic families. J Allergy Clin Immunol 100:261–5

Dhand R, Goode M, Reid R, Fink JB, Fahey PJ, Tobin MJ (1999) Preferential pulmonary retention of (S)-albuterol after inhalation of racemic albuterol. Am J Respir Crit Care Med 160:1136–41

Dickinson K, Richardson A, Nahorski SR (1981) Homogeneity of beta 2-adrenoceptors on rat erythrocytes and reticulocytes. A comparison with heterogeneous rat lung beta-adrenoceptors. Mol Pharmacol 19:194–204

Drury DE, Chong LK, Ghahramani P, Peachell PT (1998) Influence of receptor reserve on beta-adrenoceptor-mediated responses in human lung mast cells. Br J Pharmacol 124:711–8

Eda R, Sugiyama H, Hopp RJ, Okada C, Bewtra AK, Townley RG (1993) Inhibitory effects of formoterol on platelet-activating factor induced eosinophil chemotaxis and degranulation. Int Arch Allergy Immunol 102:391–8

Engel G, Hoyer D, Berthold R, Wagner H (1981) (+/-)[^{125}Iodo]-cyanopindolol, a new ligand for β-adrenoceptors. Identification and quantitation of subclasses of β adrenoceptors in guinea-pig. Naunyn Schmiedebergs Arch Pharmacol 317:277–85

Erjefalt I, Persson CG (1991) Long duration and high potency of antiexudative effects of formoterol in guinea-pig tracheobronchial airways. Am Rev Respir Dis 144:788–91

Erjefalt IA, Wagner ZG, Strand SE, Persson CG (1985) A method for studies of tracheobronchial microvascular permeability to macromolecules. J Pharmacol Methods 14:275–83

Finkel MS, Quirion R, Pert C, Patterson RE (1984) Characterization and autoradiographic distribution of the beta-adrenergic receptor in the rat lung. Pharmacology 29:247–54

Fuller RW, O'Malley G, Baker AJ, MacDermot J (1988) Human alveolar macrophage activation: inhibition by forskolin but not beta-adrenoceptor stimulation or phosphodiesterase inhibition. Pulm Pharmacol 1:101–6

Galant SP, Duriseti L, Underwood S, Allred S, Insel PA (1980) Beta adrenergic receptors of polymorphonuclear particulates in bronchial asthma. J Clin Invest 65:577–85

Gawchik SM, Saccar CL, Noonan M, Reasner DS, DeGraw SS (1999) The safety and efficacy of nebulized levalbuterol compared with racemic albuterol and placebo in the treatment of asthma in pediatric patients. J Allergy Clin Immunol 103:615–21

Gentilini G, Grazia di Bello M, Raspanti S, Bindi D, Mugnai S, Zilletti L (1994) Salmeterol inhibits anaphylactic histamine release from guinea-pig isolated mast cells. J Pharm Pharmacol 46:76–7

Gilman AG (1987) G proteins: transducers of receptor-generated signals. Annu Rev Biochem 56:615–49

Goldie RG, Papadimitriou JM, Paterson JW, Rigby PJ, Spina D (1986a) Autoradiographic localization of beta-adrenoceptors in pig lung using [^{125}I]-iodocyanopindolol. Br J Pharmacol 88:621–8

Goldie RG, Paterson JW, Lulich KM (1991) Pharmacology and therapeutics of beta-adrenoceptor agonists. In: Page CP, Barnes PJ (eds) Handbook of Experimental Pharmacology. Vol 98. Springer-Verlag, Berlin, pp 167–205

Goldie RG, Paterson JW, Wale JL (1982) Pharmacological responses of human and porcine lung parenchyma, bronchus and pulmonary artery. Br J Pharmacol 76:515–21

Goldie RG, Spina D, Henry PJ, Lulich KM, Paterson JW (1986b) In vitro responsiveness of human asthmatic bronchus to carbachol, histamine, beta-adrenoceptor agonists and theophylline. Br J Clin Pharmacol 22:669–76

Green SA, Spasoff AP, Coleman RA, Johnson M, Liggett SB (1996) Sustained activation of a G protein-coupled receptor via "anchored" agonist binding. Molecular localization of the salmeterol exosite within the 2-adrenergic receptor. J Biol Chem 271:24029–35

Green SA, Turki J, Innis M, Liggett SB (1994) Amino-terminal polymorphisms of the human beta 2-adrenergic receptor impart distinct agonist-promoted regulatory properties. Biochemistry (Mosc) 33:9414–9

Greening AP, Ind PW, Northfield M, Shaw G (1994) Added salmeterol versus higher-dose corticosteroid in asthma patients with symptoms on existing inhaled corticosteroid. Allen & Hanburys Limited UK Study Group. Lancet 344:219–24

Greiff L, Wollmer P, Andersson M, Svensson C, Persson CG (1998) Effects of formoterol on histamine induced plasma exudation in induced sputum from normal subjects. Thorax 53:1010–3

Gross N, Tashkin D, Miller R, Oren J, Coleman W, Linberg S (1998) Inhalation by nebulization of albuterol-ipratropium combination (Dey combination) is superior to either agent alone in the treatment of chronic obstructive pulmonary disease. Dey Combination Solution Study Group. Respiration 65:354–62

Gross NJ, Petty TL, Friedman M, Skorodin MS, Silvers GW, Donohue JF (1989) Dose response to ipratropium as a nebulized solution in patients with chronic obstructive pulmonary disease. A three-center study. Am Rev Respir Dis 139:1188–91

Gumbhir-Shah K, Kellerman DJ, DeGraw S, Koch P, Jusko WJ (1998) Pharmacokinetic and pharmacodynamic characteristics and safety of inhaled albuterol enantiomers in healthy volunteers. J Clin Pharmacol 38:1096–106

Guyatt GH, Townsend M, Pugsley SO, Keller JL, Short HD, Taylor DW, Newhouse MT (1987) Bronchodilators in chronic air-flow limitation. Effects on airway function, exercise capacity, and quality of life. Am Rev Respir Dis 135:1069–74

Haahtela T, Jarvinen M, Kava T, Kiviranta K, Koskinen S, Lehtonen K, Nikander K, Persson T, Reinikainen K, Selroos O (1991) Comparison of a beta 2-agonist, terbutaline, with an inhaled corticosteroid, budesonide, in newly detected asthma. N Engl J Med 325:388–92

Hall IP, Wheatley A, Wilding P, Liggett SB (1995) Association of Glu 27 beta 2-adrenoceptor polymorphism with lower airway reactivity in asthmatic subjects. Lancet 345:1213–4

Hancox RJ, Sears MR, Taylor DR (1998) Polymorphism of the beta2-adrenoceptor and the response to long-term beta2-agonist therapy in asthma. Eur Respir J 11:589–93

Handley DA, McCullough JR, Crowther SD, Morley J (1998) Sympathomimetic enantiomers and asthma. Chirality 10:262–72

Harms HH (1976) Isoproterenol antagonism of cardioselective beta adrenergic receptor blocking agents: a comparative study of human and guinea-pig cardiac and bronchial beta adrenergic receptors. J Pharmacol Exp Ther 199:329–35

Hawksworth RJ, Sykes AP, Faris M, Mant T, Lee TH (2002) Albuterol HFA is as effective as albuterol CFC in preventing exercise-induced bronchoconstriction. Ann Allergy Asthma Immunol 88:473–7

Hekking PR, Maesen F, Greefhorst A, Prins J, Tan Y, Zweers P (1990) Long-term efficacy of formoterol compared to salbutamol. Lung 168:76–82

Henry PJ, Rigby PJ, Goldie RG (1990) Distribution of beta 1- and beta 2-adrenoceptors in mouse trachea and lung: a quantitative autoradiographic study. Br J Pharmacol 99:136–44

Higgins BG, Powell RM, Cooper S, Tattersfield AE (1991) Effect of salbutamol and ipratropium bromide on airway calibre and bronchial reactivity in asthma and chronic bronchitis. Eur Respir J 4:415–20

Hill AT, Bayley D, Stockley RA (1999) The interrelationship of sputum inflammatory markers in patients with chronic bronchitis. Am J Respir Crit Care Med 160:893–8

Holen E, Elsayed S (1998) Effects of beta2 adrenoceptor agonists on T-cell subpopulations. APMIS 106:849–57

Hughes JM, Seale JP, Temple DM (1983) Effect of fenoterol on immunological release of leukotrienes and histamine from human lung in vitro: selective antagonism by beta-adrenoceptor antagonists. Eur J Pharmacol 95:239–45

Iizuka K, Shimizu Y, Tsukagoshi H, Yoshii A, Harada T, Dobashi K, Murozono T, Nakazawa T, Mori M (2000) Evaluation of Y-27632, a rho-kinase inhibitor, as a bronchodilator in guinea pigs. Eur J Pharmacol 406:273–9

Inoue H, Aizawa H, Matsumoto K, Shigyo M, Takata S, Hara M, Hara N (1997) Effect of beta 2-agonists on histamine-induced airway microvascular leakage in ozone-exposed guinea pigs. Am J Respir Crit Care Med 156:723–7

Israel E, Drazen JM, Liggett SB, Boushey HA, Cherniack RM, Chinchilli VM, Cooper DM, Fahy JV, Fish JE, Ford JG, Kraft M, Kunselman S, Lazarus SC, Lemanske RF, Jr., Martin RJ, McLean DE, Peters SP, Silverman EK, Sorkness CA, Szefler SJ, Weiss ST, Yandava CN, National Heart L, Blood Institute's Asthma Clinical Research N (2001) Effect of polymorphism of the beta(2)-adrenergic receptor on response to regular use of albuterol in asthma. Int Arch Allergy Immunol 124:183–6

Jeffery PK (1998) Structural and inflammatory changes in COPD: a comparison with asthma. Thorax 53:129–36

Jenkins SC, Heaton RW, Fulton TJ, Moxham J (1987) Comparison of domiciliary nebulized salbutamol and salbutamol from a metered-dose inhaler in stable chronic airflow limitation. Chest 91:804–7

Jones PW, Bosh TK (1997) Quality of life changes in COPD patients treated with salmeterol. Am J Respir Crit Care Med 155:1283–9

Juniper EF, Kline PA, Vanzieleghem MA, Ramsdale EH, O'Byrne PM, Hargreave FE (1990) Effect of long-term treatment with an inhaled corticosteroid (budesonide) on airway hyperresponsiveness and clinical asthma in nonsteroid-dependent asthmatics. Am Rev Respir Dis 142:832–6

Kankaanranta H, Lindsay MA, Giembycz MA, Zhang X, Moilanen E, Barnes PJ (2000) Delayed eosinophil apoptosis in asthma. J Allergy Clin Immunol 106:77–83

Keatings VM, Collins PD, Scott DM, Barnes PJ (1996) Differences in interleukin-8 and tumor necrosis factor-alpha in induced sputum from patients with chronic obstructive pulmonary disease or asthma. Am J Respir Crit Care Med 153:530–4

Kimura K, Ito M, Amano M, Chihara K, Fukata Y, Nakafuku M, Yamamori B, Feng J, Nakano T, Okawa K, Iwamatsu A, Kaibuchi K (1996) Regulation of myosin phosphatase by Rho and Rho-associated kinase (Rho-kinase). Science 273:245–8

Kips JC, Pauwels RA (2001) Long-acting inhaled beta(2)-agonist therapy in asthma. Am J Respir Crit Care Med 164:923–32

Koch-Weser J (1968) Beta adrenergic blockade and circulating eosinophils. Arch Intern Med 121:255–8

Koeter GH, Meurs H, Kauffman HF, de Vries K (1982) The role of the adrenergic system in allergy and bronchial hyperreactivity. Eur J Respir Dis Suppl 121:72–8

Kottakis J, Cioppa GD, Creemers J, Greefhorst L, Lecler V, Pistelli R, Overend T, Till D, Rapatz G, Le Gros V, Bouros D, Siafakas N (2002) Faster onset of bronchodilation with formoterol than with salmeterol in patients with stable, moderate to severe COPD: results of a randomized, double-blind clinical study. Can Respir J 9:107–15

Kraan J, Koeter GH, vd Mark TW, Sluiter HJ, de Vries K (1985) Changes in bronchial hyperreactivity induced by 4 weeks of treatment with antiasthmatic drugs in patients with allergic asthma: a comparison between budesonide and terbutaline. J Allergy Clin Immunol 76:628–36

Langeler EG, van Hinsbergh VW (1991) Norepinephrine and iloprost improve barrier function of human endothelial cell monolayers: role of cAMP. Am J Physiol 260:C1052–9

Langley SJ, Sykes AP, Batty EP, Masterson CM, Woodcock A (2002) A comparison of the efficacy and tolerability of single doses of HFA 134a albuterol and CFC albuterol in mild-to-moderate asthmatic patients. Ann Allergy Asthma Immunol 88:488–93

Larsson S, Svedmyr N (1977) Bronchodilating effect and side effects of beta2- adrenoceptor stimulants by different modes of administration (tablets, metered aerosol, and combinations thereof). A study with salbutamol in asthmatics. Am Rev Respir Dis 116:861–9

Lau HY, Wong PL, Lai CK (1994) Effects of beta 2-adrenergic agonists on isolated guinea pig lung mast cells. Agents Actions 42:92–4

Lawrence AJ, Krstew E, Jarrott B (1995) Functional dopamine D2 receptors on rat vagal afferent neurones. Br J Pharmacol 114:1329–34

Lazarus SC, Boushey HA, Fahy JV, Chinchilli VM, Lemanske RF, Jr., Sorkness CA, Kraft M, Fish JE, Peters SP, Craig T, Drazen JM, Ford JG, Israel E, Martin RJ, Mauger EA, Nachman SA, Spahn JD, Szefler SJ, Asthma Clinical Research Network for the National Heart L, Blood I (2001) Long-acting beta2-agonist monotherapy vs continued therapy with inhaled corticosteroids in patients with persistent asthma: a randomized controlled trial. JAMA 285:2583–93

Leblanc P, Knight A, Kreisman H, Borkhoff CM, Johnston PR (1996) A placebo-controlled, crossover comparison of salmeterol and salbutamol in patients with asthma. American Journal of Respiratory & Critical Care Medicine 154:324–8

Liggett SB (2000a) The pharmacogenetics of beta2-adrenergic receptors: relevance to asthma. J Allergy Clin Immunol 105:S487–92

Liggett SB (2000b) Pharmacogenetics of beta-1- and beta-2-adrenergic receptors. Pharmacology 61:167–73

Linden A, Bergendal A, Ullman A, Skoogh BE, Lofdahl CG (1993) Salmeterol, formoterol, and salbutamol in the isolated guinea pig trachea: differences in maximum relaxant effect and potency but not in functional antagonism. Thorax 48:547–53

Linden M (1992) The effects of beta 2-adrenoceptor agonists and a corticosteroid, budesonide, on the secretion of inflammatory mediators from monocytes. Br J Pharmacol 107:156–60

Lipworth BJ, Hall IP, Aziz I, Tan KS, Wheatley A (1999) Beta2-adrenoceptor polymorphism and bronchoprotective sensitivity with regular short- and long-acting beta2-agonist therapy. Clin Sci 96:253–9

Lofdahl CG, Svedmyr N (1982) Effect of prenalterol in asthmatic patients. Eur J Clin Pharmacol 23:297–302

Lopez-Vidriero MT, Jacobs M, Clarke SW (1985) The effect of isoprenaline on the ciliary activity of an in vitro preparation of rat trachea. Eur J Pharmacol 112:429–32

Lotvall J (2001) Pharmacological similarities and differences between beta2-agonists. Respir Med 95:S7–11

Mack JA, Nielson CP, Stevens DL, Vestal RE (1986) Beta-adrenoceptor-mediated modulation of calcium ionophore activated polymorphonuclear leucocytes. Br J Pharmacol 88:417–23

Mahler DA, Donohue JF, Barbee RA, Goldman MD, Gross NJ, Wisniewski ME, Yancey SW, Zakes BA, Rickard KA, Anderson WH (1999) Efficacy of salmeterol xinafoate in the treatment of COPD. Chest 115:957–65

Mak JC, Nishikawa M, Haddad EB, Kwon OJ, Hirst SJ, Twort CH, Barnes PJ (1996) Localisation and expression of beta-adrenoceptor subtype mRNAs in human lung. Eur J Pharmacol 302:215–21

Martin CA, Naline E, Bakdach H, Advenier C (1994) Beta 3-adrenoceptor agonists, BRL 37344 and SR 58611A, do not induce relaxation of human, sheep and guinea-pig airway smooth muscle in vitro. Eur Respir J 7:1610–5

Martinez FD, Graves PE, Baldini M, Solomon S, Erickson R (1997) Association between genetic polymorphisms of the beta2-adrenoceptor and response to albuterol in children with and without a history of wheezing. J Clin Invest 100:3184–8

Mazzoni L, Naef R, Chapman ID, Morley J (1994) Hyperresponsiveness of the airways following exposure of guinea-pigs to racemic mixtures and distomers of beta 2-selective sympathomimetics. Pulm Pharmacol 7:367–76

McEvoy JD, Vall-Spinosa A, Paterson JW (1973) Assessment of orciprenaline and isoproterenol infusions in asthmatic patients. Am Rev Respir Dis 108:490–500

McGraw DW, Forbes SL, Kramer LA, Witte DP, Fortner CN, Paul RJ, Liggett SB (1999) Transgenic overexpression of beta(2)-adrenergic receptors in airway smooth muscle alters myocyte function and ablates bronchial hyperreactivity. J Biol Chem 274:32241–7

Miller WC, Rice DL (1980) A comparison of oral terbutaline and fenoterol in asthma. Ann Allergy 44:15–8

Mirza ZN, Tokuyama K, Arakawa H, Kato M, Mochizuki H, Morikawa A (1998) Inhaled procaterol inhibits histamine-induced airflow obstruction and microvascular leakage in guinea-pig airways with allergic inflammation. Clin Allergy 28:644–52

Mossberg B, Strandberg K, Philipson K, Camner P (1976) Tracheobronchial clearance in bronchial asthma: response to beta-adrenoceptor stimulation. Scand J Respir Dis 57:119–28

Moy AB, Shasby SS, Scott BD, Shasby DM (1993) The effect of histamine and cyclic adenosine monophosphate on myosin light chain phosphorylation in human umbilical vein endothelial cells. J Clin Invest 92:1198–206

Mueller R, Chanez P, Campbell AM, Bousquet J, Heusser C, Bullock GR (1996) Different cytokine patterns in bronchial biopsies in asthma and chronic bronchitis. Respir Med 90:79–85

Munoz NM, Rabe KF, Vita AJ, McAllister K, Mayer D, Weiss M, Leff AR (1995) Paradoxical blockade of beta adrenergically mediated inhibition of stimulated eosinophil secretion by salmeterol. J Pharmacol Exp Ther 273:850–4

Nakahara T, Moriuchi H, Yunoki M, Sakamato K, Ishii K (2000) Y-27632 potentiates relaxant effects of beta 2-adrenoceptor agonists in bovine tracheal smooth muscle. Eur J Pharmacol 389:103–6

Naline E, Zhang Y, Qian Y, Mairon N, Anderson GP, Grandordy B, Advenier C (1994) Relaxant effects and durations of action of formoterol and salmeterol on the isolated human bronchus. Eur Respir J 7:914–20

Nelson HS, Bensch G, Pleskow WW, DiSantostefano R, DeGraw S, Reasner DS, Rollins TE, Rubin PD (1998) Improved bronchodilation with levalbuterol compared with racemic albuterol in patients with asthma. J Allergy Clin Immunol 102:943–52

Newman KB, Milne S, Hamilton C, Hall K (2002) A comparison of albuterol administered by metered-dose inhaler and spacer with albuterol by nebulizer in adults presenting to an urban emergency department with acute asthma. Chest 121:1036–41

Nials AT, Ball DI, Butchers PR, Coleman RA, Humbles AA, Johnson M, Vardey CJ (1994) Formoterol on airway smooth muscle and human lung mast cells: a comparison with salbutamol and salmeterol. Eur J Pharmacol 251:127–35

Nielson CP (1987) Beta-adrenergic modulation of the polymorphonuclear leukocyte respiratory burst is dependent upon the mechanism of cell activation. J Immunol 139:2392–7

Oddera S, Silvestri M, Testi R, Rossi GA (1998) Salmeterol enhances the inhibitory activity of dexamethasone on allergen-induced blood mononuclear cell activation. Respiration 65:199–204

Orsida BE, Ward C, Li X, Bish R, Wilson JW, Thien F, Walters EH (2001) Effect of a long-acting beta2-agonist over three months on airway wall vascular remodeling in asthma. Am J Respir Crit Care Med 164:117–21

Partridge MR (1994) Metered-dose inhalers and CFCs: what respiratory physicians need to know. Respir Med 88:645–7

Paterson JW, Woolcock AJ, Shenfield GM (1979) Bronchodilator drugs. Am Rev Respir Dis 120:1149–88

Pearlman DS, Chervinsky P, LaForce C, Seltzer JM, Southern DL, Kemp JP, Dockhorn RJ, Grossman J, Liddle RF, Yancey SW (1992) A comparison of salmeterol with albuterol in the treatment of mild-to-moderate asthma. N Engl J Med 327:1420–5

Perrin-Fayolle M, Blum PS, Morley J, Grosclaude M, Chambe MT (1996) Differential responses of asthmatic airways to enantiomers of albuterol. Implications for clinical treatment of asthma. Clin Rev Allergy Immunol 14:139–47

Persson CG (1986) Role of plasma exudation in asthmatic airways. Lancet 2:1126–9

Persson CG (1987) Leakage of macromolecules from the tracheobronchial microcirculation. Am Rev Respir Dis 135:S71–5

Persson CG, Erjefalt I, Andersson P (1986) Leakage of macromolecules from guinea-pig tracheobronchial microcirculation. Effects of allergen, leukotrienes, tachykinins, and anti-asthma drugs. Acta Physiol Scand 127:95–105

Pesci A, Balbi B, Majori M, Cacciani G, Bertacco S, Alciato P, Donner CF (1998) Inflammatory cells and mediators in bronchial lavage of patients with chronic obstructive pulmonary disease. Eur Respir J 12:380–6

Peters SP, Schulman ES, Schleimer RP, MacGlashan DW, Jr., Newball HH, Lichtenstein LM (1982) Dispersed human lung mast cells. Pharmacologic aspects and comparison with human lung tissue fragments. Am Rev Respir Dis 126:1034–9

Pfitzer G (2001) Invited review: regulation of myosin phosphorylation in smooth muscle. J Appl Physiol 91:497–503

Phipps RJ, Nadel JA, Davis B (1980) Effect of alpha-adrenergic stimulation on mucus secretion and on ion transport in cat trachea in vitro. Am Rev Respir Dis 121:359–65

Planquois JM, Mottin G, Artola M, Lagente V, Payne A, Dahl S (1998) Effects of phosphodiesterase inhibitors and salbutamol on microvascular leakage in guinea-pig trachea. Eur J Pharmacol 344:59–66

Pratt HF (1982) Abuse of salbutamol inhalers in young people. Clin Allergy 12:203–9

Prior C, Leonard MB, McCullough JR (1998) Effects of the enantiomers of R,S-salbutamol on incompletely fused tetanic contractions of slow- and fast-twitch skeletal muscles of the guinea-pig. Br J Pharmacol 123:558–64

Rabe KF, Giembycz MA, Dent G, Perkins RS, Evans P, Barnes PJ (1993) Salmeterol is a competitive antagonist at beta-adrenoceptors mediating inhibition of respiratory burst in guinea-pig eosinophils. Eur J Pharmacol 231:305–8

Reed CE (1985) Adrenergic bronchodilators: pharmacology and toxicology. J Allergy Clin Immunol 76:335–41

Reed CE, Cohen M, Enta T (1970) Reduced effect of epinephrine on circulating eosinophils in asthma and after beta-adrenergic blockade or Bordetella pertussis vaccine. With a note on eosinopenia after methacholine. J Allergy 46:90–102

Rippe B, Grega GJ (1978) Effects of isoprenaline and cooling on histamine induced changes of capillary permeability in the rat hindquarter vascular bed. Acta Physiol Scand 103:252–62

Rugg EL, Barnett DB, Nahorski SR (1978) Coexistence of beta1 and beta2 adrenoceptors in mammalian lung: evidence from direct binding studies. Mol Pharmacol 14:996–1005

Sano Y, Watt G, Townley RG (1983) Decreased mononuclear cell beta-adrenergic receptors in bronchial asthma: parallel studies of lymphocyte and granulocyte desensitization. J Allergy Clin Immunol 72:495–503

Schenkelaars EJ, Bonta IL (1984) Beta 2-adrenoceptor agonists reverse the leukotriene C4-induced release response of macrophages. Eur J Pharmacol 107:65–70

Schmekel B, Rydberg I, Norlander B, Sjosward KN, Ahlner J, Andersson RG (1999) Stereoselective pharmacokinetics of S-salbutamol after administration of the racemate in healthy volunteers. Eur Respir J 13:1230–5

Schmitz AA, Govek EE, Bottner B, Van Aelst L (2000) Rho GTPases: signaling, migration, and invasion. Exp Cell Res 261:1–12

Seasholtz TM, Majumdar M, Brown JH (1999) Rho as a mediator of G protein-coupled receptor signaling. Mol Pharmacol 55:949–56

Sekut L, Champion BR, Page K, Menius JA, Jr., Connolly KM (1995) Anti-inflammatory activity of salmeterol: down-regulation of cytokine production. Clin Exp Immunol 99:461–6

Shrewsbury S, Pyke S, Britton M (2000) Meta-analysis of increased dose of inhaled steroid or addition of salmeterol in symptomatic asthma (MIASMA). BMJ 320:1368–73

Siflinger-Birnboim A, Bode DC, Malik AB (1993) Adenosine 3',5'-cyclic monophosphate attenuates neutrophil-mediated increase in endothelial permeability. Am J Physiol 264:H370–5

Simons FE (1997) A comparison of beclomethasone, salmeterol, and placebo in children with asthma. Canadian Beclomethasone Dipropionate-Salmeterol Xinafoate Study Group. N Engl J Med 337:1659–65
Smith SR, Kendall MJ (1984) Metabolic responses to beta 2 stimulants. J R Coll Physicians Lond 18:190–4
Somlyo AP, Somlyo AV (2000) Signal transduction by G-proteins, rho-kinase and protein phosphatase to smooth muscle and non-muscle myosin II. J Physiol (Lond) 2:177–85
Spina D, Rigby PJ, Paterson JW, Goldie RG (1989) Autoradiographic localization of beta-adrenoceptors in asthmatic human lung. Am Rev Respir Dis 140:1410–5
Stiles GL, Caron MG, Lefkowitz RJ (1984) Beta-adrenergic receptors: biochemical mechanisms of physiological regulation. Physiol Rev 64:661–743
Suzuki H, Shindo K, Ueno A, Miura T, Takei M, Sakakibara M, Fukamachi H, Tanaka J, Higa T (1999) S1319: a novel beta2-andrenoceptor agonist from a marine sponge Dysidea sp. Bioorg Med Chem Lett 9:1361–4
Svedmyr NL, Larsson SA, Thiringer GK (1976) Development of "resistance" in beta-adrenergic receptors of asthmatic patients. Chest 69:479–83
Szentivanyi A (1979) The conformational flexibility of adrenoceptors and the constitutional basis of atopy. Triangle 18:109–15
Tai E, Read J (1967) Response of blood gas tensions to aminophylline and isoprenaline in patients with asthma. Thorax 22:543–9
Tan S, Hall IP, Dewar J, Dow E, Lipworth B (1997) Association between beta 2-adrenoceptor polymorphism and susceptibility to bronchodilator desensitisation in moderately severe stable asthmatics. Lancet 350:995–9
Taylor DR, Buick B, Kinney C, Lowry RC, McDevitt DG (1985) The efficacy of orally administered theophylline, inhaled salbutamol, and a combination of the two as chronic therapy in the management of chronic bronchitis with reversible air-flow obstruction. Am Rev Respir Dis 131:747–51
Taylor DR, Town GI, Herbison GP, Boothman-Burrell D, Flannery EM, Hancox B, Harre E, Laubscher K, Linscott V, Ramsay CM, Richards G (1998) Asthma control during long-term treatment with regular inhaled salbutamol and salmeterol. Thorax 53:744–52
Teixeira MM, Hellewell PG (1997) Evidence that the eosinophil is a cellular target for the inhibitory action of salmeterol on eosinophil recruitment in vivo. Eur J Pharmacol 323:255–60
Teixeira MM, Williams TJ, Hellewell PG (1995) Anti-inflammatory effects of a short-acting and a long-acting beta 2-adrenoceptor agonist in guinea pig skin. Eur J Pharmacol 272:185–93
Thirstrup S (2000) Control of airway smooth muscle tone: II-pharmacology of relaxation. Respir Med 94:519–28
Turki J, Pak J, Green SA, Martin RJ, Liggett SB (1995) Genetic polymorphisms of the beta 2-adrenergic receptor in nocturnal and nonnocturnal asthma. Evidence that Gly16 correlates with the nocturnal phenotype. J Clin Invest 95:1635–41
Ullman A, Svedmyr N (1988) Salmeterol, a new long acting inhaled beta 2 adrenoceptor agonist: comparison with salbutamol in adult asthmatic patients. Thorax 43:674–8
Ulrik CS (1995) Efficacy of inhaled salmeterol in the management of smokers with chronic obstructive pulmonary disease: a single centre randomised, double blind, placebo controlled, crossover study. Thorax 50:750–4
Van den Berg NJ, Ossip MS, Hederos CA, Anttila H, Ribeiro BL, Davies PI (2000) Salmeterol/fluticasone propionate (50/100 microg) in combination in a Diskus inhaler (Seretide) is effective and safe in children with asthma. Pediatr Pulmonol 30:97–105
van Essen-Zandvliet EE, Hughes MD, Waalkens HJ, Duiverman EJ, Pocock SJ, Kerrebijn KF (1992) Effects of 22 months of treatment with inhaled corticosteroids and/or beta-2-agonists on lung function, airway responsiveness, and symptoms in children

with asthma. The Dutch Chronic Non-specific Lung Disease Study Group. Am Rev Respir Dis 146:547-54

van Noord JA, de Munck DR, Bantje TA, Hop WC, Akveld ML, Bommer AM (2000) Long-term treatment of chronic obstructive pulmonary disease with salmeterol and the additive effect of ipratropium. Eur Respir J 15:878-85

Vathenen AS, Britton JR, Ebden P, Cookson JB, Wharrad HJ, Tattersfield AE (1988) High-dose inhaled albuterol in severe chronic airflow limitation. Am Rev Respir Dis 138:850-5

Verberne AA, Frost C, Roorda RJ, van der Laag H, Kerrebijn KF (1997) One year treatment with salmeterol compared with beclomethasone in children with asthma. The Dutch Paediatric Asthma Study Group. Am J Respir Crit Care Med 156:688-95

Verdugo P, Johnson NT, Tam PY (1980) beta-Adrenergic stimulation of respiratory ciliary activity. J Appl Physiol: Respir, Environ Exercise Physiol 48:868-71

Verleden GM, Belvisi MG, Rabe KF, Miura M, Barnes PJ (1993) Beta 2-adrenoceptor agonists inhibit NANC neural bronchoconstrictor responses in vitro. J Appl Physiol 74:1195-9

Whelan CJ, Johnson M (1992) Inhibition by salmeterol of increased vascular permeability and granulocyte accumulation in guinea-pig lung and skin. Br J Pharmacol 105:831-8

Whelan CJ, Johnson M, Vardey CJ (1993) Comparison of the anti-inflammatory properties of formoterol, salbutamol and salmeterol in guinea-pig skin and lung. Br J Pharmacol 110:613-8

Williams LT, Snyderman R, Lefkowitz RJ (1976) Identification of beta-adrenergic receptors in human lymphocytes by (-) (3H) alprenolol binding. J Clin Invest 57:149-55

Woodcock A (1995) Continuing patient care with metered-dose inhalers. J Aerosol Med 8:S5-10

Woolcock A, Lundback B, Ringdal N, Jacques LA (1996) Comparison of addition of salmeterol to inhaled steroids with doubling of the dose of inhaled steroids. Am J Respir Crit Care Med 153:1481-8

Xie GX, Jones K, Peroutka SJ, Palmer PP (1998) Detection of mRNAs and alternatively spliced transcripts of dopamine receptors in rat peripheral sensory and sympathetic ganglia. Brain Res 785:129-35

Yamaguchi H, McCullough JR (1996) S-albuterol exacerbates calcium responses to carbachol in airway smooth muscle cells. Clin Rev Allergy Immunol 14:47-55

Yamamoto C, Yoneda T, Yoshikawa M, Fu A, Tokuyama T, Tsukaguchi K, Narita N (1997) Airway inflammation in COPD assessed by sputum levels of interleukin-8. Chest 112:505-10

Young WS, 3rd, Kuhar MJ (1979) A new method for receptor autoradiography. Brain Res 179:255-70

Zhang XY, Zhu FX, Olszewski MA, Robinson NE (1998) Effects of enantiomers of beta 2-agonists on ACh release and smooth muscle contraction in the trachea. Am J Physiol 274:L32-8

Anticholinergic Bronchodilators

N. J. Gross

Departments of Medicine and Molecular Biochemistry,
Loyola University Stritch School of Medicine, Room A319, Bldg 1,
Edward Hines Jr. VA Hospital, Roosevelt and 1st Avenues, PO Box 1485,
Hines, IL 60141, USA
e-mail: nicholas.gross@med.va.gov

1	Introduction	38
2	Rationale	38
2.1	Muscarinic Receptor Subtypes in Airways	39
3	Pharmacology	40
3.1	Subsensitivity	41
3.2	Pharmacokinetics	41
4	Clinical Efficacy	41
4.1	Dose–Response	41
4.2	Against Specific Stimuli	42
4.3	In Stable Asthma	42
4.4	Acute Severe Asthma	43
4.5	Paediatric Airways Disease	43
4.6	Stable COPD	44
4.7	Effects on Sleep Quality	46
4.8	Acute Exacerbations of COPD	46
4.9	Combinations with Other Bronchodilators	46
5	Side Effects	47
6	Clinical Recommendations	48
	References	49

Abstract Anticholinergic agents produce bronchodilation by relaxing peribronchial smooth muscle. Unlike other agents, they have no other therapeutic actions on airways. Their main use is in chronic obstructive pulmonary disease (COPD) in which condition they are first-line therapy for stable disease. The new, long-acting anticholinergic agent tiotropium has the advantages that it has both a very long duration of action, making it appropriate for once-daily therapy, and is selective for the muscarinic receptor subtypes that mediate smooth muscle activity. Anticholinergic agents may have a role in asthma as adjunctive treatment in stable asthma, and in combination with an adrenergic agent in acute severe asthma. The currently available agents, ipratropium, oxitropium and tiotropium are poorly absorbed when taken by inhalation and, consequently, have a very wide margin of safety, dry mouth being the only common adverse effect.

Keywords Antimuscarinic agents · Muscarinic receptors · Vagal activity · Bronchodilators · Ipratropium · Tiotropium · Oxitropium · Stable COPD · Asthma

1
Introduction

Anticholinergic agents such as atropine are present in many plants indigenous to tropical and temperate regions and have been used in herbal remedies for many centuries. There are accounts from the seventeenth century of the use of *Datura stramonium* leaves for the treatment of asthma in India. In the early nineteenth century this plant was brought to Europe by British colonists and quickly became widely used to treat breathing difficulties (Courty 1859; Gandevia 1975).

Naturally-occurring anticholinergics such as atropine and scopolamine are well absorbed from the respiratory and gastrointestinal tracts and produce many side effects and, therefore, are poorly accepted by patients. Thus, when effective alternatives were discovered such as adrenaline in the 1920s and xanthines in the 1930s, anticholinergic agents fell out of favour. But anticholinergic bronchodilators returned to clinical use with better understanding of the role of the parasympathetic system in controlling airway tone and with the synthesis of congeners of atropine that are much less prone to produce side effects (Gross and Skorodin 1984a).

2
Rationale

Parasympathetic, cholinergic nerves supply most of the autonomic innervation to the human airways via branches of the vagus nerve (Richardson 1982). These travel along the airways and synapse at peribronchial ganglia with short postganglionic nerves which supply smooth muscle cells and mucous glands, predominantly in the central airways. The varicosities and terminals of the postganglionic nerves release acetylcholine which activates muscarinic receptors on these structures, stimulating smooth muscle contraction, releasing mucus from mucus glands, and accelerating ciliary beat frequency. At rest, a low level of ongoing vagal tone results in tonic smooth muscle contraction. In addition, phasic augmentation of cholinergic activity can result from a variety of stimuli by means of neural reflex pathways. Afferent activity from irritant receptors and C fibres located anywhere in the upper and lower airways, and probably also from the oesophagus and carotid bodies, is transmitted along vagal afferents, through the vagal nuclei in the brain-stem, and then through vagal efferents to the larger airways that receive vagal innervation. Stimuli which elicit such reflex vagal activity include mechanical irritation, many irritant gases, aerosols, particles, cold dry air, and a variety of specific mediators (Widdicombe 1979; Nadel 1980). Some bronchospastic events in humans are mediated at least partly by this reflex vagal activity. There is also evidence that baseline cholinergic bronchomotor tone is increased in both asthma (Shah et al. 1990) and chronic obstructive pulmonary disease (COPD) (Gross et al. 1989). Both tonic and phasic mechanisms are amenable to inhibition by anticholinergic agents.

Anticholinergic agents compete with acetylcholine at muscarinic receptors and thus inhibit cholinergic activity, resulting in airways dilatation. However, as vagal activity probably accounts for only a portion of the airflow obstruction in patients with asthma or COPD, airflow is not usually normalized by anticholinergic agents, nor do they inhibit other mediators of airways obstruction (such as leukotrienes) or other mechanisms of airways inflammation.

2.1
Muscarinic Receptor Subtypes in Airways

The physiological effects of acetylcholine are mediated by interaction with muscarinic receptors. There are at least five (and possibly more) closely related muscarinic receptor genes in the human genome, of which three (called M_1, M_2, and M_3) are expressed in the lung. These muscarinic receptor subtypes are found in different lung structures and appear to have different physiologic functions. Muscarinic receptors in the airways are predominantly located in central airways in contrast to β-adrenergic receptors which are located throughout the airways (Carstairs et al. 1985). Current understanding is that M_1 receptors, which are located in peribronchial ganglia, facilitate the amplification and transmission of preganglionic nerve traffic to postganglionic nerves; they thus enhance bronchoconstriction and mucus release. M_3 receptors are located on smooth muscle cells and submucosal glands and mediate smooth muscle contraction and mucus secretion (Gross and Barnes 1988). Inhibition of M_1 and M_3 receptors is thus desirable in alleviating airways obstruction. M_2 receptors, by contrast, are autoreceptors located on postganglionic fibres whose stimulation provides feedback inhibition of further acetylcholine release from postganglionic nerve terminals. They thus tend to limit vagal bronchoconstriction and their inhibition would be undesirable in treating airways obstruction.

This scheme has important physiologic implications for the use of anticholinergic agents in airways disease.

First, all naturally occurring anticholinergic agents as well as the synthetic anticholinergic agents such as ipratropium and oxitropium are not selective for muscarinic subtypes—they inhibit M_2 receptors as well as M_1 and M_3 receptors. Second, M_2 receptors appear to be selectively damaged by certain viruses as well as by some eosinophil products. It has been suggested that this may account for the bronchospasm associated with viral infections particularly in children (Fryer and Jacoby 1991, 1993). Attempts to develop selective anticholinergic agents have resulted in tiotropium bromide, a congener of ipratropium that binds strongly to M_1 and M_3 receptors, but which dissociates rapidly from M_2 receptors, rendering it functionally selective for M_1 and M_3 receptors (Barnes et al. 1995; Maesen et al. 1995; O'Connor et al. 1996; Disse et al. 1999).

3
Pharmacology

Anticholinergic agents are classified as either tertiary or quaternary ammonium alkaloids, depending on the valence of the nitrogen atom in the tropane ring, Fig. 1. In anticholinergic agents found in nature, e.g. atropine and scopolamine, the nitrogen atom is 3-valent, and these agents are thus tertiary ammonium compounds. They are soluble in water and lipids and are well absorbed from mucosal surfaces and the skin. Consequently they are widely distributed in the body and cross the blood–brain barrier, producing wide-spread, dose-related systemic effects. Thus, atropine in the dose that results in bronchodilatation (1.0–2.5 mg in adults) commonly results in flushing of the skin, dry mouth and some tachycardia. In only slightly higher doses it may produce blurred vision, urinary retention in males and mental effects such as irritability and confusion. The therapeutic margin of atropine and its natural congeners is small, making these agents less than optimal for clinical use.

Quaternary ammonium compounds, which are all synthetic, carry a charge associated with the 5-valent nitrogen atom that renders these molecules poorly absorbable from mucosal surfaces. Such agents retain their anticholinergic action at the site of deposition and will, for example, dilate the pupil if delivered to the eye or dilate the airways when inhaled. However, their systemic absorption from these sites is insufficient to produce systemic effects, even when delivered in much higher dosage than recommended (Gross and Skorodin 1985). Quaternary agents can thus be regarded for practical purposes as topically active forms of atropine. This group includes ipratropium bromide (Atrovent), oxitropium bromide (Oxivent), atropine methonitrate, glycopyrrolate bromide (Robinul) and tiotropium bromide (Spiriva).

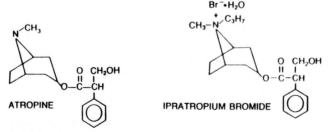

Fig. 1 Structures of atropine, a tertiary ammonium anticholinergic agent, and ipratropium bromide, a quaternary ammonium agent

Tiotropium is unique in that it is functionally selective for the muscarinic receptor subtypes that are believed to mediate bronchoconstriction (see above) and is also extremely long acting (Maesen et al. 1995; O'Connor et al. 1996), which allows for once-daily dosing as a regular bronchodilator. Its prolonged action also provides protection against nocturnal bronchoconstriction which is largely mediated by cholinergic mechanisms. For these reasons, tiotropium may prove preferable to alternative currently available anticholinergic agents. Indeed, in several randomized trials involving many patients with stable COPD, tiotropium was more effective than ipratropium in improving trough, average, and peak forced expiratory volume in 1 s (FEV_1) over periods of 3 months or more (van Noord et al. 2000, 2002; Vincken et al. 2002).

3.1
Subsensitivity

In general, subsensitivity or tachyphylaxis does not occur with receptor antagonists, even following prolonged or intensive use (in contrast with receptor agonists). In keeping with this principle, subsensitivity has been sought but not found in the case of ipratropium. This is expected to be the case with other members of this class.

3.2
Pharmacokinetics

Atropine is quantitatively absorbed from the airways, and peak blood levels are reached in about 1 h. Its plasma half-life is about 3 h in most adults, possibly longer in children and the elderly (Gross and Skorodin 1984a). Small amounts are found in the faeces and in breast milk. Studies of radiolabelled ipratropium following oral or inhaled administration show that serum levels are very low, with a peak at about 1–2 h and a half-life of about 4 h. Most of the drug is excreted unchanged in the urine. Very little reaches the central nervous system. Its bronchodilator effect following inhalation is somewhat longer than that of atropine, probably because it is not removed from the airways by absorption. Most of an oral dose is recovered unchanged in the faeces, a small amount is recovered as inactive metabolites in the urine. A similar distribution is likely for tiotropium, however the half-life of tiotropium is in excess of 1 day (Disse et al. 1999).

4
Clinical Efficacy

4.1
Dose–Response

Dose–response data for a variety of anticholinergic agents given by various inhalational methods has been provided in a previous review (Gross and Skorodin 1987).

For ipratropium by nebulized solution, the optimal dose is 500 µg in adults and 125–250 µg in children. By metered dose inhaler (MDI), its optimal dose is 40–80 µg in young adults with asthma and 160 µg in older patients with stable COPD in whom airways obstruction is generally more severe. The optimal dose by dry powder inhaler (DPI) may be lower than that by MDI. Thus 10 µg of ipratropium delivered as DPI was equipotent to 20 µg delivered by MDI (Bollert et al. 1997).

In dose-ranging studies, the optimal dose of tiotropium by DPI in patients with stable COPD was 18 µgm (Maesen et al. 1995, Littner et al 2000). Tiotropium will not be available as an MDI. It may also become available by an alternative device (Respimat) that delivers a "soft mist".

4.2
Against Specific Stimuli

In laboratory studies in asthmatic humans, anticholinergic agents provided variable protection against bronchospastic stimuli (Gross and Skorodin 1984a). They protected well against cholinergic agonists such as methacholine and against bronchospasm induced by β-blocking agents and by psychogenic factors. Against bronchospasm due to most other stimuli, e.g. histamine, prostaglandins, non-specific dusts and irritant aerosols, exercise and hyperventilation with cold, dry air, they provide only partial protection at best (Ayala and Ahmed 1989; Azevedo et al. 1990). In most of the latter instances, adrenergic agents usually provide greater protection. Ipratropium had no prophylactic effect against leukotriene-induced bronchospasm (Ayala et al. 1988).

4.3
In Stable Asthma

A very large number of studies have compared the bronchodilator effect of anticholinergic agents with those of adrenergic agents in patients with stable asthma (Ruffin et al. 1977). Most of these studies used recommended doses rather than optimal doses of each class of agent. In general, the anticholinergic agent takes longer to reach peak bronchodilator effect, e.g. 30–60 min in the case of ipratropium compared with about 15 min for most adrenergic agents. At peak effect, they almost invariably result in less bronchodilation than following an adrenergic agent, but an anticholinergic agent such as ipratropium tends to be slightly longer-acting than an adrenergic agent such as salbutamol.

There is substantial variation in responsiveness among asthmatic patients; however, a few patients respond as well to an anticholinergic agent as to adrenergic agents. Attempts to identify which asthmatic patients are likely to respond well to anticholinergics have not been successful. In general, the bronchodilator effect of ipratropium is maintained or may even increase with age, in contrast to the effect of adrenergic agents, which tends to decrease with age (Ullah et al. 1981). However, children aged 10–18 years have been shown to respond to anticholinergic use (Vichyanond et al. 1990). Individuals with intrinsic asthma and those with longer duration of asthma may also respond better to an anticholinergic agent than individuals with extrinsic asthma (Jolobe 1984). An individual trial probably remains the best way to identify responsiveness (Brown et al. 1984). Postnasal drip may provoke asthma symptoms. Ipratropium nasal spray is commercially available and effective in reducing rhinorrhoea (Baroody et al. 1992) and, in these patients, may reduce asthma symptoms.

4.4
Acute Severe Asthma

Studies suggest that β-adrenergic agonists are the bronchodilators of choice in acute severe asthma. Although an anticholinergic agent should not be used as the sole initial bronchodilator, it may add to the bronchodilatation achieved by the adrenergic agent. The combination of 500 μg nebulized ipratropium with 1.25 mg nebulized fenoterol resulted in significantly more bronchodilatation over the first 90 min of treatment than either agent alone (Rebuck et al. 1987), and patients with more severe airway obstruction obtained the greatest benefit from the combination. In a meta-analysis of 10 such studies (total of 1,377 patients), the addition of ipratropium to a β-agonist reduced hospital admissions by 27% and increased FEV_1 by 100 ml more than in patients receiving adrenergic agents alone (Stoodley et al. 1999)—benefits which were both statistically and clinically significant.

It is thus appropriate to administer both classes of bronchodilators in acute severe asthma, especially during the initial hours of treatment (Brophy et al. 1998) and particularly in patients with more severe airflow obstruction.

4.5
Paediatric Airways Disease

In children with acute severe asthma, the addition of ipratropium to a β-adrenergic agent accelerated the rate of improvement of airflow in some studies (Beck et al. 1985; Reisman et al. 1988), but not in others (Storr and Lenney 1986; Boner et al. 1987; Schuh et al. 1995; Ducharme and Davis 1998; Qureshi et al. 1998; Zorc et al. 1999). A systematic review of ten studies concluded that the addition of multiple doses of ipratropium to conventional adrenergic therapy was safe, improved lung function, and reduced hospitalization rates in children with severe asthma (Plotnick and Ducharme 1998). As for adults with status asthmaticus, therefore, the combination of ipratropium with an adrenergic agent is probably more effective than adrenergic monotherapy, particularly in severe exacerbations.

In stable paediatric asthma, the role of ipratropium is less clear. Two consensus reports reviewed the evidence, which is limited, and concluded that ipratropium was safe for the paediatric population, but its bronchodilator action was not as potent as that of an adrenergic agent (Warner et al. 1989; Hargreave et al. 1990).

There are reports of the use of ipratropium in other paediatric conditions such as cystic fibrosis, viral bronchiolitis, exercise-induced bronchospasm and bronchopulmonary dysplasia, but these provide only inconsistent evidence for the benefit of ipratropium over alternative bronchodilators.

4.6
Stable COPD

The main role of anticholinergic agents is in the management of stable COPD, in which a large number of studies have compared anticholinergic agents with other bronchodilators (Thiessen and Petersen 1982; Passamonte and Martinez 1984). Although patients with COPD, as a group, do not usually respond as much to any bronchodilator as do patients with asthma, most patients with this diagnosis are capable of significant bronchodilation and many studies show that an anticholinergic agent is a more potent bronchodilator than other agents in COPD (Brown et al. 1984; Tashkin et al. 1986; Braun et al. 1989; Bleeker and Britt 1991). This generalization applies equally to short- and long-acting bronchodilators (Donohue et al. 2002). Large cumulative doses of an anticholinergic agent alone achieve all the available bronchodilatation in the COPD population (Gross and Skorodin 1984b), in which respect COPD patients contrast markedly with asthmatic patients. In studies where bronchodilator responsiveness was compared between patients with asthma and COPD who had similar baseline airflows, patients with bronchitis had a better response to ipratropium than to various adrenergic therapies, whereas asthmatics invariably responded better to adrenergic agents (Lefcoe et al. 1982), Fig. 2. The reasons for this are unclear but one speculates it may be related to the fact that airways obstruction in asthma is due to numerous factors such as airway inflammation, mucosal oedema, etc., that are at least partially modified by adrenergic agents but not by anticholinergics. In COPD, the major reversible component is bronchomotor tone, which is best inhibited by an anticholinergic agent.

Accordingly, ipratropium is currently recommended as first-line treatment for maintenance therapy of stable COPD in almost all guidelines and statements (American Thoracic Society 1995; Siafakis et al. 1995).

Apart from transient bronchodilation, anticholinergic agents have no proven beneficial effects in COPD. A small increase in the baseline FEV_1 was suggested in a retrospective analysis of 7 large 3-month studies (Rennard et al. 1996); however, the 5-year. U.S. Lung Health Study found no discernible effect on the age-related decline in lung function of smokers (Anthonisen et al. 1994). Some studies with tiotropium suggest that its long-term use may decrease the rate of decline of FEV_1 (Casaburi et al 2002; Fig. 3), and further studies of this are in progress. For the present, anticholinergic agents should be regarded as providing transient bronchodilation, i.e. "reliever effects", but no long-term "controller" effects.

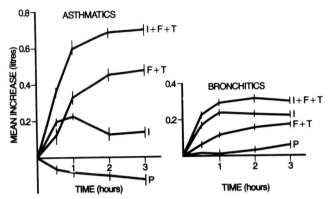

Fig. 2 Increase in forced expiratory volume in 1 s (FEV_1) of 15 patients with asthma (*left panel*) and 15 patients with chronic bronchitis (*right panel*). Baseline FEV_1 was similar in both groups. *P*, placebo metered dose inhaler (MDI); *I*, ipratropium 40 μg MDI; *F+T*, fenoterol 5 mg MDI plus oxtriphylline 400 mg oral. (Reproduced from Lefcoe et al. 1982 with permission)

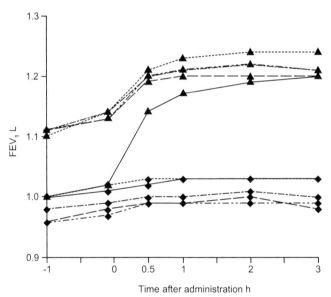

Fig. 3 Mean response to tiotropium (*triangles*) or placebo (*diamonds*) on days 1 (*continuous lines*), day 8 (*short dashes*), day 92 (*alternating one long and one short dash*), day 176 (*alternating two short dashes with long dash*), and day 344 (*long dashes*). FEV_1 (forced expiratory volume in 1 s) is given in litres. The SEM for the mean differences between groups ranged between 0.01–0.02 l. Note that the elevation over baseline prior to drug administration on days subsequent to the first treatment day in the tiotropium group is evidence of tiotropium 24-h duration of action. Note also that the pre-bronchodilator FEV_1 of the tiotropium subjects did not change between days 8 and 344, while those of the placebo group declined by about 40 ml over the same period, suggesting (but not concluding) that regular tiotropium use may decrease the age-related decline in airflow in COPD. (Reproduced from Casaburi et al. 2002 with permission)

Other possibly beneficial aspects of long-term tiotropium use in COPD include improvements in quality of life (mainly due to improvement in the impacts component) and improvement in the dyspnea scores (transitional dyspnea index) (Casaburi et al. 2002; Donohue et al. 2002; Vincken et al. 2002; Tashkin and Kesten 2003).

4.7
Effects on Sleep Quality

Sleep disturbance is common among patients with chronic bronchitis and asthma, up to 41% of patients with obstructive airways disease reporting symptoms of disturbed sleep (Klink and Quan 1987). In sleep studies, patients with COPD frequently experience nocturnal oxygen desaturation which may be profound during rapid eye movement (REM) sleep-stages, even in the absence of obstructive sleep apnoea (Douglas et al. 1998). It is not known if this contributes to the development of pulmonary hypertension, polycythaemia or cardiac problems. A randomized, double-blind study of patients with moderate-to-severe COPD showed that ipratropium increased total sleep time, decreased the severity of nocturnal desaturation, and improved perceptions of sleep quality (Martin et al. 1999).

4.8
Acute Exacerbations of COPD

Studies comparing the efficacy of bronchodilators in the management of acute exacerbations of COPD have not shown a difference among adrenergic agents, anticholinergic agents, or their combination (Rebuck et al. 1987; Karpel et al. 1990; Patrick et al. 1990; Koutsogiannis and Kelly 2000). Adrenergic agents should be included in the management. Guidelines from the American Thoracic Society, European Respiratory Society and British Thoracic Society recommend combination therapy with both adrenergic and anticholinergic agents in the initial management of acute exacerbations of COPD (American Thoracic Society 1995; Siafakis et al. 1995; British Thoracic Society 1996).

Two recent studies suggest that regular use of tiotropium may reduce the frequency of acute exacerbations of COPD by between 14%–24% (Casaburi et al. 2002; Vincken et al. 2002), an outcome that has not been consistently found with other bronchodilators. If confirmed by longer-term studies, which are in progress, the potential impact of this on quality of life and savings in health care costs would be substantial.

4.9
Combinations with Other Bronchodilators

The efficacy of combinations of different classes of bronchodilators has been studied and found often to provide more bronchodilatation than single agents.

Bearing in mind that the recommended doses of bronchodilators are rarely the optimal doses, the effects of combinations of agents may simply be additive rather than potentiating. Nonetheless, anticholinergic, adrenergic and methylxanthine agents work by different mechanisms, have different pharmacokinetic and pharmacodynamic properties, and affect different-sized airways. Their combination is thus rational and is likely to result in improved bronchodilatation. Unfavourable interactions among these classes of agents have not been reported, so the greater bronchodilation resulting from their use in combination is achieved without increasing the risk of side effects.

In practice, the use of two or even all of these agents simultaneously to manage severe airways obstruction is endorsed by all current Guidelines. Single MDIs combining different classes of bronchodilators were used in the 1950s, but fell out of favour. Since the early 1980s, combinations of ipratropium and the β-agonist fenoterol (Berodual and DuoVent) have been widely used in many countries but not marketed in the USA.

More recently, a combination MDI containing ipratropium and salbutamol was developed (Combivent). For patients who need more than one bronchodilator, a single MDI containing both agents has distinct advantages. It is less expensive than two separate MDIs as well as more convenient for patients to use and is therefore likely to improve patient compliance. Clinical trials with this combination in patients with stable COPD show that it has the rapid onset of action of the β-adrenergic component, the slightly longer duration of the anticholinergic component, plus the increased potency of the combination at intermediate times, without any increase in side effects (Petty 1994; Ikeda et al. 1995). A post hoc review of 3-month trials involving 1,067 patients in the USA concluded that the combination was cost-effective (Friedman et al. 1999). A nebulizer formulation of this combination (DuoNeb) showed similar clinical benefits (Gross et al. 1999).

5
Side Effects

Atropine has a very narrow therapeutic margin and produces numerous systemic side effects at or only slightly above the bronchodilating dose. There is no place for its use as a bronchodilator in the current era. The principal advantage of quaternary anticholinergic agents is that they are very poorly absorbed from mucosal surfaces, so the risk of systemic side effects is quite small. Even massive, inadvertent overdosage of one such agent resulted in only trivial effects (Gross and Skorodin 1985). Ipratropium, the most widely studied quaternary anticholinergic, has very few systemic atropine-like side effects (Gross 1988). It can, for example, be given to patients with narrow-angle glaucoma without affecting intraocular tension (Watson et al. 1994), provided care is taken to avoid spraying it directly into the eye. It has negligible effects on urinary flow in older men. Nor does it alter the viscosity or elasticity of respiratory mucus, or mucociliary clearance, as does atropine (Pavia et al. 1979). It has only trivial effects on ventilation, haemodynamics (Tobin et al. 1984), and the pulmonary circulation (Chapman et al. 1985). Nor do quaternary anticholinergic agents carry the risk of worsening hypoxaemia, as do adrenergic agents (Gross and Bankwala 1987; Ashotosh et al. 1995; Khoukhaz and Gross 1999)—a potentially important consideration in exacerbations of airways obstruction.

In research studies, the only side effects that occurred more commonly with ipratropium than with placebo were dry mouth and, sometimes, a brief coughing spell. Neither effect resulted in patient withdrawal from the studies. Paradoxical bronchoconstriction has been reported with ipratropium in about 0.3% of patients and has been variously attributed to hypotonicity of the nebulized solution, idiosyncrasy to the bromine radical, and idiosyncrasy to the benzalkonium preservative (Boucher et al. 1992; Beasley et al. 1998). This rare side effect, which has also occurred with oxitropium, warrants withdrawal of the drug from that patient. Other than these, extensive clinical investigation and the worldwide use of ipratropium for over two decades demonstrate a remarkably low incidence of untoward reactions.

The safety profile of tiotropium appears to be very similar to that of ipratropium (Hvizdos and Goa 2002), although, having only been introduced for general use in 2002, experience with tiotropium is not as great.

6
Clinical Recommendations

Current guidelines recommend ipratropium as first-line therapy in the long-term management of stable COPD in which condition it was probably the most efficacious single bronchodilator before the availability of tiotropium. Because of its slow onset of action it is best used on a regular, maintenance basis, rather than as needed. The conventional dose of ipratropium, two 20-μg puffs, is probably suboptimal for many patients with COPD (Gross et al. 1989) and can safely be doubled or quadrupled (Leak and O'Connor 1988). The double-strength formulation (Atrovent forte), available in many countries, is preferable.

Tiotropium has become available in many countries since the current guidelines were published, so consensus on its role in COPD has not been established. One's expectation is that tiotropium will soon be regarded as the first-line therapy for patients with COPD of all but the mildest degrees of COPD.

Combinations of ipratropium with salbutamol, Combivent MDI or DuoNeb nebulized solution, are appropriate for patients with COPD who require more than one bronchodilator agent.

Anticholinergic agents can and probably should be used in combination with adrenergic agents in the treatment of acute exacerbations of COPD.

Anticholinergic agents are sometimes useful in stable asthma as adjuncts to other bronchodilator therapy, and have a demonstrated role in combination with adrenergic agents in the treatment of acute severe asthma (status asthmaticus).

References

American Thoracic Society (1995) Standards for the diagnosis and care of patients with chronic obstructive pulmonary disease. *Am J Respir Crit Care Med* 152:S77–S121

Anthonisen NR, Connett JE, Kiley JP, Altose MD, Bailey WC, Buist AS (1994) Effects of smoking intervention and the use of an inhaled anticholinergic bronchodilator on the rate of decline of FEV1, the Lung Health Study. *JAMA* 272:1497–1505

Ashutosh K, Dev G, Steele D (1995) Nonbronchodilator effects of pirbuterol and ipratropium in chronic obstructive pulmonary disease. *Chest* 107:173–178

Ayala LE, Ahmed T (1989) Is there loss of protective muscarinic receptor mechanism in asthma? *Chest* 96:1285–1291

Ayala LE, Choudry NB, Fuller RW (1988) LTD4-induced bronchoconstriction in patients with asthma: lack of a vagal reflex. *Br J Clin Pharmacol* 26:110–112

Azevedo M, da Costa JT, Fontes P, da Silva JP, Araujo O (1990) Effect of terfenadine and ipratropium bromide on ultrasonically nebulized distilled water-induced asthma. *J Int Med Res* 18:37–49

Barnes PJ, Belvisi MG, Mak JC, Haddad EB, O'Connor B (1995) Tiotropium bromide (Ba 679 BR), a novel long-acting muscarinic antagonist for the treatment of obstructive airways disease. *Life Sci* 56:853–859

Baroody FM, Majchel AM, Roecker MM, Roszko PJ, Zegarelli EC, Wood CC (1992) Ipratropium bromide (Atrovent nasal spray) reduces the nasal response to methacholine. *J Allergy Clin Immunol* 89:1065–1075

Beasley R, Fishwick D, Miles JF, Hendeles L (1998) Preservatives in nebulizer solutions: risks without benefit. *Pharmacotherapy* 18:130–139

Beck R, Robertson C, Galdes-Sebaldt M, Levison H (1985) Combined salbutamol and ipratropium bromide by inhalation in the treatment of severe acute asthma. *J Pediatr* 107:605–608

Bleecker ER, Britt EJ (1991) Acute bronchodilating effects of ipratropium bromide and theophylline in chronic obstructive pulmonary disease. *Am J Med* 91:24S-27S.

Bollert FG, Matusiewicz SP, Dewar MH, Brown GM, McLean A, Greening AP (1997) Comparative efficacy and potency of ipratropium via Turbuhaler and pressurized metered-dose inhaler in reversible airflow obstruction. *Eur Respir J* 10:1824–1828

Boner AL, De Stefano G, Niero E, Vallone G, Gaburro D (1987) Salbutamol and pratropium bromide solution in the treatment of bronchospasm in asthmatic children. *Ann Allergy* 58:54–58

Boucher M, Roy MT, Henderson J (1992) Possible association of benzalkonium chloride in nebulizer solutions with respiratory arrest. *Ann Pharmacother* 26:772–774

Braun SR, McKenzie WN, Copeland C, Knight L, Ellersieck MA (1989) A comparison of the effect of ipratropium and albuterol in the treatment of chronic obstructive airway disease [published erratum appears in Arch Intern Med 1990 50:1242] *Arch Intern Med* 149:544–547

British Thoracic Society (1996) Guidelines for the management of chronic obstructive pulmonary disease. *Thorax* 52:S1-S28

Brophy C, Ahmed B, Bayston S, Arnold A, McGivern D, Greenstone M (1998) How long should Atrovent be given in acute asthma? *Thorax* 53:363–367

Brown IG, Chan CS, Kelly CA, Dent AG, Zimmerman PV (1984) Assessment of the clinical usefulness of nebulised ipratropium bromide in patients with chronic airflow limitation. *Thorax* 39:272–276

Carstairs JR, Nimmo AJ, Barnes PJ. (1985) Autoradiographic visualization of beta-adrenoceptor subtype in human lung. Am Rev Respir Dis 132:541–547

Casaburi R, Mahler DA, Jones PW, Wanner A, San Pedro G, ZuWallack RL, Menjoge SS, Serby CW, Witek TJ. (2002) A long-term evaluation of once daily inhaled tiotropium in chronic obstructive pulmonary disease. *Eur Respir J* 19:217–224

Chapman KR, Smith DL, Rebuck AS, Leenen FH (1985) Haemodynamic effects of inhaled ipratropium bromide, alone and combined with an inhaled beta 2-agonist. *Am Rev Respir Dis* 132:845–847

Courty MA (1859) Treatment of asthma. Edinburgh Medical Journal 5:665

Disse B, Speck GA, Rominger KL, Witek TJ, Hammer R. (1999) Tiotropium: Mechanistic considerations and clinical profile in obstructive lung disease. Life Sci 64:457–464

Donohue JF, van Noord JA, Bateman ED, Langley SJ, Lee A, Witek TJ, Kesten S, Towse L (2002) A 6-month placebo controlled study comparing lung function and health status changes in COPD patients treated with tiotropium or salmeterol. Chest 122:47–55

Douglas NJ (1998) Sleep in patients with chronic obstructive pulmonary disease. *Clin Chest Med* 19:115–125

Douglas NJ, Calverley PM, Leggett RJ, Brash HM, Flenley DC, Brezinova V (1979) Transient hypoxaemia during sleep in chronic bronchitis and emphysema. *Lancet* 1:1–4.

Ducharme FM, Davis GM (1998) Randomized controlled trial of ipratropium bromide and frequent low doses of salbutamol in the management of mild and moderate acute pediatric asthma. *J Pediatr* 133:479–485

Dweik RA (1999) Anticholinergic therapy in COPD. *UpToDate* 7:1-15

Friedman M, Serby CW, Menjoge SS, Wilson JD, Hilleman DE, Witek TJ (1999) Pharmacoeconomic evaluation of a combination of ipratropium plus albuterol compared with ipratropium alone and albuterol alone in COPD. *Chest* 115:635–641

Fryer AD, Jacoby DB (1991) Parainfluenza virus infection damages inhibitory M2 muscarinic receptors on pulmonary parasympathetic nerves in the guinea-pig. Br J Pharmacol 102:267–271

Fryer AD, Jacoby DB (1993) Effect of inflammatory cell mediators on M2 muscarinic receptors in the lungs. Life Sci 52:529–536

Gandevia B (1975) Historical review of the use of parasympatholytic agents in the treatment of respiratory disorders. Postgrad Med J 51(7Suppl):13–20

Gross NJ (1988) Ipratropium bromide. *N Engl J Med* 319:486–494

Gross NJ, Bankwala Z (1987) Effects of an anticholinergic bronchodilator on arterial blood gases of hypoxaemic patients with chronic obstructive pulmonary disease, comparison with a beta-adrenergic agent. *Am Rev Respir Dis* 136:1091–1094

Gross NJ, Barnes PJ 1988) A short tour around the muscarinic receptor. Am Rev Respir Dis 138:765–767

Gross NJ, Petty TL, Friedman M, Skorodin MS, Silvers GW, Donohue JF (1989) Dose response to ipratropium as a nebulized solution in patients with chronic obstructive pulmonary disease, a three-center study. *Am Rev Respir Dis* 139:1188–1191

Gross NJ, Co E, Skorodin MS (1989) Cholinergic bronchomotor tone in COPD, estimates, estimates of its amount in comparison with that in normal subjects. Chest 96:984–987

Gross NJ, Skorodin SM (1984a) Anticholinergic, antimuscarinic bronchodilators. Am Rev Respir Dis 129:856–870

Gross NJ, Skorodin MS (1984b) Role of the parasympathetic system in airway obstruction due to emphysema. *N Engl J Med* 311:421–425

Gross NJ, Skorodin MS (1985) Massive overdose of atropine methonitrate with only slight untoward effects [letter]. *Lancet* 2:386

Gross NJ, Skorodin M (1987) Anticholinergic agents. In: Jenne JW, Murphy S (eds) *Drug therapy for asthma*. Marcel Dekker, New York pp615–668

Gross NJ, Tashkin D, Miller R, Oren J, Coleman W, Linberg S (1998) Inhalation by nebulization of albuterol-ipratropium combination (Dey combination) is superior to either agent alone in the treatment of chronic obstructive pulmonary disease. Dey Combination Solution Study Group. *Respiration* 65:354–36

Hargreave FE, Dolovich J, Newhouse MT (1990) The assessment and treatment of asthma, a conference report. *J Allergy Clin Immunol* 85:1098–1111

Hvizdos KM, Goa KL (2002) Tiotropium bromide. Drugs 62:1195–1203.

Ikeda A, Nishimura K, Koyama H, Izumi T (1995) Bronchodilating effects of combined therapy with clinical dosages of ipratropium bromide and salbutamol for stable COPD: comparison with ipratropium bromide alone. *Chest* 107:401–405

Jolobe OM (1984) Asthma vs. non-specific reversible airflow obstruction: clinical features and responsiveness to anticholinergic drugs. *Respiration* 45:237–242

Karpel JP, Pesin J, Greenberg D, Gentry E (1990) A comparison of the effects of ipratropium bromide and metaproterenol sulfate in acute exacerbations of COPD. *Chest* 98:835–839

Khoukhaz G, Gross NJ (1999) Effects of salmeterol on arterial blood gases in patients with stable chronic obstructive pulmonary disease, comparison with albuterol and ipratropium. *Am J Respir Crit Care Med* 160:1028–1030

Klink M, Quan SF (1987) Prevalence of reported sleep disturbances in a general adult population and their relationship to obstructive airways diseases. *Chest* 91:540–546

Koutsogiannis Z, Kelly A (2000) Does high dose ipratropium bromide added to salbutamol improve pulmonary function for patients with chronic obstructive airways disease in the emergency department? *Australian New Zealand J Med* 30:38–40

Leak A, O'Connor T (1988) High dose ipratropium bromide is it safe? *Practitioner* 232:9–10

Lefcoe NM, Toogood JH, Blennerhassett G, Baskerville J, Paterson NA (1982) The addition of an aerosol anticholinergic to an oral beta agonist plus theophylline in asthma and bronchitis, a double-blind single dose study. *Chest* 82:300–305

Littner MR, Ilowite JS, Tashkin DP, Friedman, Serby CW, Menjoge SS, Witek TJ (2000) Long-acting bronchodilation with once-daily dosing of tiotropium in stable chronic obstructive pulmonary disease. *Am J Respir Crit Care Med* 161:1136–1142

Maesen FP, Smeets JJ, Sledsens TJ, Wald FD, Cornelissen PJ (1995) Tiotropium bromide, a new long-acting antimuscarinic bronchodilator, a pharmacodynamic study in patients with chronic obstructive pulmonary disease (COPD). Dutch Study Group. *Eur Respir J* 8:1506–1513

Martin RJ, Bucher-Bartleson BL, Smith P, Hudgel DW, Lewis D, Pohl G, Koker P, Souhrada JF (1999) Effect of ipratropium bromide treatment on oxygen saturation and sleep quality in COPD. *Chest* 115:1338–1345

Nadel JA (1980) Autonomic regulation of airway smooth muscle. In: Nadel JA (ed) Physiology and Pharmacology of the Airways. Marcel Dekker, New York pp 217–257

O'Connor BJ, Towse LJ, Barnes PJ (1996) Prolonged effect of tiotropium bromide on methacholine-induced bronchoconstriction in asthma. Am J Respir Crit Care Med 154:876–880

Passamonte PM, Martinez AJ (1984) Effect of inhaled atropine or metaproterenol in patients with chronic airway obstruction and therapeutic serum theophylline levels. *Chest* 85:610–615

Patrick DM, Dales RE, Stark RM, Laliberte G, Dickinson G. (1990) Severe exacerbations of COPD and asthma, incremental benefit of adding ipratropium to usual therapy. *Chest* 98:295–297

Pavia D, Bateman JR, Sheahan NF, Clarke SW (1979) Effect of ipratropium bromide on mucociliary clearance and pulmonary function in reversible airways obstruction. *Thorax* 34:501–507

Petty TL (1994) In chronic obstructive pulmonary disease, a combination of ipratropium and albuterol is more effective than either agent alone, an 85-day multicenter trial. COMBIVENT Inhalation Aerosol Study Group. *Chest* 105:1411–1419

Plotnick L, Ducharme F (1998) Should inhaled anticholinergics be added to $\beta 2$ agonists for treating acute childhood and adolescent asthma? A systematic review (1998) *Brit Med J* 317:971–977

Qureshi F, Pestian J, Davis P, Zaritsky A (1998) Effect of nebulized ipratropium on the hospitalization rates of children with asthma. *N Engl J Med* 339:1030–1035

Rebuck AS, Chapman KR, Abboud R, Pare PD, Kreisman H, Wolkove N (1987) Nebulized anticholinergic and sympathomimetic treatment of asthma and chronic obstructive airways disease in the emergency room. *Am J Med* 82:59–64

Reisman J, Galdes-Sebalt M, Kazim F, Canny G, Levison H (1988) Frequent administration by inhalation of salbutamol and ipratropium bromide in the initial management of severe acute asthma in children. J Allergy Clin Immunol 81:16–20

Rennard SI, Serby CW, Ghafouri M, Johnson PA, Friedman M. (1996) Extended therapy with ipratropium is associated with improved lung function in patients with COPD, a retrospective analysis of data from seven clinical trials. Chest 10:62–70

Richardson JB (1982) The innervation of the lung. Eur J Respir Dis (Suppl) 117:13–31

Ruffin RE, Fitzgerald JD, Rebuck AS (1977) A comparison of the bronchodilator activity of Sch 1000 and salbutamol. J Allergy Clin Immunol 59:136–141

Schuh S, Johnson DW, Callahan S, Canny G, Levison H (1995) Efficacy of frequent nebulized ipratropium bromide added to frequent high-dose albuterol therapy in severe childhood asthma. J Pediatr 126:639–645

Shah PK, Lakhotia M, Mehta S, Jain SK, Gupta GL (1990) Clinical dysautonomia in patients with bronchial asthma, study with seven autonomic function tests. Chest 98:1408–1413

Siafakas NM, Vermeire P, Pride NB, Paoletti P, Gibson J, Howard P (1995) Optimal assessment and management of chronic obstructive pulmonary disease (COPD). The European Respiratory Society Task Force. Eur Respir J 8:1398–1420

Stoodley RG, Aaron SD, Dales RE (1999) The role of ipratropium bromide in the emergency management of acute asthma exacerbation: a meta-analysis of randomized clinical trials. Ann Emerg Med 34:8–18

Storr J, Lenney W (1986) Nebulised ipratropium and salbutamol in asthma. Arch Dis Child 61:602–603

Tashkin DP, Ashutosh K, Bleecker ER, Britt EJ, Cugell DW, Cummiskey JM, Gross NJ (1986) Comparison of the anticholinergic bronchodilator ipratropium bromide with metaproterenol in chronic obstructive pulmonary disease, a 90-day multi-center study. Am J Med 81:81–90

Tashkin DP, Kesten S (2003) Long term treatment benefits with tiotropium in COPD patients with and without acute bronchodilator responses. Chest 123 (in press)

Thiessen B, Pedersen OF (1982) Maximal expiratory flows and forced vital capacity in normal, asthmatic and bronchitic subjects after salbutamol and ipratropium bromide. Respiration 43:304–316

Tobin MJ, Hughes JA, Hutchison DC (1984)Effects of ipratropium bromide and fenoterol aerosols on exercise tolerance. Eur J Respir Dis 65:441–446

Ullah MI, Newman GB, Saunders KB (1981) Influence of age on response to ipratropium and salbutamol in asthma. Thorax 36:523–529

Van Noord JA, Bantje TA, Eland ME (2000) A randomised controlled comparison of tiotropium and ipratropium in the treatment of chronic obstructive pulmonary disease. Thorax 55:289–294

Vichyanond P, Sladek WA, Sur S, Hill MR, Szefler SJ, Nelson HS (1990) Efficacy of atropine methylnitrate alone and in combination with albuterol in children with asthma. Chest 98:637–642

Vincken W, van Noord JA, Greenfhorst APM, Bantje TA, Kesten S, Korducki L, Cornelissen PJG. (2002) Improved health outcomes in patients with COPD during 1 year's treatment with tiotropium. Eur Respir J. 19:209–216

Warner JO, Gotz M, Landau LI, Levison H, Milner AD, Pedersen S (1989) Management of asthma, a consensus statement. Arch Dis Child 64:1065–1079

Watson WT, Shuckett EP, Becker AB, Simons FE (1994) Effect of nebulized ipratropium bromide on intraocular pressures in children. Chest 105:1439–1441

Widdicombe JG (1979) The parasympathetic nervous system in airways disease. Scand J Respir Dis Suppl 103:38–43

Zorc JJ, Pusic MV, Ogborn CJ, Lebet R, Duggan AK (1999) Ipratropium bromide added to asthma treatment in the pediatric emergency department. Pediatrics 103:748–752

Theophylline in the Treatment of Respiratory Disease

N. A. Jones · C. P. Page

The Sackler Institute of Pulmonary Pharmacology,
Pharmacology and Therapeutics Division, GKT School of Biomedical Sciences,
King's College London, 5th Floor Hodgkin Building, Guy's Campus,
London, SE1 9RT, UK
e-mail: clive.page@kcl.ac.uk

1	Background	54
2	The Effect of Theophylline on Inflammatory Cell Function	60
2.1	Mast Cells and Basophils	60
2.2	Neutrophil	61
2.3	Eosinophil	61
2.4	T Lymphocyte	62
2.5	Monocytes	62
2.6	Macrophages	63
3	Mechanism of Action of Theophylline	63
4	Conclusion	66
	References	66

Abstract Theophylline has now been in clinical use, most notably in the treatment of respiratory disease, for more than a century. Traditionally classified as a bronchodilator drug, it is becoming increasingly apparent that theophylline has a range of other pharmacological effects of potential therapeutic value in the treatment of respiratory diseases. Anti-inflammatory and immunomodulatory actions of theophylline have been observed both in the laboratory with respect to inflammatory cell function and in the clinic in patient populations suffering with a range of respiratory diseases. Many of the biological effects of theophylline have been suggested to occur via an inhibitory effect on the phosphodiesterase (PDE) family of enzymes; however, studies have also shown theophylline to antagonise adenosine receptors, inhibit NF-κB, a transcription factor important in regulation of inflammatory cell cytokine activity, inhibit interleukin (IL)-5 or granulocyte-macrophage colony-stimulating factor (GM-CSF)-induced eosinophil survival, enhance histone deacetylase (HDAC) activity and affect lipid kinase and protein kinase activities in inflammatory cells. Research will no doubt continue in order to elucidate these mechanisms in an attempt to produce a theophylline-type drug with greater selectivity and higher therapeutic index, which in turn may lead to improved patient compliance and greater control of airway inflammation.

Keywords Theophylline · Phosphodiesterase (PDE) · Anti-inflammatory · Bronchodilator · Immunomodulatory · Asthma · COPD

1
Background

Theophylline has now been in clinical use for more than a century, although it is only during the last 50 years that this drug has been in regular use for the treatment of respiratory diseases. In 1886, Henry Hyde Salter described the efficacious use of strong coffee taken on an empty stomach as a treatment for asthma (Persson et al. 1991). The principal agent in coffee producing this effect was the methylxanthine caffeine. Theophylline has a similar chemical structure to caffeine and was first used in the treatment of asthma as early as 1922, when it was found to be effective in the treatment of three asthmatic subjects (Becker et al. 1984). In 1937, theophylline was administered i.v. for the treatment of acute asthma, and in 1940 theophylline was first used orally in combination with ephedrine. There are now many studies in the literature describing the effects of theophylline in the treatment of both asthma (Weinberger et al. 1996) and COPD (Ashutosh et al. 1997). Theophylline continues to be used in various slow-release formulations to overcome rapid metabolism and to maintain constant plasma levels. However, over the last decade the number of prescriptions being written for theophylline has continued to decline as newer medications with an improved safety record have been introduced for the treatment of respiratory disease. This decline has mainly come about due to concerns raised over the narrow therapeutic window of theophylline, which has typically been classified as being 10–20 μg/ml in plasma (Weinberger et al. 1996), although this has recently been revised to 5–15 μg/ml in plasma to reflect recent work with lower doses of theophylline (Jarjour et al. 1998).

Whilst theophylline has traditionally been classified as a bronchodilator drug, it is becoming increasingly apparent that this drug has a range of other pharmacological effects of potential therapeutic value in the treatment of respiratory diseases (Spina et al. 1998a) that occur independently of the bronchodilator actions, including anti-inflammatory, immunomodulatory actions (Ward et al. 1993; Kidney et al. 1995) and increased respiratory drive (Ashutosh et al. 1997). These effects often occur at plasma levels below 10 μg/ml, suggesting that lower levels of theophylline than have previously been used to obtain bronchodilation may be of benefit in the treatment of lung diseases, thus reducing the side effect profile and improving the safety margin of this drug. Indeed, these actions have caused the recommended therapeutic window to be reduced to 5–15 μg/ml in plasma in many countries.

A number of studies have reported that i.v. administration of theophylline or the related xanthine enprofylline prior to allergen challenge can inhibit the development of the late asthmatic response without any significant effect on the acute bronchoconstrictor response (Pauwels et al. 1985; Crescioli et al. 1991; Ward et al. 1993; Hendeles et al. 1995) or associated bronchial hyperresponsiveness to methacholine (Cockcroft et al. 1989). Thus, neither functional antagonism of airway smooth muscle shortening nor inhibition of mast cell degranulation accounted for the attenuated late asthmatic response caused by theophylline or enprofylline, although, in allergic rhinitis, 1 week treatment with theophylline reduced histamine release during pollen exposure (Naclerio et al. 1986), which indicated that either (1) theophylline inhibited mast cell or basophil degranulation in this disorder or (2) reduced the number of mast cells as has been shown with glucocorticosteroids (Mygind 1993).

Individuals exposed for long periods of time to certain industrial chemicals develop asthma-like symptoms that can be duplicated in the clinical laboratory following aerosol challenge with the inciting agent. For example, susceptible individuals demonstrate acute bronchospasm, late asthmatic responses and bronchial hyperresponsiveness following inhalation of toluene di-isocyanate (TDI) (Mapp et al. 1987). The inflammatory nature of this response has been confirmed by its sensitivity to inhibition by the glucocorticosteroid beclomethasone. Theophylline partially modified the acute response and attenuated the late asthmatic response induced by TDI but was ineffective against bronchial hyperresponsiveness (Mapp et al. 1987; Crescioli et al. 1991). This latter finding is consistent with the inability of theophylline to modulate allergen-induced bronchial hyperresponsiveness in asthmatics (Cockcroft et al. 1989; Ward et al. 1993).

The late asthmatic response to allergen is known to be accompanied by an influx of inflammatory cells into the airways (De Monchy et al. 1985) and this allergen-induced infiltration of activated eosinophils into the airways (assessed as the total number of eosinophils and as an increase in the number of EG_2+ eosinophils in biopsies) was also reduced significantly by 6 weeks of treatment with theophylline (Sullivan et al. 1994), an effect that occurred at plasma levels well below the 10–20 $\mu g/ml$ plasma levels required for bronchodilation (mean plasma levels of 6.7 $\mu g/ml$). More recent clinical studies have confirmed these anti-inflammatory properties of theophylline in patients with asthma. In two randomised, placebo controlled studies (Ward et al. 1993; Hendeles et al. 1995), the effect of theophylline or placebo was investigated on various inflammatory indices following once and twice daily treatment for 1 and 5 weeks respectively. The late asthmatic response was reduced in those subjects treated with theophylline after 5 weeks (Ward et al. 1993) despite a mean plasma concentration of only 7.8 $\mu g/ml$. The lack of effect of theophylline on the acute response is presumably due to the low plasma levels in these subjects. Inhibition of the late asthmatic response, therefore, was unlikely to be due to functional antagonism of airway smooth muscle shortening or inhibition of mast cell degranulation (Ward et al. 1993). Similarly, various surrogate markers of nasal inflammation in response to allergen challenge, including the late phase response and the accumulation/activation of eosinophils in the nose, was significantly attenuated in allergic rhinitis subjects following chronic treatment with theophylline (Aubier et al. 1998), again consistent with an anti-inflammatory rather than bronchodilatory property of this drug.

The mechanism whereby theophylline inhibits the recruitment of activated eosinophils into the airways is not known, but several mechanisms have been put forward to explain this observation. The first mechanism relates to an immunomodulatory action of theophylline whereby it is thought that the inhibitory effect of theophylline may be a consequence of a restoration of T suppressor cell function, since it has long been recognised that theophylline can increase T suppressor cell function (Limatibul et al. 1978; Shohat et al. 1983; Pardi et al. 1984; Zocchi et al. 1985; Scordamaglia et al. 1988), impair cellular immune response in vitro (Fink et al. 1987), and inhibit graft rejection in vivo (Guillou et al. 1984). Individuals who do not develop a late asthmatic response have been shown to recruit a greater proportion of $CD8^+$ (suppressor) than $CD4^+$ (helper) T lymphocytes in bronchoalveolar lavage (BAL) fluid (Gonzalez et al. 1987). It is recognised that T lymphocytes play a central role in the pathogenesis of allergic asthma, in particular the orchestration of eosinophil migration into the airways, via the release of cytokines such as interleukin (IL)-5 (Hamid et al. 1991). Regular treatment with theophylline has also been reported to inhibit allergen-induced recruitment of T lymphocytes into the airway (Jaffar et al. 1996) and to increase the number of suppressor CD8 cells in peripheral blood (Shohat et al. 1983; Fink et al. 1987; Ward et al. 1993). Furthermore, withdrawal of theophylline from asthmatics has been shown to cause a significant increase in asthma symptoms (Brenner et al. 1988; Kidney et al. 1995), which was associated with an increase in T lymphocytes in the airways (Kidney et al. 1995), an immunomodulatory effect that again occurred at plasma levels below 10 μg/ml. Regular treatment with theophylline has also been reported to reduce the number of inflammatory cells expressing IL-4 in the airway (Finnerty et al. 1996) and to induce the production of IL-10 from peripheral blood mononuclear cells obtained from asthmatics (Mascali et al. 1996), an observation of considerable interest as IL-10 can shorten eosinophil survival (Punnonen et al. 1993) and induce tolerance in T cells (Enk et al. 1993).

Another suggested mechanism of action of theophylline that occurs at clinically relevant concentrations is the ability of theophylline to alter eosinophil survival. A number of cytokines, including IL-5 have been shown to prolong eosinophil survival (Yamaguchi et al. 1988). Theophylline has been shown to inhibit IL-5-mediated survival of human eosinophils and to accelerate apoptosis, again at concentrations below 10 μg/ml (Yasui et al. 1997). Analysis of bronchial biopsies taken from mild asthmatics treated with low-dose theophylline over 6 weeks revealed a significant reduction in $EG2^+$ staining cells (activated eosinophils) and in the total number of eosinophils (Sullivan et al. 1994), which may be a consequence of the ability of theophylline to induce apoptosis of human eosinophils in this way (Yasui et al. 1997; Ohta et al. 1999). Similarly, a reduction in $CD3^+$ T lymphocytes and expression of various activation markers on $CD4^+$ T lymphocytes including HLA-DR and VLA-1 was observed in BAL fluid (Jaffar et al. 1996). Furthermore, a reduction in $CD4^+$, $CD8^+$ T lymphocytes and IL-4- and IL-5-containing cells was observed in bronchial biopsies from asthmatics who were taking theophylline over a 6-week period (Finnerty et al. 1996) and a fall in circulating levels of T helper (Th)2 cytokines, IL-4 and IL-5, was observed after a single low dose of theophylline (Kosmas et al. 1999).

Regular theophylline treatment has also been demonstrated to produce anti-inflammatory activity in patients having natural exacerbations of their asthma, in the form of nocturnal asthma. Theophylline treatment significantly improved the overnight deterioration in lung function associated with nocturnal asthma compared with placebo treatment (Kraft et al. 1996b), a finding consistent with previous studies using theophylline for the treatment of asthma (D'Alonzo et al. 1990). Theophylline also inhibited the ability of neutrophils to migrate into the airways of patients undergoing nocturnal attacks of asthma (Kraft et al. 1996b), associated with a reduction in the ability of polymorphonuclear neutrophils (PMNs) to release leukotriene B4 (LTB4). This work not only extends the anti-inflammatory actions of theophylline, but also supports earlier work that regular treatment with theophylline can reduce PMN activation (Nielson et al. 1986; Nielson et al. 1988) in addition to the actions of theophylline on eosinophils and lymphocytes discussed above. Theophylline treatment has also been reported to reduce the steepness of methacholine dose–response curves in asthmatics versus placebo treatment (Magnussen et al. 1987; Page et al. 1998), a change also seen with glucocorticosteroids (Bel et al. 1991a), but not with B_2 agonists (Bel et al. 1991b), which actually steepen the curve. Again, such observations would suggest that theophylline behaves more like an anti-inflammatory drug than a bronchodilator.

The clinical relevance of these anti-inflammatory actions of theophylline is now being evaluated and a number of recent clinical studies lend weight to the suggestion that such activities may offer clinical benefit. Two separate studies have demonstrated that in asthmatics who were poorly controlled on existing glucocorticosteroid therapy, a significant improvement in a number of clinical outcomes including peak expiratory flow, forced expiratory volume in 1 s (FEV_1), and symptom scores; and reduced rescue medication was observed when patients were taking theophylline together with low-dose glucocorticosteroid compared with high-dose glucocorticosteroid treatment (Evans et al. 1997; Ukena et al. 1997). In both studies, the plasma levels of theophylline measured were unlikely to be sufficient to induce bronchodilation [median 7.8 µg/ml (Evans et al. 1997) and mean 10.1 µg/ml (Ukena et al. 1997)]. In other studies, withdrawing theophylline from asthmatics who continued taking glucocorticosteroids resulted in a significant deterioration of their disease (Brenner et al. 1988; Kidney et al. 1995), together with a concomitant rise in the number of $CD4^+$ and $CD8^+$ T lymphocytes in bronchial biopsies (Kidney et al. 1995). These results suggest that theophylline may offer additional benefit to glucocorticosteroids as has been previously suggested from other clinical studies by the use of different types of protocol (Kidney et al. 1995; Rivington et al. 1995; Ukena et al. 1997).

Other studies in paediatric asthma have shown that there is a clear effect of theophylline in the treatment of asthma that is comparable to low doses of glucocorticosteroids (Tinkelman et al. 1993). This observation is of particular interest given that theophylline is an orally active drug and has been shown to have a better compliance rate than inhaled medications (Kelloway et al. 1994), which is particularly relevant to the treatment of asthmatic children. Given the relatively low cost of theophylline in comparison to other anti-asthma medications (Barnes et al. 1996) and the fact that it is still one of the few drugs available for use orally in the treatment of this common disease, the growing body of evidence suggesting that theophylline has anti-inflammatory and immunomodulatory efficacy at lower than conventional plasma levels suggests that it is timely to reconsider the wider use of theophylline in the overall management of asthma (Barnes and Pauwels 1994).

Chronic obstructive pulmonary disease (COPD), a clinical condition characterised by progressive, irreversible airflow limitation, is a major cause of morbidity and mortality worldwide. COPD has been described as chronic airflow obstruction due to a mixture of emphysema and peripheral airway obstruction from chronic bronchitis (Barnes 2000). This is reflected by a reduction in maximum expiratory flow and slow forced emptying of the lungs. Extensive pulmonary damage occurs before the patient is aware of symptoms, such as exertional dyspnoea, due to the slowly progressive nature of airflow obstruction and various coping manoeuvres (Barnes 2000). There have been relatively few developments in the management of COPD despite the significant advances in asthma treatment in recent years. Smoking cessation appears to be the only strategy that reduces the relentless decline in lung function observed in COPD; however, theophylline has been widely used in the management of COPD for many years, although it is considered to be a third-line agent after the use of β_2-agonists and anticholinergics (ZuWallack et al. 2001).

A range of clinical studies have demonstrated the ability of theophylline to reduce symptoms, increase lung function and improve diaphragmatic muscle function in COPD patients. There are a number of reports of meaningful symptomatic responses in patients treated with bronchodilators or theophylline, even in the absence of significant changes in FEV_1 and forced vital capacity (FVC) (Mahler et al. 1985; Wolkove et al. 1989; Hay et al. 1992). Studies measuring dyspnea, wheezing, cough, sputum, walking and other general feelings of wellbeing have reported modest to significant improvements in dyspnea (Dullinger et al. 1986; Guyatt et al. 1987; Chrystyn et al. 1988) and improvements in both wheezing and shortness of breath with theophylline (Marvin et al. 1983; Thomas et al. 1992).

Various mechanisms have been proposed to explain how theophylline could improve symptoms and reduce breathlessness in patients with COPD. An investigation of the effects of theophylline in 33 patients with stable COPD at a dose which resulted in serum concentrations of 15–20 mg/l showed a minor increase in FEV_1 (13%–130 mls), but a significant 64% decrease in trapped gas volume (from 1.84 l to 0.67 l) (Chrystyn et al. 1988). These results suggest that improvements in lung function observed with theophylline may be due to dilatation of the small airways with a consequent reduction in gas trapping. A fall in trapped gas volume, thus functional residual capacity (FRC), is likely to improve the mechanical advantage of the diaphragm and chest wall muscles which may explain many of the observed effects of theophylline on respiratory muscles (Murciano et al. 1984). Theophylline has also been demonstrated to increase the pressure generated by respiratory muscles (Umut et al. 1992) and increase diaphragmatic strength (Kongragunta et al. 1988). Its effect has been shown to be greater in fatigued diaphragm (Murciano et al. 1984; Aubier 1986) and in one study, theophylline was observed to increase transdiaphragmatic pressure by 16%, an increase that persisted even after 30 days of treatment with theophylline (Murciano et al. 1984). In therapeutic doses, theophylline is known to increase respiratory drive, independent of its effect on lung function (Ashutosh et al. 1997) and increase respiratory muscle function in both normal (Sherman et al. 1996) and COPD subjects (Umut et al. 1992), an effect detected by measuring increases in maximal inspiratory and expiratory pressures. One trial demonstrated an improvement in respiratory muscle performance as indicated by a decline in the ratio of inspiratory plural pressure during quiet breathing to the maximal pleural pressure (Murciano et al. 1989), suggesting theophylline reduces breathlessness by improving diaphragmatic contractility.

In many cases respiratory rate is increased in COPD and this may be observed in conjunction with shallow breathing that is further pronounced by carbon dioxide retention. It is known that theophylline improves minute ventilation in humans (Darnall-Jr 1995) and also alters the ventilatory response in COPD, detected as improved ventilatory capacity measured as increased VO_{2max}. This ventilatory response leads to increased tidal volume and thus an improvement in blood gas tensions. Both of changes could be related to a direct positive inotropic effect of theophylline on the respiratory muscles (Okubo et al. 1987; Kongragunta et al. 1988; Landsberg et al. 1990; Marsh et al. 1993) or due to the action of theophylline via a central stimulatory pathway (Cooper et al. 1987; Javaheri et al. 1990).

A recent study has also shown a useful additive effect between theophylline and long-acting β_2-agonists. This is of interest, as patients with COPD often require multiple therapies to improve lung function and decrease symptoms and exacerbations. Combination treatment with salmeterol and theophylline consistently provided significantly greater improvements in pulmonary function, significantly greater decreases in symptoms, dyspnea and albuterol use, and significantly fewer COPD exacerbations than treatment with either drug alone (ZuWallack et al. 2001).

Unfortunately, theophylline has a relatively narrow therapeutic index and many clinical trials have reported a number of adverse effects, even when serum concentrations are within the "therapeutic range" of 10–20 mg/l. Nausea is most commonly reported by significantly more patients treated with theophylline compared with placebo, another more extreme and less common unwanted effect of theophylline in elderly patients with COPD is supraventricular arrhythmias (Levine 1985; Varriale et al. 1993); however, the benefits of theophylline treatment in stable COPD have to be weighed against the risks of adverse effects.

Theophylline may also have anti-inflammatory effects in COPD. It is known that chronic obstructive bronchitis is associated with structural narrowing of bronchioles as a result of inflammatory changes represented by T lymphocytes and macrophages (Thompson et al. 1989), while emphysema is due to enzymatic destruction of alveolar walls with both macrophages and T lymphocytes, particularly $CD8^+$ cells, found to be prominent at sites of destruction (Finkelstein et al. 1995). However, the cellular mechanisms of inflammation in COPD are poorly understood. It is thought that cigarette smoke and other inhaled irritants initiate an inflammatory response in the peripheral airways by activating macrophages, causing the release of neutrophil chemotactic factors, cytokines, lipid mediators (LTB_4) and proteolytic enzymes, the levels of which correlate with the number of neutrophils observed in bronchoalveolar lavage and in turn, with the degree of airflow obstruction (Keatings et al. 1996). Neutrophils play an important role in COPD through the release of neutrophil elastase and other proteases, such as proteinase 3, cathepsin G and matrix metalloproteinases (MMPs). These contribute to parenchymal destruction and, in the airways, stimulate mucus hypersecretion. Macrophage numbers are increased five- to tenfold in bronchoalveolar lavage fluid of patients with COPD and are concentrated in the centriacinar zones where emphysema is most marked (Finkelstein et al. 1995). In a recent study, 25 patients with COPD were treated for 4 weeks with low-dose theophylline (plasma level 9–11 mg/l). Induced sputum inflammatory cells, neutrophils, IL-8, myeloperoxidase and lactoferrin were all significantly reduced by theophylline, while neutrophils from subjects treated with theophylline showed reduced chemotaxis (Culpitt et al. 2002). These data suggest that unlike corticosteroids, low-dose theophylline could reduce neutrophil recruitment and neutrophil-mediated damage in the airways. In this way, theophylline may prove useful in the long-term management of COPD by significantly affecting the rate of decline in lung function observed in patients.

2
The Effect of Theophylline on Inflammatory Cell Function

Given the clear indication that theophylline can act as an anti-inflammatory drug clinically, the ability of theophylline to influence the function of inflammatory cells will now be discussed.

2.1
Mast Cells and Basophils

Theophylline was the archetypal phosphodiesterase (PDE) inhibitor that could lead to the elevation of cyclic adenosine monophosphate (cAMP) in cells, and it has been recognised for over 25 years that cAMP-elevating drugs inhibit mast cell degranulation (Lichtenstein et al. 1968; Orange et al. 1971). The suppression of mast cell and basophil degranulation in response to different stimuli by a range of non-selective PDE inhibitors has been well documented in rodents and man (Frossard et al. 1981; Pearce et al. 1982; Louis et al. 1990, 1992; Peachell et al. 1992). Isobutylmethylxanthine (IBMX) decreases basophil histamine release induced by platelet-activating factor (PAF) (Columbo et al. 1993), and theophylline, enprophylline and IBMX have been shown to inhibit (1) anti-IgE-induced histamine release by both human lung mast cells and basophils (Peachell et al. 1988, 1992; Weston et al. 1997) and (2) cytokine release in human basophils (Shichijo et al. 1997), although whether this occurs in vivo at clinically relevant concentrations is still a matter of debate (Ward et al. 1993).

2.2
Neutrophil

Theophylline has been shown to inhibit phagocytosis of latex particles (Bessler et al. 1986), superoxide anion production (Bessler et al. 1986; Nielson et al. 1986; Carletto et al. 1997; Yasui et al. 2000b), chemotaxis (Rivkin et al. 1977; Harvath et al. 1991; Ferretti et al. 1994; Elferink et al. 1997; Yasui et al. 2000b), aggregation (Schmeichel et al. 1987), adhesion (Franzini et al. 1995), PAF-induced CD11b upregulation and L-selectin shedding (Spoelstra et al. 1998), degranulation (Schmeichel et al. 1987; Paul et al. 1995; Jones et al. 2001), apoptosis (Yasui et al. 1997), and PAF biosynthesis in neutrophils (Fonteh et al. 1993). It is thought that the effects of theophylline on neutrophil function were associated with an increase in the level of intracellular cAMP, as similar effects are observed with respect to neutrophil adhesion (Bloemen et al. 1997; Derian et al. 1995), chemotaxis (Harvath et al. 1991), apoptosis (Rossi et al. 1995; Yasui et al. 1997; Ottonello et al. 1998; Niwa et al. 1999), superoxide anion production, and degranulation (Nourshargh et al. 1986) when cAMP analogues or cAMP-elevating agents are applied. These observations are of interest clinically since regular theophylline treatment in patients with either COPD or asthma can influence PMN effects (Kraft et al. 1996a; Culpitt et al. 2002).

2.3
Eosinophil

Theophylline (Griswold et al. 1993) inhibits zymosan-induced superoxide anion generation by guinea-pig eosinophils and will also inhibit the C5a-stimulated formation of reactive oxygen species in intact human eosinophils (Hatzelmann et al. 1995; Yasui et al. 2000b). Interestingly, low doses of theophylline augmented superoxide anion generation secondary to adenosine (A2)-receptor antagonism (Griswold et al. 1993). Theophylline has also been observed to decrease the viability of eosinophils in culture (Hossain et al. 1994), attenuate Ig-induced (Kita et al. 1991) and C5a-induced secretion of cationic proteins (Hatzelmann et al. 1995), inhibit PAF and C5a-induced release of LTC_4 (Tenor et al. 1996), reduce granulocyte-macrophage colony-stimulating factor (GM-CSF) and IL-8 release in response to sIgA-coated beads (Shute et al. 1998) and suppress PAF-induced upregulation of Mac-1 (Sagara et al. 1996). Theophylline also inhibits PAF- and C5a-induced chemotaxis of eosinophils (Tenor et al. 1996), an effect that was substantially reversed by addition of Rp-cAMPs, which in turn suggests protein kinase A (PKA) dependence and is thus likely to occur via a PDE inhibitory mechanism. Suppression of eosinophil chemotaxis in vitro by PDE inhibitors may be due to inhibition of adhesion molecule expression as theophylline has been seen to inhibit PAF-induced CD11b upregulation on the eosinophil cell surface (Momose et al. 1998). Such observations may account for the ability of theophylline to inhibit the recruitment and activation of eosinophils in airway tissues clinically (Sullivan et al. 1994; Aubier et al. 1998).

2.4
T Lymphocyte

Cyclic AMP-elevating agents can modulate development, proliferation, cytokine generation, expression of cytokine receptors, chemotaxis and antibody production in T lymphocytes (Kammer 1988; Scherer et al. 1994; van der Pouw et al. 1995; Hidi et al. 2000). It is perhaps not surprising, therefore, that theophylline has been shown to inhibit lymphocyte proliferation in response to a variety of stimuli, including phytohemagglutinin (PHA) and anti-CD3 (Scordamaglia et al. 1988; Crocker et al. 1996; Banner et al. 1997a; Banner et al. 1999; Landells et al. 2000a), which may be secondary to inhibition of IL-2 synthesis (Mary et al. 1987; Scordamaglia et al. 1988) and downregulation of IL-2 receptor expression (Hancock et al. 1988). Theophylline has also been observed to inhibit PAF- or IL-8-induced human T lymphocyte chemotaxis in vitro (Hidi et al. 2000), lymphocyte migration through human endothelium—an effect thought to be mediated via inhibition of lymphocyte motility (Lidington et al. 1996)—and the release of both IL-4 and IL-5 by PMA- and anti-CD3 stimulated Th2 cells (Crocker et al. 1996). Furthermore, it has been suggested that theophylline may stimulate a subpopulation of T lymphocytes with suppressor cell activity (Zocchi et al. 1985). These observations perhaps explain the ability of theophylline to influence T cell effects in vivo, including in patients with pulmonary disease (Jaffar et al. 1996).

2.5
Monocytes

In human monocytes, theophylline, will inhibit the release of arachidonic acid (Hichami et al. 1995), superoxide anion generation (Elliott et al. 1989), tumour necrosis factor (TNF)-α production at the level of gene transcription (Endres et al. 1991; Spatafora et al. 1994; Verghese et al. 1995; Souness et al. 1996; Landells et al. 2000b), complement component C2 (Lappin et al. 1984), phagocytosis (Bessler et al. 1986), IL-2R expression (Hancock et al. 1988), production of IL-12 (van der Pouw et al. 1995) and the generation of LTB$_4$ (Jeurgens et al. 1993). Theophylline has also been reported to prevent adherence-dependent expression of platelet-derived growth factor (PDGF)β mRNA (Kotecha et al. 1994) and facilitate the production of IL-10 (Platzer et al. 1995; van der Pouw et al. 1995). Some studies have demonstrated that non-selective phosphodiesterase inhibitors and cAMP-elevating drugs have either no effect (Endres et al. 1991), inhibited (Knudsen et al. 1986) or enhanced (Kassis et al. 1989; Sung et al. 1991; Lorenz et al. 1995) IL-1 production in monocytes. These discrepancies may be accounted for by a number of observations. First, cAMP inhibited the release but had no effect on the intracellular concentration of IL-1β in monocytes (Viherluoto et al. 1991; Verghese et al. 1995). Second, the inhibition of IL-1 production by methylxanthines is not due to a reduction in the level of IL-1 mRNA but to a reduction in IL-1 activity (Knudsen et al. 1986).

2.6
Macrophages

Functional studies have shown that elevation of intracellular cAMP via inhibition of PDE can also affect the inflammatory response of macrophages. Theophylline and enprofylline inhibited lipoprotein lipase activity, a consequence of reduced synthesis and increased lysosomal acid hydrolase activity in human monocyte-derived macrophages (Gardette et al. 1987). Furthermore, these drugs inhibited TNF-α release from alveolar macrophages (Spatafora et al. 1994), superoxide anion production from guinea-pig (Turner et al. 1993) and rat (Lim et al. 1983) peritoneal and human alveolar macrophages, respectively (Baker et al. 1992), and to a lesser extent, attenuated thromboxane (TXB)$_2$ release from human alveolar macrophages (Baker et al. 1992). Theophylline was also observed to suppress human alveolar macrophage respiratory burst, an effect reversed by PKA inhibition, suggesting that the functional effect observed here was mediated through elevation of cAMP as a result of PDE inhibition.

3
Mechanism of Action of Theophylline

Many of the biological effects of theophylline have been suggested to be via an inhibitory effect on the PDE family of enzymes (Barnes et al. 1994; Spina et al. 1998b). However, the effect of theophylline on apoptosis of eosinophils was not shared by the selective PDE4 inhibitor rolipram, suggesting that this anti-inflammatory effect of theophylline may not be via inhibition of PDE4 (Yasui et al. 1997). This observation supports other work carried out in mononuclear cells obtained from asthmatics where theophylline was able to inhibit mononuclear cell proliferation via mechanisms distinct from selective PDE4 inhibitors (Banner et al. 1997b) and data with the related xanthine, pentoxifylline, showing that this drug can inhibit proliferation of fibroblasts via a mechanism unrelated to cAMP generation (Peterson et al. 1998). In the clinic, theophylline is able to exert therapeutic effects at serum levels of less than 10 mg/l; however, it has been observed that serum levels of 10 mg/l have a negligible effect on cAMP hydrolysis in vivo (Giembycz 2000).These data suggest that some or all of theophylline's anti-inflammatory effects occur via mechanisms independent of PDE inhibition, a hypothesis supported by the anti-inflammatory effect observed in patients with asthma of alkyl-xanthine structures devoid of adenosine or PDE receptor affinity. Arofylline is a relatively weak PDE4 inhibitor currently in phase 3 clinical trials for the treatment of pulmonary disease that displays an anti-inflammatory effect equivalent to that of the PDE4 inhibitor rolipram (Ferrer et al. 1997).

Another prominent action of theophylline is the ability of this drug to antagonise adenosine receptors (Persson et al. 1989). However, for more than a decade this suggestion was questioned as the related drug enprofylline had similar effects to theophylline clinically (Pauwels et al. 1985), yet was claimed to lack adenosine receptor antagonism (Persson et al. 1989). However, studies have now reported that enprofylline can act as a selective A_{2b} receptor antagonist on human mast cells (Feoktistov et al. 1995), a property shared by theophylline, which has been suggested to be of potential importance for the clinical activities of theophylline (Feoktistov et al. 1998). However, other studies have shown that whilst asthmatics are very sensitive to inhaled adenosine (Cushley et al. 1983), an effect that is blocked by theophylline (Cushley et al. 1984; Mann et al. 1985), there is no evidence to date that this effect is mediated via activation of A_{2b} receptors. Rather, there is evidence from experimental animals and man that it is the A_1 receptor that is upregulated as a result of allergic sensitisation (Bjorck et al. 1992; Ali et al. 1994; el Hashim et al. 1996), an observation supported by the study of Nyce and Metzger (Nyce et al. 1997) that an anti-sense oligonucleutide to A_1 receptors blocks allergen-reduced eosinophilia and allergen-induced bronchial hyperresponsiveness in allergic rabbits. In more recent studies it has been suggested that theophylline has an immunomodulatory effect on neutrophil apoptosis through A_{2a} receptor antagonism at relevant therapeutic concentrations (Yasui et al. 2000a). In contrast to this, inhibition by theophylline of complement C5a-induced degranulation of human eosinophils was significantly reversed by the selective A_3 antagonist MRS 1220, but not A_1 or A_2 antagonists, suggesting that therapeutic concentrations of theophylline inhibit human eosinophil degranulation by acting as an $A_{(3)}$ agonist (Ezeamuzie 2001). Conversely, a study investigating the anti-proliferative effects of theophylline on human peripheral blood mixed mononuclear cells (HPBMC) in vitro showed that theophylline was only capable of reducing proliferation at higher concentrations than are required to significantly antagonise A_{2b} receptors (Landells et al. 2000a). This study also demonstrated that exogenous or endogenous adenosine has little impact on HPBMC proliferation, as neither adenosine receptor agonists, antagonists nor adenosine deaminase had a significant effect on the proliferation of HPBMC from either healthy or asthmatic subjects (Landells et al. 2000a). Whilst it remains plausible that adenosine may be involved in a number of anti-inflammatory functions, which can also be modulated by theophylline, results such as these suggest the anti-inflammatory effects of theophylline to be mediated through mechanisms other than adenosine antagonism.

One interesting observation concerning the potential mechanisms of action of theophylline is the apparent ability of theophylline to inhibit the activation of the transcription factor nuclear factor-κB (NF-κB), in human mast cells at concentrations of 6–18 mg/l (Coward et al. 1998). NF-κB is a transcription factor that, when activated, stimulates transcription of TNF-α, GM-CSF and IL-8 in a variety of cell types. NF-κB is a member of a group of DNA-binding proteins that are a target of activated glucocorticoid receptors and control the transcription of numerous inflammatory mediators. This is a particularly relevant finding as the binding of NF-κB to its consensus DNA-binding site is two- to threefold higher in the sputum of patients with asthma when compared to the sputum of healthy subjects (Hart et al. 1998) and increased activation of NF-κB has been observed in alveolar macrophages from patients with COPD (Di Stefano et al. 2002). Also, high levels of TNF-α, the transcription of which is stimulated upon activation of NF-κB, are observed both in the circulation and in induced sputum of COPD patients and thought to play a major role in the characteristic inflammation and accompanying weight loss exhibited in patients with this and similar pulmonary diseases (Barnes 2003).

In inflammatory cells, acetylation of core histones by coactivator proteins, such as cAMP-response element binding protein (CREB), facilitate transcription of inflammatory genes (Imhof et al. 1998). This increased inflammatory gene transcription can, in turn, be suppressed by proteins which possess histone deacetylase (HDAC) activity (Ito et al. 2002a). In a recent study, low-dose theophylline has been shown to enhance HDAC activity in epithelial cells and macrophages, both in vitro and in vivo (Ito et al. 2002b). This potential anti-inflammatory mechanism of action of theophylline is of great interest as reduced HDAC activity is observed in alveolar macrophages and peripheral lung tissue of cigarette smokers and COPD patients (Ito et al. 2000, 2001). In COPD it appears that there is an active resistance to corticosteroids (Barnes 2003) and in patients with mild to moderate asthma, the addition of low-dose theophylline to ongoing inhaled corticosteroid therapy gives a greater improvement in asthma control than that achieved by doubling the dose of inhaled corticosteroid (Evans et al. 1997; Ukena et al. 1997; Lim et al. 2001). As the anti-inflammatory actions of corticosteroids are mediated through the activated glucocorticoid receptor recruiting corepressor proteins that have HDAC activity (Ito et al. 2001), the reduction in HDAC activity observed in COPD patients may prevent the anti-inflammatory action of corticosteroids in this condition (Barnes2003), while low-dose theophylline may exert an anti-asthma effect through increasing activation of HDAC which can be subsequently recruited by corticosteroids to suppress inflammatory genes, if both drugs are administered concomitantly (Ito et al. 2002b).

It has been recognised for more than 25 years that certain methylxanthines have an inhibitory effect on phosphoinositide metabolism (Buckley 1977; Honeyman et al. 1983; Steele et al. 1988; Cox et al. 1990), although up until recently the molecular basis of this effect had not been established. In a recent study, theophylline was identified as a novel phosphoinositide (PI)-3-kinase inhibitor, with the ability to inhibit both class I and class II PI-3-kinase and the potential for isoform selectivity towards p110δ (Foukas et al. 2002). The p110δ isoform of class IA PI-3-kinases is found in a limited sub population of cells such as leukocytes and melanoma cells (Vanhaesebroeck et al. 1997) and a specific inhibitor of this isoform might prevent inflammation or cell migration (Vanhaesebroeck et al. 1999). These novel inhibitory effects of theophylline occur within a physiologically achievable concentration range which suggest potential therapeutic benefit in inflammatory conditions; however, long-term administration could have adverse effects, as animals in which the p100δ gene is deleted show reduced B and T cell function and also develop inflammatory bowel disease (Okkenhaug et al. 2002).

4
Conclusion

The anti-inflammatory effects of theophylline are well documented and in tolerant patients it is thus still the preferred long-term control medication for the treatment of asthma or COPD. However, despite an increased knowledge of how to best use theophylline clinically, the molecular mechanisms of action of theophylline are still poorly understood and may occur through a range of actions at different sites. These may include inhibition of phosphodiesterase enzymes, antagonism of adenosine receptors, inhibition of NF-κB—a transcription factor important in regulation of inflammatory cell cytokine activity—inhibition of IL-5 or GM-CSF-induced eosinophil survival, enhancement of HDAC activity, or effects on lipid kinase and protein kinase activities in inflammatory cells. Research will continue to elucidate these mechanisms in an attempt to produce a theophylline-type drug with greater selectivity and higher therapeutic index which in turn, may lead to improved patient compliance and greater control of airway inflammation.

References

Ali, S., Mustafa, S. J., and Metzger, W. J. Adenosine receptor-mediated bronchoconstriction and bronchial hyperresponsiveness in allergic rabbit model. Am.J.Physiol 266(3 Pt 1), L271-L277. 1994

Ashutosh, K., Sedat, M., and Fragale-Jackson, J. Effects of theophylline on respiratory drive in patients with chronic obstructive pulmonary disease. J.Clin.Pharmacol. 37(12), 1100–1107. 1997

Aubier, M. Effect of theophylline on diaphragmatic and other skeletal muscle function. J Allergy Clin.Immunol. 78(4 Pt 2), 787–792. 1986

Aubier, M., Neukirch, C., Maachi, M., Boucara, D., Engelstatter, R., Steinijans, V., Samoyeau, R., and Dehoux, M. Effect of slow-release theophylline on nasal antigen challenge in subjects with allergic rhinitis. European Respiratory Journal 11(5), 1105–1110. 1998

Baker, A. J. and Fuller, R. W. Effect of cyclic adenosine monophosphate, 5'-(N- ethylcarboxyamido)-adenosine and methylxanthines on the release of thromboxane and lysosomal enzymes from human alveolar macrophages and peripheral blood monocytes in vitro. Eur.J.Pharmacol. 211(2), 157–161. 1992

Banner, K. H., Harbinson, P., Costello, J. F., and Page, C. P. Effect of PDE inhibitors on the proliferation of human peripheral blood mononuclear cells (HPBM) from mild asthmatics and normals. Am.J.Resp.Crit.Care Med. 155(4). 1997a

Banner, K. H., Hoult, J. R., Taylor, M. N., Landells, L. J., and Page, C. P. Possible Contribution of Prostaglandin E2 to the antiproliferative effect of phosphodiesterase 4 inhibitors in human mononuclear cells. Biochem.Pharmacol. 58(9), 1487–1495. 1-11-1999

Banner, K. H. and Page, C. Prostaglandins contribute to the anti-proliferative effect of isoenzyme selective phosphodiesterase 4 inhibitors but not theophylline in human mononuclear cells. Br.J.Pharmacol. 120. 1997b

Barnes, P. J. Managing Chronic Obstructive Pulmonary Disease. 2nd. 2000. Science Press Ltd

Barnes, P. J. New concepts in chronic obstructive pulmonary disease. Annu.Rev.Med. 54, 113–129. 2003

Barnes, P. J., Jonsson, B., and Klim, J. B. The costs of asthma. Eur.Respir.J. 9(4), 636–642. 1996

Barnes, P. J. and Pauwels, R. A. Theophylline in the management of asthma: time for reappraisal? Eur.Respir.J. 7(3), 579–591. 1994

Becker, A. B., Simons, K. J., Gillespie, C. A., and Simons, F. E. The bronchodilator effects and pharmacokinetics of caffeine in asthma. N.Engl.J.Med. 310, 743–746. 1984

Bel, E. H., Timmers, M. C., Zwinderman, A. H., Dijkman, J. H., and Sterk, P. J. The effect of inhaled corticosteroids on the maximal degree of airway narrowing to methacholine in asthmatic subjects. Am.Rev.Respir.Dis. 143(1), 109–113. 1991a

Bel, E. H., Zwinderman, A. H., Timmers, M. C., Dijkman, J. H., and Sterk, P. J. The protective effect of a beta 2 agonist against excessive airway narrowing in response to bronchoconstrictor stimuli in asthma and chronic obstructive lung disease. Thorax 46(1), 9–14. 1991b

Bessler, H., Gilgal, R., Djaldetti, M., and Zahavi, I. Effect of pentoxifylline on the phagocytic activity, cAMP levels, and superoxide anion production by monocytes and polymorphonuclear cells. J.Leukoc.Biol. 40(6), 747–754. 1986

Bjorck, T., Gustafsson, L. E., and Dahlen, S. E. Isolated bronchi from asthmatics are hyperresponsive to adenosine, which apparently acts indirectly by liberation of leukotrienes and histamine. Am.Rev.Respir.Dis. 145(5), 1087–1091. 1992

Bloemen, P. G., van-den-Tweel, M. C., Henricks, P. A., Engels, F., Kester, M. H., van-de-Loo, P. G., Blomjous, F. J., and Nijkamp, F. P. Increased cAMP levels in stimulated neutrophils inhibit their adhesion to human bronchial epithelial cells. Am.J.Physiol. 272(4 Pt 1), L580-L587. 1997

Brenner, M., Berkowitz, R., Marshall, N., and Strunk, R. C. Need for theophylline in severe steroid-requiring asthmatics. Clin.Allergy 18(2), 143–150. 1988

Buckley, J. T. Properties of human erythrocyte phosphatidylinositol kinase and inhibition by adenosine, ADP and related compounds. Biochim.Biophys.Acta 498(1), 1–9. 23-6-1977

Carletto, A., Biasi, D., Bambara, L. M., Caramaschi, P., Bonazzi, M. L., Lussignoli, S., Andrioli, G., and Bellavite, P. Studies of skin-window exudate human neutrophils: increased resistance to pentoxifylline of the respiratory burst in primed cells. Inflammation 21(2), 191–203. 1997

Chrystyn, H., Mulley, B. A., and Peake, M. D. Dose response relation to oral theophylline in severe chronic obstructive airways disease. BMJ 297(6662), 1506–1510. 10-12-1988

Cockcroft, D. W., Murdock, K. Y., Gore, B. P., O'Byrne, P. M., and Manning, P. Theophylline does not inhibit allergen-induced increase in airway responsiveness to methacholine. J.Allergy Clin.Immunol. 83(5), 913–920. 1989

Columbo, M., Horowitz, E. M., McKenzie, White, Kagey, Sobotka, and Lichtenstein, L. M. Pharmacologic control of histamine release from human basophils induced by platelet-activating factor. Int.Arch.Allergy Immunol. 102(4), 383–390. 1993

Cooper, C. B., Davidson, A. C., and Cameron, I. R. Aminophylline, respiratory muscle strength and exercise tolerance in chronic obstructive airway disease. Bull.Eur Physiopathol.Respir 23(1), 15–22. 1987

Coward, W. R., Sagara, H., and Church, M. K. Asthma, adenosine, mast cells and theophylline. Clin.Exp.Allergy 28 Suppl 3, 42–46. 1998

Cox, L. R., Murphy, S. K., and Ramos, K. Modulation of phosphoinositide metabolism in aortic smooth muscle cells by allylamine. Exp.Mol.Pathol. 53(1), 52–63. 1990

Crescioli, S., Spinazzi, A., Plebani, M., Pozzani, M., Mapp, C. E., Boschetto, P., and Fabbri, L. M. Theophylline inhibits early and late asthmatic reactions induced by allergens in asthmatic subjects. Ann.Allergy 66(3), 245–251. 1991

Crocker, I. C., Townley, R. G., and Khan, M. M. Phosphodiesterase inhibitors suppress proliferation of peripheral blood mononuclear cells and interleukin-4 and −5 secretion by human T-helper type 2 cells. Immunopharmacology 31, 223–235. 1996

Culpitt, S. V., de Matos, C., Russell, R. E., Donnelly, L. E., Rogers, D. F., and Barnes, P. J. Effect of theophylline on induced sputum inflammatory indices and neutrophil chemotaxis in chronic obstructive pulmonary disease. Am.J Respir Crit Care Med. 165(10), 1371–1376. 15-5-2002

Cushley, M. J., Tattersfield, A. E., and Holgate, S. T. Inhaled adenosine and guanosine on airway resistance in normal and asthmatic subjects. Br.J.Clin.Pharmacol. 15(2), 161–165. 1983

Cushley, M. J., Tattersfield, A. E., and Holgate, S. T. Adenosine-induced bronchoconstriction in asthma. Antagonism by inhaled theophylline. Am.Rev.Respir.Dis. 129(3), 380–384. 1984

D'Alonzo, G. E., Smolensky, M. H., Feldman, S., Gianotti, L. A., Emerson, M. B., Staudinger, H., and Steinijans, V. W. Twenty-four hour lung function in adult patients with asthma. Chronoptimized theophylline therapy once-daily dosing in the evening versus conventional twice-daily dosing. Am.Rev.Respir.Dis. 142(1), 84–90. 1990

Darnall-Jr, R. A. Aminophylline reduces hypoxic ventilatory depression: possible role of adenosine. Pediatr.Res. 19(7), 706–710. 1995

De Monchy, J. G., Kauffman, H. F., Venge, P., Koeter, G. H., Jansen, H. M., Sluiter, H. J., and de Vries, K. Bronchoalveolar eosinophilia during allergen-induced late asthmatic reactions. Am.Rev.Respir.Dis. 131(3), 373–376. 1985

Derian, C. K., Santulli, R. J., Rao, P. E., Solomon, H. F., and Barrett, J. A. Inhibition of chemotactic peptide-induced neutrophil adhesion to vascular endothelium by cAMP modulators. J.Immunol. 154(1), 308–317. 1995

Di Stefano, A., Caramori, G., Oates, T., Capelli, A., Lusuardi, M., Gnemmi, I., Ioli, F., Chung, K. F., Donner, C. F., Barnes, P. J., and Adcock, I. M. Increased expression of nuclear factor-kappaB in bronchial biopsies from smokers and patients with COPD. Eur Respir J 20(3), 556–563. 2002

Dullinger, D., Kronenberg, R., and Niewoehner, D. E. Efficacy of inhaled metaproterenol and orally-administered theophylline in patients with chronic airflow obstruction. Chest 89(2), 171–173. 1986

el Hashim, A., D'Agostino, B., Matera, M. G., and Page, C. Characterization of adenosine receptors involved in adenosine-induced bronchoconstriction in allergic rabbits. Br.J.Pharmacol. 119(6), 1262–1268. 1996

Elferink, J. G., Huizinga, T. W., and de-Koster, B. M. The effect of pentoxifylline on human neutrophil migration: a possible role for cyclic nucleotides. Biochem.Pharmacol. 54(4), 475–480. 15-8-1997

Elliott, K. R. and Leonard, E. J. Interactions of formylmethionyl-leucyl-phenylalanine, adenosine, and phosphodiesterase inhibitors in human monocytes. Effects on superoxide release, inositol phosphates and cAMP. FEBS Lett. 254(1–2), 94–98. 1989

Endres, S., Fulle, H. J., Sinha, B., Stoll, D., Dinarello, C. A., Gerzer, R., and Weber, P. C. Cyclic nucleotides differentially regulate the synthesis of tumour necrosis factor-alpha and interleukin-1 beta by human mononuclear cells. Immunology 72(1), 56–60. 1991

Enk, A. H., Angeloni, V. L., Udey, M. C., and Katz, S. I. Inhibition of Langerhans cell antigen-presenting function by IL-10. A role for IL-10 in induction of tolerance. J.Immunol. 151(5), 2390–2398. 1-9-1993

Evans, D. J., Taylor, D. A., Zetterstrom, O., Chung, K. F., O'Connor, B. J., and Barnes, P. J. A comparison of low-dose inhaled budesonide plus theophylline and high-dose inhaled budesonide for moderate asthma. N.Engl.J.Med 337, 1412–1418. 1997

Ezeamuzie, C. I. Involvement of A(3) receptors in the potentiation by adenosine of the inhibitory effect of theophylline on human eosinophil degranulation: possible novel mechanism of the anti-inflammatory action of theophylline. Biochem.Pharmacol. 61(12), 1551–1559. 15-6-2001

Feoktistov, I and Biaggioni, I. Adenosine A2b receptors evoke interleukin-8 secretion in human mast cells. An enprofylline-sensitive mechanism with implications for asthma. J.Clin.Invest. 96(4), 1979–1986. 1995

Feoktistov, I., Polosa, R., Holgate, S. T., and Biaggioni, I. Adenosine A2B receptors: a novel therapeutic target in asthma? Trends Pharmacol.Sci. 19(4), 148–153. 1998

Ferrer, P., Dihn-Xuan, T., and Chanal, I. Bronchodilator activity of LAS 31025, a new selective phosphodiesterase inhibitor. Am.J Respir Crit Care Med. 155. 1997

Ferretti, M. E., Spisani, S., Pareschi, M. C., Buzzi, M., Cavallaro, R., Traniello, S., Reali, E., Torrini, I, Paradisi, M. P., and Zecchini, G. P. Two new formulated peptides able to activate chemotaxis and respiratory burst selectively as tools for studying human neutrophil responses. Cell Signal. 6(1), 91–101. 1994

Fink, G., Mittelman, M., Shohat, B., and Spitzer, S. A. Theophylline-induced alterations in cellular immunity in asthmatic patients. Clin.Allergy 17(4), 313–316. 1987

Finkelstein, R., Fraser, R. S., and Ghezzo, H. Alveolar inflammation and its relation to emphysema in smokers. Am.J.Resp.Crit.Care Med. 152, 1666–1672. 1995

Finnerty, J. P., Lee, C., Wilson, S., Madden, J., Djukanovic, R., and Holgate, S. T. Effects of theophylline on inflammatory cells and cytokines in asthmatic subjects: a placebo-controlled parallel group study. Eur.Respir.J. 9, 1672–1677. 1996

Fonteh, A. N., Winkler, J. D., Torphy, T. J., Heravi, J., Undem, B. J., and Chilton, F. H. Influence of isoproterenol and phosphodiesterase inhibitors on platelet-activating factor biosynthesis in the human neutrophil. J.Immunol. 151(1), 339–350. 1993

Foukas, L. C., Daniele, N., Ktori, C., Anderson, K. E., Jensen, J., and Shepherd, P. R. Direct effects of caffeine and theophylline on p110 delta and other phosphoinositide 3-kinases. Differential effects on lipid kinase and protein kinase activities. J Biol.Chem. 277(40), 37124–37130. 4-10-2002

Franzini, E., Sellak, H., Babin, Chevaye C., Hakim, J., and Pasquier, C. Effects of pentoxifylline on the adherence of polymorphonuclear neutrophils to oxidant-stimulated human endothelial cells: involvement of cyclic AMP. J.Cardiovasc.Pharmacol. 25 Suppl 2, S92–S95. 1995

Frossard, N., Landry, Y., Pauli, G., and Ruckstuhl, M. Effects of cyclic AMP- and cyclic GMP- phosphodiesterase inhibitors on immunological release of histamine and on lung contraction. Br.J.Pharmacol. 73(4), 933–938. 1981

Gardette, J., Margelin, D., Maziere, J. C., Bertrand, J., and Picard, J. Effect of dibutyryl cyclic AMP and theophylline on lipoprotein lipase secretion by human monocyte-derived macrophages. FEBS Lett. 225(1–2), 178–182. 1987

Giembycz, M. A. Phosphodiesterase 4 inhibitors and the treatment of asthma: where are we now and where do we go from here? Drugs 59(2), 193–212. 2000

Gonzalez, M. C., Diaz, P., Galleguillos, F. R., Ancic, P., Cromwell, O., and Kay, A. B. Allergen-induced recruitment of bronchoalveolar helper (OKT4) and suppressor (OKT8) T-cells in asthma. Relative increases in OKT8 cells in single early responders compared with those in late- phase responders. Am.Rev.Respir.Dis. 136(3), 600–604. 1987

Griswold, D. E., Webb, E. F., Breton, J., White, J. R., Marshall, P. J., and Torphy, T. J. Effect of selective phosphodiesterase type IV inhibitor, rolipram, on fluid and cellular phases of inflammatory response. Inflammation 17(3), 333–344. 1993

Guillou, P. J., Ramsden, C., Kerr, M., Davison, A. M., and Giles, G. R. A prospective controlled clinical trial of aminophylline as an adjunctive immunosuppressive agent. Transplant.Proc. 16(5), 1218–1220. 1984

Guyatt, G. H., Townsend, M., Pugsley, S. O., Keller, J. L., Short, H. D., Taylor, D. W., and Newhouse, M. T. Bronchodilators in chronic air-flow limitation. Effects on airway function, exercise capacity, and quality of life. Am.Rev.Respir Dis. 135(5), 1069–1074. 1987

Hamid, Q., Azzawi, M., Ying, S., Moqbel, R., Wardlaw, A. J., Corrigan, C. J., Bradley, B., Durham, S. R., Collins, J. V., Jeffery, P. K., and . Expression of mRNA for interleukin-5 in mucosal bronchial biopsies from asthma. J.Clin.Invest 87(5), 1541–1546. 1991

Hancock, W. W., Pleau, M. E., and Kobzik, L. Recombinant granulocyte-macrophage colony-stimulating factor down-regulates expression of IL-2 receptor on human mononuclear phagocytes by induction of prostaglandin E. J.Immunol. 140(9), 3021–3025. 1988

Hart, L. A., Krishnan, V. L., Adcock, I. M., Barnes, P. J., and Chung, K. F. Activation and localization of transcription factor, nuclear factor- kappaB, in asthma. Am.J Respir Crit Care Med. 158(5 Pt 1), 1585–1592. 1998

Harvath, L., Robbins, J. D., Russell, A. A., and Seamon, K. B. cAMP and human neutrophil chemotaxis. Elevation of cAMP differentially affects chemotactic responsiveness. J.Immunol. 146(1), 224–232. 1991

Hatzelmann, A., Tenor, H., and Schudt, C. Differential effects of non-selective and selective phosphodiesterase inhibitors on human eosinophil functions. Br.J.Pharmacol. 114(4), 821–831. 1995

Hay, J. G., Stone, P., Carter, J., Church, S., Eyre-Brook, A., Pearson, M. G., Woodcock, A. A., and Calverley, P. M. Bronchodilator reversibility, exercise performance and breathlessness in stable chronic obstructive pulmonary disease. Eur Respir J 5(6), 659–664. 1992

Hendeles, L., Harman, E., Huang, D., O'Brien, R., Blake, K., and Delafuente, J. Theophylline attenuation of airway responses to allergen: comparison with cromolyn metered-dose inhaler. J.Allergy Clin.Immunol. 95(2), 505–514. 1995

Hichami, A., Boichot, E., Germain, N., Legrand, A., Moodley, I, and Lagente, V. Involvement of cyclic AMP in the effects of phosphodiesterase IV inhibitors on arachidonate release from mononuclear cells. Eur.J.Pharmacol. 291(2), 91–97. 1995

Hidi, R., Timmermans, S., Liu, E., Schudt, C., Dent, G., Holgate, S. T., and Djukanovic, R. Phosphodiesterase and cyclic adenosine monophosphate-dependent inhibition of T-lymphocyte chemotaxis. Eur Respir J 15(2), 342–349. 2000

Honeyman, T. W., Strohsnitter, W., Scheid, C. R., and Schimmel, R. J. Phosphatidic acid and phosphatidylinositol labelling in adipose tissue. Relationship to the metabolic effects of insulin and insulin-like agents. Biochem.J 212(2), 489–498. 15-5-1983

Hossain, M., Okubo, Y., and Sekiguchi, M. Effects of various drugs (staurosporine, herbimycin A, ketotifen, theophylline, FK506 and cyclosporin A) on eosinophil viability. Arerugi. 43(6), 711–717. 1994

Imhof, A. and Wolffe, A. P. Transcription: gene control by targeted histone acetylation. Curr.Biol. 8(12), R422-R424. 4-6-1998

Ito, K. and Adcock, I. M. Histone acetylation and histone deacetylation. Mol.Biotechnol. 20(1), 99–106. 2002a

Ito, K., Barnes, P. J., and Adcock, I. M. Glucocorticoid receptor recruitment of histone deacetylase 2 inhibits interleukin-1beta-induced histone H4 acetylation on lysines 8 and 12. Mol.Cell Biol. 20(18), 6891–6903. 2000

Ito, K., Lim, S., Caramori, G., Chung, K. F., Barnes, P. J., and Adcock, I. M. Cigarette smoking reduces histone deacetylase 2 expression, enhances cytokine expression, and inhibits glucocorticoid actions in alveolar macrophages. FASEB J 15(6), 1110–1112. 2001

Ito, K., Lim, S., Caramori, G., Cosio, B., Chung, K. F., Adcock, I. M., and Barnes, P. J. A molecular mechanism of action of theophylline: Induction of histone deacetylase activity to decrease inflammatory gene expression. Proc.Natl.Acad.Sci.U.S.A 99(13), 8921–8926. 25-6-2002b

Jaffar, Z. H., Sullivan, P., Page, C. P., and Costello, J. Low-dose theophylline modulates T-lymphocyte activation in allergen-challenged asthmatics. Eur.Respir.J. 9, 456–462. 1996

Jarjour, N. N., Lacouture, P. G., and Busse, W. W. Theophylline inhibits the late asthmatic response to nighttime antigen challenge in patients with mild atopic asthma. Ann.Allergy Asthma Immunol. 81(3), 231–236. 1998

Javaheri, S. and Guerra, L. Lung function, hypoxic and hypercapnic ventilatory responses, and respiratory muscle strength in normal subjects taking oral theophylline. Thorax 45(10), 743–747. 1990

Jeurgens, U. R., Overlack, A., and Vetter, H. Theophylline inhibits the formation of leukotriene B_4 (LTB_4) by enhancement of cyclic-AMP and prostaglandin E_2 (PGE_2) production in normal human monocytes in vitro. Eur.Respr.J. 17S, 3685. 1993

Jones, N. A., Page, C., and Lever, R. The effect of selective phosphodiesterase (PDE) isoenzyme inhibition on F-MET-LEU-PHE (fMLP) and tumor necrosis factor-alpha induced human neutrophil elastase release. Am.J.Resp.Crit.Care Med. 163. 2001

Kammer, G. M. The adenylate cyclase-cAMP-protein kinase A pathway and regulation of the immune response. Immunol.Today 9(7–8), 222–229. 1988

Kassis, S., Lee, J. C., and Hanna, N. Effects of prostaglandins and cAMP levels on monocyte IL-1 production. Agents Actions 27(3–4), 274–276. 1989

Keatings, V. M., Collins, P. D., and Scott, D. M. Differences in interleukin-8 and tumour necrosis factor-alpha in induced sputum from patients with chronic obstructive pulmonary disease or asthma. Am.J.Resp.Crit.Care Med. 153, 530–534. 1996

Kelloway, J. S., Wyatt, R. A., and Adlis, S. A. Comparison of patients' compliance with prescribed oral and inhaled asthma medications. Arch.Intern.Med. 154(12), 1349–1352. 27-6-1994

Kidney, J., Dominguez, M., Taylor, P. M., Rose, M., Chung, K. F., and Barnes, P. J. Immunomodulation by theophylline in asthma. Demonstration by withdrawal of therapy. American Journal of Respiratory and Critical Care Medicine 151(6), 1907–1914. 1995

Kita, H., Abu, Ghazaleh, Gleich, G. J., and Abraham, R. T. Regulation of Ig-induced eosinophil degranulation by adenosine 3',5'-cyclic monophosphate. J.Immunol. 146(8), 2712–2718. 1991

Knudsen, P. J., Dinarello, C. A., and Strom, T. B. Prostaglandins posttranscriptionally inhibit monocyte expression of interleukin 1 activity by increasing intracellular cyclic adenosine monophosphate. J.Immunol. 137(10), 3189–3194. 1986

Kongragunta, V. R., Druz, W. S., and Sharp, J. T. Dyspnea and diaphragmatic fatigue in patients with chronic obstructive pulmonary disease. Responses to theophylline. Am.Rev.Respir Dis. 137(3), 662–667. 1988

Kosmas, E. N., Michaelides, S. A., Polychronaki, A., Roussou, T., Toukmatzi, S., Polychronopoulos, V., and Baxevanis, C. N. Theophylline induces a reduction in circulating interleukin-4 and interleukin-5 in atopic asthmatics [In Process Citation]. Eur Respir J 13(1), 53–58. 1999

Kotecha, S., Taylor, I. K., and Shaw, R. J. Pharmacological modulation of platelet-derived growth factor (B) mRNA expression in alveolar macrophages and adherent monocytes. Pulm.Pharmacol. 7(6), 383–391. 1994

Kraft, M., Pak, J., Borish, L., and Martin, R. J. Theophylline's effect on neutrophil function and the late asthmatic response. Journal of Allergy & Clinical Immunology 98(2), 251–257. 1996a

Kraft, M., Torvik, J. A., Trudeau, J. B., Wenzel, S. E., and Martin, R. J. Theophylline: potential antiinflammatory effects in nocturnal asthma. J.Allergy Clin.Immunol. 97(6), 1242–1246. 1996b

Landells, L. J., Jensen, M. W., Orr, L. M., Spina, D., O'Connor, B. J., and Page, C. P. The role of adenosine receptors in the action of theophylline on human peripheral blood mononuclear cells from healthy and asthmatic subjects. Br.J.Pharmacol. 129(6), 1140–1144. 2000a

Landells, L. J., Spina, D., Souness, J. E., O'Connor, B. J., and Page, C. P. A biochemical and functional assessment of monocyte phosphodiesterase activity in healthy and asthmatic subjects. Pulm.Pharmacol.Ther. 13(5), 231–239. 2000b

Landsberg, K. F., Vaughan, L. M., and Heffner, J. E. The effect of theophylline on respiratory muscle contractility and fatigue. Pharmacotherapy 10(4), 271–279. 1990

Lappin, D., Riches, D. W., Damerau, B., and Whaley, K. Cyclic nucleotides and their relationship to complement- component-C2 synthesis by human monocytes. Biochem.J. 222(2), 477–486. 1984

Lichtenstein, L. M. and Margolis, S. Histamine release in vitro: inhibition by catecholamines and methylxanthines. Science 161(844), 902–903. 1968

Lidington, E., Nohammer, C., Dominguez, M., Ferry, B., and Rose, M. L. Inhibition of the transendothelial migration of human lymphocytes but not monocytes by phosphodiesterase inhibitors. Clinical and Experimental Immunology 104, 66–71. 1996

Lim, L. K., Hunt, N. H., and Weidemann, M. J. Reactive oxygen production, arachidonate metabolism and cyclic AMP in macrophages. Biochem.Biophys.Res.Commun. 114(2), 549–555. 1983

Lim, S., Tomita, K., Carramori, G., Jatakanon, A., Oliver, B., Keller, A., Adcock, I., Chung, K. F., and Barnes, P. J. Low-dose theophylline reduces eosinophilic inflammation but not exhaled nitric oxide in mild asthma. Am.J Respir Crit Care Med. 164(2), 273–276. 15-7-2001

Limatibul, S., Shore, A., Dosch, H. M., and Gelfand, E. W. Theophylline modulation of E-rosette formation: an indicator of T-cell maturation. Clinical and Experimental Immunology 33(3), 503–513. 1978

Lorenz, J. J., Furdon, P. J., Taylor, J. D., Verghese, M. W., Chandra, G., Kost, T. A., Haneline, S. A., Roner, L. A., and Gray, J. G. A cyclic adenosine 3',5'-monophosphate signal is required for the induction of IL-1 beta by TNF-alpha in human monocytes. J.Immunol. 155(2), 836–844. 1995

Louis, R., Bury, T., Corhay, J. L., and Radermecker, M. LY 186655, a phosphodiesterase inhibitor, inhibits histamine release from human basophils, lung and skin fragments. Int.J.Immunopharmacol. 14(2), 191–194. 1992

Louis, R. E. and Radermecker, M. F. Substance P-induced histamine release from human basophils, skin and lung fragments: effect of nedocromil sodium and theophylline. Int.Arch.Allergy Appl.Immunol. 92(4), 329–333. 1990

Magnussen, H., Reuss, G., and Jorres, R. Theophylline has a dose-related effect on the airway response to inhaled histamine and methacholine in asthmatics. Am.Rev.Respir.-Dis. 136(5), 1163–1167. 1987

Mahler, D. A., Matthay, R. A., Snyder, P. E., Wells, C. K., and Loke, J. Sustained-release theophylline reduces dyspnea in nonreversible obstructive airway disease. Am.Rev.-Respir Dis. 131(1), 22–25. 1985

Mann, J. S. and Holgate, S. T. Specific antagonism of adenosine-induced bronchoconstriction in asthma by oral theophylline. Br.J.Clin.Pharmacol. 19(5), 685–692. 1985

Mapp, C., Boschetto, P., dal Vecchio, L., Crescioli, S., de Marzo, N., Paleari, D., and Fabbri, L. M. Protective effect of antiasthma drugs on late asthmatic reactions and increased airway responsiveness induced by toluene diisocyanate in sensitized subjects. Am.-Rev.Respir.Dis. 136(6), 1403–1407. 1987

Marsh, G. D., McFadden, R. G., Nicholson, R. L., Leasa, D. J., and Thompson, R. T. Theophylline delays skeletal muscle fatigue during progressive exercise. Am.Rev.Respir Dis. 147(4), 876–879. 1993

Marvin, P. M., Baker, B. J., Dutt, A. K., Murphy, M. L., and Bone, R. C. Physiologic effects of oral bronchodilators during rest and exercise in chronic obstructive pulmonary disease. Chest 84(6), 684–689. 1983

Mary, D., Aussel, C., Ferrua, B., and Fehlmann, M. Regulation of interleukin 2 synthesis by cAMP in human T cells. J.Immunol. 139(4), 1179–1184. 1987

Mascali, J. J., Cvietusa, P., Negri, J., and Borish, L. Anti-inflammatory effects of theophylline: modulation of cytokine production. Ann.Allergy Asthma Immunol. 77(1), 34–38. 1996

Momose, T., Okubo, Y., Horie, S., Suzuki, J., Isobe, M., and Sekiguchi, M. Effects of intracellular cyclic AMP modulators on human eosinophil survival, degranulation and CD11b expression. International Archives of Allergy & Immunology 117(2), 138–145. 1998

Murciano, D., Aubier, M., Lecocguic, Y., and Pariente, R. Effects of theophylline on diaphragmatic strength and fatigue in patients with chronic obstructive pulmonary disease. N.Engl.J Med. 311(6), 349–353. 9-8-1984

Murciano, D., Auclair, M. H., Pariente, R., and Aubier, M. A randomized, controlled trial of theophylline in patients with severe chronic obstructive pulmonary disease. N.Engl.J Med. 320(23), 1521–1525. 8-6-1989

Mygind, N. Glucocorticosteroids and rhinitis. Allergy 48(7), 476–490. 1993

Naclerio, R. M., Bartenfelder, D., Proud, D., Togias, A. G., Meyers, D. A., Kagey, Sobotka, Norman, P. S., and Lichtenstein, L. M. Theophylline reduces histamine release during pollen-induced rhinitis. J.Allergy Clin.Immunol. 78(5 Pt 1), 874–876. 1986

Nielson, C. P., Crowley, J. J., Cusack, B. J., and Vestal, R. E. Therapeutic concentrations of theophylline and enprofylline potentiate catecholamine effects and inhibit leukocyte activation. J.Allergy Clin.Immunol. 78(4 Pt 1), 660–667. 1986

Nielson, C. P., Crowley, J. J., Morgan, M. E., and Vestal, R. E. Polymorphonuclear leukocyte inhibition by therapeutic concentrations of theophylline is mediated by cyclic-3',5'-adenosine monophosphate. Am.Rev.Respir.Dis. 137(1), 25–30. 1988

Niwa, M., Hara, A., Kanamori, Y., Matsuno, H., Kozawa, O., Yoshimi, N., Mori, H., and Uematsu, T. Inhibition of tumor necrosis factor-alpha induced neutrophil apoptosis by cyclic AMP: involvement of caspase cascade. Eur.J.Pharmacol. 371(1), 59–67. 23-4-1999

Nourshargh, S. and Hoult, J. R. Inhibition of human neutrophil degranulation by forskolin in the presence of phosphodiesterase inhibitors. Eur.J.Pharmacol. 122(2), 205–212. 1986

Nyce, J. W. and Metzger, W. J. DNA antisense therapy for asthma in an animal model. Nature 385(6618), 721–725. 20-2-1997

Ohta, K. and Yamashita, N. Apoptosis of eosinophils and lymphocytes in allergic inflammation. J Allergy Clin Immunol 104(1), 14–21. 1999

Okkenhaug, K., Bilancio, A., Farjot, G., Priddle, H., Sancho, S., Peskett, E., Pearce, W., Meek, S. E., Salpekar, A., Waterfield, M. D., Smith, A. J., and Vanhaesebroeck, B. Impaired B and T cell antigen receptor signaling in p110delta PI 3- kinase mutant mice. Science 297(5583), 1031–1034. 9-8-2002

Okubo, S., Konno, K., Ishizaki, T., Kubo, M., Suganuma, T., and Takizawa, T. Effect of theophylline on respiratory neuromuscular drive. Eur J Clin.Pharmacol. 33(1), 85–88. 1987

Orange, R. P., Kaliner, M. A., Laraia, P. J., and Austen, K. F. Immunological release of histamine and slow reacting substance of anaphylaxis from human lung. II. Influence of cellular levels of cyclic AMP. Fed.Proc. 30(6), 1725-1729. 1971

Ottonello, L., Gonella, R., Dapino, P., Sacchetti, C., and Dallegri, F. Prostaglandin E2 inhibits apoptosis in human neutrophilic polymorphonuclear leukocytes: role of intracellular cyclic AMP levels. Exp.Hematol. 26(9), 895-902. 1998

Page, C. P., Cotter, T., Kilfeather, S., Sullivan, P., Spina, D., and Costello, J. F. Effect of chronic theophylline treatment on the methacholine dose-response curve in allergic asthmatic subjects. European Respiratory Journal 12(1), 24-29. 1998

Pardi, R., Zocchi, M. R., Ferrero, E., Ciboddo, G. F., Inverardi, L., and Rugarli, C. In vivo effects of a single infusion of theophylline on human peripheral blood lymphocytes. Clinical and Experimental Immunology 57(3), 722-728. 1984

Paul, Eugene, Pene, J., Bousquet, J., and Dugas, B. Role of cyclic nucleotides and nitric oxide in blood mononuclear cell IgE production stimulated by IL-4. Cytokine 7(1), 64-69. 1995

Pauwels, R., van Renterghem, D., van der Straeten, M., Johannesson, N., and Persson, C. G. The effect of theophylline and enprofylline on allergen-induced bronchoconstriction. J.Allergy Clin.Immunol. 76(4), 583-590. 1985

Peachell, P. T., MacGlashan, D. W., Jr., Lichtenstein, L. M., and Schleimer, R. P. Regulation of human basophil and lung mast cell function by cyclic adenosine monophosphate. J.Immunol. 140(2), 571-579. 1988

Peachell, P. T., Undem, B. J., Schleimer, R. P., MacGlashan, D. W., Jr., Lichtenstein, L. M., Cieslinski, L. B., and Torphy, T. J. Preliminary identification and role of phosphodiesterase isozymes in human basophils. J.Immunol. 148(8), 2503-2510. 1992

Pearce, F. L., Befus, A. D., Gauldie, J., and Bienenstock, J. Mucosal mast cells. II. Effects of anti-allergic compounds on histamine secretion by isolated intestinal mast cells. J.Immunol. 128(6), 2481-2486. 1982

Persson, C. G. and Pauwels, R. Pharmacology of Anti-Asthma Xanthines. (7), 207-225. 1989. Academic Press London. Pharmacology of Asthma. Page, C. and Barnes, P. J

Persson, C. G. A. and Pauwels, R. Pharmacology of anti-asthma xanthines. Page, C. P. and Barnes, P. J. Pharmacology of asthma. (7), 207-225. 1991. Berlin, Springer-Verlag. Handbook of Experimental Pharmacology. Born, G. V. R., Cuatrecasas, P., and Herken, H

Peterson, T. C., Slysz, G., and Isbrucker, R. The inhibitory effect of ursodeoxycholic acid and pentoxifylline on platelet derived growth factor-stimulated proliferation is distinct from an effect by cyclic AMP. Immunopharmacology 39(3), 181-191. 1998

Platzer, C., Meisel, C., Vogt, K., Platzer, M., and Volk, H. D. Up-regulation of monocytic IL-10 by tumor necrosis factor-alpha and cAMP elevating drugs. Int.Immunol. 7(4), 517-523. 1995

Punnonen, J., Punnonen, K., Jansen, C. T., and Kalimo, K. Interferon (IFN)-alpha, IFN-gamma, interleukin (IL)-2, and arachidonic acid metabolites modulate IL-4-induced IgE synthesis similarly in healthy persons and in atopic dermatitis patients. Allergy 48(3), 189-195. 1993

Rivington, R. N., Boulet, L. P., Cote, J., Kreisman, H., Small, D. I., Alexander, M., Day, A., Harsanyi, Z., and Darke, A. C. Efficacy of Uniphyl, salbutamol, and their combination in asthmatic patients on high-dose inhaled steroids. Am.J.Respir.Crit Care Med. 151(2 Pt 1), 325-332. 1995

Rivkin, I and Neutze, J. A. Influence of cyclic nucleotides and a phosphodiesterase inhibitor on in vitro human blood neutrophil chemotaxis. Arch.Int.Pharmacodyn.Ther. 228(2), 196-204. 1977

Rossi, A. G., Cousin, J. M., Dransfield, I., Lawson, M. F., Chilvers, E. R., and Haslett, C. Agents that elevate cAMP inhibit human neutrophil apoptosis. Biochem.Biophys.Res.Commun. 217(3), 892-899. 26-12-1995

Sagara, H., Fuiuda, T., Okada, T., Ishikawa, A., and Makino, S. Theophylline at therapeutic concentration suppresses PAF-induced upregulation of Mac-1 on human eosinophils. Clinical and Experimental Allergy, Supplement 26, 16–21. 1996

Scherer, L. J., Diamond, R. A., and Rothenberg, E. V. Developmental regulation of cAMP signaling pathways in thymocyte development. Thymus 23, 231–257. 1994

Schmeichel, C. J. and Thomas, L. L. Methylxanthine bronchodilators potentiate multiple human neutrophil functions. J.Immunol. 138(6), 1896–1903. 15-3-1987

Scordamaglia, A., Ciprandi, G., Ruffoni, S., Caria, M., Paolieri, F., Venuti, D., and Canonica, G. W. Theophylline and the immune response: in vitro and in vivo effects. Clin.Immunol.Immunopathol. 48(2), 238–246. 1988

Sherman, M. S., Lang, D. M., Matityahu, A., and Campbell, D. Theophylline improves measurements of respiratory muscle efficiency. Chest 110(6), 1437–1442. 1996

Shichijo, M., Shimizu, Y., Hiramatsu, K., Inagaki, N., Tagaki, K., and Nagai, H. Cyclic AMP-elevating agents inhibit mite-antigen-induced IL-4 and IL-13 release from basophil-enriched leukocyte preparation. Int.Arch.Allergy Immunol. 114(4), 348–353. 1997

Shohat, B., Volovitz, B., and Varsano, I. Induction of suppressor T cells in asthmatic children by theophylline treatment. Clin.Allergy 13(5), 487–493. 1983

Shute, J. K., Tenor, H., Church, M. K., and Holgate, S. T. Theophylline inhibits the release of eosinophil survival cytokines—Is Raf-1 the protein kinase A target? Clinical & Experimental Allergy, Supplement 28(3), 47–52. 1998

Souness, J. E., Griffin, M., Maslen, C., Ebsworth, K., Scott, L. C., Pollock, K., Palfreyman, M. N., and Karlsson, J. A. Evidence that cyclic AMP phosphodiesterase inhibitors suppress TNFα generation from human monocytes by interacting with a 'low-affinity' phosphodiesterase 4 conformer. Br.J.Pharmacol. 118, 649–658. 1996

Spatafora, M., Chiappara, G., Merendino, A. M., D'Amico, D., Bellia, V, and Bonsignore, G. Theophylline suppresses the release of tumour necrosis factor- alpha by blood monocytes and alveolar macrophages. Eur.Respir.J. 7(2), 223–228. 1994

Spina, D., Ferlenga, P., Biasini, I., Moriggi, E., Marchini, F., Semeraro, C., and Page, C. P. The effect duration of selective phosphodiesterase inhibitors in the guinea-pig. Life Sci. 11, 953–965. 1998a

Spina, D., Landells, L. J., and Page, C. P. The role of phosphodiesterase isoenzymes in health and in atopic disease. August, T. Advances in Pharmacology. 33–89. 1998b. San Diego, Academic Press Inc

Spoelstra, F. M., Berends, C., Dijkhuizen, B., De Monchy, J. G. R., and Kauffman, H. F. Effect of theophylline on CD11b and L-selectin expression and density of eosinophils and neutrophils in vitro. European Respiratory Journal 12(3), 585–591. 1998

Steele, T. A. and Brahmi, Z. Phosphatidylinositol metabolism accompanies early activation events in tumor target cell-stimulated human natural killer cells. Cell.Immunol. 112, 402–443. 1988

Sullivan, P., Bekir, S., Jaffar, Z., Page, C., Jeffery, P., and Costello, J. Anti-inflammatory effects of low-dose oral theophylline in atopic asthma [published erratum appears in Lancet 1994 Jun 11; 343(8911):1512]. Lancet 343(8904), 1006–1008. 1994

Sung, S. S. and Walters, J. A. Increased cyclic AMP levels enhance IL-1 alpha and IL-1 beta mRNA expression and protein production in human myelomonocytic cell lines and monocytes. J.Clin.Invest. 88(6), 1915–1923. 1991

Tenor, H., Hatzelmann, A., Church, M. K., Schudt, C., and Shute, J. K. Effects of theophylline and rolipram on leukotriene C4 (LTC4) synthesis and chemotaxis of human eosinophils from normal and atopic subjects. Br.J.Pharmacol. 118, 1727–1735. 1996

Thomas, P., Pugsley, J. A., and Stewart, J. H. Theophylline and salbutamol improve pulmonary function in patients with irreversible chronic obstructive pulmonary disease. Chest 101(1), 160–165. 1992

Thompson, P. B., Daughton, D., and Robbins, G. A. Intramural airway inflammation in chronic bronchitis. Characterization and correlation with clinical parameters. Am.-Rev.Respir.Dis. 140, 1527–1537. 1989

Tinkelman, D. G., Reed, C. E., Nelson, H. S., and Offord, K. P. Aerosol beclomethasone dipropionate compared with theophylline as primary treatment of chronic, mild to moderately severe asthma in children. Pediatrics 92(1), 64–77. 1993

Turner, C. R., Esser, K. M., and Wheeldon, E. B. Therapeutic intervention in a rat model of ARDS: IV. Phosphodiesterase IV inhibition. Circ.Shock 39(3), 237–245. 1993

Ukena, D., Harnest, U., Sakalauskas, R., Magyar, P., Vetter, N., Steffen, H., Leichtl, S., Rathgeb, F., Keller, A., and Steinijans, V. W. Comparison of addition of theophylline to inhaled steroid with doubling of the dose of inhaled steroid in asthma. Eur.Respir.J. 10, 2754–2760. 1997

Umut, S., Gemicioglu, B., Yildirim, N., Barlas, A., and Ozuner, Z. Effect of theophylline in chronic obstructive lung disease. Int.J Clin.Pharmacol.Ther.Toxicol. 30(5), 149–152. 1992

van der Pouw, Kraan, Boeije, L. C., Smeenk, R. J., Wijdenes, J., and Aarden, L. A. Prostaglandin-E2 is a potent inhibitor of human interleukin 12 production. J.Exp.Med. 181(2), 775–779. 1995

Vanhaesebroeck, B., Jones, G. E., Allen, W. E., Zicha, D., Hooshmand-Rad, R., Sawyer, C., Wells, C., Waterfield, M. D., and Ridley, A. J. Distinct PI(3)Ks mediate mitogenic signalling and cell migration in macrophages. Nat.Cell Biol. 1(1), 69–71. 1999

Vanhaesebroeck, B., Welham, M. J., Kotani, K., Stein, R., Warne, P. H., Zvelebil, M. J., Higashi, K., Volinia, S., Downward, J., and Waterfield, M. D. P110delta, a novel phosphoinositide 3-kinase in leukocytes. Proc.Natl.Acad.Sci.U.S.A 94(9), 4330–4335. 29-4-1997

Varriale, P. and Ramaprasad, S. Aminophylline induced atrial fibrillation. Pacing Clin.Electrophysiol. 16(10), 1953–1955. 1993

Verghese, M. W., McConnell, R. T., Strickland, A. B., Gooding, R. C., Stimpson, S. A., Yarnall, D. P., Taylor, J. D., and Furdon, P. J. Differential regulation of human monocyte-derived TNF alpha and IL-1 beta by type IV cAMP-phosphodiesterase (cAMP-PDE) inhibitors. J.Pharmacol.Exp.Ther. 272(3), 1313–1320. 1995

Viherluoto, J., Palkama, T., Silvennoinen, O., and Hurme, M. Cyclic adenosine monophosphate decreases the secretion, but not the cell-associated levels, of interleukin-1 beta in lipopolysaccharide-activated human monocytes. Scand.J.Immunol. 34(1), 121–125. 1991

Ward, A. J., McKenniff, M., Evans, J. M., Page, C. P., and Costello, J. F. Theophylline—an immunomodulatory role in asthma? Am.Rev.Respir.Dis. 147(3), 518–523. 1993

Weinberger, M. and Hendeles, L. Theophylline in asthma. N.Engl.J.Med. 334(21), 1380–1388. 23-5-1996

Weston, M. C., Anderson, N., and Peachell, P. T. Effects of phosphodiesterase inhibitors on human lung mast cell and basophil function. Br.J.Pharmacol. 121(2), 287–295. 1997

Wolkove, N., Dajczman, E., Colacone, A., and Kreisman, H. The relationship between pulmonary function and dyspnea in obstructive lung disease. Chest 96(6), 1247–1251. 1989

Yamaguchi, Y., Hayashi, Y., Sugama, Y., Miura, Y., Kasahara, T., Kitamura, S., Torisu, M., Mita, S., Tominaga, A., and Takatsu, K. Highly purified murine interleukin 5 (IL-5) stimulates eosinophil function and prolongs in vitro survival. IL-5 as an eosinophil chemotactic factor. J.Exp.Med. 167(5), 1737–1742. 1-5-1988

Yasui, K., Agematsu, K., Shinozaki, K., Hokibara, S., Nagumo, H., Nakazawa, T., and Komiyama, A. Theophylline induces neutrophil apoptosis through adenosine A2A receptor antagonism. J.Leukoc.Biol. 67(4), 529–535. 2000a

Yasui, K., Agematsu, K., Shinozaki, K., Hokibara, S., Nagumo, H., Yamada, S., Kobayashi, N., and Komiyama, A. Effects of theophylline on human eosinophil functions: comparative study with neutrophil functions. J.Leukoc.Biol. 68(2), 194–200. 2000b

Yasui, K., Hu, B., Nakazawa, T., Agematsu, K., and Komiyama, A. Theophylline accelerates human granulocyte apoptosis not via phosphodiesterase inhibition. Journal of Clinical Investigation 100(7), 1677–1684. 1997

Zocchi, M. R., Pardi, R., Gromo, G., Ferrero, E., Ferrero, M. E., Besana, C., and Rugarli, C. Theophylline induced non specific suppressor activity in human peripheral blood lymphocytes. J.Immunopharmacol. 7(2), 217–234. 1985

ZuWallack, R. L., Mahler, D. A., Reilly, D., Church, N., Emmett, A., Rickard, K., and Knobil, K. Salmeterol plus theophylline combination therapy in the treatment of COPD. Chest 119(6), 1661–1670. 2001

Corticosteroids

P. J. Barnes

Department of Thoracic Medicine, National Heart and Lung Institute,
Dovehouse St, London, SW3 6LY, UK
e-mail: p.j.barnes@ic.ac.uk

1	Introduction	81
2	**Molecular Mechanisms**	82
2.1	Glucocorticoid Receptors	82
2.2	Increased Gene Transcription	82
2.3	Decreased Gene Transcription	85
2.4	Histone Deacetylation	86
2.5	Non-transcriptional Effects	87
2.6	Target Genes in Inflammation Control	87
2.6.1	Anti-inflammatory Proteins	87
2.6.2	β_2-Adrenoceptors	87
2.6.3	Cytokines	88
2.6.4	Inflammatory Enzymes	88
2.6.5	Inflammatory Receptors	89
2.6.6	Adhesion Molecules	89
2.6.7	Apoptosis	89
3	**Effects on Cell Function**	89
3.1	Macrophages	90
3.2	Eosinophils	90
3.3	T Lymphocytes	91
3.4	Mast Cells	91
3.5	Dendritic Cells	91
3.6	Neutrophils	91
3.7	Endothelial Cells	91
3.8	Epithelial Cells	92
3.9	Mucus Secretion	92
4	**Effects on Airway Inflammation in Asthma**	93
4.1	Effects on Airway Hyperresponsiveness	93
5	**Clinical Efficacy of Inhaled Corticosteroids in Asthma**	93
5.1	Studies in Adults	94
5.2	Studies in Children	95
5.3	Dose–Response Studies	95
5.4	Prevention of Irreversible Airway Changes	96
5.5	Reduction in Mortality	96
5.6	Comparison Between Inhaled Corticosteroids	97

6		**Clinical Use of Inhaled Corticosteroids in Asthma**	97
6.1		Add-on Therapy	98
	6.1.1	Long-Acting Inhaled β_2-Agonists	98
	6.1.2	Theophylline	98
	6.1.3	Anti-leukotrienes	99
	6.1.4	Mechanisms	99
6.2		Cost Effectiveness	99
6.3		Corticosteroid-Sparing Therapy	99
7		**Pharmacokinetics**	100
8		**Side Effects of Inhaled Corticosteroids**	101
8.1		Local Side Effects	101
	8.1.1	Dysphonia	102
	8.1.2	Oropharyngeal Candidiasis	102
	8.1.3	Other Local Complications	102
8.2		Systemic Side Effects	103
	8.2.1	Effect of Delivery Systems	103
	8.2.2	Hypothalamic–Pituitary–Adrenal Axis	104
	8.2.3	Effects on Bone Metabolism	105
	8.2.4	Effects on Connective Tissue	106
	8.2.5	Ocular Effects	106
	8.2.6	Growth	106
	8.2.7	Metabolic Effects	107
	8.2.8	Haematological Effects	107
	8.2.9	Central Nervous System Effects	108
	8.2.10	Safety in Pregnancy	108
9		**Systemic Corticosteroids**	108
9.1		Acute Severe Asthma	109
10		**Corticosteroid-Resistant Asthma**	110
10.1		Mechanisms of Corticosteroid Resistance	110
11		**Corticosteroids in COPD**	111
11.1		Effect on Inflammation	111
11.2		Clinical Studies	112
12		**Future Directions**	112
12.1		New Corticosteroids	113
12.2		Dissociated Steroids	113
References			114

Abstract Corticosteroids are by far the most effective treatments currently available for the treatment of asthma. Inhaled corticosteroids have revolutionised the management of asthma and have been the most important advance in therapy over the last 30 years. Initially, inhaled corticosteroids were introduced to reduce the requirement for oral corticosteroids in patients with more severe disease, but now their use has extended to patients with much milder disease, including children. This is in part because it is recognised that airway inflammation is present even in patients with mild asthma when they may have no symptoms, and in part because of their great safety at low doses. The molecular mechanisms of corticosteroids are now much better understood, and it is now known that corticosteroids suppress the increased expression of multiple inflammatory genes that are over-expressed in asthmatic airways, accounting for their clinical efficacy. The mechanism whereby corticosteroids switch off multiple inflammatory genes appears to involve reversal of acetylation of core histones, particularly through the recruitment of histone deacetylases. Inhaled corticosteroids are now recommended as first-line treatment for all patients (adults and children) with persistent asthma. The early use of inhaled steroids in asthma has been shown to reduce symptoms, improve the quality of life of patients, prevent exacerbations and reduce airway hyperresponsiveness. There is increasing evidence that early use of inhaled corticosteroids may reduce the irreversible changes in airway function that occur in some patients with asthma. Several studies now suggest that inhaled corticosteroids have a relatively flat dose–response curve and that most benefit is achieved at rather low doses. Increasing the dose of inhaled corticosteroids has less benefit than adding another class of drug, such as an inhaled long-acting β_2-agonist or theophylline in patients not controlled on low doses of inhaled corticosteroids. Systemic side effects are not an important issue when low doses of inhaled corticosteroids are used. There is no evidence for growth suppression in children treated with low doses of inhaled corticosteroids and no evidence for osteoporosis in adults. Some patients with asthma are resistant or relatively resistant to the anti-inflammatory effects of corticosteroids, and several molecular mechanisms for this steroid resistance have been elucidated. Patients with chronic obstructive pulmonary disease (COPD) show a poor response to corticosteroids and there may be an active resistance, due to impaired function of histone deacetylases. Several new corticosteroids are in development with the view of improving the therapeutic ratio.

Keywords Inflammation · Transcription factor · Histone acetylation · Steroid resistance

1
Introduction

Corticosteroids are by far the most effective therapy currently available for asthma, and improvement with corticosteroids is one of the hallmarks of asthma. Inhaled corticosteroids have revolutionised asthma treatment and have become the mainstay of therapy for patients with chronic disease (Barnes 1995, 1998b). By contrast, corticosteroids are much less ineffective in chronic obstructive pulmonary disease (COPD) (Barnes 2000a). We now have a much better understanding of the molecular mechanisms whereby corticosteroids suppress inflammation in asthma and why they may be ineffective in COPD and in rare patients with asthma who are resistant to corticosteroids. This chapter discusses the current understanding of the mechanism of action of corticosteroids and how corticosteroids are used in the management of airway diseases.

2
Molecular Mechanisms

Corticosteroids are highly effective anti-inflammatory therapy in asthma, and the molecular mechanisms involved in suppression of airway inflammation in asthma are now better understood (Barnes 1998a, 2001). Corticosteroids are effective because they block many of the inflammatory pathways that are abnormally activated in asthma and they have a wide spectrum of anti-inflammatory actions.

2.1
Glucocorticoid Receptors

Corticosteroids cross the cell membrane and bind to glucocorticoid receptors (GR) in the cytoplasm that are normally bound to chaperone proteins, such as the 90-kDa heat shock protein (hsp90), that prevent nuclear localisation of the receptor. There is a single gene encoding GR but several variants are now recognised (Yudt and Cidlowski 2002). GRα binds corticosteroids, whereas GRβ is an alternatively spliced form that binds to DNA but is not activated by corticosteroids. GRβ has been implicated in steroid-resistance in asthma (Leung et al. 1997), although whether GRβ has any functional significance has been questioned (Hecht et al. 1997). GR is also subject to post-translational modification by phosphorylation, nitration and ubiquitination, and this may lead to differential sensitivity to corticosteroids in different cell types. For example, there are a number of serine/threonines in the N-terminal domain where GR may be phosphorylated by various kinases and this may change corticosteroid binding affinity, nuclear import and export, receptor stability and *trans*-activating efficacy (Bodwell et al. 1998).

Once corticosteroids bind to the N-terminal end of GR, conformational changes in the receptor structure result in dissociation of molecular chaperone molecules, thereby exposing nuclear localisation signals on GR. This results in rapid transport of the activated GR–corticosteroid complex into the nucleus, where it binds to DNA at specific glucocorticoid response elements in steroid-responsive genes (Fig. 1). Two GR molecules bind to glucocorticoid recognition element (GRE) sites as a dimer, resulting in changes in gene transcription. The crystal structure of the GR ligand-binding domain has recently been determined in the presence of dexamethasone (Bledsoe et al. 2002). This reveals a novel dimerisation interface and a region that is important for coactivator binding.

2.2
Increased Gene Transcription

Corticosteroids produce their effect on responsive cells by activating GR to directly or indirectly regulate the transcription of target genes (Reichardt et al. 1998). The number of genes per cell directly regulated by corticosteroids is esti-

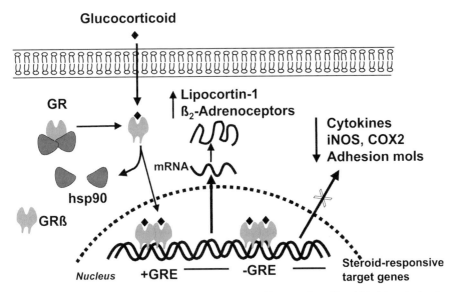

Fig. 1 Classical model of corticosteroid action. Corticosteroids enter the cell and bind to cytoplasmic glucocorticoid receptors (*GR*) that are complexed with two molecules of a 90-kDa heat shock protein (*hsp90*). GR translocates to the nucleus where, as a dimer, it binds to a glucocorticoid recognition elements (*GRE*) on the 5'-upstream promoter sequence of steroid-responsive genes. GREs increase transcription, whereas nGREs may decrease transcription, resulting in increased or decreased messenger RNA (*mRNA*) and protein synthesis. An isoform of GR, GR-β, binds to DNA but is not activated by corticosteroids

mated to be between 10 and 100, but many genes are indirectly regulated through an interaction with other transcription factors and coactivators. GR dimers bind to DNA at consensus GRE sites in the 5'-upstream promoter region of steroid-responsive genes. Interaction of the activated GR homodimer with GRE usually increases transcription, resulting in increased protein synthesis. GR may increase transcription by interacting with coactivator molecules, such as CREB-binding protein (CBP), thus switching on histone acetylation and gene transcription. For example, relatively high concentrations of corticosteroids increase the secretion of the antiprotease secretory leukoprotease inhibitor (SLPI) from epithelial cells (Ito et al. 2000).

Gene transcription is regulated by acetylation of core histones and a change in the structure of chromatin (Urnov and Wolffe 2001). The activation of genes by corticosteroids is associated with a selective acetylation of lysine residues 5 and 16 on histone-4, resulting in increased gene transcription (Ito et al. 2000, 2001a) (Fig. 2). Activated GR may bind to coactivator molecules, such as CBP as well as steroid-receptor coactivator-1 (SRC-1), which all have histone acetyltransferase (HAT) activity (Yao et al. 1996; Kurihara et al. 2002). However, SRC-1 does not appear to be involved in nuclear factor (NF)-κB-activated HAT activity (Ito et al. 2000), but it is likely that other similar p160 coactivator mole-

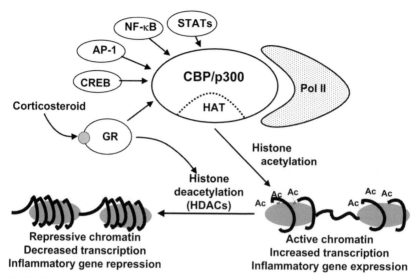

Fig. 2 Effect of corticosteroids on chromatin structure. Transcription factors, such as STATs, AP-1 and NF-κB bind to co-activator molecules, such as CREB-binding protein (*CBP*) or p300, which have intrinsic histone acetyltransferase (*HAT*) activity, resulting in acetylation (*Ac*) of histone residues. This leads to unwinding of DNA and this allows increased binding of transcription factors, resulting in increased gene transcription. Glucocorticoid receptors (*GR*) after activation by corticosteroids bind to a glucocorticoid receptor co-activator which is bound to CBP. This results in deacetylation of histone, with increased coiling of DNA around histone, thus preventing transcription factor binding leading to gene repression

cules are involved. Corticosteroids may suppress inflammation by increasing the synthesis of anti-inflammatory proteins, such as annexin-1, interleukin (IL)-10 and the inhibitor of NF-κB, IκB-α. However, therapeutic doses of inhaled corticosteroids have not been shown to increase annexin-1 concentrations in bronchoalveolar lavage fluid (Hall et al. 1999) and an increase in IκB-α has not been shown in most cell types, including epithelial cells (Heck et al. 1997; Newton et al. 1998a). It seems highly unlikely that the widespread anti-inflammatory actions of corticosteroids could be explained by increased transcription of small numbers of anti-inflammatory genes, particularly as high concentrations of corticosteroids are usually required for this response, whereas in clinical practice corticosteroids are able to suppress inflammation at much lower concentrations.

Little is known about the molecular mechanisms of corticosteroid side effects, such as osteoporosis, growth retardation in children, skin fragility and metabolic effects. The systemic side effects of corticosteroids may be due to gene activation. Some insight into this has been provided by mutant GR which do not dimerise and therefore cannot bind to GRE to switch on genes. In transgenic mice expressing these mutant GR, corticosteroids show no loss in their anti-inflammatory effects and are able to suppress NF-κB-activated genes in the normal way (Reichardt et al. 2001).

2.3
Decreased Gene Transcription

In controlling inflammation, the major effect of corticosteroids is to inhibit the synthesis of multiple inflammatory proteins through suppression of the genes that encode them. This was originally believed to be through interaction of GR with negative GREs, resulting in repression of transcription. However, negative GREs have only very rarely been demonstrated and are not a feature of the promoter region of the inflammatory genes that are suppressed by steroids in the treatment of asthma. In patients with asthma, the increased expression of multiple inflammatory genes includes those encoding cytokines, chemokines, adhesion molecules, inflammatory enzymes and inflammatory receptors (Table 1).

Activated GRs have been shown to interact functionally with other activated transcription factors. Most of the inflammatory genes that are activated in asthma do not have GREs in their promoter regions, yet are potently repressed by corticosteroids. There is persuasive evidence that corticosteroids inhibit the effects of pro-inflammatory transcription factors, such as AP-1 and NF-κB, that regulate the expression of genes that code for inflammatory proteins, such as cytokines, inflammatory enzymes, adhesion molecules and inflammatory receptors (Barnes and Karin 1997; Barnes and Adcock 1998). The activated GR can interact directly with activated transcription factors by protein–protein interaction, but this may be a feature of transfected cells and over-expression, rather

Table 1 Effect of corticosteroids on gene transcription

Increased transcription
Lipocortin-1 (phospholipase A_2 inhibitor)
β_2-Adrenoceptor
Secretory leukoprotease inhibitor
Clara cell protein (CC10, phospholipase A_2 inhibitor)
IL-1 receptor antagonist
IL-1R2 (decoy receptor)
IκB-α (inhibitor of NF-κB)
Decreased transcription
Cytokines
IL-1, IL-2, IL-3, IL-4, IL-5, IL-6, IL-9, IL-11, IL-12, IL-13, IL-16, IL-17, IL-18, TNF-α, GM-CSF, SCF
Chemokines
IL-8, RANTES, MIP-1α, MCP-1, MCP-3, MCP-4, eotaxin
Inflammatory enzymes
Inducible nitric oxide synthase (iNOS), inducible cyclo-oxygenase (COX-2)
Cytoplasmic phospholipase A_2 (cPLA$_2$)
Inflammatory receptors
NK$_1$-receptors, NK$_2$-receptors, bradykinin B$_2$-receptors
Adhesion molecules
ICAM-1, E-selectin, VCAM-1
Inflammatory peptides
Endothelin-1

than a property of primary cells. Treatment of asthmatic patients with high doses of inhaled corticosteroids that suppress airway inflammation is not associated with any reduction in NF-κB binding to DNA (Hart et al. 2000). This suggests that corticosteroids are more likely to be acting downstream of the binding of proinflammatory transcription factors to DNA and attention has now focused on their effects on chromatin structure and histone acetylation.

2.4
Histone Deacetylation

Repression of genes occurs through reversal of the histone acetylation that switches on inflammatory genes (Imhof and Wolffe 1998). Activated GR may bind to CBP or other coactivators directly to inhibit their HAT activity (Ito et al. 2000), thus reversing the unwinding of DNA round-core histones and thereby repressing inflammatory genes. More importantly, particularly at low concentrations that are likely to be relevant therapeutically in asthma treatment, activated GR recruits histone deacetylases (HDACs) to the activated transcriptional complex, resulting in deacetylation of histones, and thus a decrease in inflammatory gene transcription (Ito et al. 2000) (Fig. 2). At least 11 HDACs have now been identified and these are differentially expressed and regulated in different cell types (Gao et al. 2002). There is now evidence that the different HDACs target different patterns of acetylation (Peterson 2002). These differences in HDACs may contribute to differences in responsiveness to corticosteroids between different genes and cells.

An important question is why corticosteroids only switch off inflammatory genes, as they clearly do not suppress all activated genes and are well tolerated as a therapy. It is likely that GR only binds to coactivators that are activated by proinflammatory transcription factors, such as NF-κB and AP-1, although we do not understand how this specific recognition occurs. It is likely that there are several specific coactivators that interact with GR. AP-1 and NF-κB repression is normal in mice which express a form of GR that does not dimerise (dim$^{-/-}$), indicating that GR monomers are able to mediate the anti-inflammatory effects of corticosteroids, whereas dimerisation is needed for gene activation (Reichardt et al. 1998).

Methylation of histones, particularly histone-3, by histone methyltransferases, results in gene suppression (Bannister et al. 2002). The anti-inflammatory effects of corticosteroids are reduced by a methyltransferase inhibitor, 5-aza-2'-deoxycytidine, suggesting that this may be an additional mechanism whereby corticosteroids suppress genes (Kagoshima et al. 2001). Indeed there may be an interaction between acetylation, methylation and phosphorylation of histones, so that the sequence of chromatin modifications may give specificity to expression of particular genes (Jenuwein and Allis 2001).

2.5
Non-transcriptional Effects

Although most of the actions of corticosteroids are mediated by changes in transcription through chromatin remodelling, it is increasingly recognised that they may also affect protein synthesis by reducing the stability of mRNA. Some inflammatory genes, such as the gene encoding granulocyte-macrophage colony-stimulating factor (GM-CSF), produce mRNA that is rich in AU nucleotides at the 3′-untranslated end. It is this region that interacts with ribonucleases that break down mRNA, thus switching off protein synthesis. Corticosteroids may have effects on proteins that stabilise the AU-rich region, leading to more rapid breakdown of mRNA and consequent reduction in protein expression (Bergmann et al. 2000; Reichardt et al. 2001).

2.6
Target Genes in Inflammation Control

Corticosteroids may control inflammation by inhibiting many aspects of the inflammatory process in asthma through increasing the transcription of anti-inflammatory genes and decreasing the transcription of inflammatory genes (Table 1).

2.6.1
Anti-inflammatory Proteins

Corticosteroids may suppress inflammation by increasing the synthesis of anti-inflammatory proteins. For example, corticosteroids increase the synthesis of lipocortin-1, a 37-kDa protein that has an inhibitory effect on phospholipase A_2 (PLA_2), and therefore may inhibit the production of lipid mediators. Corticosteroids induce the formation of lipocortin-1 in several cells, and recombinant lipocortin-1 has acute anti-inflammatory properties. However, lipocortin-1 does not appear to be increased by inhaled corticosteroid treatment in asthma (Hall et al. 1999). Corticosteroids increase the expression of other potentially anti-inflammatory proteins, such as IL-1 receptor antagonist (which inhibits the binding of IL-1 to its receptor), secretory leukoprotease inhibitor (which inhibits proteases, such as tryptase), neutral endopeptidase (which degrades bronchoactive peptides such as kinins), CC-10 (an immunomodulatory protein), an inhibitor of NF-κB (IκB-α) and IL-10 (an anti-inflammatory cytokine).

2.6.2
β_2-Adrenoceptors

Corticosteroids increase the expression of β_2-adrenoceptors by increasing the rate of transcription, and the human β_2-receptor gene has three potential GREs. Corticosteroids double the rate of β_2-receptor gene transcription in human lung in vitro, resulting in increased expression of β_2-receptors (Mak et al. 1995a). This also occurs in vivo in nasal mucosa with treatment with topical corticosteroids (Baraniuk et al. 1997). This may be relevant in asthma as corticosteroids may prevent down-regulation of β-receptors in response to prolonged treatment with β_2-agonists. In rats, corticosteroids prevent down-regulation and reduced transcription of β_2-receptors in response to chronic β-agonist exposure (Mak et al. 1995b).

2.6.3
Cytokines

The inhibitory effect of corticosteroids on cytokine synthesis is likely to be of particular importance in the control of inflammation in asthma. Corticosteroids inhibit the transcription of many cytokines and chemokines that are relevant in asthma (Table 1). These inhibitory effects are due, at least in part, to an inhibitory effect on the transcription factors that regulate induction of these cytokine genes, including AP-1 and NF-κB. For example, eotaxin which is important in selective attraction of eosinophils from the circulation into the airways is regulated in part by NF-κB and its expression in airway epithelial cells is inhibited by corticosteroids (Lilly et al. 1997). Many transcription factors are likely to be involved in the regulation of inflammatory genes in asthma in addition to AP-1 and NF-κB. IL-4 and IL-5 expression in T lymphocytes plays a critical role in allergic inflammation, but NF-κB does not play a role, whereas the transcription factor nuclear factor of activated T cells (NF-AT) is important (Rao et al. 1997). AP-1 is a component of the NF-AT transcription complex, so that corticosteroids inhibit IL-5, at least in part, by inhibiting the AP-1 component of NF-AT.

There may be marked differences in the response of different cells and of different cytokines to the inhibitory action of corticosteroids, and this may be dependent on the relative abundance of transcription factors within different cell types. Thus in alveolar macrophages and peripheral blood monocytes, GM-CSF secretion is more potently inhibited by corticosteroids than IL-1β or IL-6 secretion.

2.6.4
Inflammatory Enzymes

Corticosteroids inhibit the synthesis of several inflammatory mediators implicated in asthma through an inhibitory effect on the induction of enzymes such as cyclo-oxygenase-2 and cytosolic PLA_2 (Newton et al. 1997, 1998b). Corticosteroids markedly inhibit the induction of inducible nitric oxide synthase (iNOS), which is induced in asthma and COPD (Saleh et al. 1998), but this does not appear to be a direct effect on iNOS expression (Donnelly and Barnes 2002). It is more likely due to suppression of the inducing cytokines.

2.6.5
Inflammatory Receptors

Corticosteroids also decrease the transcription of genes coding for certain receptors. Thus the gene for the NK_1-receptor, which mediates the inflammatory effects of tachykinins in the airways, has an increased expression in asthma and is inhibited by corticosteroids, probably via an inhibitory effect on AP-1 (Adcock et al. 1993). Corticosteroids also inhibit the transcription of the NK_2-receptor which mediates the bronchoconstrictor effects of tachykinins (Katsunuma et al. 1998) and bradykinin B_1 and B_2 receptors (Haddad et al. 2000).

2.6.6
Adhesion Molecules

Adhesion molecules play a key role in the trafficking of inflammatory cells to sites of inflammation. The expression of many adhesion molecules on endothelial cells is induced by cytokines, and corticosteroids may lead indirectly to a reduced expression via their inhibitory effects on cytokines, such as IL-1β and tumour necrosis factor (TNF)-α. Corticosteroids may also have a direct inhibitory effect on the expression of adhesion molecules, such as intercellular adhesion molecule (ICAM)-1 and E-selectin at the level of gene transcription. ICAM-1 and vascular cell adhesion molecule (VCAM)-1 expression in bronchial epithelial cell lines and monocytes is inhibited by corticosteroids (Atsuta et al. 1999).

2.6.7
Apoptosis

Corticosteroids markedly reduce the survival of certain inflammatory cells, such as eosinophils. Eosinophil survival is dependent on the presence of certain cytokines, such as IL-5 and GM-CSF. Exposure to corticosteroids blocks the effects of these cytokines and leads to programmed cell death or apoptosis, although the corticosteroid-sensitive molecular pathways have not yet been defined (Walsh 1997). By contrast, corticosteroids decrease apoptosis in neutrophils and thus prolong their survival (Meagher et al. 1996). This may contribute to the lack of anti-inflammatory effects of corticosteroids in COPD where neutrophilic inflammation is predominant.

3
Effects on Cell Function

Corticosteroids may have direct inhibitory actions on several inflammatory cells and structural cells that are implicated in asthma (Fig. 3).

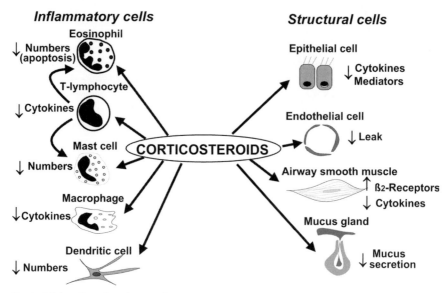

Fig. 3 Cellular effect of corticosteroids

3.1
Macrophages

Corticosteroids inhibit the release of inflammatory mediators and cytokines from alveolar macrophages in vitro. Inhaled corticosteroids reduce the secretion of chemokines and proinflammatory cytokines from alveolar macrophages from asthmatic patients, whereas the secretion of IL-10 is increased (John et al. 1998).

3.2
Eosinophils

Corticosteroids have a direct inhibitory effect on mediator release from eosinophils, although they are only weakly effective in inhibiting secretion of reactive oxygen species and eosinophil basic proteins. More importantly corticosteroids induce apoptosis by inhibiting the prolonged survival due to IL-3, IL-5 and GM-CSF (Walsh 1997), resulting in an increased number of apoptotic eosinophils in induced sputum of asthmatic patients (Woolley et al. 1996). There is a delay in the apoptosis of eosinophils in asthma, which is reversed by treatment with corticosteroids (Kankaanranta et al. 2000). One of the best-described actions of corticosteroids in asthma is a reduction in circulating eosinophils, which may reflect an action on eosinophil production in the bone marrow.

3.3
T Lymphocytes

T helper 2 lymphocytes (Th2) play an important orchestrating role in asthma through the release of the cytokines IL-4, IL-5, IL-9 and IL-13 and may be an important target for corticosteroids in asthma therapy. Corticosteroids increase apoptosis in T cells, although the molecular mechanisms are not yet certain.

3.4
Mast Cells

While corticosteroids do not appear to have a direct inhibitory effect on mediator release from lung mast cells, chronic corticosteroid treatment is associated with a marked reduction in mucosal mast cell numbers. This may be linked to a reduction in IL-3 and stem cell factor (SCF) production, which are necessary for mast cell expression at mucosal surfaces. Mast cells also secrete various cytokines (TNF-α, IL-4, IL-5, IL-6 and IL-8), and this may also inhibited by corticosteroids (Williams and Galli 2000).

3.5
Dendritic Cells

Dendritic cells in the epithelium of the respiratory tract appear to play a critical role in antigen presentation in the lung, as they have the capacity to take up allergen, process it into peptides and present it via MHC molecules on the cell surface for presentation to uncommitted T lymphocytes. In experimental animals, the number of dendritic cells is markedly reduced by systemic and inhaled corticosteroids, thus dampening the immune response in the airways (Nelson et al. 1995).

3.6
Neutrophils

Neutrophils, which are not prominent in the biopsies of asthmatic patients, are not sensitive to the effects of corticosteroids. Indeed, systemic corticosteroids increase peripheral neutrophil counts, which may reflect an increased survival time due to an inhibitory action of neutrophil apoptosis (Meagher et al. 1996). High doses of inhaled corticosteroids have no effect on airway neutrophilia induced by ozone (Nightingale et al. 2000)

3.7
Endothelial Cells

GR gene expression in the airways is most prominent in endothelial cells of the bronchial circulation and airway epithelial cells (Adcock et al. 1996). Corticosteroids do not appear to directly inhibit the expression of adhesion molecules, although they may inhibit cell adhesion indirectly by suppression of cytokines involved in the regulation of adhesion molecule expression. Corticosteroids may have an inhibitory action on airway microvascular leak induced by inflammatory mediators. This appears to be a direct effect on postcapillary venular epithelial cells. Although there have been no direct measurements of the effects of corticosteroids on airway microvascular leakage in asthmatic airways, regular treatment with inhaled corticosteroids decreases the elevated plasma proteins found in bronchoalveolar lavage fluid of patients with stable asthma.

3.8
Epithelial Cells

Epithelial cells may be an important source of many inflammatory mediators in asthmatic airways and may drive and amplify the inflammatory response in the airways through the secretion of proinflammatory cytokines, chemokines and inflammatory peptides. Airway epithelium may be one of the most important cellular targets for inhaled corticosteroids in asthma (Barnes 1996b; Schweibert et al. 1996) (Fig. 4). Inhaled corticosteroids inhibit the increased expression of many inflammatory proteins in airway epithelial cells (Barnes 1996b).

3.9
Mucus Secretion

Corticosteroids inhibit mucus secretion in airways and this may be a direct action of corticosteroids on submucosal gland cells. Corticosteroids may also inhibit the expression of mucin genes, such as MUC2 and MUC5AC (Kai et al. 1996). In addition there are indirect inhibitory effects due to the reduction in inflammatory mediators that stimulate increased mucus secretion.

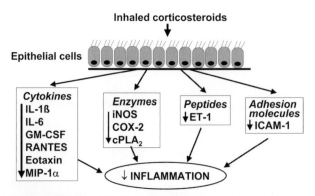

Fig. 4 Inhaled corticosteroids may inhibit the transcription of several "inflammatory" genes in airway epithelial cells and thus reduce inflammation in the airway wall

4
Effects on Airway Inflammation in Asthma

Corticosteroids are remarkably effective in controlling the inflammation in asthmatic airways, and it is likely that they have multiple cellular effects. Biopsy studies in patients with asthma have now confirmed that inhaled corticosteroids reduce the number and activation of inflammatory cells in the airway mucosa and in bronchoalveolar lavage (Barnes 1996b). These effects may be due to inhibition of cytokine synthesis in inflammatory and structural cells and suppression of adhesion molecules. The disrupted epithelium is restored and the ciliated-to-goblet cell ratio is normalised after 3 months of therapy with inhaled corticosteroids. There is also some evidence for a reduction in the thickness of the basement membrane, although in asthmatic patients taking inhaled corticosteroids for over 10 years the characteristic thickening of the basement membrane was still present.

4.1
Effects on Airway Hyperresponsiveness

By reducing airway inflammation, inhaled corticosteroids consistently reduce airway hyperresponsiveness (AHR) in asthmatic adults and children (Barnes 1990). Chronic treatment with inhaled corticosteroids reduces responsiveness to histamine, cholinergic agonists, allergen (early and late responses), exercise, fog, cold air, bradykinin, adenosine and irritants (such as sulphur dioxide and metabisulphite). The reduction in AHR takes place over several weeks and may not be maximal until several months of therapy. The magnitude of reduction is variable between patients and is in the order of one to two doubling dilutions for most challenges and often fails to return to the normal range. This may reflect less suppression of the inflammation than persistence of structural changes which cannot be reversed by corticosteroids. Inhaled corticosteroids not only make the airways less sensitive to spasmogens, but they also limit the maximal airway narrowing in response to spasmogens.

5
Clinical Efficacy of Inhaled Corticosteroids in Asthma

Inhaled corticosteroids are very effective in controlling asthma symptoms in asthmatic patients of all ages and severity (Kamada et al. 1996; Barnes 1999). Inhaled corticosteroids improve the quality of life of patients with asthma and allow many patients to lead normal lives, improve lung function, reduce the frequency of exacerbations and may prevent irreversible airway changes. They were first introduced to reduce the requirement for oral corticosteroids in patients with severe asthma and many studies have confirmed that the majority of patients can be weaned off oral corticosteroids (Barnes et al. 1998).

5.1
Studies in Adults

As experience has been gained with inhaled corticosteroids they have been introduced in patients with milder asthma, with the recognition that inflammation is present even in patients with mild asthma. Inhaled anti-inflammatory drugs have now become first-line therapy in any patient who needs to use a β_2-agonist inhaler more than once a day, and this is reflected in national and international guidelines for the management of chronic asthma. In patients with newly diagnosed asthma, inhaled corticosteroids (budesonide 600 µg twice daily) reduced symptoms and β_2-agonist inhaler usage, and improved lung function. These effects persisted over the 2 years of the study, whereas in a parallel group treated with inhaled β_2-agonists alone there was no significant change in symptoms or lung function (Haahtela et al. 1991). In another study patients with mild asthma treated with a low dose of inhaled corticosteroid (budesonide 400 µg daily) showed less symptoms and a progressive improvement in lung function over several months, and many patients became completely asymptomatic (Juniper et al. 1990). There was also a significant reduction in the number of exacerbations. Although the effects of inhaled corticosteroids on AHR may take several months to reach a plateau, the reduction in asthma symptoms occurs more rapidly (Vathenen et al. 1991).

High-dose inhaled corticosteroids have now been introduced for the control of more severe asthma. This markedly reduces the need for maintenance oral corticosteroids and has revolutionised the management of more severe and unstable asthma. Inhaled corticosteroids are the treatment of choice in nocturnal asthma, which is a manifestation of inflamed airways, reducing night time awakening and reducing the diurnal variation in airway function.

High doses of inhaled corticosteroids may also substitute for a course of oral steroids in controlling acute exacerbations of asthma. High-dose fluticasone propionate (2,000 µg daily) was as effective as a course of oral prednisolone in controlling acute exacerbations of asthma in general practice (Levy et al. 1996). Although doubling the dose of inhaled corticosteroids is recommended for mild exacerbation of asthma, this does not appear to be useful, but a fourfold increase in dose is effective (Foresi et al. 2000).

Inhaled corticosteroids effectively control asthmatic inflammation but must be taken regularly. When inhaled corticosteroids are discontinued there is usually a gradual increase in symptoms and airway responsiveness back to pretreatment values (Vathenen et al. 1991), although in patients with mild asthma who have been treated with inhaled corticosteroids for a long time symptoms may not recur in some patients (Juniper et al. 1991). Reduction in the dose of inhaled corticosteroids is associated with an increase in symptoms and this is preceded by an increase in exhaled NO and sputum eosinophils (Jatakanon et al. 2000).

5.2
Studies in Children

Inhaled corticosteroids are equally effective in children. In an extensive study of children aged 7–17 years there was a significant improvement in symptoms, peak flow variability and lung function compared to a regular inhaled β_2-agonist which was maintained over the 22 months of the study (van Essen-Zandvliet et al. 1992), but asthma deteriorated when the inhaled corticosteroids were withdrawn (Waalkens et al. 1993). There was a high proportion of drop-outs (45%) in the group treated with inhaled β_2-agonist alone. Inhaled corticosteroids are more effective than a long-acting β_2-agonist in controlling asthma in children (Simons 1997). Inhaled corticosteroids are also effective in younger children. Nebulised budesonide reduces the need for oral corticosteroids and also improved lung function in children under the age of three (Ilangovan et al. 1993). Inhaled corticosteroids given via a large volume spacer improve asthma symptoms and reduce the number of exacerbations in preschool children and in infants.

5.3
Dose–Response Studies

Surprisingly, the dose–response curve for the clinical efficacy of inhaled corticosteroids is relatively flat and—while all studies have demonstrated a clinical benefit of inhaled corticosteroids—it has been difficult to demonstrate differences between doses, with most benefit obtained at the lowest doses used (Kamada et al. 1996; Barnes et al. 1998; Busse et al. 1998; Adams et al. 2001). This is in contrast to the steeper dose–response for systemic effects, implying that while there is little clinical benefit from increasing doses of inhaled corticosteroids, the risk of adverse effects is increased. However, the dose–response effect of inhaled corticosteroids may depend on the parameters measured and, while it is difficult to discern a dose response when traditional lung function parameters are measured, there may be a dose–response effect in prevention of asthma exacerbations. Thus, there is a significantly greater effect of budesonide 800 μg daily compared to 200 μg daily in preventing severe and mild asthma exacerbations (Pauwels et al. 1997). Normally, a fourfold or greater difference in dose has been required to detect a statistically significant (but often small) difference in effect on commonly measured outcomes such as symptoms, peak expiratory flow (PEF), use of rescue β_2-agonist, and lung function; and even such large differences in dose are not always associated with significant differences in response. These findings suggest that pulmonary function tests or symptoms may have a rather low sensitivity in the assessment of the effects of inhaled corticosteroids. This is obviously important for the interpretation of clinical comparisons between different inhaled corticosteroids or inhalers. It is also important to consider the type of patient included in clinical studies. Patients with relatively mild asthma may have relatively little room for improvement with inhaled corticosteroids, so that maximal improvement is obtained with relatively low doses. Patients with more severe asthma or with unstable asthma may have more room for improvement and may therefore show a greater response to increasing doses, but it is often difficult to include such patients in controlled clinical trials.

More studies are needed to assess whether other outcome measures such as AHR or more direct measurements of inflammation, such as sputum eosinophils or exhaled NO, may be more sensitive than traditional outcome measures such as symptoms or lung function tests (Jatakanon et al. 1998, 1999; Lim et al. 1999; Green et al. 2002). Higher doses of inhaled corticosteroids are needed to control AHR than to improve symptoms and lung function, and this may have a better long-term outcome in terms of reduction in structural changes of the airways (Sont et al. 1999).

5.4
Prevention of Irreversible Airway Changes

Some patients with asthma develop an element of irreversible airflow obstruction, but the pathophysiological basis of these changes is not yet understood. It is likely that they are the result of chronic airway inflammation and that they may be prevented by treatment with inhaled corticosteroids. There is some evidence that the annual decline in lung function may be slowed by the introduction of inhaled corticosteroids (Dompeling et al. 1992). Increasing evidence also suggests that delay in starting inhaled corticosteroids may result in less overall improvement in lung function in both adults and children (Agertoft and Pedersen 1994; Haahtela et al. 1994; Selroos et al. 1995). These studies suggest that introduction of inhaled corticosteroids at the time of diagnosis is likely to have the greatest impact (Agertoft and Pedersen 1994; Selroos et al. 1995). Several large studies are now underway to assess the benefit of very early introduction of inhaled corticosteroids in children and adults. So far there is no evidence that early use of inhaled corticosteroids is curative and even when inhaled corticosteroids are introduced at the time of diagnosis, symptoms and lung function revert to pretreatment levels when corticosteroids are withdrawn (Haahtela et al. 1994).

5.5
Reduction in Mortality

Inhaled corticosteroids may reduce the mortality from asthma, but prospective studies are almost impossible to conduct. In a retrospective review of the risk of mortality and prescribed anti-asthma medication, there was a significant protection provided by regular inhaled corticosteroid therapy (Suissa et al. 2000).

5.6
Comparison Between Inhaled Corticosteroids

Several inhaled corticosteroids are currently available for the treatment of asthma, although their availability varies between countries. There have been relatively few studies comparing efficacy of the different inhaled corticosteroids, and it is important to take into account the delivery system and the type of patient under investigation when such comparisons are made. Because of the relatively flat dose–response curve for the clinical parameters normally used in comparing doses of inhaled corticosteroids, it may be difficult to see differences in efficacy of inhaled corticosteroids. Most comparisons have concentrated in differences in systemic effects at equally efficacious doses, although it has often proved difficult to establish dose equivalence (Martin et al. 2002). There are few studies comparing different doses of inhaled corticosteroids in asthmatic patients. Budesonide has been compared with BDP, and in adults and children it appears to have comparable anti-asthma effects at equal doses, whereas FP appears to be approximately twice as potent as BDP and budesonide. Studies have consistently shown that fluticasone propionate (FP) and budesonide have less systemic effects than BDP, triamcinolone and flunisolide (Lipworth 1999). A new inhaled corticosteroid mometasone also has less systemic effects (Nathan et al. 2001).

6
Clinical Use of Inhaled Corticosteroids in Asthma

Inhaled corticosteroids are now recommended as first-line therapy for all patients with persistent symptoms. Inhaled corticosteroids should be started in any patient who needs to use a β_2-agonist inhaler for symptom control more than two to three times weekly. It is conventional to start with a low dose of inhaled corticosteroid and to increase the dose until asthma control is achieved. However, this may take time and a preferable approach is to start with a dose of corticosteroids in the middle of the dose range (400 µg twice daily) to establish control of asthma more rapidly (Barnes 1996a). Once control is achieved (defined as normal or best possible lung function and infrequent need to use an inhaled β_2-agonist) the dose of inhaled corticosteroid should be reduced in a step-wise manner to the lowest dose needed for optimal control. It may take as long as 3 months to reach a plateau in response, and any changes in dose should be made at intervals of 3 months or more. When daily doses of greater than or equal to 800 µg daily are needed, a large volume spacer device should be used with a metered dose inhaler (MDI) and mouth washing with a dry powder inhaler in order to reduce local and systemic side effects. Inhaled corticosteroids are usually given as a twice-daily dose in order to increase compliance. When asthma is unstable, a four-times daily dosage is preferable (Malo et al. 1989). For patients who require less than or equal to 400 µg daily, once-daily dosing appears to be as effective as twice-daily dosing, at least for budesonide (Jones et al. 1994).

The dose of inhaled corticosteroid should be increased to 2,000 µg daily if necessary, but higher doses may result in systemic effects. It may be preferable to add a low dose of oral corticosteroid, since higher doses of inhaled corticosteroids are expensive and have a high incidence of local side effects. Nebulised budesonide has been advocated in order to give an increased dose of inhaled corticosteroid and to reduce the requirement for oral corticosteroids (Otulana et al. 1992), but this treatment is expensive and may achieve its effects largely via systemic absorption.

6.1
Add-on Therapy

Conventional advice was to increase the dose of inhaled corticosteroids if asthma was not controlled, on the assumption that there was residual inflammation of the airways. However, it is now apparent that the dose–response effect of inhaled corticosteroids is relatively flat, so that there is little improvement in lung function after doubling the dose of inhaled corticosteroids. The preferred strategy for most patients is to add some other class of controller drug.

6.1.1
Long-Acting Inhaled β_2-Agonists

In patients in general practice who are not controlled on BDP 200 µg twice daily, addition of salmeterol 50 µg twice daily was more effective than increasing the dose of inhaled corticosteroid to 500 µg twice daily, in terms of lung function improvement, use of rescue β_2-agonist use and symptom control (Greening et al. 1994). This has been confirmed in several other studies (Shrewsbury et al. 2000). Similar results have been found with another long-acting inhaled β_2-agonist formoterol, which in addition reduced the frequency of mild and severe asthma exacerbations (Pauwels et al. 1997). This has led to the development of fixed combinations of corticosteroids and long-acting β_2-agonists, such as FP and salmeterol (Seretide/Advair/Vianni) and budesonide with formoterol (Symbicort), which may be more convenient for patients (Chapman et al. 1999; Shapiro et al. 2000; Zetterstrom et al. 2001). These fixed combination inhalers also ensure that patients do not discontinue their inhaled corticosteroids when a long-acting bronchodilator is used.

6.1.2
Theophylline

Addition of low doses of theophylline (giving plasma concentrations of <10 mg/l) are more effective than doubling the dose of inhaled budesonide, either in mild or severe asthma (Evans et al. 1997; Ukena et al. 1997; Lim et al. 2000). However, this is less effective as an add-on therapy than long-acting β_2-agonists (Wilson et al. 2000)

6.1.3
Anti-leukotrienes

Anti-leukotrienes are also used as add-on therapies (Laviolette et al. 1999), although this is less effective than addition of long-acting β_2-agonists (Nelson et al. 2000), particularly in patients with severe asthma (Robinson et al. 2001). A systematic review has suggested that anti-leukotrienes provide relatively little benefit as add-on therapy (Ducharme 2002)

6.1.4
Mechanisms

The reason why add-on controller treatments are more effective than higher doses of inhaled corticosteroids remains to be elucidated. The add-on effect suggests that there is some reversible component of asthma that is not treatable with steroids. As discussed above, the dose–response curve for inhaled corticosteroids efficacy is relatively shallow, and control of inflammation may be achieved at low doses in most patients. The add-on therapies may be working on some other component if asthma that is not sensitive to inhibition by inhaled corticosteroids. This may be an abnormality in airway smooth muscle itself (as a result of remodelling) or oedema of the airway wall. In the case of long-acting β_2-agonists there may also be a positive effect of β_2-agonists on the anti-inflammatory effects of corticosteroids (Barnes 2002). β_2-Agonists increase the nuclear translocation of GR and this might enhance the anti-inflammatory effects of corticosteroids with enhanced suppression of cytokines.

6.2
Cost Effectiveness

Although inhaled corticosteroids may be more expensive than short-acting inhaled β_2-agonists, they are the most cost-effective way of controlling asthma, since reducing the frequency of asthma attacks will save on total costs (Barnes et al. 1996). Inhaled corticosteroids also improve the quality of life of patients with asthma and allow many patients a normal lifestyle, thus saving costs indirectly (van Schayk et al. 1995).

6.3
Corticosteroid-Sparing Therapy

In patients who have serious side effects with maintenance corticosteroid therapy there are several treatments which have been shown to reduce the requirement for oral corticosteroids (Hill and Tattersfield 1995). These treatments are commonly termed corticosteroid sparing, although this is a misleading description that could be applied to any additional asthma therapy (including bronchodilators). The amount of corticosteroid sparing with these therapies is not impressive.

Several immunosuppressive agents have been shown to have corticosteroid effects, including methotrexate, oral gold and cyclosporin A. These therapies all have side effects that may be more troublesome than those of oral corticosteroids and are therefore only indicated as an additional therapy to reduce the requirement of oral corticosteroids (Davies et al. 2000). None of these treatments is very effective, but there are occasional patients who appear to show a good response. Because of side effects, these treatments cannot be considered as a way to reduce the requirement for inhaled corticosteroids. Several other therapies, including azathioprine, dapsone and hydroxychloroquine have not been found to be beneficial. The macrolide antibiotic troleandomycin is also reported to have corticosteroid-sparing effects, but this is only seen with methylprednisolone and is due to reduced metabolism of this corticosteroid, so that there is little therapeutic gain (Nelson et al. 1993).

7
Pharmacokinetics

The pharmacokinetics of inhaled corticosteroids is important in determining the concentration of drug reaching target cells in the airways and in the fraction of drug reaching the systemic circulation and therefore causing side effects (Barnes et al. 1998). Beneficial properties in an inhaled corticosteroid are a high topical potency, a low systemic bioavailability of the swallowed portion of the dose and rapid metabolic clearance of any corticosteroid reaching the systemic circulation. After inhalation a large proportion of the inhaled dose (80%–90%) is deposited on the oropharynx and is then swallowed and therefore available for absorption via the liver into the systemic circulation (Fig. 5). This fraction is markedly reduced by using a large volume spacer device with a metered dose inhaler (MDI) or by mouth washing and discarding the washing with dry powder inhalers. Between 10% and 20% of inhaled drug enters the respiratory tract, where it is deposited in the airways and this fraction is available for absorption into the systemic circulation. Most of the early studies on the distribution of inhaled corticosteroids were conducted in healthy volunteers, and it is not certain what effect inflammatory disease, airway obstruction, age of the patient or concomitant medication may have on the disposition of the inhaled dose. There may be important differences in the metabolism of different inhaled corticosteroids. BDP is metabolised to its more active metabolite beclomethasone monopropionate in many tissues including lung, but there is no information about its absorption or metabolism of this metabolite in humans. Flunisolide and budesonide are subject to extensive first-pass metabolism in the liver so that less reaches the systemic circulation. Little is known about the distribution

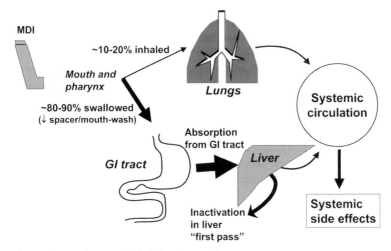

Fig. 5 Pharmacokinetics of inhaled corticosteroids

of triamcinolone. FP is almost completely metabolised by first-pass metabolism, which reduces systemic effects.

When inhaled corticosteroids were first introduced it was recommended that they should be given four times daily, but several studies have now demonstrated that twice daily administration gives comparable control, although four times daily administration may be preferable in patients with more severe asthma. However, patients may find it difficult to comply with such frequent administration unless they have troublesome symptoms. For patients with mild asthma who require less than or equal to 400 μg daily, once-daily therapy may be sufficient.

8
Side Effects of Inhaled Corticosteroids

The efficacy of inhaled corticosteroids is now established in short- and long-term studies in adults and children, but there are still concerns about side effects, particularly in children and when high inhaled doses are needed. Several side effects have been recognised (Table 2).

8.1
Local Side Effects

Side effects due to the local deposition of the inhaled corticosteroid in the oropharynx may occur with inhaled corticosteroids, but the frequency of complaints depends on the dose and frequency of administration and on the delivery system used.

Table 2 Side-effects of inhaled corticosteroids

Local side-effects
Dysphonia
Oropharyngeal candidiasis
Cough
Systemic side-effects
Adrenal suppression
Growth suppression
Bruising
Osteoporosis
Cataracts
Glaucoma
Metabolic abnormalities (glucose, insulin, triglycerides)
Psychiatric disturbances

8.1.1
Dysphonia

The commonest complaint is of hoarseness of the voice (dysphonia) and may occur in over 50% of patients using high doses of MDI. Dysphonia is not appreciably reduced by using spacers, but may be less with dry powder devices. Dysphonia may be due to myopathy of laryngeal muscles and is reversible when treatment is withdrawn (Williamson et al. 1995). For most patients it is not troublesome but may be disabling in singers and lecturers.

8.1.2
Oropharyngeal Candidiasis

Oropharyngeal candidiasis (thrush) may be a problem in some patients, particularly in the elderly, with concomitant oral corticosteroids and more than twice-daily administration (Toogood et al. 1980). Large volume spacer devices protect against this local side effect by reducing the dose of inhaled corticosteroid that deposits in the oropharynx.

8.1.3
Other Local Complications

There is no evidence that inhaled corticosteroid, even in high doses, increases the frequency of infections, including tuberculosis, in the lower respiratory tract. There is no evidence for atrophy of the airway epithelium, and even after 10 years of treatment with inhaled corticosteroids there is no evidence for any structural changes in the epithelium. Cough and throat irritation, sometimes accompanied by reflex bronchoconstriction, may occur when inhaled corticosteroids are given via a metered dose inhaler. These symptoms are likely to be due to surfactants in pressurised aerosols as they disappear after switching to a dry powder corticosteroid inhaler device.

8.2
Systemic Side Effects

The efficacy of inhaled corticosteroids in the control of asthma is undisputed, but there are concerns about systemic effects of inhaled corticosteroids, particularly as they are likely to be used over long periods and in children of all ages (Kamada et al. 1996; Lipworth 1999). The safety of inhaled corticosteroids has been extensively investigated since their introduction 30 years ago (Barnes et al. 1998). One of the major problems is to decide whether a measurable systemic effect has any significant clinical consequence, and this necessitates careful long-term follow-up studies. As biochemical markers of systemic corticosteroid effects become more sensitive, then systemic effects may be seen more often, but this does not mean that these effects are clinically relevant. There are several case reports of adverse systemic effects of inhaled corticosteroids, and these may be idiosyncratic reactions, which may be due to abnormal pharmacokinetic handing of the inhaled corticosteroid. The systemic effect of an inhaled corticosteroid will depend on several factors, including the dose delivered to the patient, the site of delivery (gastrointestinal tract and lung), the delivery system used and individual differences in the patient's response to the corticosteroid. Studies now suggest that systemic effects of inhaled corticosteroid are less in patients with more severe asthma, presumably as less drug reaches the lung periphery (Brutsche et al. 2000; Harrison et al. 2001).

8.2.1
Effect of Delivery Systems

The systemic effect of an inhaled corticosteroid is dependent on the amount of drug absorbed into the systemic circulation. Approximately 90% of the inhaled dose from an MDI deposits in the oropharynx and is swallowed and subsequently absorbed from the gastrointestinal tract. Use of a large volume spacer device markedly reduces the oropharyngeal deposition, and therefore the systemic effects of inhaled corticosteroids, although thus is less important when oral bioavailability is minimal, as with FP. For dry powder inhalers, similar reductions in systemic effects may be achieved with mouth washing and discarding the fluid. All patients using a daily dose of greater than or equal to 800 µg of an inhaled corticosteroid should therefore use either a spacer or mouth washing to reduce systemic absorption. Approximately 10% of an MDI enters the lung and this fraction (which presumably exerts the therapeutic effect) may be absorbed into the systemic circulation. As the fraction of inhaled corticosteroid deposited in the oropharynx is reduced, the proportion of the inhaled dose entering the lungs is increased. More efficient delivery to the lungs is therefore accompanied by increased systemic absorption, but this is offset by a reduction in the dose needed for optimal control of airway inflammation. For example, a multiple dry powder delivery system, the Turbuhaler, delivers approximately twice as much corticosteroid to the lungs as other devices, and therefore has increased systemic effects. However, this is compensated for by the fact that only half the dose is required. BDP MDI in hydrofluoroalkane propellant gives smaller particle sizes which appears to deliver a higher dose to the lungs than in the conventional chlorofluorocarbon propellant (Shaw 1999).

8.2.2
Hypothalamic–Pituitary–Adrenal Axis

Corticosteroids may cause hypothalamic–pituitary–adrenal (HPA) axis suppression by reducing corticotrophin (ACTH) production, which reduces cortisol secretion by the adrenal gland. The degree of HPA suppression is dependent on dose, duration, frequency and timing of corticosteroid administration. There is no evidence that cortisol responses to the stress of an asthma exacerbation or insulin-induced hypoglycaemia are impaired, even with high doses of inhaled corticosteroids. However, measurement of HPA axis function provides evidence for systemic effects of an inhaled corticosteroid. Basal adrenal cortisol secretion may be measured by a morning plasma cortisol, 24 h urinary cortisol or by plasma cortisol profile over 24 h. Other tests measure the HPA response following stimulation with tetracosactrin (which measures adrenal reserve) or stimulation with metyrapone and insulin (which measure the response to stress).

There are many studies of HPA axis function in asthmatic patients with inhaled corticosteroids, but the results are inconsistent, as they have often been uncontrolled and patients have also been taking courses of oral corticosteroids (which may affect the HPA axis for weeks) (Barnes et al. 1998). BDP, budesonide and FP at high doses by conventional MDI (>1,600 µg daily) give a dose-related decrease in morning serum cortisol levels and 24 h urinary cortisol, although values still lie well within the normal range. However, when a large volume spacer is used, doses of 2,000 µg daily of BDP or budesonide have little effect on 24 h urinary cortisol excretion. Stimulation tests of HPA axis function similarly show no consistent effects of doses of 1,500 µg or less of inhaled corticosteroid. At high doses (>1,500 µg daily) budesonide and FP have less effect than BDP on HPA axis function. In children, no suppression of urinary cortisol is seen with doses of BDP of 800 µg or less. In studies where plasma cortisol has been measured at frequent intervals there was a significant reduction in cortisol peaks with doses of inhaled BDP as low as 400 µg daily, although this does not appear to be dose-related in the range 400–1,000 µg. The clinical significance of these effects is not certain, however.

Overall, the studies which are not confounded by concomitant treatment with oral corticosteroids have consistently shown that there are no significant suppressive effects on HPA axis function at doses at or below 1,500 µg in adults and 400 µg in children.

8.2.3
Effects on Bone Metabolism

Corticosteroids lead to a reduction in bone mass by direct effects on bone formation and resorption and indirectly by suppression of the pituitary–gonadal and HPA axes, effects on intestinal calcium absorption, renal tubular calcium reabsorption and secondary hyperparathyroidism (Efthimou and Barnes 1998). The effects of oral corticosteroids on osteoporosis and increased risk of vertebral and rib fractures are well known, but there are no reports suggesting that long-term treatment with inhaled corticosteroids is associated with an increased risk of fractures. Bone densitometry has been used to assess the effect of inhaled corticosteroids on bone mass. Although there is evidence that bone density is less in patients taking high-dose inhaled corticosteroids, interpretation is confounded by the fact that these patients are also taking intermittent courses of oral corticosteroids.

Changes in bone mass occur very slowly and several biochemical indices have been used to assess the short-term effects of inhaled corticosteroids on bone metabolism. Bone formation has been measured by plasma concentrations of bone-specific alkaline phosphatase, serum osteocalcin or procollagen peptides. Bone resorption may be assessed by urinary hydroxyproline after a 12-h fast, urinary calcium excretion and pyridinium cross-link excretion. It is important to consider the age, diet, time of day and physical activity of the patient in interpreting any abnormalities. It is also necessary to choose appropriate control groups, as asthma itself may have an effect on some of the measurements, such as osteocalcin. Inhaled corticosteroids, even at doses up to 2,000 μg daily, have no significant effect on calcium excretion, but acute and reversible dose-related suppression of serum osteocalcin has been reported with BDP and budesonide when given by conventional MDI in several studies. Budesonide consistently has less effect than BDP at equivalent doses and only BDP increases urinary hydroxyproline at high doses. With a large volume spacer, even doses of 2,000 μg daily of either BDP or budesonide are without effect on plasma osteocalcin concentrations, however. Urinary pyridinium and deoxypyridinoline cross-links, which are a more accurate and stable measurement of bone and collagen degradation, are not increased with inhaled corticosteroids (BDP >1,000 μg daily), even with intermittent courses of oral corticosteroids. It is important to monitor changes in markers of bone formation as well as bone degradation, as the net effect on bone turnover is important.

There has been particular concern about the effect of inhaled corticosteroids on bone metabolism in growing children. A very low dose of oral corticosteroids (prednisolone 2.5 mg) causes significant changes in serum osteocalcin and urinary hydroxyproline excretion, whereas daily BDP and budesonide at doses up to 800 μg daily have no effect. It is important to recognise that the changes in biochemical indices of bone metabolism are less than those seen with even low doses of oral corticosteroids. This suggests that even high doses of inhaled corticosteroids, particularly when used with a spacer device, are unlikely to have any long-term effect on bone structure. Careful long-term follow-up studies in patients with asthma are needed.

There is no evidence that inhaled corticosteroids increase the frequency of fractures. Long-term treatment with high-dose inhaled corticosteroids has not been associated with any consistent change in bone density (Egan et al. 1999), apart from triamcinolone, which has greater systemic absorption than other inhaled corticosteroids (Israel et al. 2001). A decrease in bone density with inhaled BDP has been observed in post-menopausal women who are a high-risk group (Fujita et al. 2001).

8.2.4
Effects on Connective Tissue

Oral and topical corticosteroids cause thinning of the skin, telangiectasiae and easy bruising, probably as a result of loss of extracellular ground substance within the dermis, due to an inhibitory effect on dermal fibroblasts. There are reports of increased skin bruising and purpura in patients using high doses of inhaled BDP, but the amount of intermittent oral corticosteroids in these patients is not known. Easy bruising in association with inhaled corticosteroids is more frequent in elderly patients (Roy et al. 1996) and there are no reports of this problem in children. Long-term prospective studies with objective measurements of skin thickness are needed with different inhaled corticosteroids.

8.2.5
Ocular Effects

Long-term treatment with oral corticosteroids increases the risk of posterior subcapsular cataracts and there are several case reports describing cataracts in individual patients taking inhaled corticosteroids (Barnes et al. 1998). In a recent cross-sectional study in patients aged 5–25 years taking either inhaled BDP or budesonide, no cataracts were found on slit-lamp examination, even in patients taking 2,000 µg daily for over 10 years (Simons et al. 1993). However, epidemiological studies have identified an increased risk of cataracts in patients taking high-dose inhaled steroids over prolonged periods (Cumming et al. 1997), although the increased risk is slight and applies mainly to elderly patients (Jick et al. 2001). A slight increase in the risk of glaucoma in patients taking very high does of inhaled corticosteroids has also been identified (Garbe et al. 1997).

8.2.6
Growth

There has been particular concern that inhaled corticosteroids may cause stunting of growth, and several studies have addressed this issue. Asthma itself (as with other chronic diseases) may have an effect on the growth pattern and has been associated with delayed onset of puberty and deceleration of growth velocity that is more pronounced with more severe disease. However, asthmatic children appear to grow for longer, so that their final height is normal. The effect of asthma on growth makes it difficult to assess the effects of inhaled corticosteroids on growth in cross-sectional studies, particularly as courses of oral corticosteroids is a confounding factor (Pedersen 2001). Longitudinal studies have demonstrated that there is no significant effect of inhaled corticosteroids on statural growth in doses of up to 800 µg daily and for up to 5 years of treatment (Barnes et al. 1998). A meta-analysis of 21 studies, including over 800 children, showed no effect of inhaled BDP on statural height, even with higher doses and long duration of therapy (Allen et al. 1994) and in a large study of asthmatics treated with inhaled corticosteroids during childhood there was no difference in statural height compared to normal children (Silverstein et al. 1997). Another long-term follow-up study showed no effect of corticosteroids on final height in children treated over several years (Agertoft and Pedersen 2000).

Short-term growth measurements (knemometry) have demonstrated that even a low dose of an oral corticosteroid (prednisolone 2.5 mg) is sufficient to give complete suppression of lower leg growth. However, inhaled budesonide up to 400 µg is without effect, although some suppression is seen with 800 µg and with 400 µg BDP. The relationship between knemometry measurements and final height are uncertain since low doses of oral corticosteroid that have no effect on final height cause profound suppression.

8.2.7
Metabolic Effects

Several metabolic effects have been reported after inhaled corticosteroids, but there is no evidence that these are clinically relevant at therapeutic doses. In fasting adults, glucose and insulin are unchanged after doses of BDP up to 2000 µg daily and in children with inhaled budesonide up to 800 µg daily. In normal individuals, high-dose inhaled BDP may slightly increase resistance to insulin. However, in patients with poorly controlled asthma, high doses of BDP and budesonide paradoxically decrease insulin resistance and improve glucose tolerance, suggesting that the disease itself may lead to abnormalities in carbohydrate metabolism. Neither BDP 2000 µg daily in adults nor budesonide 800 µg daily in children have any effect on plasma cholesterol or triglycerides.

8.2.8
Haematological Effects

Inhaled corticosteroids may reduce the numbers of circulating eosinophils in asthmatic patients, possibly due to an effect on local cytokine generation in the airways. Inhaled corticosteroids may cause a small increase in circulating neutrophil counts.

8.2.9
Central Nervous System Effects

There are various reports of psychiatric disturbance, including emotional lability, euphoria, depression, aggressiveness and insomnia, after inhaled corticosteroids. Only eight such patients have so far been reported, suggesting that this is very infrequent and a causal link with inhaled corticosteroids has usually not been established.

8.2.10
Safety in Pregnancy

Based on extensive clinical experience inhaled corticosteroids appear to be safe in pregnancy, although no controlled studies have been performed. There is no evidence for any adverse effects of inhaled corticosteroids on the pregnancy, the delivery or on the foetus (Schatz et al. 1997; Schatz 1999). It is important to recognise that poorly controlled asthma may increase the incidence of perinatal mortality and retard intra-uterine growth, so that more effective control of asthma with inhaled corticosteroids may reduce these problems.

9
Systemic Corticosteroids

Oral or intravenous corticosteroids may be indicated in several situations. Prednisolone, rather than prednisone, is the preferred oral corticosteroid, as prednisone has to be converted in the liver to the active prednisolone. In pregnant patients, prednisone may be preferable, as it is not converted to prednisolone in the foetal liver, thus diminishing the exposure of the foetus to corticosteroids. Enteric-coated preparations of prednisolone are used to reduce side effects (particularly gastric side effects) and give delayed and reduced peak plasma concentrations, although the bioavailability and therapeutic efficacy of these preparations is similar to uncoated tablets. Prednisolone and prednisone are preferable to dexamethasone, betamethasone or triamcinolone, which have longer plasma half-lives and therefore an increased frequency of adverse effects.

Short courses of oral corticosteroids (30–40 mg prednisolone daily for 1–2 weeks or until the peak flow values return to best attainable) are indicated for exacerbations of asthma, and the dose may be tailed off over 1 week once the exacerbation is resolved. The tail-off period is not strictly necessary, but some patients find it reassuring.

Maintenance oral corticosteroids are only needed in a small proportion of asthmatic patients with the most severe asthma that cannot be controlled with maximal doses of inhaled corticosteroids (2,000 µg daily) and additional bronchodilators. The minimal dose of oral corticosteroid needed for control should be used and reductions in the dose should be made slowly in patients who have been on oral corticosteroids for long periods (e.g. by 2.5 mg per month for doses down to 10 mg daily and thereafter by 1 mg per month). Oral corticosteroids are usually given as a single morning dose, as this reduces the risk of adverse effects, since it coincides with the peak diurnal concentrations. There is some evidence that administration in the afternoon may be optimal for some patients who have severe nocturnal asthma (Beam et al. 1992). Alternate day administration may also reduce adverse effects, but control of asthma may not be as good on the day when the oral dose is omitted in some patients.

Intramuscular triamcinolone acetonide (80 mg monthly) has been advocated in patients with severe asthma as an alternative to oral corticosteroids (McLeod et al. 1985; Ogirala et al. 1991). This may be considered in patients in whom compliance is a particular problem, but the major concern is the high frequency of proximal myopathy associated with this fluorinated corticosteroid. Some patients who do not respond well to prednisolone are reported to respond to oral betamethasone, presumably because of pharmacokinetic handling problems with prednisolone.

9.1
Acute Severe Asthma

Intravenous hydrocortisone is given in acute severe asthma, with a recommended dose of 200 mg (i.v.). While the value of corticosteroids in acute severe asthma has been questioned, others have found that they speed the resolution of attacks (Engel and Heinig 1991). There is no apparent advantage in giving very high doses of intravenous corticosteroids (such as methylprednisolone 1 g). Indeed, intravenous corticosteroids have occasionally been associated with an acute severe myopathy (Decramer et al. 1995). No difference in recovery from acute severe asthma was seen whether i.v. hydrocortisone in doses of 50, 200 or 500 mg 6 hourly were used (Bowler et al. 1992), and another placebo-controlled study showed no beneficial effect of i.v. corticosteroids (Morell et al. 1992). Intravenous corticosteroids are indicated in acute asthma if lung function is less than 30% predicted and in whom there is no significant improvement with nebulised β_2-agonist. Intravenous therapy is usually given until a satisfactory response is obtained and then oral prednisolone may be substituted. Oral prednisolone (40–60 mg) has a similar effect to intravenous hydrocortisone and is easier to administer (Harrison et al. 1986; Engel and Heinig 1991). Oral prednisolone is the preferred treatment for acute severe asthma, providing there are no contraindications to oral therapy (British Thoracic Society 1997). There is some evidence that high does of nebulised corticosteroids may also be effective in acute exacerbations of asthma, with a more rapid onset of action (Devidayal et al. 1999).

10
Corticosteroid-Resistant Asthma

Although corticosteroids are highly effective in the control of asthma and other chronic inflammatory or immune diseases, a small proportion of patients with asthma fail to respond even to high doses of oral glucocorticoids (Barnes et al. 1995; Szefler and Leung 1997; Barnes 2000b). Resistance to the therapeutic effects of corticosteroids is also recognised in other inflammatory and immune diseases, including rheumatoid arthritis and inflammatory bowel disease. Corticosteroid-resistant patients, although uncommon, present considerable management problems. Recently, new insights into the mechanisms whereby corticosteroids suppress chronic inflammation have shed new light on the molecular basis of corticosteroid-resistant asthma.

Corticosteroid-resistant asthma is defined as a failure to improve forced expiratory volume in 1 s (FEV_1) or PEF by greater than 15% after treatment with oral prednisolone 30–40 mg daily for 2 weeks, providing the oral steroid is taken (verified by plasma prednisolone level or a reduction in early morning cortisol level). These patients are not addisonian and they do not suffer from the abnormalities in sex hormones described in the very rare familial glucocorticoid resistance. Plasma cortisol and adrenal suppression in response to exogenous cortisol is normal in these patients, so they suffer from side effects of corticosteroids.

Complete corticosteroid resistance in asthma is very rare, with a prevalence of less than 1:1000 asthmatic patients. Much more common is a reduced responsiveness to corticosteroids, so that large inhaled or oral doses are needed to control asthma adequately (corticosteroid-dependent asthma). It is likely that there is a range of responsiveness to corticosteroids and that corticosteroid resistance is at one extreme of this range.

It is important to establish that the patient has asthma, rather than COPD, "pseudoasthma" (a hysterical conversion syndrome involving vocal cord dysfunction), left ventricular failure or cystic fibrosis that do not respond to corticosteroids. Asthmatic patients are characterised by a variability in PEF and, in particular, a diurnal variability of greater than 15% and episodic symptoms. It is also important to identify provoking factors (allergens, drugs, psychological problems) that may increase the severity of asthma and its resistance to therapy. Biopsy studies have demonstrated the typical eosinophilic inflammation of asthma in these patients (Szefler and Leung 1997).

10.1
Mechanisms of Corticosteroid Resistance

There may be several mechanisms for resistance to the effects of corticosteroids. Certain cytokines (particularly IL-2, IL-4 and IL-13) may induce a reduction in affinity of glucocorticoid receptors in inflammatory cells such as T-lymphocytes, resulting in local resistance to the anti-inflammatory actions of corticosteroids (Szefler and Leung 1997). This may be mediated via activation of p38 mitogen-activated protein (MAP) kinase pathways which result in reduced nuclear translocation of GR (Irusen et al. 2002). Another mechanism is an increased activation of the transcription factor AP-1 by inflammatory cytokines, so that AP-1 may consume activated glucocorticoid receptors and thus reduce their availability for suppression of inflammation at inflamed sites (Adcock et al. 1994). There is an increased expression of c-Fos, one of the components of AP-1 (Lane et al. 1998). The reasons for this excessive activation of AP-1 by activating enzymes is currently unknown, but may be genetically determined. Another proposed mechanism is an increase in expression of GR-β which then interferes with DNA binding of GR (Hamid et al. 1999; Sousa et al. 2000), but any increase in GR-β is insufficient to account for reduced responsiveness to corticosteroids (Gagliardo et al. 2000). Recently a subgroup for patients with steroid resistance that have a defect in acetylation of histone-4 has been described (Matthews et al. 2000). This suggests that there are probably several different molecular mechanisms contributing to corticosteroid resistance in asthma and that different therapeutic approaches may be needed in the future.

11
Corticosteroids in COPD

Although inhaled corticosteroids are highly effective in asthma, they provide little benefit in COPD, despite the fact that airway and lung inflammation is present (Barnes 2000a).

11.1
Effect on Inflammation

The resistance to corticosteroids in COPD may reflect that the inflammation in COPD is not suppressed by corticosteroids, with no reduction in inflammatory cells, cytokines or proteases in induced sputum even with oral corticosteroids (Keatings et al. 1997; Culpitt et al. 1999). Corticosteroids do not suppress neutrophilic inflammation in the airways and corticosteroids may prolong the survival of neutrophils (Nightingale et al. 2000). This may be an active resistance mechanism as cytokines, such as IL-8 and TNF-α, that are normally suppressed by corticosteroids, are not reduced even by maximal does of inhaled corticosteroids (Keatings et al. 1997; Culpitt et al. 1999). This resistance to the anti-inflammatory effects of corticosteroids is also seen at a cellular level and alveolar macrophages from patients with COPD show little or no suppression of inflammatory cytokines, such as IL-8 and matrix metalloproteinase (MMP)-9, compared with cells from normal smokers and non-smokers (Culpitt et al. 2002; Russell et al. 2002). This lack of response to corticosteroids may be explained in part by an inhibitory effect of cigarette smoking on histone deacetylases, thus interfering with an important anti-inflammatory action of corticosteroids (Ito et al. 2001b). This mechanism may be mediated via oxidative stress which impairs the activity of certain HDACs.

11.2
Clinical Studies

Four large studies conducted over 3 years have demonstrated no beneficial effect of inhaled corticosteroids on the decline in lung function in patients from mild to moderate COPD (Pauwels et al. 1999; Vestbo et al. 1999; Burge et al. 2000; Lung Health Study Research Group 2000). There is some evidence for a reduction in more severe exacerbations with high does of inhaled corticosteroids (Paggiaro et al. 1998; Burge et al. 2000), but the effect is small and similar to the effect of bronchodilators in this respect. Some patients with COPD (approximately 10%) show some response to inhaled corticosteroids, and it is likely that these are patients who have concomitant asthma. Indeed, the corticosteroid responders are more likely to have sputum eosinophils and an increase in exhaled NO, which are features of asthmatic inflammation (Papi et al. 2000). These patients should be treated as if they have asthma. The remaining majority of patients are unlikely to derive much benefit from inhaled corticosteroids and there are good reasons for not prescribing these drugs. They are often given in high doses as this has a risk of systemic side effects in a vulnerable patient population, who are elderly, relatively immobile, may have a poor diet and have comorbid conditions, all of which increase the risk of side effects, such as osteoporosis and cataracts. In addition high does of inhaled corticosteroids are relatively expensive.

In the management of acute exacerbations there is evidence that oral corticosteroids increase the rate of recovery, although the effects are relatively small (Davies et al. 1999; Niewoehner et al. 1999).

12
Future Directions

Inhaled corticosteroids are now used as first-line therapy for the treatment of persistent asthma in adults and children in many countries, as they are the most effective treatments for asthma currently available (Barnes et al. 1998). While may patients, particularly with more severe asthma, remain undertreated, there is also a danger of overtreatment, and may patients with mild asthma who may require very low doses of inhaled corticosteroids are inappropriately treated with high doses. It is essential that inhaled corticosteroids are slowly reduced to the minimal dose required to control asthma. An important clinical development is the recognition that asthma is better controlled by addition of an alternative class of treatment (long-acting inhaled β_2-agonists, low-dose theophylline, anti-leukotrienes) than increasing the dose of inhaled steroid. The recent introduction of fixed combination inhalers with long-acting β_2-agonists is an important advance, as it greatly simplifies asthma management and provides very effective control. Improvement in techniques for the non-invasive monitoring of airway inflammation may be valuable in the future for assessing the requirement for inhaled corticosteroids (Kharitonov and Barnes 2001).

12.1
New Corticosteroids

Budesonide and FP have been important advances in inhaled corticosteroid therapy, as they have reduced systemic effects because of greater first-pass hepatic metabolism than BDP. New inhaled corticosteroids in development, such as mometasone, show a similar improved profile (Prakash and Benfield 1998). However, all currently available corticosteroids are absorbed from the lungs into the systemic circulation and therefore inevitably have some systemic component.

A class of steroids was developed that was metabolised in the lung, but such so-called soft steroids, such as tipredane and butixocort, did not prove to be clinically effective, probably because they were metabolised too rapidly in the airways. A new steroid ciclesonide is a prodrug that releases active corticosteroids in the lungs after enzymatic action (Dent 2002). In addition the active metabolite as a very high degree of protein binding in plasma, so that the concentrations of free steroid are low. This gives ciclesonide improved therapeutic ration. Ciclesonide appears to have good efficacy and is now in clinical development (Taylor et al. 1999). Steroids that are metabolised by enzymes in the circulation may be the safest type of inhaled corticosteroid, and novel esterified corticosteroids are now in clinical development. However, it is still not certain whether the anti-inflammatory effects of inhaled corticosteroids in asthma are mediated entirely by local anti-inflammatory effects in the airways, and it is possible that there is a systemic component, for example on bone marrow eosinophil precursors or on regional lymph nodes. Furthermore, it is not clear whether inhaled corticosteroids are distributed from their point of deposition in the airways to more peripheral airways via the local circulation. If this is the case then corticosteroids that are degraded by enzymes in the circulation may not reach small airways that are inflamed in asthma.

12.2
Dissociated Steroids

Understanding the molecular mechanisms of action of corticosteroids has led to the development of a new generation of corticosteroids. As discussed above, a major mechanism of the anti-inflammatory effect of corticosteroids appears to be direct inhibition of transcription factors, such as NF-κB and AP-1 that are activated by proinflammatory cytokines (*trans*-repression). By contrast, the endocrine and metabolic effects of steroids that are responsible for the systemic side effects of corticosteroids are likely to be mediated via DNA binding (*trans*-activation). This has led to a search for novel corticosteroids that selectively *trans*-repress, thus reducing the potential risk of systemic side effects. Since corticosteroids bind to the same GR, this seems at first to be an unlikely possibility, but while DNA binding involves a GR homodimer, interaction with transcription factors AP-1 and NF-κB involves only a single GR. A separation of *trans*-activation and *trans*-repression has been demonstrated using reporter gene constructs in transfected cells using selective mutations of the glucocorticoid receptor. Furthermore, some steroids, such as the antagonist RU486, have a greater *trans*-repression than *trans*-activation effect. Indeed, the topical steroids used in asthma therapy today, such as FP and budesonide, appear to have more potent *trans*-repression than *trans*-activation effects, which may account for their selection as potent anti-inflammatory agents (Adcock et al. 1999). Recently, a novel class of steroids has been described in which there is potent *trans*-repression with relatively little *trans*-activation. These "dissociated" steroids, including RU24858 and RU40066, have anti-inflammatory effects in vitro (Vayssiere et al. 1997), although there is little separation of anti-inflammatory effects and systemic side effects in vivo (Belvisi et al. 2001). This suggests that the development of steroids with a greater margin of safety is possible and may even lead to the development of oral steroids that do not have significant adverse effects (Belvisi Brown et al. 2001).

References

Adams N, Bestall J, Jones PW (2001) Inhaled fluticasone propionate for chronic asthma (Cochrane Review). Cochrane.Database.Syst.Rev 3:CD003135

Adcock IM, Brown CR, Shirasaki H, Barnes PJ (1994) Effects of dexamethasone on cytokine and phorbol ester stimulated c-Fos and c-Jun DNA binding and gene expression in human lung. Eur Resp J 7:2117–2123

Adcock IM, Gilbey T, Gelder CM, Chung KF, Barnes PJ (1996) Glucocorticoid receptor localization in normal human lung and asthmatic lung. Am J Respir Crit Care Med 154:771–782

Adcock IM, Nasuhara Y, Stevens DA, Barnes PJ (1999) Ligand-induced differentiation of glucocorticoid receptor trans-repression and transactivation: preferential targetting of NF-κB and lack of I-κB involvement. Br J Pharmacol 127:1003–1011

Adcock IM, Peters M, Gelder C, Shirasaki H, Brown CR, Barnes PJ (1993) Increased tachykinin receptor gene expression in asthmatic lung and its modulation by steroids. J Mol Endocrinol 11:1–7

Agertoft L, Pedersen S (1994) Effects of long-term treatment with an inhaled corticosteroid on growth and pulmonary function in asthmatic children. Resp Med 5:369–372

Agertoft L, Pedersen S (2000) Effect of long-term treatment with inhaled budesonide on adult height in children with asthma. New Engl J Med 343:1064–1069

Allen DB, Mullen M, Mullen B (1994) A meta-analysis of the effects of oral and inhaled corticosteroids on growth. J Allergy Clin Immunol 93:967–976

Atsuta J, Plitt J, Bochner BS, Schleimer RP (1999) Inhibition of VCAM-1 expression in human bronchial epithelial cells by glucocorticoids. Am.J.Respir.Cell Mol.Biol. 20:643–650

Bannister AJ, Schneider R, Kouzarides T (2002) Histone methylation: dynamic or static? Cell 109:801–806

Baraniuk JN, Ali M, Brody D, Maniscalco J, Gaumond E, Fitzgerald T, Wonk G, Mak JCW, Bascom R, Barnes PJ, Troost T (1997) Glucocorticoids induce β_2-adrenergic receptor function in human nasal mucosa. Am J Respir Crit Care Med 155:704–710

Barnes PJ (1990) Effect of corticosteroids on airway hyperresponsiveness. Am Rev Respir Dis 141:S70-S76

Barnes PJ (1995) Inhaled glucocorticoids for asthma. New Engl J Med 332:868–875

Barnes PJ (1996a) Inhaled glucocorticoids: new developments relevant to updating the Asthma Management Guidelines. Resp Med 90:379–384

Barnes PJ (1996b) Mechanism of action of glucocorticoids in asthma. Am J Respir Crit Care Med 154:S21-S27

Barnes PJ (1998a) Anti-inflammatory actions of glucocorticoids: molecular mechanisms. Clin Sci 94:557–572

Barnes PJ (1998b) Efficacy of inhaled corticosteroids in asthma. J.Allergy Clin.Immunol. 102:531–538

Barnes PJ (1999) Therapeutic strategies for allergic diseases. Nature 402:B31-B38

Barnes PJ (2000a) Inhaled corticosteroids are not helpful in chronic obstructive pulmonary disease. Am J Resp Crit Care Med 161:342–344

Barnes PJ (2000b) Steroid-resistant asthma. Eur Resp Rev 10:74–78

Barnes PJ (2001) Molecular mechanisms of corticosteroids in allergic diseases. Allergy 56:928–936

Barnes PJ (2002) Scientific rationale for combination inhalers with a long-acting $\beta 2$-agonists and corticosteroids. Eur Respir J 19:182–191

Barnes PJ, Adcock IM (1998) Transcription factors and asthma. Eur Respir J 12:221–234

Barnes PJ, Greening AP, Crompton GK (1995) Glucocorticoid resistance in asthma. Am J Respir Crit Care Med 152:125S-140S

Barnes PJ, Jonsson B, Klim J (1996) The costs of asthma. Eur Respir J 9:636–642

Barnes PJ, Karin M (1997) Nuclear factor-κB: a pivotal transcription factor in chronic inflammatory diseases. New Engl J Med 336:1066–1071

Barnes PJ, Pedersen S, Busse WW (1998) Efficacy and safety of inhaled corticosteroids: an update. Am J Respir Crit Care Med 157:S1-S53

Beam WR, Ballard RD, Martin RJ (1992) Spectrum of corticosteroid sensitivity in nocturnal asthma. Am Rev Respir Dis 145:1082–1086

Belvisi MG, Wicks SL, Battram CH, Bottoms SE, Redford JE, Woodman P, Brown TJ, Webber SE, Foster ML (2001) Therapeutic benefit of a dissociated glucocorticoid and the relevance of in vitro separation of transrepression from transactivation activity. J.Immunol. 166:1975–1982

Bergmann M, Barnes PJ, Newton R (2000) Molecular regulation of granulocyte macrophage colony-stimulating factor in human lung epithelial cells by interleukin (IL)-1β, IL-4, and IL-13 involves both transcriptional and post-transcriptional mechanisms. Am J Respir Cell Mol.Biol. 22:582–589

Bledsoe RK, Montana VG, Stanley TB, Delves CJ, Apolito CJ, McKee DD, Consler TG, Parks DJ, Stewart EL, Willson TM, Lambert MH, Moore JT, Pearce KH, Xu HE (2002) Crystal structure of the glucocorticoid receptor ligand binding domain reveals a novel mode of receptor dimerization and coactivator recognition. Cell 110:93–105

Bodwell JE, Webster JC, Jewell CM, Cidlowski JA, Hu JM, Munck A (1998) Glucocorticoid receptor phosphorylation: overview, function and cell cycle-dependence. J Steroid Biochem Mol Biol 65:91–99

Bowler SD, Mitchell CA, Armstrong JG (1992) Corticosteroids in acute severe asthma: effectiveness of low doses. Thorax 47:584–587

British Thoracic Society (1997) The British guidelines on asthma management. Thorax 52(Suppl 1):S1-S21

Brutsche MH, Brutsche IC, Munawar M, Langley SJ, Masterson CM, Daley-Yates PT, Brown R, Custovic A, Woodcock A (2000) Comparison of pharmacokinetics and systemic effects of inhaled fluticasone propionate in patients with asthma and healthy volunteers: a randomised crossover study. Lancet 356:556–561

Burge PS, Calverley PMA, Jones PW, Spencer S, Anderson JA, Maslen T (2000) Randomised, double-blind, placebo-controlled study of fluticasone propionate in patients with moderate to severe chronic obstructive pulmonary disease; the ISOLDE trial. Br Med J 320:1297–1303

Busse WW, Chervinsky P, Condemi J, Lumry WR, Petty TL, Rennard S, Townley RG (1998) Budesonide delivered by Turbuhaler is effective in a dose- dependent fashion when used in the treatment of adult patients with chronic asthma. J.Allergy Clin.Immunol. 101:457–463

Chapman KR, Ringdal N, Backer V, Palmqvist M, Saarelainen S, Briggs M (1999) Salmeterol and fluticasone propionate (50/250 mg) administered via combination diskus inhaler: As effective as when given via separate diskus inhalers. Can.Respir.J. 6:45–51

Culpitt SV, Nightingale JA, Barnes PJ (1999) Effect of high dose inhaled steroid on cells, cytokines and proteases in induced sputum in chronic obstructive pulmonary disease. Am J Respir Crit Care Med 160:1635–1639

Culpitt SV, Rogers DF, Shah P, de Matos C, Russell R, Donnelly LE, Barnes PJ (2002) Impaired inhibition by dexamethasone of cytokine release by alveolar macrophages from COPD patients. Am J Respir.Crit Care Med

Cumming RG, Mitchell P, Leeder SR (1997) Use of inhaled corticosteroids and the risk of cataracts. New Engl J Med 337:8–14

Davies H, Olson L, Gibson P (2000) Methotrexate as a steroid sparing agent for asthma in adults. Cochrane.Database.Syst.Rev. 2:CD000391:CD000391

Davies L, Angus RM, Calverley PM (1999) Oral corticosteroids in patients admitted to hospital with exacerbations of chronic obstructive pulmonary disease: a prospective randomised controlled trial. Lancet 354:456–460

Decramer M, Lacquet LM, Fagard R, Rogiers P (1995) Corticosteroids contribute to muscle weakness in chronic airflow obstruction. Am J Respir Crit Care Med 150:11–16

Dent G (2002) Ciclesonide (Byk Gulden). Curr Opin Investig.Drugs 3:78–83

Devidayal, Singhi S, Kumar L, Jayshree M (1999) Efficacy of nebulized budesonide compared to oral prednisolone in acute bronchial asthma. Acta Paediatr. 88:835–840

Dompeling E, Van Schayck CP, Molema J, Folgering H, van Grusven PM, van Weel C (1992) Inhaled beclomethasone improves the course of asthma and COPD. Eur Resp J 5:945–952

Donnelly LE, Barnes PJ (2002) Expression and regulation of inducible nitric oxide synthase from human primary airway epithelial cells. Am J Respir Cell Mol.Biol. 26:144–151

Ducharme F (2002) Anti-leukotrienes as add-on therapy to inhaled glucocorticoids in patients with asthma: systematic review of current evidence. Br Med J 324:1545–1548

Efthimou J, Barnes PJ (1998) Effect of inhaled corticosteroids on bone and growth. Eur Respir J 11:1167–1177

Egan JJ, Maden C, Kalra S, Adams JE, Eastell R, Woodcock AA (1999) A randomized, double-blind study comparing the effects of beclomethasone and fluticasone on bone density over two years. Eur Respir J 13:1267–1275

Engel T, Heinig JH (1991) Glucocorticoid therapy in acute severe asthma—a critical review. Eur Respir J 4:881–889

Evans DJ, Taylor DA, Zetterstrom O, Chung KF, O'Connor BJ, Barnes PJ (1997) A comparison of low-dose inhaled budesonide plus theophylline and high-dose inhaled budesonide for moderate asthma. New Engl J Med 337:1412–1418

Foresi A, Morelli MC, Catena E (2000) Low-dose budesonide with the addition of an increased dose during exacerbations is effective in long-term asthma control. On behalf of the Italian Study Group. Chest 117:440–446

Fujita K, Kasayama S, Hashimoto J, Nagasaka Y, Nakano N, Morimoto Y, Barnes PJ, Miyatake A (2001) Inhaled corticosteroids reduce bone mineral density in early postmenopausal but not premenopausal asthmatic women. J Bone Miner.Res 16:782–787

Gagliardo R, Chanez P, Vignola AM, Bousquet J, Vachier I, Godard P, Bonsignore G, Demoly P, Mathieu M (2000) Glucocorticoid receptor α and β in glucocorticoid dependent asthma. Am.J.Respir.Crit Care Med. 162:7–13

Gao L, Cueto MA, Asselbergs F, Atadja P (2002) Cloning and functional characterization of HDAC11, a novel member of the human histone deacetylase family. J Biol Chem. 277:25748–25755

Garbe E, LeLorier J, Boivin J-F, Suissa S (1997) Inhaled and nasal glucocorticoids and the risks of ocular hypertension or open-angel glaucoma. JAMA 227:722–727

Green RH, Brightling CE, McKenna S, Hargadon B, Parker D, Bradding P, Wardlaw AJ, Pavord ID (2002) Asthma exacerbations and sputum eosinophil counts: a randomised controlled trial. Lancet 360:1715–1721

Greening AP, Ind PW, Northfield M, Shaw G (1994) Added salmeterol versus higher-dose corticosteroid in asthma patients with symptoms on existing inhaled corticosteroid. Lancet 344:219–224

Haahtela T, Jarvinen M, Kava T, Kiviranta K, Koskinen S, Lehtonen K, Nivander K, Persson T, Reinikainen R, Selroos O, Sovijarvi A, Stenius-Aarniala B, Svahn T, Tammivaara R, Laitinen LA (1991) Comparison of a b2-agonist terbutaline with an inhaled steroid in newly detected asthma. New Engl J Med 325:388–392

Haahtela T, Järvinsen M, Kava T, Kiviranta K, Koskinen S, Lemtonen K, Nikander K, Person T, Selroos O, Sovijäri A, Stenius-Aarniala B, Svahn T, Tammivaara R, Laitinen LA (1994) Effects of reducing or discontinuing inhaled budesonide in patients with mild asthma. New Engl J Med 331:700–705

Haddad EB, Fox AJ, Rousell J, Burgess G, McIntyre P, Barnes PJ, Chung KF (2000) Post-transcriptional regulation of bradykinin B_1 and B_2 receptor gene expression in human lung fibroblasts by tumor necrosis factor-a: modulation by dexamethasone. Mol.Pharmacol. 57:1123–1131

Hall SE, Lim S, Witherden IR, Tetley TD, Barnes PJ, Kamal AM, Smith SF (1999) Lung type II cell and macrophage annexin I release: differential effects of two glucocorticoids. Am.J.Physiol. 276:L114-L121

Hamid QA, Wenzel SE, Hauk PJ, Tsicopoulos A, Wallaert B, Lafitte JJ, Chrousos GP, Szefler SJ, Leung DY (1999) Increased glucocorticoid receptor beta in airway cells of glucocorticoid-insensitive asthma. Am.J.Respir.Crit.Care Med. 159:1600–1604

Harrison BDN, Stokes TC, Hart GJ, Vaughan DA, Ali NJ, Robinson AA (1986) Need for intravenous hydrocortisone in addition to oral prednisolone in patients admitted to hospital with severe asthma without ventilatory failure. Lancet i:181–184

Harrison TW, Wisniewski A, Honour J, Tattersfield AE (2001) Comparison of the systemic effects of fluticasone propionate and budesonide given by dry powder inhaler in healthy and asthmatic subjects. Thorax 56:186–191

Hart L, Lim S, Adcock I, Barnes PJ, Chung KF (2000) Effects of inhaled corticosteroid therapy on expression and DNA-binding activity of nuclear factor-κB in asthma. Am J Respir Crit Care Med 161:224–231

Hecht K, Carlstedt-Duke J, Stierna P, Gustaffson J-Å, Bronnegard M, Wilkstrom A-C (1997) Evidence that the β-isoform of the human glucocorticoid receptor does not act as a physiologically significant repressor. J Biol Chem 272:26659–26664

Heck S, Bender K, Kullmann M, Gottlicher M, Herrlich P, Cato AC (1997) IKBα-independent downregulation of NF-κB activity by glucocorticoid receptor. EMBO J. 16:4698–4707

Hill SJ, Tattersfield AE (1995) Corticosteroid sparing agents in asthma. Thorax 50:577–582

Ilangovan P, Pedersen S, Godfrey S, Nikander K, Novisky N, Warner JO (1993) Nebulised budesonide suspension in severe steroid-dependent preschool asthma. Arch Dis Child 68:356–359

Imhof A, Wolffe AP (1998) Transcription: gene control by targeted histone acetylation. Curr.Biol. 8:R422–4

Irusen E, Matthews JG, Takahashi A, Barnes PJ, Chung KF, Adcock IM (2002) p38 Mitogen-activated protein kinase-induced glucocorticoid receptor phosphorylation reduces its activity: Role in steroid-insensitive asthma. J Allergy Clin Immunol 109:649–657

Israel E, Banerjee TR, Fitzmaurice GM, Kotlov TV, LaHive K, LeBoff MS (2001) Effects of inhaled glucocorticoids on bone density in premenopausal women. N.Engl.J Med 345:941–947

Ito K, Barnes PJ, Adcock IM (2000) Glucocorticoid receptor recruitment of histone deacetylase 2 inhibits IL-1β-induced histone H4 acetylation on lysines 8 and 12. Mol Cell Biol 20:6891–6903

Ito K, Jazwari E, Cosio B, Barnes PJ, Adcock IM (2001a) p65-activated histone acetyl-transferase activity is repressed by glucocorticoids: Mifepristone fails to recruit HDAC2 to the p65/HAT complex. J Biol.Chem. 276:30208–30215

Ito K, Lim S, Caramori G, Chung KF, Barnes PJ, Adcock IM (2001b) Cigarette smoking reduces histone deacetylase 2 expression, enhances cytokine expression and inhibits glucocorticoid actions in alveolar macrophages. FASEB J 15:1100–1102

Jatakanon A, Kharitonov S, Lim S, Barnes PJ (1999) Effect of differing doses of inhaled budesonide on markers of airway inflammation in patients with mild asthma. Thorax 54:108–114

Jatakanon A, Lim S, Barnes PJ (2000) Changes in sputum eosinophils predict loss of asthma control. Am.J.Respir.Crit.Care Med. 161:64–72

Jatakanon A, Lim S, Chung KF, Barnes PJ (1998) An inhaled steroid improves markers of inflammation in asymptomatic steroid-naive asthmatic patients. Eur Respir J 12:1084–1088

Jenuwein T, Allis CD (2001) Translating the histone code. Science 293:1074–1080

Jick SS, Vasilakis-Scaramozza C, Maier WC (2001) The risk of cataract among users of inhaled steroids. Epidemiology 12:229–234

John M, Lim S, Seybold J, Robichaud A, O'Connor B, Barnes PJ, Chung KF (1998) Inhaled corticosteroids increase IL-10 but reduce MIP-1α, GM-CSF and IFN-γ release from alveolar macrophages in asthma. Am J Respir Crit Care Med 157:256–262

Jones AH, Langdon CG, Lee PS, Lingham SA, Nankani JP, Follows RMA, Tollemar U, Richardson PDI (1994) Pulmicort Turbohaler once daily as initial prophylactic therapy for asthma. Resp Med 88:293–299

Juniper EF, Kline PA, Vanzieleghem MA, Ramsdale EH, O'Byrne PM, Hargreave FE (1990) Effect of long-term treatment with an inhaled corticosteroid (budesonide) on airway hyperresponsiveness and clinical asthma in nonsteroid-dependent asthmatics. Am Rev Respir Dis 142:832–836

Juniper EF, Kline PA, Vanzielegmem MA, Hargreave FE (1991) Reduction of budesonide after a year of increased use: a randomized controlled trial to evaluate whether improvements in airway responsiveness and clinical asthma are maintained. J Allergy Clin Immunol 87:483–489

Kagoshima M, Wilcke T, Ito K, Tsaprouni L, Barnes PJ, Punchard N, Adcock IM (2001) Glucocorticoid-mediated transrepression is regulated by histone acetylation and DNA methylation. Eur J Pharmacol 429:327–334

Kai H, Yoshitake K, Hisatsune A, Kido T, Isohama Y, Takahama K, Miyata T (1996) Dexamethasone suppresses mucus production and MUC-2 and MUC-5AC gene expression by NCI-H292 cells. Am J Physiol 271:L484-L488

Kamada AK, Szefler SJ, Martin RJ, Boushey HA, Chinchilli VM, Drazen JM, Fish JE, Israel E, Lazarus SC, Lemanske RF (1996) Issues in the use of inhaled steroids. Am J Respir Crit Care Med 153:1739–1748

Kankaanranta H, Lindsay MA, Giembycz MA, Zhang X, Moilanen E, Barnes PJ (2000) Delayed eosinophil apoptosis in asthma. J Allergy Clin Immunol 106:77–83

Katsunuma T, Mak JCW, Barnes PJ (1998) Glucocorticoids reduce tachykinin NK_2-receptor expression in bovine tracheal smooth muscle. Eur J Pharmacol 344:99–107

Keatings VM, Jatakanon A, Worsdell YM, Barnes PJ (1997) Effects of inhaled and oral glucocorticoids on inflammatory indices in asthma and COPD. Am J Respir Crit Care Med 155:542–548

Kharitonov SA, Barnes PJ (2001) Exhaled markers of pulmonary disease. Am J Respir Crit Care Med 163:1693–1772

Kurihara I, Shibata H, Suzuki T, Ando T, Kobayashi S, Hayashi M, Saito I, Saruta T (2002) Expression and regulation of nuclear receptor coactivators in glucocorticoid action. Mol Cell Endocrinol. 189:181–189

Lane SJ, Adcock IM, Richards D, Hawrylowicz C, Barnes PJ, Lee TH (1998) Corticosteroid-resistant bronchial asthma is associated with increased c-Fos expression in monocytes and T-lymphocytes. J Clin Invest 102:2156–2164

Laviolette M, Malmstrom K, Lu S, Chervinsky P, Pujet JC, Peszek I, Zhang J, Reiss TF (1999) Montelukast added to inhaled beclomethasone in treatment of asthma. Am J Respir Crit Care Med 160:1862–1868

Leung DYM, Hamid Q, Vottero A, Szefler SJ, Surs W, Minshall E, Chrousos GP, Klemm DJ (1997) Association of glucocorticoid insensitivity with increased expression of glucocorticoid receptor b. J Exp Med 186:1567–1574

Levy ML, Stevenson C, Maslen T (1996) Comparison of short courses of oral prednisolone and fluticasone propionate in the treatment of adults with acute exacerbations of asthma in primary care. Thorax 51:1087–1092

Lilly CM, Nakamura H, Kesselman H, Nagler Anderson C, Asano K, Garcia Zepeda EA, Rothenberg ME, Drazen JM, Luster AD (1997) Expression of eotaxin by human lung epithelial cells: induction by cytokines and inhibition by glucocorticoids. J.Clin.Invest. 99:1767–1773

Lim S, Jatakanon A, Gordon D, Macdonald C, Chung KF, Barnes PJ (2000) Comparison of high dose inhaled steroids, low dose inhaled steroids plus low dose theophylline, and low dose inhaled steroids alone in chronic asthma in general practice. Thorax 55:837–841

Lim S, Jatakanon A, John M, Gilbey T, O'Connor BJ, Chung KF, Barnes PJ (1999) Effect of inhaled budesonide on lung function and airway inflammation. Assessment by various inflammatory markers in mild asthma. Am.J.Respir.Crit.Care Med. 159:22–30

Lipworth BJ (1999) Systemic adverse effects of inhaled corticosteroid therapy: A systematic review and meta-analysis. Arch.Intern.Med 159:941–955

Lung Health Study Research Group (2000) Effect of inhaled triamcinolone on the decline in pulmonary function in chronic obstructive pulmonary disease. New Engl J Med 343:1902–1909

Mak JCW, Nishikawa M, Barnes PJ (1995a) Glucocorticosteroids increase β_2-adrenergic receptor transcription in human lung. Am J Physiol 12:L41-L46

Mak JCW, Nishikawa M, Shirasaki H, Miyayasu K, Barnes PJ (1995b) Protective effects of a glucocorticoid on down-regulation of pulmonary β_2-adrenergic receptors in vivo. J Clin Invest 96:99–106

Malo J-L, Cartier A, Merland N, Ghezzo H, Burke A, Morris J, Jennings BH (1989) Four-times-a-day dosing frequency is better than twice-a-day regimen in subjects requiring a high-dose inhaled steroid, budesonide, to control moderate to severe asthma. Am Rev Respir Dis 140:624–628

Martin RJ, Szefler SJ, Chinchilli VM, Kraft M, Dolovich M, Boushey HA, Cherniack RM, Craig TJ, Drazen JM, Fagan JK, Fahy JV, Fish JE, Ford JG, Israel E, Kunselman SJ,

Lazarus SC, Lemanske RF, Jr., Peters SP, Sorkness CA (2002) Systemic effect comparisons of six inhaled corticosteroid preparations. Am J Respir Crit Care Med 165:1377–1383

Matthews JG, Ito K, Barnes PJ, Adcock IM (2000) Corticosteroid-resistant and corticosteroid-dependent asthma: two clinical phenotypes can be associated with the same in vitro defects in nuclear translocation and acetylation of histone 4. Am J Respir Crit Care Med 161:A189

McLeod DT, Capewell SJ, Law J, MacLaren W, Seaton A (1985) Intramuscular triamcinolone acetonide in chronic severe asthma. Thorax 40:840–845

Meagher LC, Cousin JM, Seckl JR, Haslett C (1996) Opposing effects of glucocorticoids on the rate of apoptosis in neutrophilic and eosinophilic granulocytes. J Immunol 156:4422–4428

Morell F, Orkiols R, de Gracia J, Curul V, Pujol A (1992) Controlled trial of intravenous corticosteroids in severe acute asthma. Thorax 47:588–591

Nathan RA, Nayak AS, Graft DF, Lawrence M, Picone FJ, Ahmed T, Wolfe J, Vanderwalker ML, Nolop KB, Harrison JE (2001) Mometasone furoate: efficacy and safety in moderate asthma compared with beclomethasone dipropionate. Ann.Allergy Asthma Immunol. 86:203–210

Nelson DJ, McWilliam AS, Haining S, Holt PG (1995) Modulation of airway intraepithelial dendritic cells following exposure to steroids. Am J Respir Crit Care Med 151:475–481

Nelson HS, Busse WW, Kerwin E, Church N, Emmett A, Rickard K, Knobil K (2000) Fluticasone propionate/salmeterol combination provides more effective asthma control than low-dose inhaled corticosteroid plus montelukast. J.Allergy Clin.Immunol. 106:1088–1095

Nelson HS, Hamilos DL, Corsello PR, Levesque NV, Buchameier AD, Bucher BL (1993) A double-blind study of troleandamycin and methylprednisolone in asthmatic patients who require daily corticosteroids. Am Rev Respir Dis 147:398–404

Newton R, Hart LA, Stevens DA, Bergmann M, Donnelly LE, Adcock IM, Barnes PJ (1998a) Effect of dexamethasone on interleukin-1b-(IL-1b)-induced nuclear factor-κB (NF-κB) and kB-dependent transcription in epithelial cells. Eur.J.Biochem. 254:81–89

Newton R, Kuitert LM, Slater DM, Adcock IM, Barnes PJ (1997) Cytokine induction of cytosolic phosholipase A_2 and cyclooxygenase-2 mRNA by proinflammatory cytokines is suppressed by dexamethasone in human epithelial cells. Life Sci 60:67–78

Newton R, Seybold J, Kuitert LME, Bergmann M, Barnes PJ (1998b) Repression of cyclooxygenase-2 and prostaglandin E_2 release by dexamethasone occurs by transcriptional and post-transcriptional mechanisms Involving loss of polyadenylated mRNA. J.Biol.Chem. 273:32312–32321

Niewoehner DE, Erbland ML, Deupree RH, Collins D, Gross NJ, Light RW, Anderson P, Morgan NA (1999) Effect of systemic glucocorticoids on exacerbations of chronic obstructive pulmonary disease. N.Engl.J.Med. 340:1941–1947

Nightingale JA, Rogers DF, Chung KF, Barnes PJ (2000) No effect of inhaled budesonide on the response to inhaled ozone in normal subjects. Am J Respir Crit Care Med 161:479–486

Ogirala RG, Aldrich TK, Prezant DJ, Sinnett MJ, Enden JB, Williams MH (1991) High dose intramuscular triamcinolone in severe life-threatening asthma. New Engl J Med 329:585–589

Otulana BA, Varma N, Bullock A, Higenbottam T (1992) High dose nebulized steroid in the treatment of chronic steroid-dependent asthma. Resp Med 86:105–108

Paggiaro PL, Dahle R, Bakran I, Frith L, Hollingworth K, Efthimou J (1998) Multicentre randomised placebo-controlled trial of inhaled fluticasone propionate in patients with chronic obstructive pulmonary disease. Lancet 351:773–780

Papi A, Romagnoli M, Baraldo S, Braccioni F, Guzzinati I, Saetta M, Ciaccia A, Fabbri LM (2000) Partial reversibility of airflow limitation and increased exhaled NO and sputum eosinophilia in chronic obstructive pulmonary disease. Am J Respir Crit Care Med 162:1773–1777

Pauwels RA, Lofdahl C-G, Postma DS, Tattersfield AE, O'Byrne PM, Barnes PJ, Ullman A (1997) Effect of inhaled formoterol and budesonide on exacerbations of asthma. New Engl J Med 337:1412–1418

Pauwels RA, Lofdahl CG, Laitinen LA, Schouten JP, Postma DS, Pride NB, Ohlsson SV (1999) Long-term treatment with inhaled budesonide in persons with mild chronic obstructive pulmonary disease who continue smoking. N.Engl.J.Med. 340:1948–1953

Pedersen S (2001) Do inhaled corticosteroids inhibit growth in children? Am J Respir Crit Care Med 164:521–535

Peterson CL (2002) HDAC's at work: everyone doing their part. Mol Cell 9:921–922

Prakash A, Benfield P (1998) Topical mometasone. A review of its pharmacological properties and therapeutic use in the treatment of dermatological disorders. Drugs 55:145–163

Rao A, Luo C, Hogan PG (1997) Transcription factors of the NFAT family: regulation and function. Annu.Rev.Immunol. 15:707–47:707–747

Reichardt HM, Kaestner KH, Tuckermann J, Kretz O, Wessely O, Bock R, Gass P, Schmid W, Herrlich P, Angel P, Schutz G (1998) DNA binding of the glucocorticoid receptor is not essential for survival. Cell 93:531–541

Reichardt HM, Tuckermann JP, Gottlicher M, Vujic M, Weih F, Angel P, Herrlich P, Schutz G (2001) Repression of inflammatory responses in the absence of DNA binding by the glucocorticoid receptor. EMBO J 20:7168–7173

Robinson DS, Campbell DA, Barnes PJ (2001) Addition of an anti-leukotriene to therapy in chronic severe asthma in a clinic setting:: a double-blind, randomised, placebo-controlled study. Lancet 357:2007–2011

Roy A, Leblanc C, Paquette L, Ghezzo H, Cote J, Cartier A, Malo J-L (1996) Skin bruising in asthmatic subjects treated with high doses of inhaled steroids: frequency and association with adrenal function. Eur Respir J 9:226–231

Russell RE, Culpitt SV, DeMatos C, Donnelly L, Smith M, Wiggins J, Barnes PJ (2002) Release and activity of matrix metalloproteinase-9 and tissue inhibitor of metalloproteinase-1 by alveolar macrophages from patients with chronic obstructive pulmonary disease. Am J Respir Cell Mol Biol 26:602–609

Saleh D, Ernst P, Lim S, Barnes PJ, Giaid A (1998) Increased formation of the potent oxidant peroxynitrite in the airways of asthmatic patients is associated with induction of nitric oxide synthase: effect of inhaled glucocorticoid. FASEB J. 12:929–937

Schatz M (1999) Asthma and pregnancy. Lancet 353:1202–1204

Schatz M, Zeiger RS, Harden K, Hoffman CC, Chilingar L, Petitti D (1997) The safety of asthma and allergy medications during pregnancy. J.Allergy Clin.Immunol. 100:301–306

Schweibert LM, Stellato C, Schleimer RP (1996) The epithelium as a target for glucocorticoid action in the treatment of asthma. Am J Respir Crit Care Med 154:S16-S20

Selroos O, Pietinalcho A, Lofroos A-B, Riska A (1995) Effect of early and late intervention with inhaled corticosteroids in asthma. Chest 108:1228–1234

Shapiro G, Lumry W, Wolfe J, Given J, White MV, Woodring A, Baitinger L, House K, Prillaman B, Shah T (2000) Combined salmeterol 50 μg and fluticasone propionate 250 μg in the Diskus device for the treatment of asthma. Am J Respir Crit Care Med 161:527–534

Shaw RJ (1999) Inhaled corticosteroids for adult asthma: impact of formulation and delivery device on relative pharmacokinetics, efficacy and safety. Respir Med 93:149–160

Shrewsbury S, Pyke S, Britton M (2000) Meta-analysis of increased dose of inhaled steroid or addition of salmeterol in symptomatic asthma (MIASMA). BMJ. 320:1368–1373

Silverstein MD, Yunginger JW, Reed CE, Petterson T, Zimmerman D, Li JT, O'Fallon WM (1997) Attained adult height after childhood asthma: effect of glucocorticoid therapy. J.Allergy Clin.Immunol. 99:466–474

Simons FE (1997) A comparison of beclomethasone, salmeterol, and placebo in children with asthma. N.Engl.J.Med. 337:1659–1665

Simons FER, Persaud MP, Gillespie CA, Cheang M, Shuckett EP (1993) Absence of posterior subcapsular cataracts in young patients treated with inhaled glucocorticoids. Lancet 342:736–738

Sont JK, Willems LN, Bel EH, van Krieken JH, Vandenbroucke JP, Sterk PJ (1999) Clinical control and histopathologic outcome of asthma when using airway hyperresponsiveness as an additional guide to long-term treatment. The AMPUL Study Group. Am.J.Respir.Crit.Care Med. 159:1043–1051

Sousa AR, Lane SJ, Cidlowski JA, Staynov DZ, Lee TH (2000) Glucocorticoid resistance in asthma is associated with elevated in vivo expression of the glucocorticoid receptor β-isoform. J.Allergy Clin.Immunol. 105:943–950

Suissa S, Ernst P, Benayoun S, Baltzan M, Cai B (2000) Low-dose inhaled corticosteroids and the prevention of death from asthma. N.Engl.J.Med. 343:332–336

Szefler SJ, Leung DY (1997) Glucocorticoid-resistant asthma: pathogenesis and clinical implications for management. Eur Respir J 10:1640–1647

Taylor DA, Jensen MW, Kanabar V, Englestatter R, Steinjans VW, Barnes PJ, O'Connor BJ (1999) A dose-dependent effect of the novel inhaled corticosteroid ciclesonide on airway responsiveness to adenosine-5'-monophosphate in asthmatic patients. Am J Respir Crit Care Med 160:237–243

Toogood JA, Jennings B, Greenway RW, Chung L (1980) Candidiasis and dysphonia complicating beclomethasone treatment of asthma. J Allergy Clin Immunol 65:145–153

Ukena D, Harnest U, Sakalauskas R, Magyar P, Vetter N, Steffen H, Leichtl S, Rathgeb F, Keller A, Steinijans VW (1997) Comparison of addition of theophylline to inhaled steroid with doubling of the dose of inhaled steroid in asthma. Eur.Respir.J. 10:2754–2760

Urnov FD, Wolffe AP (2001) Chromatin remodeling and transcriptional activation: the cast (in order of appearance). Oncogene 20:2991–3006

van Essen-Zandvliet EE, Hughes MD, Waalkens HJ, Duiverman EJ, Pocock SJ, Kerrebijn KF (1992) Effects of 22 months of treatment with inhaled corticosteroids and/or b_2-agonists on lung function, airway responsiveness and symptoms in children with asthma. Am Rev Respir Dis 146:547–554

van Schayk CP, Dompeling E, Rutten MP, Folgering H, van den Boom G, van Weel C (1995) The influence of an inhaled steroid on quality of life in patients with asthma or COPD. Chest 107:1199–1205

Vathenen AS, Knox AJ, Wisniewski A, Tattersfield AE (1991) Time course of change in bronchial reactivity with an inhaled corticosteroid in asthma. Am Rev Respir Dis 143:1317–1321

Vayssiere BM, Dupont S, Choquart A, Petit F, Garcia T, Marchandeau C, Gronemeyer H, Resche-Rigon M (1997) Synthetic glucocorticoids that dissociate transactivation and AP-1 transrepression exhibit antiinflammatory activity in vivo. Mol Endocrinol 11:1245–1255

Vestbo J, Sorensen T, Lange P, Brix A, Torre P, Viskum K (1999) Long-term effect of inhaled budesonide in mild and moderate chronic obstructive pulmonary disease: a randomised controlled trial. Lancet 353:1819–1823

Waalkens HJ, van Essen-Zandvliet EE, Hughes MD, Gerritsen J, Duiverman EJ, Knol K, Kerrebijn KF (1993) Cessation of long-term treatment with inhaled corticosteroids (budesonide) in children with asthma results in deterioration. Am Rev Respir Dis 148:1252–1257

Walsh GM (1997) Mechanisms of human eosinophil survival and apoptosis. Clin.Exp.Allergy 27:482–487

Williams CM, Galli SJ (2000) The diverse potential effector and immunoregulatory roles of mast cells in allergic disease. J.Allergy Clin.Immunol. 105:847–859

Williamson IJ, Matusiewicz SP, Brown PH, Greening AP, Crompton GK (1995) Frequency of voice problems and cough in patients using pressurised aersosol inhaled steroid preparations. Eur Resp J 8:590–592

Wilson AJ, Gibson PG, Coughlan J (2000) Long acting beta-agonists versus theophylline for maintenance treatment of asthma. Cochrane Database Syst RevCD001281

Woolley KL, Gibson PG, Carty K, Wilson AJ, Twaddell SH, Woolley MJ (1996) Eosinophil apoptosis and the resolution of airway inflammation in asthma. Am.J.Respir.Crit.-Care Med. 154:237–243

Yao TP, Ku G, Zhou N, Scully R, Livingston DM (1996) The nuclear hormone receptor coactivator SRC-1 is a specific target of p300. Proc.Natl.Acad.Sci.U.S.A. 93:10626–10631

Yudt MR, Cidlowski JA (2002) The glucocorticoid receptor: coding a diversity of proteins and responses through a single gene. Mol Endocrinol. 16:1719–1726

Zetterstrom O, Buhl R, Mellem H, Perpina M, Hedman J, O'Neill S, Ekstrom T (2001) Improved asthma control with budesonide/formoterol in a single inhaler, compared with budesonide alone. Eur Respir J 18:262–2688

Mediator Antagonists and Anti-Allergic Drugs

N. C. Thomson

Department of Respiratory Medicine, Western Infirmary, University of Glasgow, Glasgow, G11 6NT, UK
e-mail: n.c.thomson@clinmed.gla.ac.uk

1	Introduction	126
2	Leukotriene-Receptor Antagonists and Leukotriene Synthesis Inhibitors	127
2.1	Leukotrienes and Asthma	127
2.2	Drugs Acting on the 5-Lipoxygenase Pathway	127
2.3	Effect of Drugs Modifying the 5-Lipoxygenase Pathway on Inflammation in Man	128
2.4	Bronchial Challenge Studies	129
2.5	Effect on Baseline Lung Function	131
2.6	Therapeutic Effects in Chronic Asthma	131
2.6.1	Efficacy	131
2.6.2	Adverse Effects	135
2.7	Conclusions	135
3	Histamine Receptor Antagonists	136
3.1	Histamine and Asthma	136
3.2	Histamine Receptor Antagonists	137
3.3	Effect of Histamine Receptor Antagonists on Inflammation in Asthma	137
3.4	Bronchial Challenge Studies	137
3.5	Effect on Baseline Lung Function	138
3.6	Therapeutic Effects in Chronic Asthma	138
3.6.1	Efficacy	138
3.6.2	Adverse Effects	139
3.7	Conclusions	139
4	Cromones	139
4.1	Mode of Action	139
4.2	Effect of Cromones on Inflammation	140
4.3	Bronchial Challenge Studies	141
4.4	Effect on Baseline Lung Function	141
4.5	Therapeutic Effects in Chronic Asthma	141
4.5.1	Efficacy	141
4.5.2	Adverse Effects	143
4.6	Conclusions	143
	References	144

Abstract Many inflammatory mediators have been implicated in the pathogenesis of asthma, and this has led the pharmaceutical industry to develop specific mediator antagonists and non-steroidal anti-allergy drugs as potential therapeutic agents for asthma. The role of these drugs in the treatment of chronic obstructive pulmonary disease (COPD) is largely untested. The leukotriene-receptor antagonists and the 5-lipoxygenase inhibitors have been shown to have a mild bronchodilator effect in asthmatic patients with airflow obstruction and to attenuate bronchoconstriction induced by exercise, allergen and aspirin. There is also some evidence to indicate that they have anti-inflammatory actions. Several therapeutic studies in mild to moderate asthma have shown evidence of efficacy. In clinical trials, the leukotriene-receptor antagonists zafirlukast and montelukast and the 5-lipoxygenase inhibitor zileuton have been well tolerated. The clinical effectiveness of the 5-lipoxygenase inhibitors and the leukotriene-receptor antagonists appears to be quite variable between individuals. The place of the leukotriene-receptor antagonists and 5-lipoxygenase inhibitors in asthma management has not been fully established. Current published evidence suggests a role as an alternative, but less effective first-line prophylactic agent to inhaled corticosteroids and as an alternative second-line add-on drug to long-acting β_2-agonists for patients with persistent symptoms despite low-dose inhaled corticosteroid therapy. H_1-receptor antagonists have been shown to be mild bronchodilators and to attenuate bronchoconstriction induced by exercise and allergen. Current evidence would suggest that the antihistamines do not have a place in the treatment of chronic persistent asthma. The cromones, sodium cromoglycate and nedocromil sodium, have a limited role in the management of chronic asthma and are less effective anti-inflammatory agents than inhaled corticosteroids.

Keywords Asthma · COPD · Leukotriene receptor antagonists · Leukotriene synthesis inhibitors · Histamine receptor antagonists · Cromones

1
Introduction

Many inflammatory mediators have been implicated in the pathogenesis of asthma and this has led the pharmaceutical industry to develop specific mediator antagonists as potential therapeutic agents for asthma. There has also been interest in identifying non-steroidal anti-allergic drugs for asthma. Of particular therapeutic importance is whether these agents can be used as alternative therapies to low-dose inhaled corticosteroids and/or as add-on treatments for patients in whom asthma control remains inadequate despite inhaled corticosteroid therapy. The role of these drugs in the treatment of chronic obstructive pulmonary disease (COPD) is largely untested. This chapter will review the role of currently licensed mediator antagonists and non-steroidal anti-allergic drugs in the treatment of asthma.

2
Leukotriene-Receptor Antagonists and Leukotriene Synthesis Inhibitors

2.1
Leukotrienes and Asthma

The leukotrienes are inflammatory mediators that play a role in the pathogenesis of asthma (Hay et al. 1995). The cysteinyl leukotrienes, leukotriene C_4, leukotriene D_4 and leukotriene E_4, are synthesised from arachidonic acid by the 5-lipoxygenase pathway in the wall of inflammatory cells such as eosinophils, alveolar macrophages and mast cells (Fig. 1). The cysteinyl leukotrienes have several properties that might contribute to the pathogenesis of asthma, including bronchial smooth muscle contraction, mucus hypersecretion and oedema formation. In addition, the leukotrienes can cause eosinophil recruitment and smooth muscle proliferation. Leukotriene A_4, which is the unstable precursor of the cysteinyl leukotrienes, is generated by neutrophils and converted to leukotriene B_4, which is in turn a potent chemotactic factor for neutrophils. In view of the potential involvement of leukotrienes in airway diseases, a number of drugs have been identified that block their effects or synthesis in the hope that these agents will be useful for the treatment of asthma.

2.2
Drugs Acting on the 5-Lipoxygenase Pathway

The effects of the leukotrienes on the lungs can be inhibited in two ways (Fig. 1):

a. Leukotriene-receptor antagonists
   There are two subtypes of cysteinyl leukotriene (Cys-LT) receptors termed the Cys-LT_1 receptor and the Cys-LT_2 receptor. In human airway smooth

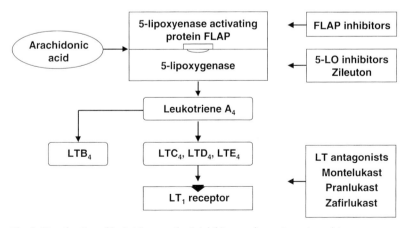

Fig. 1 Site of action of leukotriene synthesis inhibitors and receptor antagonists

muscle, leukotriene C_4, leukotriene D_4 and leukotriene E_4 all act on the Cys-LT_1 receptor. Montelukast, pranlukast and zafirlukast are orally active leukotriene-receptor antagonists that act at the Cys-LT_1 receptor (Fig. 1). The chemical structure of each of these drugs is different, although they have similar receptor-binding affinities. They are metabolised by cytochrome P450 enzymes in the liver and metabolites are excreted largely in bile. Pranlukast and zafirlukast are administered twice daily, whereas montelukast is administered once daily at bedtime. Montelukast and zafirlukast are licensed worldwide for the treatment of asthma. Panlukast is approved only in Japan. Cys-LT_2 receptors are present in pulmonary vascular tissue and are not thought to be involved in the pathogenesis of asthma. Leukotriene B_4 acts on a seven transmembrane-spanning receptor termed the B leukotriene-receptor (BLT) and mediates chemotaxis. Several LTB_4-receptor antagonists have been synthesised, but none are licensed for the treatment of asthma or COPD.

b. Leukotriene synthesis inhibitors

  Leukotriene synthesis can be blocked by inhibition of either 5-lipoxygenase or 5-lipoxygenase-activating protein (FLAP). The 5-lipoxygenase inhibitors block the enzyme directly, whereas the FLAP inhibitors prevent 5-lipoxygenase binding with FLAP on the nuclear membrane (Fig. 1). The orally active 5-lipoxygenase inhibitor zileuton is to date the only compound licensed for use in asthma and is approved only in the United States. In addition to inhibiting the production of the cysteinyl leukotrienes, the 5-lipoxygenase inhibitors also prevent the formation of leukotriene B_4 and other 5-lipoxygenase products. The clinical importance of the additional pharmacological effects of 5-lipoxygenase inhibitors compared to leukotriene-receptor antagonists is uncertain.

2.3
Effect of Drugs Modifying the 5-Lipoxygenase Pathway on Inflammation in Man

In a mouse model of allergen-induced chronic inflammation and fibrosis, leukotriene-receptor antagonist treatment reduced airway eosinophil infiltration and mucus plugging as well as changes of airway remodelling including smooth muscle hyperplasia and subepithelial fibrosis (Henderson et al. 2002). A limited number of studies have investigated the effects of drugs modifying the 5-lipoxygenase pathway on inflammatory responses in asthma. The leukotriene-receptor antagonist montelukast had no effect on allergen-induced increases in sputum eosinophil counts when administered for 36 h before allergen exposure' (Diamant et al. 1999), whereas more prolonged treatment for 4 weeks decreased sputum eosinophil counts induced by allergen challenge (Pizzichini et al. 1999). The leukotriene-receptor antagonist zafirlukast, when administered at a dosage of 20 mg twice daily for 1 week, reduced bronchoalveolar lavage lymphocyte and basophil counts, but not eosinophil influx, 48 h after segmental allergen challenge (Calhoun et al. 1998). A preliminary report of a study that used a similar protocol, but with zafirlukast administered at a higher dosage of 160 mg twice daily, found reduced eosinophil influx 48 h following segmental allergen challenge (Calhoun et al. 1997). Raised exhaled nitric oxide concentrations in children with asthma are reduced by montelukast (Bisgaard et al. 1999).

The 5-lipoxygenase inhibitor zileuton decreases nocturnal bronchoalveolar lavage eosinophil counts in patients with nocturnal asthma (Wenzel et al. 1995) and the influx of eosinophils following segmental allergen challenge (Kane et al. 1996), particularly in a subset of asthmatics in whom leukotriene levels are raised within the airways by allergen (Hasday et al. 2000). Furthermore, chronic treatment with leukotriene-receptor antagonists or zileuton reduces circulating blood eosinophil counts (Liu et al. 1996; Reiss et al. 1998; Pizzichini et al. 1999; Simons et al. 2001). A number of studies have found that zileuton suppresses urinary leukotriene E_4 excretion (Hui et al. 1991; Israel et al. 1993a,b).

Taken together these findings indicate that drugs modifying the 5-lipoxygenase pathway have some anti-inflammatory activity in asthma.

2.4
Bronchial Challenge Studies

Both leukotriene-receptor antagonists and 5-lipoxygenase inhibitors attenuate the bronchoconstrictor response to a number of trigger factors including allergen, exercise, cold air and aspirin (Chung 1995; McGill and Busse 1996; Drazen et al. 1999a). In addition, a single dose of zileuton attenuates bronchial reactivity to histamine and to ultrasonically nebulised distilled water (Dekhuijzen et al. 1997).

a. Allergen-induced asthma
   The leukotriene-receptor antagonist zafirlukast administered in a single oral dose 2 h before allergen challenge inhibited the early response by 80% and the late response by 50% (Taylor et al. 1991). Zafirlukast, 20 mg twice daily for 1 week, demonstrated a significant protective effect on symptoms of asthma and alterations in pulmonary function induced by natural cat exposure (Corren et al. 2001). Similar results to those obtained with zafirlukast on acute allergen challenge have been reported with montelukast (Diamant et al. 1999). A single 800-mg dose of the 5-lipoxygenase inhibitor zileuton had no effect on either the early or late response to allergen (Hui et al. 1991).
b. Exercise-induced asthma
   Montelukast and zafirlukast attenuate exercise-induced asthma in both children and adults (Finnerty et al. 1992; Kemp et al. 1998; Leff et al. 1998; Edelman et al. 2000). The mean maximal percentage fall in forced expiratory volume in 1 s (FEV_1) after exercise following a single dose of zafirlukast was 22% compared to 36% after placebo, although the degree of protection against exercise-induced asthma varied between individuals (Finnerty et al. 1992). Following 3 months treatment with montelukast, the maximal fall in

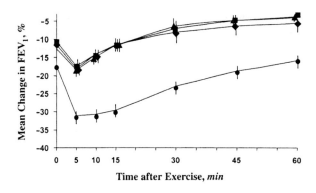

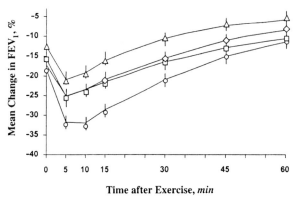

Fig. 2 Effect of montelukast (*top*) or salmeterol (*bottom*) on FEV_1 after exercise. The mean response curves are shown for percentage change in FEV_1 from pre-challenge FEV_1 at baseline (*circles*), days 1 and 3 (*triangles*), week 4 (*squares*), and week 8 (*diamonds*) after study treatment. (Reproduced from Edelman et al. 2000)

FEV_1 after exercise was 22% compared to 32% after placebo (Leff et al. 1998). The results remained consistent throughout the study and tolerance did not develop. In patients with mild asthma who were not receiving inhaled corticosteroid therapy, the inhaled long-acting β_2-agonist salmeterol, but not montelukast, showed attenuation of bronchoprotection after 4 and 8 weeks of treatment (Edelman et al. 2000) (Fig. 2). Acute and long-term treatment with zileuton has been shown to decrease the response to isocapnic hyperventilation (Fischer et al. 1995).

c. Aspirin-induced asthma

Aspirin-intolerant asthma is associated with elevated formation of the cysteinyl leukotrienes and this may be related to up-regulation of leukotriene C_4 synthase. Both 5-lipoxygenase inhibitors and leukotriene-receptor antagonists effectively inhibit acute aspirin-induced bronchoconstriction (Chung 1995). Pre-treatment with zileuton for 1 week not only prevented the fall in FEV_1 after aspirin challenge but also reduced urinary leukotriene E_4 levels at baseline and after aspirin challenge (Israel et al. 1993a). Pretreatment with zileuton also blocked nasal, gastrointestinal and dermal symptoms.

2.5
Effect on Baseline Lung Function

The finding that drugs that modify the 5-lipoxygenase pathway can cause mild bronchodilation suggests that leukotrienes released within the airways contribute to bronchoconstriction in asthma (Hui and Barnes 1991; Israel et al. 1993b; Reiss et al. 1997; Dockhorn et al. 2000). A single 40-mg oral dose of zafirlukast produced a small bronchodilator effect, increasing the mean FEV_1 value by 8% (Hui and Barnes 1991). In this study the increase in FEV_1 after nebulised salbutamol and zafirlukast was 26% compared to 18% after nebulised salbutamol and placebo. This finding suggests that the bronchodilator effect of β_2-agonists and leukotriene-receptor antagonists might be additive. Single oral doses of montelukast cause bronchodilation irrespective of the concurrent use of inhaled corticosteroids in asthmatic subjects with airflow limitation (Reiss et al. 1997). The finding that intravenous montelukast produces a rapid onset of bronchodilation raises the possibility that leukotriene-receptor antagonists might have a role as a treatment for acute asthma (Dockhorn et al. 2000). In a group of 139 asthmatic patients whose baseline FEV_1 values were approximately 60% of predicted, a single 600-mg dose of zileuton increased mean FEV_1 values by 14.6% at 1 h, which was significantly greater than the change with placebo (Israel et al. 1993b). Further improvements in FEV_1 values occurred during the following 4 weeks of chronic dosing with zileuton.

2.6
Therapeutic Effects in Chronic Asthma

2.6.1
Efficacy

a. Comparison with placebo
   The leukotriene-receptor antagonists, montelukast and zafirlukast have been shown to be effective anti-asthma agents in both children and adults (Spector et al. 1994; Suissa et al. 1997; Knorr et al. 1998; Noonan et al. 1998; Reiss et al. 1998; Barnes et al. 2001; Knorr et al. 2001). In a 6-week trial of zafirlukast, significant improvements in symptoms and lung function were seen in the highest dose group (40 mg daily) (Spector et al. 1994). Treatment with zafirlukast in a dosage of 40 mg daily for 3 months was also found to reduce the rate of exacerbations of asthma (Suissa et al. 1997). In a group of 681 adult patients with chronic stable asthma, 23% of whom were receiving inhaled corticosteroids, montelukast, compared with placebo, significantly improved asthma control (FEV_1, morning and evening PEF, asthma symptoms and exacerbation rates) during a 3-month treatment period

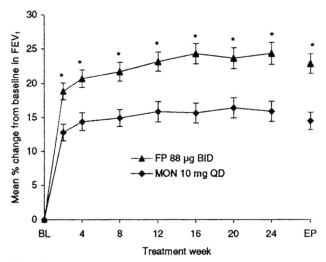

Fig. 3 The mean percentage change from baseline in FEV_1 in patients receiving inhaled fluticasone (*FP; triangles*) or oral montelukast (*MON; diamonds*) during a 24-week treatment period. * $p < 0.001$, FP vs. montelukast. *BL*, baseline; *EP*, endpoint. (Reproduced from Busse et al. 2001a)

(Reiss et al. 1998). Montelukast produces dose-related improvements in asthma control in patients with mild persistent asthma (Noonan et al. 1998; Barnes et al. 2001). Montelukast has also been shown to improve asthma control in 6- to 14-year-old children with chronic asthma (Knorr et al. 1998) as well as in younger asthmatic children aged 2 to 5 years (Knorr et al. 2001).

Three large multicentre trials using zileuton in mild to moderate chronic asthma have shown evidence of efficacy (Israel et al. 1993b; Israel et al. 1996; Liu et al. 1996). The duration of the trials were 4, 13 and 26 weeks, respectively, and each employed a double-blind, parallel group, placebo-controlled study design. Zileuton was shown to improve daily symptoms of asthma, night waking and peak expiratory flow measurements, and to reduce the use of rescue inhaled β_2-agonists and the number of exacerbations of asthma requiring corticosteroids. In general, the 5-lipoxygenase inhibitors and leukotriene-receptor antagonists appear to produce similar clinical effects, although to date there have been no comparative studies.

b. Comparison with inhaled corticosteroids

The clinical efficacy of montelukast and zafirlukast appears to be slightly less than that seen with low doses of inhaled corticosteroids (equivalent to 400 mcg daily of inhaled beclomethasone) (Laviolette et al. 1999; Malmström et al. 1999; Bleeker et al. 2000; Busse et al. 2001a,b; Ducharme and Hicks 2001) (Fig. 3). In a group of 895 adult patients with chronic stable asthma, montelukast 10 mg once daily compared with inhaled beclomethasone 200 µg twice daily, increased mean FEV_1 values by 7.5% compared to the larger increase of 13.3% after the inhaled corticosteroid following a 3-month treatment period (Malmstrom et al. 1999). Multi-centre randomised controlled trial of zafirlukast (20 mg twice daily) or montelukast (10 mg once daily) for 3 and 6 months, respectively, when compared to inhaled fluticasone (100 μg twice daily) found that the improvement in asthma control was greater in the inhaled corticosteroid groups (Bleeker et al. 2000; Busse et al. 2001a,b). A Cochrane systematic review of randomised controlled trials identified up until 1999 concluded that leukotriene-receptor antagonist treatment had a similar rate of exacerbations compared to inhaled corticosteroids. Inhaled corticosteroids, however, produced better lung function and quality of life, as well as a reduction in both symptoms and rescue β_2-agonist usage (Ducharme and Hicks 2001).

c. As an add-on therapy

The efficacy of leukotriene-receptor antagonists as add-on treatment to inhaled corticosteroids has been assessed using a number of different protocols.

1. *As an add-on to low dose inhaled corticosteroids:* In a randomised, placebo-controlled, parallel group trial lasting 16 weeks, of 642 adult asthmatic patients not adequately controlled on inhaled beclomethasone (400 mcg daily), the addition of montelukast 10 mg daily resulted in an improvement in mean FEV_1 of 5.43% compared to 1.04% for inhaled corticosteroids alone and also caused modest improvements in peak expiratory flow (PEF) and reductions in β_2-agonist use (Laviolette et al. 1999). Montelukast 5 mg daily when added to inhaled budesonide (400 mcg daily) in children with persistent asthma produced modest improvements in PEF and reductions in β_2-agonist use over a 4-week treatment period (Simons et al. 2001). The results of these published studies suggest a modest effect of licensed doses of montelukast in symptomatic children and adults. However, a Cochrane systematic review of published and unpublished randomised controlled trials identified up until 2001 of leukotriene-receptor antagonists as add-on therapy in chronic asthma, concluded that there was insufficient evidence to firmly support the use of leukotriene-receptor antagonists as add-on therapy to inhaled corticosteroids (Ducharme et al. 2002).

2. *To allow tapering of inhaled corticosteroids:* Several short-term studies have reported modest inhaled corticosteroid sparing effects with the addition of montelukast daily for 12 weeks (Löfdahl et al. 1999) or pranlukast administered daily for 6 weeks (Tamaoki et al. 1997). In general, the addition of leukotriene-receptor antagonists may be associated with superior asthma control after corticosteroid tapering, but a corticosteroid-sparing effect cannot be quantified at present (Ducharme et al. 2002).

3. *As an add-on to low-dose inhaled corticosteroids versus double dose of inhaled corticosteroid:* A Cochrane systematic review of randomised controlled trials identified up until 2001 concluded that the addition of a leukotriene-receptor antagonist cannot be recommended as a substitute for increasing the dose of inhaled corticosteroid (Ducharme et al. 2002).

4. *As an add-on to low-dose inhaled corticosteroids versus the addition of long-acting β_2-agonist to inhaled corticosteroids:* A randomised controlled trial of inhaled salmeterol (50 mcg twice daily) compared to zafirlukast (20 mg twice daily) found that the long-acting inhaled β_2-agonists was more effective in terms of improving pulmonary function and symptom control in a group of 301 patients aged above 12 years with persistent asthma (Busse et al. 1999). The majority of patients in this study were receiving inhaled corticosteroids. A 12-week study in 447 asthmatic patients comparing the combination of fluticasone 100 mcg plus salmeterol 50 mcg twice daily, with fluticasone 100 mcg twice daily plus montelukast 10 mg daily, found the former combination produced greater improvement in asthma control (Nelson et al. 2000). A study in a small group of 20 patients with persistent asthma found that the addition of montelukast or inhaled salmeterol over a 2-week treatment period produced similar improvements in the asthma control of patients not controlled with inhaled corticosteroids, but only montelukast therapy reduced blood eosinophil counts (Wilson et al. 2001).
5. *As an add-on to low-dose inhaled corticosteroids versus other add-on therapies:* The efficacy and adverse effect profile of anti-leukotrienes with other add-on therapies such as oral slow release β_2-agonists or oral slow-release xanthines has not been reported.
6. *As an add-on to high-dose inhaled corticosteroids plus other add-on therapies:* In a group of 72 patients with symptomatic chronic persistent asthma already taking high-dose inhaled corticosteroids and other add-on therapies such as long-acting inhaled β_2-agonists the addition of montelukast for 2 weeks did not improve symptoms or PEF recordings (Robinson et al. 2001). This result suggests that leukotriene-receptor antagonists might be less effective or ineffective in patients with moderate or severe asthma. However, zafirlukast administered for 6 weeks, at an unlicensed dose of 80 mg twice daily, was found to improve asthma control as assessed by reduced exacerbations, improvements in PEF readings and reduction in both β_2-agonist usage and symptoms in patients with severe persistent asthma receiving high-dose inhaled corticosteroids (>1,200 mcg daily) (Virchow et al. 2000). It is possible that patients with chronic severe asthma may require higher than currently licensed doses of leukotriene-receptor antagonists.
d. Influence of asthmatic phenotype on efficacy
   In aspirin–sensitive asthma, chronic dosing with montelukast for 4 weeks (Dahlen et al. 2002) or with zileuton for 6 weeks (Dahlen et al. 1998) has been shown to improve asthma control over that achieved with medium to high doses of inhaled corticosteroids. These findings are of interest in view of the observation that corticosteroids do not inhibit leukotriene formation in vivo (O'Shaughnessy et al. 1993) and aspirin-intolerant asthma is associated with increased formation of the cysteinyl leukotrienes. However, the therapeutic response to montelukast did not correlate with baseline urinary leukotriene E_4 concentrations (Dahlen et al. 2002). Interestingly, naturally occurring mutations in the 5-lipoxygenase gene may influence the clinical

response to drugs modifying the 5-lipoxygenase pathway such as zileuton (Drazen et al. 1999b).

2.6.2
Adverse Effects

In clinical trials, the leukotriene-receptor antagonists zafirlukast and montelukast and the 5-lipoxygenase inhibitor zileuton have been well tolerated. The main side effects reported with the use of zafirlukast include headache and gastrointestinal disturbance. Zafirlukast can interact with other drugs including warfarin (increases the prothrombin time by approximately 35%), aspirin (increase in plasma levels of zafirlukast by approximately 45%), theophylline (decrease in plasma levels of zafirlukast by approximately 30%), and erythromycin (decrease in plasma levels of zafirlukast by approximately 40%). Cigarette smokers exhibit increased clearance of zafirlukast. In clinical trials with montelukast, abdominal pain and headache were reported in more than 1% of patients, although the incidence was only slightly higher from those receiving placebo. Zileuton has been associated with rises in liver enzymes that return to normal on stopping the drug (Israel et al. 1993b, 1996). The incidence of zileuton-induced hepatitis is approximately 3%. In the USA, the Food and Drug Administration (FDA) recommends that liver function tests be monitored every 2 weeks during the first year of treatment.

The administration of the leukotriene-receptor antagonist zafirlukast has been associated with the emergence of the Churg-Strauss syndrome (Stirling and Chung 1999). Six patients developed the Churg-Strauss syndrome while taking zafirlukast (Josefson 1997) and further cases have been reported both with the use of zafirlukast (Wechsier et al. 1998) and montelukast (Tuggery and Hosker 2000). In each case, the patient was reducing their dose of oral corticosteroids for asthma. There is no proof of a causal relationship between the Churg-Strauss syndrome and these drugs. It is likely that the reduction in oral corticosteroid dose by these patients resulted in unmasking the syndrome. Nevertheless, it is advisable that patients with chronic oral corticosteroid-dependent asthma who are receiving a leukotriene-receptor antagonist are monitored carefully when oral corticosteroid reduction is being considered.

2.7
Conclusions

In asthma, the leukotriene-receptor antagonists and the 5-lipoxygenase inhibitors have been shown to have a mild bronchodilator effect in patients with airflow obstruction and to attenuate bronchoconstriction induced by exercise, allergen and aspirin. There is also some evidence to indicate that they have anti-inflammatory actions. Several therapeutic studies in mild to moderate asthma have shown evidence of efficacy. A theoretical advantage of the oral formulation of these drugs, and for some agents the infrequent dose scheduling, is an improved concordance with therapy. Systemic administration of drugs acting on the 5-lipoxygenase pathway may also improve co-existing diseases such as allergic rhinitis. The effect of drugs that modify the 5-lipoxygenase pathway in COPD has not been reported.

The clinical effectiveness of the 5-lipoxygenase inhibitors and the leukotriene-receptor antagonists appears to be quite variable between individuals. To date, there are few known predictors of response to therapy, although naturally occurring mutations in the 5-lipoxygenase gene may confer a poor clinical response to zileuton. The involvement of leukotrienes in aspirin-intolerant asthma and the beneficial effect of these agents in clinical trials would suggest that asthmatic patients intolerant of aspirin would respond particularly well to treatment with anti-leukotrienes. However to date, the effect of 5-lypoxygenase-modifying drugs have not been compared between aspirin-intolerant and aspirin-tolerant patients with asthma of similar severity and baseline treatment.

The place of the leukotriene-receptor antagonists and 5-lipoxygenase inhibitors in asthma management has not been fully established. Current published evidence suggests a role as alternative, but less effective first-line prophylactic agents to inhaled corticosteroids and as an alternative second-line add-on drug to long-acting β_2-agonists for patients with persistent symptoms despite low-dose inhaled corticosteroid therapy. The role of the leukotriene-receptor antagonists and the 5-lipoxygenase inhibitors in moderate to severe asthma is less well established.

3
Histamine Receptor Antagonists

3.1
Histamine and Asthma

Histamine is an inflammatory mediator that plays a role in allergic reactions; particularly allergen-induced early phase reactions (Chung and Barnes 1998). The majority of histamine is stored preformed in cytoplasmic granules of mast cells and basophils. Many stimuli, including antigen, cause mast cell degranulation and histamine release. Histamine has several properties that might contribute to the pathogenesis of asthma including bronchial smooth muscle contraction, increased vascular permeability leading to mucosal oedema, and mucus hypersecretion. In view of the potential involvement of histamine in airway disease, a number of histamine-receptor antagonists have been assessed for the treatment of asthma.

3.2
Histamine Receptor Antagonists

Four subtypes of histamine receptor (H_1, H_2, H_3, H_4) have been identified (Alexander et al. 2001). Clinical studies evaluating the efficacy of histamine-receptor antagonists in asthma have concentrated predominately on drugs acting at the H_1 receptor, which mediates bronchoconstriction and microvascular leakage. Non-sedating antihistamines, such as astemizole, azelastine, cetirizine, loratadine and terfenadine act as histamine H_1-receptor antagonists. A number of these compounds possess additional pharmacological properties (Chung and Barnes 1998; Holgate 1999; Leurs et al. 2002). For example, cetirizine has been reported to inhibit eosinophil chemotaxis in vitro and adhesion molecule expression by epithelial cells in vivo (Walsh 1997) and loratadine inhibits leukotriene release (Temple et al. 1988). H_1-receptor antagonists have also been shown to down-regulate nuclear factor (NF)-κB expression (Leurs et al. 2002). Ketotifen inhibits in vitro mediator release from mast cells and possibly from other inflammatory cells and reverses β-adrenergic tachyphylaxis. The exact mode of action of ketotifen in vivo is uncertain, although the main pharmacological effects are likely to be due to H_1-receptor antagonism (Craps 1985; Grant et al. 1990).

3.3
Effect of Histamine Receptor Antagonists on Inflammation in Asthma

The importance of the possible anti-inflammatory actions of antihistamines demonstrated in vitro is unclear. There have been very few studies that have examined the effects of H_1-receptor antagonists on inflammation in asthma using either invasive or non-invasive assessments of airway pathology. In one small study, pre-treatment with cetirizine for 8 days reduced bronchoalveolar lavage eosinophil numbers in late asthmatic responders, 24 h after allergen challenge (Redier et al. 1992).

3.4
Bronchial Challenge Studies

Anti-histamines have been shown to attenuate early and late responses to allergen, although the combination of H_1-receptor antagonists, loratadine and the leukotriene-receptor antagonist zafirlukast was found to be more effective than either drug alone (Roquet et al. 1997). H_1-receptor antagonists also inhibit exercise-induced asthma (Patel 1991) and bronchoconstriction mediated by cold-air challenge (O'Byrne et al. 1983). Pre-treatment with oral terfenadine attenuates adenosine-induced bronchoconstriction in smokers with COPD (Rutgers et al. 1999).

3.5
Effect on Baseline Lung Function

H_1-receptor antagonists administered both orally or by inhalation cause modest bronchodilation in patients with asthma (Thomson and Kerr 1980; Chung and Barnes 1998).

3.6
Therapeutic Effects in Chronic Asthma

3.6.1
Efficacy

In a small number of clinical trials in chronic or seasonal asthma, cetirizine, loratadine and terfenadine have been shown to cause slight reductions in symptoms and/or in the use of rescue inhaled β_2-agonists (Simons 1999). Azelastine was shown to slightly reduce the need for inhaled corticosteroids (Busse et al. 1996). Ketotifen has been found to be of little value in the treatment of asthma in adults (Grant et al. 1990), although improvements in both symptoms and inhaled β_2-agonist usage were reported in children with asthma (Rackham et al. 1989). A meta-analysis of 19 randomised controlled trials found that antihistamines had little effect on lung function or on the use of inhaled β_2-agonists and that sedation occurred more often than with placebo (Van Ganse et al. 1997) (Fig. 4). The systematic review concluded that antihistamines were not of value in the treatment of asthma.

In a 2-week study of 117 adult patients with chronic asthma, the combination of the leukotriene-receptor antagonist montelukast (10 mg daily—licensed dose) and the H_1-receptor antagonist loratadine (20 mg daily—higher than licensed dose) compared to montelukast alone produced slight improvements in lung function, asthma symptoms and inhaled β_2-agonist use (Reicin et al. 2000).

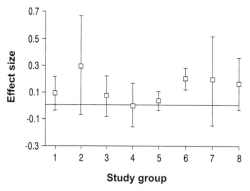

Fig. 4 Meta-analysis of the effectiveness of H_1-receptor antagonists in adults with asthma. Sensitivity analyses on mean effect size (morning PEF). *Bars* indicate the 95% confidence interval. Study groups are as: (*1*) mild asthma (6 trials); (*2*) mild-moderate asthma (4 trials); (*3*) moderate-severe asthma (5 trials); (*4*) quality scoring >70%; (3 trials); (*5*) all ketotifen studies (5 trials); (*6*) all trials with "new" antihistamines (10 trials); (*7*) all studies that excluded oral/inhaled corticosteroids (8 trials); and (*8*) all studies except corticosteroid-sparing trials (14 trials). (Reproduced from Van Ganse et al. 1997)

3.6.2
Adverse Effects

On starting therapy, drowsiness occurs in 15%–20% of patients, but this effect usually disappears. Arrhythmias have been associated with astemizole and terfenadine use as well as rare reports of sudden death. These drugs are known to act on cardiac potassium channels and to prolong the QT interval. Astemizole is no longer licensed because of the severity of cardiac adverse effects.

3.7
Conclusions

In asthma, the H_1-receptor antagonists have been shown to have a mild bronchodilator effect and to attenuate bronchoconstriction induced by exercise and allergen. There is minimal evidence to indicate that they have anti-inflammatory actions in asthma. Therapeutic studies in mild to moderate asthma have shown little evidence of efficacy. Current evidence would suggest that the antihistamines do not have a place in the treatment of chronic persistent asthma. The role of combination therapy with H_1-receptor antagonist and leukotriene-receptor antagonist in chronic asthma therapy remains to be established. There is no evidence to support a role for H_1-receptor antagonists in the treatment of COPD.

4
Cromones

The cromones, sodium cromoglycate and nedocromil sodium, are often referred to as anti-allergy or non-steroidal anti-asthma drugs. These drugs have different chemical structures, but their mode of action and therapeutic effects are very similar (Gonzalez and Brogden 1987; Thomson 1989; Norris and Alton 1996).

4.1
Mode of Action

The mode of action of the cromones has not been clearly established. Both drugs act as non-specific chloride channel blockers in a large range of cell types and through this action these compounds may reduce alterations in cell volume and function. (Norris and Alton 1996). Sodium cromoglycate and nedocromil sodium may also have inhibitory effects on sensory nerve endings in the lung, thus preventing the release of tachykinins.

4.2
Effect of Cromones on Inflammation

a. In vitro and in vivo animal studies
   In vitro studies of human inflammatory cells and in vivo studies in experimental animals have shown that both drugs inhibit functions of a variety of inflammatory cells including mast cells, eosinophils, neutrophils, platelets and alveolar macrophages (see Thomson 1989; Norris and Alton 1996; Corin 2000). Nedocromil sodium has either similar or slightly greater potency to that of sodium cromoglycate. Sodium cromoglycate inhibits degranulation of mast cells triggered by immunological and non-immunological stimuli. Pre-treatment of eosinophils and neutrophils with sodium cromoglycate inhibits certain cell responses such as complement (C3b) and IgG (Fc) receptor expression and cytotoxicity against schistosomula (Kay et al. 1987). Sodium cromoglycate also has inhibitory effects in vitro on alveolar macrophage function. Nedocromil sodium can inhibit anti-human IgE histamine release from mast cells obtained by bronchoalveolar lavage, and it has been shown to be more active against these mast cells than against parenchymal-derived mast cells. The platelets of patients with aspirin-induced asthma demonstrate increased cytotoxicity against schistosomula in the presence of aspirin and other non-steroidal anti-inflammatory drugs. This response is inhibited by nedocromil sodium, whereas sodium cromoglycate demonstrates markedly reduced potency in this system. Taken together, these data would suggest that these drugs possess properties that might reduce components of airway inflammation in asthma.
b. Clinical evidence for anti-inflammatory actions
   There is limited and conflicting evidence for anti-inflammatory effects of sodium cromoglycate and nedocromil sodium in asthma. Chronic treatment with sodium cromoglycate has been shown to significantly reduce the percentage of eosinophils in bronchial alveolar lavage specimens following 4 weeks of treatment (Diaz et al. 1984). An open study of 12 weeks treatment with inhaled sodium cromoglycate reported a reduction in the numbers of eosinophils, mast cells and T lymphocytes and in the expression of adhesion molecules in bronchial biopsy specimens from 9 patients with atopic asthma (Hoshino and Nakamura 1997). A randomised controlled trial of nedocromil sodium for a period of 16 weeks in mild to moderate asthma, however, found no significant change in bronchial biopsy eosinophil counts compared to the placebo group (Manolitsas et al. 1995).

4.3
Bronchial Challenge Studies

In both children and adults pre-treatment with sodium cromoglycate has been shown to reduce bronchoconstriction following allergen challenge and after non-allergic stimuli such as exercise (see Holgate 1996a; Thomson et al. 1978).

Sodium cromoglycate can inhibit allergen induced early and late asthmatic responses and the increase in histamine reactivity that is associated with late asthmatic responses. The seasonal increase in non-allergic bronchial reactivity can be prevented by sodium cromoglycate treatment. The results of studies examining the effects of sodium cromoglycate therapy on non-seasonal bronchial hyperreactivity have produced conflicting results; any reduction in bronchial reactivity is likely to be very small.

Nedocromil sodium also attenuates the bronchoconstrictor response induced by exercise (Roberts and Thomson 1985; Spooner et al. 2002), cold air and the early and late response to allergen (see Thomson 1989). The increase in histamine reactivity that occurs during the pollen season in pollen-sensitive patients can be attenuated by nedocromil sodium (Dorward et al. 1986). Nedocromil sodium can also produce small reductions in non-seasonal bronchial reactivity (Svendsen et al. 1989; Bel et al. 1990). Sodium cromoglycate and nedocromil sodium are equally effective in attenuating exercise-induced asthma (Kelly et al. 2000).

4.4
Effect on Baseline Lung Function

The cromones do not have an acute bronchodilator action (Thomson et al. 1981)

4.5
Therapeutic Effects in Chronic Asthma

4.5.1
Efficacy

a. Comparison with placebo
   In both paediatric and adult asthma, treatment with inhaled sodium cromoglycate has been shown to improve asthma control (Brompton Hospital/Medical Research Council Collaborative Trial 1972; Eigen et al. 1987; see Carlsen and Larsson 1996). Both atopic and non-atopic asthmatic patients have been shown to respond to sodium cromoglycate treatment. A recent systematic review of 24 randomised controlled trials, however, questioned the efficacy of sodium cromoglycate in children with asthma and concluded that it was no longer justified to recommend sodium cromoglycate as a first-line prophylactic agent in chronic childhood asthma (Tasche et al. 1997, 2000) (Fig. 5). Studies comparing nedocromil sodium with placebo have demonstrated efficacy in both children and adult asthmatic patients (Edwards and Stevens 1993; see Holgate 1996b).
b. Comparison with inhaled steroids
   The results of most short- to medium-term studies suggest that the improvement in asthma control produced by the cromones is less than that

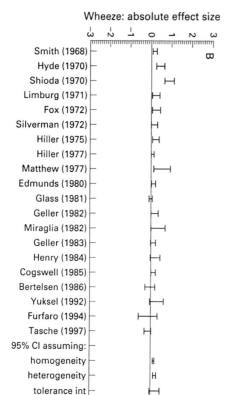

Fig. 5 Sodium cromoglycate in children: 95% confidence intervals of absolute difference for wheeze in sodium cromoglycate group compared with placebo. (Reproduced from Tasche et al. 2000)

produced by low doses of inhaled corticosteroids (equivalent to 400 mcg daily of inhaled beclomethasone) (Svendsen et al. 1989; Bel et al. 1990; Faurschou et al. 1994; Childhood Asthma Management Program Research Group 2000). The differences in results between studies may be explained in part by the dose of sodium cromoglycate or nedocromil sodium administered, e.g. nedocromil sodium 8 mg or 16 mg daily and the severity of asthma. A large long-term randomised controlled trial in 1,041 children aged 5 to 12 years with mild to moderate asthma compared the anti-asthma effects of 200 mcg of budesonide, 8 mg of nedocromil, or placebo twice daily over a 4- to 6-year period (Childhood Asthma Management Program Research Group 2000). Inhaled budesonide provided better control of asthma and improved airway responsiveness when compared with placebo or nedocromil. Neither drug was better than placebo in terms of lung function. Several studies have reported that nedocromil sodium can partially replace inhaled corticosteroids (Lal et al. 1984; Fyans et al. 1986), but not when the corticosteroid is withdrawn over a period of a few weeks (Ruffin et al. 1987). The addition of inhaled nedocromil sodium to asthmatic patients with symptoms poorly controlled by high-dose inhaled corticosteroids

(>1,000 mcg daily) has been reported to produce only minor improvements in asthma control (Svendsen and Jorgensen 1991).
c. Oral corticosteroid sparing effect
   Nedocromil sodium does not appear to have a clinical relevant oral corticosteroid sparing effect. In a study reported by Goldin and Bateman (1988), the total corticosteroid reduction in the nedocromil sodium-treated group (2.5 mg) was not significantly different from that in the placebo-treated group (3 mg). Boulet et al. (1990) reported a minor oral steroid sparing action of nedocromil sodium.

Comparison of Sodium Cromoglycate and Nedocromil Sodium. A small number of clinical trials in adult asthma have compared the two cromones and in most studies the therapeutic effects were found to be comparable (Boldy and Ayres 1993; Schwartz et al. 1996).

COPD. Both sodium cromoglycate and nedocromil sodium are considered to be ineffective in the treatment of COPD, although there are very few published studies (ERS Consensus Statement 1995). A randomised controlled study in patients with COPD reported that nedocromil sodium treatment for 10 weeks had no effect on symptoms, lung function or airway responsiveness to histamine and adenosine. However, the number of dropouts because of exacerbations was fewer with nedocromil sodium compared to placebo (de Jong et al. 1994).

4.5.2
Adverse Effects

Sodium cromoglycate is a safe drug and uncommonly causes side effects. Very occasionally it can cause acute bronchoconstriction in a sensitive individual and some patients may notice slight cough or irritation of the throat following inhalation. Occasional complaints of nausea and headaches have been reported. Nedocromil sodium has not been associated with any severe side effects. The main side effects reported include a distinctive bitter taste, headache and nausea.

4.6
Conclusions

The cromones have a limited role in the management of chronic asthma. The National Institutes of Health expert panel report on the Guidelines for the Diagnosis and Management of Asthma (1997) suggest that sodium cromoglycate or nedocromil sodium may be considered as initial prophylactic therapy for children and as preventative treatment prior to exercise or unavoidable exposure to known allergens. The report emphasises that safety is the primary advantage of these drugs. The British Guidelines on Asthma Management (2003) recommend the cromones as alternative, but less effective anti-inflammatory agents, compared to low-dose inhaled corticosteroids. Most studies suggest that the improvements in asthma control resulting from sodium cromoglycate or nedocromil sodium treatment are less than those produced by low doses of inhaled corticosteroids (equivalent to 400 mcg daily of inhaled beclomethasone). The need for 3- or 4-times daily dosing is a major disadvantage when compared to twice- or even once-daily dosing with inhaled corticosteroids. Taken together, the data would indicate that the cromones are not of clinical benefit as additional therapy for patients already receiving inhaled or systemic corticosteroids.

References

Alexander SPH, Mathie A, Peters JA. (2001) Nomenclature Supplement 12th edition TiPS Elsevier Trend Journals, pp 60.
Barnes N, Wei LX, Reiss TF, Leff JA, Shino S, Yu C, Edelman JM. (2001) Analysis of montelukast in mild persistent asthmatic patients with near-normal lung function. Respir Med 95:379–386.
Bel EH, Timmers MC, Hermans J et al (1990) The long-term effects of nedocromil sodium and beclomethasone dipropionate on bronchial responsiveness to methacholine in nonatopic asthmatic subjects. Am Rev Respir Dis 141:21–28
Bisgaard H, Loland L, Anhøj J (1999) NO in exhaled air of asthmatic children is reduced by the leukotriene-receptor antagonist montelukast. Am J Respir Crit Care Med 160:1227–1231
Bleecker ER, Welch MJ, Weinstein SF et al (2000) Low-dose inhaled fluticasone propionate versus oral zafirlukast in the treatment of persistent asthma. J Allergy Clin Immunol 105:1123–1129
Boldy DAR, Ayres JG (1993) Nedocromil sodium and sodium cromoglycate in patients aged over 50 years with asthma. Respir Med 87:517–523
Boulet L-P, Cartier A, Cockcroft DW et al (1990) Tolerance to reduction of oral steroid dosage in severely asthmatic patients receiving nedocromil sodium. Respir Med 84:317–323
British Guideline on the Management of Asthma (2003) Thorax 58 (suppl 1):i1–i94
Brompton Hospital/Medical Research Council Collaborative Trial (1972) Long term study of disodium cromoglycate in the treatment of severe extrinsic or intrinsic bronchial asthma in adults. Br Med J 4:383–388
Busse WW, Middleton E, Storms W et al (1996) Corticosteroid-sparing effect of azelastine in the management of bronchial asthma. Am J Respir Crit Care Med 153:122–7
Busse W, Nelson H, Wolfe J, Kalberg C, Yancey SW, Rickard KA. (1999) Comparison of inhaled salmeterol and oral zafirlukast in patients with asthma. J Allergy Clin Immunol 103:1075–1080
Busse W, Wolfe J, Storms W, Srebro S et al (2001) (a) Fluticasone propionate compared with zafirlukast in controlling persistent asthma: a randomised double-blind, placebo controlled trial. J Family Practice 50:595–602
Busse W, Raphael GD, Galant S, et al (2001) (b) Low-dose fluticasone propionate compared with montelukast for first-line treatment of persistent asthma: a randomised clinical trial. J Allergy Clin Immunol 107:461–468.
Calhoun WJ, Williams KL, Simonson SG et al (1997) Effect of zafirlukast (Accolate) on airway inflammation after segmental allergen challenge in patients with mild asthma. Am J Respir Crit Care Med 155:A662

Calhoun WJ, Lavins BJ, Minkwitz MC et al (1998) Effect of zafirlukast (Accolate) on cellular mediators of inflammation. Bronchoalveolar lavage fluid findings after segmental allergen challenge. Am J Respir Crit Care Med 157:1381–1389.
Carlsen K-H, Larsson K (1996) The efficacy of inhaled disodium cromoglycate and glucocorticoids. Clin Exper Allergy 26 (suppl 4): 8–17.
Childhood Asthma Management Program Research Group. (2000) Long-term effects of budesonide or nedocromil in children with asthma. New Eng J Med 343:1054–1063.
Chung KF (1995) Leukotriene-receptor antagonists and biosynthesis inhibitors: potential breakthrough in asthma therapy. Eur Respir J 6:1203–1213
Chung KF, Barnes PB (1998) In Asthma Basic Mechanisms and clinical management 3rd edition Eds Barnes PB, Rodger IW, Thomson NC. Academic Press London pp 343–368.
Corin RE. (2000) Nedocromil sodium: a review of the evidence for a dual mechanism of action. Clin Exp Allergy 30:461–468.
Corren J, Spector S, Fuller L, Minkwitz M, Mezzanotte W (2001) Effects of zafirlukast upon clinical, physiologic, and inflammatory responses to natural cat allergen exposure. Ann of Allergy, Asthma, and Immunology 87(3):211-7
Craps LP (1985) Immunologic and therapeutic aspects of ketotifen. J Allergy Clin Immunol 76:389–393
Dahlen B, Nizankowska E, Szczeklik A et al (1998) Benefits from adding the 5-lipoxygenase inhibitor zileuton to conventional therapy in aspirin–intolerant asthmatics. Am J Respir Crit Care Med 157:1187–1194
Dahlen S-E, Malmström K, Nizankowska E et al (2002) Improvement of aspirin-intolerant asthma by montelukast, a leukotriene antagonist. Am J Respir Crit Care Med 165:9–14
De Jong JW, Postma DS, van der Mark TW, Koeter GH. (1994) Effects of nedocromil sodium in the treatment of non-allergic subjects with chronic obstructive pulmonary disease. Thorax 49:1022–1224
Dekhuijzen PNR, Bootsma GP, Wieldres PLNL et al (1997) Effects of a single-dose zileuton on bronchial hyperresponsiveness in asthmatic patients treated with inhaled corticosteroids. Eur Respir J 10:2749–2753
Diaz P, Galleguillos FR, Gonazelez MC et al (1984) Bronchoalveolar lavage in asthma: The effect of sodium cromoglycate (Cromolyn) on leucocyte counts. J Allergy Clinic Immunol 74:41–48
Diamant Z, Grootendorst DC, Veselic-Charvat M, Timmers MC, De Smet M, Leff JA, Seidenberg BC, Zwinderman AH, Peszek I, Sterk PJ (1999) The effect of montelukast (MK-0476), a cysteinyl leukotriene-receptor antagonist, on allergen-induced airways responses and sputum cell counts in asthma. Clin Exper Allergy 29:42–51
Dorward, AJ, Roberts JA, Thomson NC. Effect of nedocromil sodium on histamine airway responsiveness in grass-pollen sensitive asthmatics during the pollen season. Clin Allergy 16:309–315
Drazen JM, Israel E, O'Byrne PM (1999) (a) Treatment of asthma with drugs modifying the leukotriene pathway. N Eng J Med 340:197–206
Drazen JM, Yandava, CN, Dube L, et al (1999) (b) Pharmacogenetic association between ALOX5 promoter genotype and the response to ant-asthma treatment. Nat Genet 22:168–170
Ducharme FM, Hicks GC (2001) Anti-leukotriene agents compared to inhaled corticosteroids in the management of recurrent and/or chronic asthma. (Cochrane Review) In: The Cochrane Library, Issue 4, Oxford: Update Software
Ducharme F, Hicks G, Kakuma R (2002) Addition of anti-leukotriene agents compared to inhaled corticosteroids for chronic asthma. (Cochrane Review) In: The Cochrane Library, Issue 1, Oxford: Update Software.
Edelman JM, Turpin JA, Bronsky EA, Grossman J, Kemp JP, Ghannam AF, DeLucca PT, Gormley GJ, Pearlman DS (2000) Oral montelukast compared with inhaled salmeterol

in preventing exercise-induced bronchoconstriction. A randomised, double-blind trial. Ann Int Med 132:97–104

Edwards AM, Stevens MT (1993) The clinical efficacy of inhaled nedocromil sodium (Tilade) in the treatment of asthma. Eur Respir J 6:35–41

Eigen H, Reid JJ, Dahl R et al (1987) Evaluation of the addition of cromolyn sodium to bronchodilator maintenance therapy in the long-term management of asthma. J Allergy Clin Immunol 80:612–21

ERS Consensus Statement. Optimal assessment and management of chronic obstructive pulmonary disease (COPD). (1995) Eur Respir J 8:1398–1420.

Faurschou P, Bing J, Edman G, et al (1994) Comparison between sodium cromoglycate MDI: metered dose inhaler and beclomethasone dipropionate MDI in treatment of adult patients with mild to moderate bronchial asthma. A double-blind, double dummy randomized, parallel-group study. Allergy 49:656–660

Finnerty JP, Wood-Baker R, Thomson H et al (1992) Role of leukotrienes in exercise-induced asthma. Inhibitory effect of ICI-204-219, a potent leukotriene D4 antagonist. Am Rev Respir Dis 145:746–749

Finnerty JP, Lee C, Wilson S et al (1996) Effects of theophylline on inflammatory cells and cytokines in asthmatic subjects: a placebo-controlled parallel group study. Eur Respir J 9:1672–1677

Fischer AR, McFadden, CA, Frantz R et al (1995) Effect of chronic 5-lipoxygenase inhibition on airway hyperresponsiveness in asthmatic subjects. Am J Respir Crit Care Med 152:1203–1207

Fyans PG, Chatterjee PC, Chatterjee SS (1986) A trial comparing nedocromil sodium (TILADE) and placebo in the management of bronchial asthma. Clin Allergy 16:505–511

Golden JG, Bateman ED (1988) Does nedocromil sodium have a steroid sparing effect in adult asthmatic patients requiring maintenance oral corticosteroids? Thorax 43:982–986

Gonzalez JP, Brogden RN (1987) Nedocromil sodium. A preliminary review of its pharmacolodynamic and pharmacokinetic properties and therapeutic efficacy in the treatment of reversible obstructive airways disease. Drugs 34:560–577

Grant SM, Goa KL, Fitton A et al (1990) Ketotifen. A review of its pharmacodynamic and pharmacokinetic properties, and therapeutic use in asthma and allergic disorders. Drugs 40:412–48

Hasday JD, Meltzer SS, Moore WC, Wisniewski P, Hebel JR, Lanni C, Dube LM, Bleeker ER. (2000) Anti-inflammatory effects of zileuton in a subpopulation of allergic asthmatics. Am J Respir Care Med 161:1229–1236

Hay DWP, Torphy TJ, Undem BJ (1995) Cysteinyl leukotrienes in asthma: old mediators up to new tricks. TiPS 16:304–309

Henderson WR, Tang L-O, Chu S-J, Tsao S-M, Chiang GKS, Jones F, Jonas M, Pae C, Wang H, Chi EY (2002) A role for cysteinyl leukotrienes in airway remodelling in a mouse asthma model. Am J Respir Crit Care Med 165:108–116

Holgate ST (1996) (a) Inhaled sodium cromoglycate. Respir Med, 90:387–390

Holgate ST (1996) (b) The efficacy and therapeutic position of nedocromil sodium Respir Med 90:391–394

Holgate ST (1999) Antihistamines: back to the future Clin Exper Allergy 29 (suppl 3):1–250

Hoshino M, Nakamura Y. (1997) The effects of inhaled sodium cromoglycate on cellular infiltration into the bronchial mucosa and the expression of adhesion molecules in asthmatics. Eur Respir J 10:858–865

Hui KP, Barnes NC (1991) Lung function improvement in asthma with a cysteinyl-leukotriene-receptor antagonist. Lancet 337:1062–63

Israel E, Cohn J, Dube L, Drazen JM (1996) Effect of treatment with zileuton, a 5-lipoxygenase inhibitor, in patients with asthma. A randomised controlled trial. Zileuton Clinical Trial Group. JAMA 275:931–36

Israel E, Fischer AR, Rosenberg MA et al (1993) (a). The pivotal role of 5-lipoxygenase products in the reaction of aspirin-intolerant asthmatics to aspirin. Am Rev Respir Dis 148:1447–51

Israel E, Rubin P, Kemp J et al (1993) (b). The effect of inhibition of 5-lipoxygenases by zileuton in mild to moderate asthma. Ann Intern Med 119:1059–66

Josefson D (1997) Asthma drug linked with Churg-Strauss syndrome. Br Med J 315:330.

Kane GC, Dube LM, Lancaster J et al (1996) A controlled trial of the effects of the 5-lipoxygenase inhibitor, zileuton, on lung inflammation produced by segmental antigen challenge in human beings. J Allergy Clin Immunol 97:646–54

Kay AB, Walsh GM, Davis S et al (1987) Disodium cromoglycate inhibits activation of human inflammatory cells in vitro. J Allergy Clin Immunol 80:1–8

Kelly KD, Spooner CH, Rowe BH. (2000) Nedocromil sodium versus cromoglycate for the pre-treatment of exercise induced bronchoconstriction in asthma. Cochrane Database of systematic reviews: CD002169.

Kemp JP, Dockhorn RJ, Shapiro GG, et al (1998) Montelukast once daily inhibits exercise-induced bronchoconstriction in 6- to 14-year-old children with asthma. J Pediatr 133:424–428

Kidney J, Dominguez M, Taylor PM et al (1995) Immunomodulation by theophylline in asthma-demonstration by withdrawal of therapy. Am J Respir Crit Care Med 151:1907–1914

Knorr B, Matz J, Bernstein JA et al (1998) Montelukast for chronic asthma in 6- to 14-yearold children. JAMA 279:1181–1186

Knorr B, Franchi LM, Bisgaard H, Vermeulin JH, Le Souef P, Santanello N, Michele TM, Reiss TF, Nguyen HH, Bratton DL (2001) Montelukast, a leukotriene-receptor antagonist, for the treatment of persistent asthma in children aged 2 to 5 years. Pediatrics 103:E48

Kraft M, Torvik JA, Trudeau JB et al (1996) Theophylline: potential anti-inflammatory effects in nocturnal asthma. J Allergy Clin Immunol 97:1242–1246

Laviolette M, Malmström K, Lu S, Chervinsky P, Pujet J-C, Peszek I, Zhang J, Reiss TF (1999) Montelukast added to inhaled beclomethasone in treatment of asthma. Am J Respir Crit Care Med 160:1862–1868

Lal S, Malhotra S, Gribben D, Hodder D (1984) Nedocromil sodium: a new drug for the management of bronchial asthma. Thorax 39:809–812

Leff JA, Busse WW, Pearlman D et al (1998). Montelukast, a leukotriene-receptor antagonist, for the treatment of mild asthma and exercise-induced bronchoconstriction. N Engl J Med 339:147–152

Leurs R, Church MK, Taglialatela M. (2002). H_1-antihistamines: inverse agonism, anti-inflammatory actions and cardiac effects. Clin Exp Allergy 32:489–498

Liu MC, Dube LM, Lancaster J (1996) Acute and chronic effects of a 5-lipoygenase inhibitor in asthma: a 6-month randomized trial. Zileuton Study Group. J Allergy Clin Immunol 98:859–71

Löfdahl C-G, Reiss TF, Leff JA et al (1999) Randomised, placebo controlled trial of effect of a leukotriene-receptor antagonist, montelukast, on tapering inhaled corticosteroids in asthmatic patients. Br Med J 319:87–90

Manolitsas ND, Wang JH, Devalia JL (1995) Regular albuterol, nedocromil sodium and bronchial inflammation in asthma. Am J Respir Crit Care Med 151:1925–1930

McGill KA, Busse WW (1996) Zileuton. Lancet 348:519–24

Malmström K, Rodriquez-Gomez G, Guerra J et al (1999) Oral montelukast, inhaled beclomethasone, and placebo for chronic asthma. A randomised, controlled trial. Ann Intern Med 130:487–495

Nassif EG, Weinberger M, Thompson R, Huntely W (1981) The value of maintenance theophylline in steroid-dependent asthma. N Engl J Med 304:71–75

National Institutes of Health: Highlights of the Expert Panel Report II: Guidelines for the Diagnosis and Management of Asthma. US Department of Health and Human Services Publication 1997.

Nelson HS, Busse WW, Kerwin E et al (2000) Fluticasone propionate/salmeterol combination provides more effective asthma control than low dose inhaled corticosteroid plus montelukast. J Allergy Clin Immunol 106:1088–1095

Noonan MJ, Chervinsky P, Brandon M, Zhang J, Kundu S, McBurney J, Reiss TF (1998) Montelukast, a potent leukotriene-receptor antagonist, causes dose-related improvements in chronic asthma. Eur Respir J 11:1232–1239

Norris, AA, Alton EWFW (1996) Chloride transport and the actions of sodium cromoglycate and nedocromil sodium in asthma. Clin Exp Allergy 26:250–253

O'Byrne PM, Thomson NC, Morris M, Roberts RS, Daniel EE, Hargreave FE (1983) The protective effect of inhaled chlorpheniramine and atropine on bronchoconstriction stimulated by airway cooling. Am Rev Respir Dis 128:611–6117

Patel, KR (1991) Histamine and exercise-induced asthma Eur Respir L 2:89–96

Pizzichini E, Leff JA, Reiss TF, Hedeles L, Boulet L-P, Wei LX, Efthimiadis AE, Zhang J, Hargreave FE (1997) Montelukast reduces airway eosinophilic inflammation in asthma: a randomised, controlled trial. Eur Respir J 14:12–18

Rackham A, Brown CA, Chandra RK, Ho P, Hoogerwerf PF, Kennedy RJ (1989). A Canadian multicenter study with Zaditen (ketotofen) in the treatment of bronchial asthma in children aged 5–17 years J Allergy Clin Immunol 14:267–273

Redier H, Chanez P, De Vos C, Rifai N, Clauzel AM, Michel F-B, Godard P. (1992) Inhibitory effect of cetirizine on the bronchial eosinophil recruitment induced by allergen inhalation challenge in allergic patients with asthma. J Allergy Clin Immunol 90:215–224

Reicin A, White R, Weinstein SF, et al (2000) Montelukast, a leukotriene-receptor antagonist, in combination with loratadine, a histamine receptor antagonist, in the treatment of chronic asthma Arch Intern Med 160,2481–2488

Reiss TF, Sorkness CA, Stricker W, Botto A, Busse WW, Kundu S, Zhang J. (1997) Effects of montelukast (MK-0476), a potent cysteinyl leukotriene-receptor antagonist, on bronchodilation in asthmatic subjects treated with and without inhaled corticosteroids. Thorax 52:45–48

Reiss TF, Chervinsky P, Dockhorn RJ et al (1998) Montelukast, a once-daily leukotrienereceptor antagonist in the treatment of chronic asthma. Arch Intern Med 158:1213–1220

Roberts JA, Thomson NC (1985) Attenuation of exercised-induced asthma by pre-treatment with nedocromil sodium and minocromil. Clin Allergy 15:377–381

Robinson DS, Campbell D, Barnes PJ (2001) Addition of leukotriene antagonists to therapy in chronic persistent asthma: a randomised double-blind placebo-controlled trial. Lancet 357:1991–1992

Roquet A, Dahlen B, Kumlin M, et al (1997) Combined antagonism of leukotrienes and histamine produces predominant inhibition of allergen-induced early and late phase airway obstruction in asthmatics. Am J Respir Crit Care Med 155:1856–1863

Ruffin RE, Alpers JH, Pain MCF et al (1987) The efficacy of nedocromil sodium (TILADE) in asthma. Aust NZ J Med 17:557–561

Rutgers SR, Koeter GH, Van Der Mark TW, Postma DS. (1999) Protective effect of oral terfenadine and not inhaled ipratropium bromide on adenosine 5'-monophosphate-induced bronchoconstriction in patients with COPD Clin Exp Allergy 29:1287–1292

Schwartz HJ, Blumenthal M, Brady R et al (1996) A comparative study of the clinical efficacy of nedocromil sodium and placebo. How does cromolyn sodium compare as an active control treatment? Chest 109:945–952

Simons FER. (1999) Is antihistamine (H_1-receptor antagonist) therapy useful in clinical asthma? Clin Exp Allergy 29 (suppl 3):98–104

Simons FER, Villa JR, Lee BW et al (2001) Montelukast added to budesonide in children with persistent asthma: a randomised, double-blind, crossover study. J Pediat 138:694–698

Spector SL, Smith LJ, Glass M (1994) Accolate Trialist Group. Effects of 6 weeks of therapy with oral doses of ICI 204,219, a leukotriene D_4 receptor antagonist, in subjects with bronchial asthma. Am J Respir Crit Care Med 150:618–23

Spooner CH, Saunders LD, Rowe BH. (2002) Nedocromil sodium for preventing exercise-induced bronchoconstriction (Cochrane Review) In: The Cochrane Library, Issue 1, Oxford: Update Software

Stirling RG, Chung KF (1999) Leukotriene antagonists and Churg-Strauss syndrome: the smoking gun. Thorax 54:865–866

Suissa S, Dennis R, Ernst P et al (1997) Effectiveness of the leukotriene-receptor antagonist zafirlukast for mild-to-moderate asthma. A randomised, double-blind, placebo-controlled trial. Ann Inter Med 126:177–183

Sullivan P, Bekir S, Jaffar Z et al (1994) Anti-inflammatory effects of low-dose theophylline in atopic asthma. Lancet 343:1006–1008

Svendsen UG, Frolund L, Madsen F, Nielson NH (1989) A comparison of the effects of nedocromil sodium and beclomethasone dipropionate on pulmonary function, symptoms, and bronchial responsiveness in patients with asthma. J Allergy Clin Immunol 84:224–231

Svendsen UG, Jorgensen H (1991) Inhaled nedocromil sodium as additional treatment to high dose inhaled corticosteroids in the management of bronchial asthma. Eur Respir J 4:992–999

Schwartz HJ, Blumenthal M, Brady R et al. (1996) A comparative study of the clinical efficacy of nedocromil sodium and placebo. How does cromolyn sodium compare as an active control treatment? Chest 109:945–952.

Tamaoki J, Kondo M, Sakai N et al (1997) Leukotriene antagonists prevent exacerbation of asthma during reduction of high-dose inhaled corticosteroid. Am J Respir Crit Care Med 155:1235–1240.

Tasche MJA, Van der Wouden JC, Uijen JHJM et al (1997) Randomised placebo-controlled trial of inhaled sodium cromoglycate in 1–4 year-old children with moderate asthma. Lancet 350:1060–1064

Tasche MJA, Uijen JHJM, Bernsen RMD et al (1997) Inhaled disodium cromoglycate (DSCG) as maintenance therapy in children with asthma: a systematic review. Thorax 55:913–920

Taylor IK, O'Shaughnessy KM, Fuller RW et al (1991) Effect of cysteinyl-leukotriene-receptor antagonist ICI 204-219 on allergen-induced bronchoconstriction and airway hyperreactivity in atopic subjects. Lancet 337:690–694

Temple DM, McCluskey M (1988) Loratadine, an antihistamine, blocks antigen and ionophore-induced leukotriene release from human lung, in vitro. Prostaglandins 35:549–54.

Thomson NC, Patel KR, Kerr JW (1978) Sodium, cromoglycate and ipratropium bromide in exercise-induced asthma Thorax 33:694–699

Thomson NC, Kerr JW (1980) The effect of inhaled H_1 and H_2 receptor antagonists in normal and asthmatic subjects Thorax 35:428–434

Thomson NC, Clark CJ, Boyd G, Moran F (1981) The effect of sodium cromoglycate on bronchial smooth muscle Br J Clin Pharmacol 12:440–441

Thomson, NC (1989) Nedocromil sodium: an overview. Respir Med 83:269–76

Tuggery JM, Hosker HSR (2000) Churg-Strauss syndrome associated with montelukast therapy. Thorax 55:805–806

Van Ganse E, Kaufman L, Derde MP, Yernault JC, Delaunois L, Vincken W. (1997) Effects of antihistamines in adult asthma: a meta-analysis of clinical trials. Eur Respir J 10:2216–2224

Virchow JC, Prasse A, Naya I et al (2000) Zafirlukast improves asthma control in patients receiving high-dose inhaled corticosteroids. Am J Respir Cri Care Med 162:578–585

Walsh, GM (1997) The effects of cetirizine on the function of inflammatory cells involved in the allergic response. Clin Exp Allergy 27 (suppl 2):47–53

Wechsier ME, Garpestad E, Flier SR, et al (1998). Pulmonary infiltrates, eosinophilia, and cardiomyopathy following corticosteroid withdrawal in patients with asthma receiving zafirlukast. JAMA 279:455–457

Wenzel SE, Trudeau JB, Kaminsky DA et al (1995) Effect of 5-lipoxgenase inhibition on the bronchoconstriction and airway inflammation in nocturnal asthma. Am J Respir Crit Care Med 152:897–905

Wilson AM. Dempsey OJ, Sims EJ, Lipworth BJ (2001) Evaluation of salmeterol or montelukast as second-line therapy for asthma not controlled with inhaled corticosteroids. Chest 119:1021–1026

Part 2
Future Drugs

New Bronchodilator Drugs

D. Spina

The Sackler Institute of Pulmonary Pharmacology, GKT School of Biomedical Science, Guy's Campus, King's College London, 5th floor Hodgkin Building, London, SE1 1UL, UK
e-mail: domenico.spina@kcl.ac.uk

1	Introduction	154
2	**Vasoactive Intestinal Peptide**	154
2.1	Localization	154
2.2	VIP and Airway Smooth Muscle	155
2.2.1	Functional Response	155
2.2.2	Receptor Heterogeneity	156
2.3	VIP Analogues	157
3	**Nitric Oxide**	158
3.1	Nitric Oxide Synthesis	158
3.2	NO and Airway Smooth Muscle	158
3.3	Clinical Studies	159
3.4	Pro-inflammatory Properties of NO	161
4	**Atrial Natriuretic Peptide**	162
5	**Phosphodiesterase 3 Inhibitors**	164
5.1	In Vitro Studies	164
5.2	In Vivo Studies	165
6	**Potassium Channel Openers**	165
6.1	Voltage-Dependent Channels	166
6.2	Calcium-Activated Potassium Channels	166
6.3	K_{ATP} Channels	167
6.4	Inward Rectifiers	169
7	**Conclusion**	169
	References	171

Abstract β_2-Adrenoceptor agonists are the gold standard for the symptomatic treatment of asthma, resulting in quick and effective bronchodilator relief, although they are not completely free from side-effects. A range of different pharmacological approaches is being actively pursued in order to achieve relaxation of airway smooth muscle comparable to that seen with β_2-adrenoceptor agonists and by mechanisms distinct from activation of the β-adrenoceptor without attendant side-effects. These various approaches include signalling via vasoactive intestinal polypeptide (VIP) receptors, activation of soluble and particulate guanylyl cyclase, inhibition of phosphodiesterase (PDE)3 and opening of potassium channels. The pharmacology, clinical experience and merit of each of these novel approaches are discussed.

Keywords VIP · Nitric oxide · Airway smooth muscle · Potassium channels · ANF · PDE3 · Relaxation

1
Introduction

β_2-Adrenoceptor agonists are the mainstay for the treatment of asthma, effective in providing symptomatic relief to a wide range of allergic (e.g. house dust mite) and non-allergic (e.g. cold air, fog, exercise, pollutants) stimuli. Whilst β_2-adrenoceptor agonists remain the drug of choice in the symptomatic treatment of asthma, this class of drug is not free from side-effects including tremor, hypokalaemia and tachycardia. There are also concerns that the over-reliance on the use of β_2-adrenoceptor agonists may lead to an increase in asthma morbidity (Sears 2001). Furthermore, it is recognized that polymorphisms in the β_2-adrenoceptor protein may confer a reduction in the efficacy of these agents in some patients (Martinez et al. 1997). The possibility therefore of developing non-β_2-agonist bronchodilators free from such deficiencies is currently under investigation. A number of endogenous processes exist within the lung that, when activated, appear to play a role in down regulating airway smooth muscle function (Fig. 1). This chapter will review the literature concerning the utility of exploiting a number of these mechanisms for the development of potentially new bronchodilator agents for the treatment of respiratory diseases such as asthma and chronic obstructive pulmonary disease.

2
Vasoactive Intestinal Peptide

2.1
Localization

Vasoactive intestinal peptide (VIP) is a bioactive peptide present in abundance within mammalian lung. VIP-immunoreactive nerves are present throughout the respiratory tract innervating airway smooth muscle, blood vessels, submucosal glands and airway parasympathetic ganglia with an absence of innervation within the respiratory epithelium. The density of innervation diminishes from central to peripheral airways with few VIP-immunoreactive fibres innervating the terminal bronchioles and alveolar spaces (Laitinen et al. 1985; Fischer et al. 1996). There is a significantly greater density of VIP-containing nerves compared with nerves specific for nitric oxide synthase (nNOS), suggesting the presence of different subpopulation of nerve fibres that presumably subserve different functions within the airway wall. However, there is also evidence of co-localization in some nerves, suggesting the possibility that release of both these neurotransmitters act in concert at neuroeffector sites in the regulation of cell function (Fischer and Hoffmann 1996; Fischer et al. 1996). Furthermore, co-localization of VIP in a subpopulation of parasympathetic nerves was also confirmed following immunostaining for choline acetyltransferase, an enzyme specific for parasympathetic nerves (Fischer and Hoffmann 1996). Similarly, co-localization of both VIP and nitric oxide synthase (NOS) to intrinsic parasympathetic ganglion cells was a consistent observation in human airway tissue (Fischer et al. 1996). Whilst not reported in human airways, studies of lung from primates also reveal co-localization of VIP within sensory nerves containing the neuropeptides substance P and calcitonin gene-related peptide (Nohr et al. 1995). There is also evidence for non-neuronal sources of VIP in a variety of inflammatory cells including mast cells, mononuclear cells, neutrophils and eosinophils, although to what extent the release of VIP from these sources contributes to the modulation of airway cell function remains to be determined (Groneberg et al. 2001b).

2.2
VIP and Airway Smooth Muscle

VIP plays an important physiological role in regulating various aspects of cell biology within the lung, including inhibition of neurogenic mediated mucus secretion, vasodilation, and suppression of inflammatory cell function (Said 1998; Groneberg et al. 2001b). The role of VIP in the control of airway smooth muscle function will be discussed in greater detail.

2.2.1
Functional Response

The exogenous administration of VIP to human isolated bronchial preparations in vitro induces relaxation which diminishes progressively down the respiratory tract (Palmer et al. 1986; Ellis and Undem 1992; Kinhult et al. 2000; Naline et al. 2000). Furthermore, the response to VIP is limited by metabolism from various proteolytic enzymes, including neutral endopeptidase, mast cell tryptase and chymase (Tam et al. 1990)—which may explain the relative lack of potency compared with β_2-adrenoceptor agonists in many of these studies. In contrast, it has been difficult to document relaxation of human isolated preparations in a VIP-dependent manner following nerve depolarization (Ellis and Undem 1992; Ward et al. 1995a). This would indicate that if released from airway nerves, VIP does not play a role in regulating airway smooth muscle tone. Alternatively, the use of tissue preparations from patients undergoing thoracotomy, albeit from uninvolved areas of resected lung tissue, might be characterized by unavoidable loss of VIP due to disease or enzymatic destruction following surgical removal of tissue. This would potentially confound attempts to discern a possible role, if any, of VIP as an endogenous inhibitory neurotransmitter in human airways.

A number of studies have examined the effect of VIP in man. Inhaled VIP did not produce significant bronchoprotection against exercise (Bundgaard et al. 1983) and histamine (Altiere and Diamond 1984), which is perhaps not surprising in view of the potential for enzymatic degradation following deposition within the lung. Therefore, the development of more potent and stable analogues of VIP has been investigated.

2.2.2
Receptor Heterogeneity

VIP binding sites have been documented throughout the lung and localized to airway epithelium, airway smooth muscle, vascular smooth muscle and submucosal glands (Carstairs and Barnes 1986; Robberecht et al. 1988), which correlates with the wide range of biological activity attributed to VIP (Said 1998; Groneberg et al. 2001b). This diminution in the density of VIP binding sites down the respiratory tree is consistent with the functional observations documenting a loss in agonist potency at different sites within the respiratory tract. There are currently two receptors that demonstrate high affinity for this peptide and mediate many of the biological actions of VIP, designated VPAC1 and VPAC2 (Harmar et al. 1998). Northern blot analysis of human lung tissue has revealed the expression of VPAC1 (Sreedharan et al. 1995), whilst expression of VPAC2 was sparse (Adamou et al. 1995; Wei and Mojsov 1996; Groneberg et al. 2001a). Further studies using RT-PCR have confirmed the expression of mRNA for both receptor subtypes in human lung, and receptor protein was also detected in lung membranes (Busto et al. 1999). Several methodologies have been employed to ascertain the location of receptor protein in human airways. The detection of receptor protein using an immunohistochemical approach has revealed the presence of both receptors on lymphocytes, macrophages and smooth muscle cells of small arteries (Busto et al. 2000). Using an in situ hybridization approach, mRNA for VPAC2 was detected in basal and ciliated epithelial cells, serous and mucous cells and within alveolar tissue; transcript was detected only in macrophages (Groneberg et al. 2001a). In contrast, very weak or no message was detected over airway smooth muscle and vascular smooth muscle (Groneberg et al. 2001a; Schmidt et al. 2001). The paucity of VPAC2 expression in smooth muscle suggests that receptors other than VPAC2 are responsible for mediating relaxation or that activation of VPAC2 possibly on another cell leads to the release of as-yet-unidentified factors which mediate relaxation (Groneberg et al. 2001a). It is also possible that VIP mediates relaxation of airway smooth muscle via another receptor subtype. Indeed, VIP demonstrates relatively high affinity for PAC1, which is also found in human lung (Busto et al. 1999). However, these receptors represent only a fraction of the total amount of VPAC/PAC receptor protein found in the lung, and it would therefore be unlikely that this receptor mediates relaxation to VIP. Furthermore, the selective

VPAC2 receptor agonist, Ro 25-1553, is a potent relaxant agonist of human airway smooth muscle (see below) and shows 3,000 times greater affinity for VPAC2 than PAC1 (Gourlet et al. 1997), thereby ruling out this receptor in the functional responses observed to VIP in airway tissue. Alternatively, the low levels of receptor expression detected in smooth muscle may be sufficient to stimulate relaxation of airway smooth muscle following activation by VIP- or VPAC2-selective agonists (Schmidt et al. 2001).

2.3
VIP Analogues

Enzymatically stable analogues of VIP have been evaluated for their potential as novel bronchodilator agents for the treatment of respiratory disease (Bolin et al. 1995). The cyclization of the VIP peptide yielded analogues with high affinity for VPAC2 as illustrated by the synthesis of Ro 25-1553 [Ac-Glu^8,Lys^{12},Nle^{17}, Ala^{19},Asp^{25},Leu^{26},Lys^{27,28},Gly^{29,30},Thr^{31}-vasoactive intestinal peptide(cyclo 21–25) (Gourlet et al. 1997) and Ro 25–1392 [Ac-Glu^8,OCH3-Tyr^{10},Lys^{12},Nle^{17}, Ala^{19},Asp^{25},Leu^{26},Lys^{27,28}-vasoactive intestinal peptide(cyclo 21–25)] (Xia et al. 1997) which are three orders of magnitude more selective for VPAC2 than VPAC1. In contrast, substitution of specific amino acids with alanine ([Ala^{11,22,28}]VIP) resulted in a highly selective VPAC1 agonist (Nicole et al. 2000). These studies have clear implications for selectively targeting cells expressing different VPAC subtypes.

Ro 25-1553 relaxed human bronchial tissue some 390 times more potently than native VIP and 20 times more potently than the β_2-adrenoceptor agonist, salbutamol (O'Donnell et al. 1994a). Moreover, the intrinsic activity of this agent was similar to that observed with salbutamol, and suggests that these analogues would be as effective as salbutamol in providing bronchoprotection to various bronchoconstrictor stimuli. Similarly, whilst less potent, Ro 25-1553 was as efficacious as formoterol in reversing contraction of guinea-pig isolated preparations (Kallstrom and Waldeck 2001). Furthermore, Ro 25-1553 provided significant bronchoprotection against a range of mediators in sensitized guinea-pigs, demonstrating the potential of this class of molecules as bronchoprotective agents (O'Donnell et al. 1994a,b). Moreover, administration of Ro 25-1553 by the inhaled route suppressed bronchoconstriction induced by antigen challenge, whilst native VIP was ineffective (O'Donnell et al. 1994a). It is therefore clear that Ro 25-1553 is resistant to enzymatic degradation and is a potent and efficacious relaxant agonist comparable to that observed with β_2-selective agonists. Like salbutamol, Ro 25-1553 prevented oedema formation following antigen challenge and thereby demonstrating acute anti-inflammatory activity (O'Donnell et al. 1994b). It is unlikely that relaxation was mediated by activation of VPAC1, as this molecule has 800 times lower affinity for this receptor subtype (Gourlet et al. 1997).

3
Nitric Oxide

Nitric oxide (NO) is derived from the amino acid L-arginine by the enzyme NOS, of which exist two calcium-dependent constitutive isoforms: endothelium (e)NOS or NOS3 and the neural (central and peripheral) isoform, nNOS or NOS1. These isoforms are also expressed in tissue other than their original designation. Thus, nNOS is also present in human bronchial epithelium (Kobzik et al. 1993). A third isoform is only expressed following an inflammatory insult (iNOS or NOS2), is less dependent upon calcium for activation, produces significant quantities of NO and is thought to play a role in pathophysiological situations. The release of NO from nitrergic nerves leads to relaxation of airway smooth muscle and therefore is thought to behave as an endogenous bronchoprotective agent. Whether such pathways can be utilized for the development of novel bronchodilator agents remains to be seen.

3.1
Nitric Oxide Synthesis

Nitrergic nerves have been documented throughout the respiratory tree in and around airway smooth muscle, diminishing in density from central to peripheral airways, and evidence exists of nNOS localized to parasympathetic ganglia, sensory nerves and in some instances co-localized with VIP and acetylcholine (Fischer et al. 1993; Ward et al. 1995a; Fischer et al. 1996; Fischer and Hoffmann 1996). In asthmatic airways, NOS2 is predominantly found within the airway epithelium (Hamid et al. 1993; Saleh et al. 1998) and inflammatory cells, notably macrophages, neutrophils and eosinophils (Saleh et al. 1998). Clinical studies have detected the presence of NO in exhaled air of asthmatic subjects (Kharitonov et al. 1994) which is increased during the late asthmatic response (Kharitonov et al. 1995), supporting the view that during airway inflammation the expression of NOS2 is increased leading to the generation of large quantities of NO. In the lung, NOS3 is localized to endothelial cells in the bronchial circulation and within the epithelium (Shaul et al. 1994).

3.2
NO and Airway Smooth Muscle

The functional consequence of NO in the airways has been investigated in a number of studies. NO is a potent vasodilator of the pulmonary circulation (Moya et al. 2001) and may either inhibit (Erjefalt et al. 1994) or promote (Bernareggi et al. 1997) plasma protein extravasation in the airways, this latter action being dependent upon the expression of NOS2. These studies suggest that NO generated from different NOS isoforms exhibits anti- and pro-inflammatory activities in the lung.

In human airways, it appears that NO is the neurotransmitter responsible for mediating relaxation following activation of non-adrenergic non-cholinergic nerves. The stimulation of nitrergic nerves leads to elevations in the levels of the second messenger, cyclic guanosine monophosphate (GMP) within airway smooth muscle cells. Furthermore, both relaxation of airway smooth muscle and elevation of cyclic GMP following nerve stimulation are inhibited in the presence of a NOS inhibitor (Ellis and Undem 1992; Ward et al. 1995a,b). Relaxation of airway smooth muscle is also observed following the administration of NO or donors of NO, including sodium nitroprusside and S-nitrosothiols that is also accompanied by elevations of intracellular cyclic GMP in human airway tissue (Ellis and Undem 1992; Gaston et al. 1993, 1994; Ward et al. 1995a,b).

The expression of guanylyl cyclase in airway smooth muscle is consistent with the ability of NO either generated endogenously by NOS following nerve depolarization or administered in the form of a donor to mediate relaxation of airway smooth muscle (Hamad et al. 1997). Whilst elevations in cyclic GMP in these studies are thought to be responsible for mediating the observed relaxation in reaction to NO, it has been suggested that a component of this response is not mediated by the activation of guanylyl cyclase. This conclusion is based on the observation that relaxation in response to NO donors was not inhibited by methylene blue, an agent often used as an inhibitor of guanylyl cyclase (Gaston et al. 1994). Therefore, relaxation may involve a non-cyclic GMP-dependent mechanism for example, by a direct interaction between NO and calcium-activated potassium channels, thereby leading to hyperpolarization of cell membrane and relaxation (Abderrahmane et al. 1998), although the role of these channels in mediating relaxation in response to NO donors has been questioned (Corompt et al. 1998). However, the suggestion of cyclic GMP-independent mechanisms arises from studies employing methylene blue. Unfortunately, this agent is a weak inhibitor of guanylyl cyclase and has a number of actions, including the ability to generate oxygen radicals, which would inactivate NO and thereby confound interpretation as to the lack of involvement of soluble guanylyl cyclase in the relaxation in response to NO donors. Confirmation that NO-mediated relaxation of airway smooth muscle is predominantly mediated via activation of guanylyl cyclase was obtained with the use of a selective inhibitor of this enzyme, ODQ (1H-[1,2,4]oxadiazolol[4,3-a]quinoxalin-1-one). Relaxation induced by stimulation of nitrergic nerves in the guinea-pig and relaxation of both guinea-pig and human airway preparations in response to NO donors was abolished in the presence of ODQ (Ellis 1997).

3.3
Clinical Studies

The role of NO in modulating airway smooth muscle tone has also been investigated in vivo in gene-disrupted mice lacking the various NOS isoforms. It appears that NOS1 (to a lesser extent NOS3) is an important determinant of baseline bronchial reactivity to spasmogenic agonists (De Sanctis et al. 1997, 1999).

NOS3 also appears to suppress neural responses to cholinergic stimulation in the mouse (Kakuyama et al. 1999). Hence, NO generated from constitutively expressed NOS, which produces discreet amounts of NO compared with the inducible isoform, might act as an endogenous suppressor of airway reactivity. Indeed, NOS inhibitors augment bronchoconstriction to inhaled bradykinin (Ricciardolo et al. 1996), histamine and adenosine (Taylor et al. 1998a) in asthmatic subjects and increased nasal reactivity to both histamine and bradykinin in healthy subjects (Turner et al. 2000).

A number of clinical studies have investigated the potential utility of NO and NO donors as bronchodilator agents. Inhalation of NO (80–100 ppm) provided significant bronchoprotection against the fall in specific airway conductance induced by methacholine in healthy subjects (Sanna et al. 1994), subjects with hyperresponsive airways (Hogman et al. 1993) and in asthmatic subjects (Kacmarek et al. 1996). Furthermore, NO induced modest, albeit significant bronchodilation in asthmatic subjects but not in patients with chronic obstructive pulmonary disease and was considerably less efficacious than salbutamol and therefore would not be useful in the clinical setting (Hogman et al. 1993). Furthermore, a potential problem with delivery of NO to the airways is that higher concentrations than those employed in these clinical studies may be toxic to the lung. In order to avoid such concerns, the delivery of NO in the form of NO donors may be more appropriate. However, cardiovascular side-effects are a concern with drugs of this nature. Indeed, the administration of glyceryl trinitrate by nebulization induced a modest bronchodilator response in comparison with salbutamol in subjects with acute severe asthma, but at the expense of systemic side-effects (Sharara et al. 1998). Another approach might be to deliver NO with a β_2-adrenoceptor agonist, which would provide an additive bronchodilator effect (Hogman et al. 1993). This has been taken one step further by formulating salbutamol with a NO moiety leading to an NO-donating β_2-adrenoceptor agonist, which in recent clinical trials was shown to be more effective than salbutamol in providing bronchoprotection against methacholine in the absence of untoward side-effects (Sardina et al. 2002).

Alternatively, interventions that may increase the synthesis of endogenous nitrovasodilators might be another therapeutic approach. It has been observed that the levels of S-nitrosothiols and thiols in samples of tracheal lining fluid were significantly reduced in children suffering from a severe acute exacerbation of asthma (Gaston et al. 1998). This loss in a pool of endogenous S-nitrosothiols may have important functional implications during exacerbation of the inflammatory response. Endogenous S-nitrosothiols are thought to act as a reservoir for the detoxification of NO and any reduction in the levels of these protective agents could result in an increase in the levels of free NO. Hence, the pro-inflammatory potential of NO might prevail. Furthermore, these endogenous S-nitrosothiols are also potent relaxant agonists, and an increase in bronchial reactivity might also occur in the absence of these naturally occurring bronchodilator substances within the lung (Gaston et al. 1993; Dweik et al. 2001). The supplementation of nitrosothiols in the inspired air afforded protection against hypoxic pulmonary vasoconstriction in pigs (Moya et al. 2001), and whether a similar intervention provides bronchoprotective relief in respiratory disease remains to be established.

An important consideration concerning the use of NO donors is the potential for tachyphylaxis following prolonged administration and the potential that pro-inflammatory activities of NO, e.g. nitrosation of proteins, would limit the utility of this drug class. It might be possible to avoid such potential problems by direct activation of soluble guanylyl cyclase with novel activators of this enzyme (Hobbs 2002).

3.4
Pro-inflammatory Properties of NO

The preceding section described the potential utility of NO donors as novel bronchodilator and bronchoprotective substances. However, it is also clear that the generation of NO from NOS2 during an inflammatory insult to the airways could have pro-inflammatory consequences. Therefore, the administration of NO donors may increase the NO burden within the lung, which would be detrimental. Thus, the role of NO under these conditions merits discussion.

Animal studies have documented increased NOS2 expression at various times following antigen challenge in immunized mice (De Sanctis et al. 1999; Xiong et al. 1999; Koarai et al. 2000; Iijima et al. 2001). However, in some circumstances, airway inflammation can be induced in the absence of up-regulation of NOS2 expression within the lung (Feder et al. 1997), perhaps reflecting differences in the intensity of the inflammatory response induced in these models. Nevertheless, inhibition of NOS2 activity with the use of inhibitors has highlighted the role of this isoform in the allergic inflammatory and airway response in these models. Non-selective NOS inhibitors significantly reduced the recruitment of eosinophils to the lung following antigen challenge (Feder et al. 1997; Ferreira et al. 1998), although the role of NOS2 in this response is controversial, since selective NOS2 inhibitors were either ineffective (Feder et al. 1997) or suppressed (Koarai et al. 2000; Trifilieff et al. 2000; Iijima et al. 2001) eosinophil recruitment to the lung. A potential mechanism by which NO facilitates eosinophil recruitment may be linked to increased CC chemokine expression in this model (Trifilieff et al. 2000).

Further studies in mice with a gene disruption for the various NOS isoforms has also yielded conflicting outcomes. Thus, eosinophil recruitment following antigen challenge is either attenuated (Xiong et al. 1999) or unaffected (De Sanctis et al. 1999) by disruption of the NOS2 gene. Moreover, targeted disruption of the NOS1 and NOS3 genes was without effect upon eosinophil recruitment (De Sanctis et al. 1999), and therefore NO generated from NOS2 does not appear to play an obligatory role in recruiting eosinophils to the airways of allergic mice. Indeed, in other studies eosinophil recruitment was exacerbated in allergic mice following treatment with the NOS inhibitor, *N*-(G)-nitro-L-arginine methyl ester (L-NAME), suggesting that NO may also have an anti-inflammatory role (Blease et al. 2000). While the mechanism by which NO may promote or inhibit eosinophil recruitment to the airways in these models is not clear, this was associated with suppression (Trifilieff et al. 2000) or augmentation (Blease et al. 2000) of CC chemokine expression, respectively. It is therefore far from clear as to the exact role played by NO in cellular recruitment to the airways.

In both knockout studies outlined above, bronchial hyperresponsiveness to methacholine following antigen challenge was not significantly altered in the absence of NOS2. This suggests that an alteration in airway responsiveness following an allergic insult is not a direct consequence of the overgeneration of NO during an inflammatory response. Interestingly, there was a discrepancy between the degree of baseline airway responsivity to methacholine in non-immunized mice, probably reflecting the different methods used to measure respiratory lung mechanics in these models (De Sanctis et al. 1999; Xiong et al. 1999). However, in contrast to the findings obtained in genetically modified mice, one study has reported a potential role for NO, generated by NOS2, in the induction of bronchial hyperresponsiveness. Treatment with a selective NOS2 inhibitor abolished the increase in airway responsiveness to methacholine observed following antigen challenge (Koarai et al. 2000).

Few studies have assessed the consequences of NOS inhibition in the clinical setting. The NOS inhibitor L-NAME failed to attenuate the early and late asthmatic response following antigen challenge in asthmatic subjects (Taylor et al. 1998a). The level of expired NO was significantly decreased following treatment with L-NAME, which probably reflects inhibition of NOS1 and NOS3, since there was a lack of increase in expired NO during the late asthmatic response (8–10 h) in the placebo arm. This suggests that NOS2 expression was not increased during this time period, therefore making interpretation of the data difficult. However, 21 h post allergen challenge, there was a significant increase in expired NO which was not evident in the L-NAME-treated group, possibly reflecting an increase in NOS2 expression. Whether or not the development of more selective NOS2 inhibitors will prove of greater benefit in asthma requires further investigation. The lack of effect of L-NAME on airways hyperresponsiveness might be related to the removal of a protective endogenous dilator from the airways (Ricciardolo et al. 1996; Taylor et al. 1998b; Turner et al. 2000) or loss of a negative feedback regulator, which tends to suppress the allergic inflammatory response (Thomassen et al. 1999).

4
Atrial Natriuretic Peptide

Atrial natriuretic peptide (ANP) belongs to a family of peptides with potent vasorelaxant, diuretic and natriuretic properties. In the lung, ANP is produced by alveolar type II cells, epithelial cells and vascular smooth muscle and stimulates a number of physiological responses including vasodilation, bronchial smooth muscle relaxation, the inhibition of pulmonary oedema and the stimulation of surfactant secretion (Perreault and Gutkowska 1995). ANP binds with high affinity to cell surface receptors [natriuretic peptide receptor (NPR)-A; guanylyl cyclase (GC)-A] with constitutive guanylyl cyclase activity and catalyses the synthesis of cyclic GMP. ANP also binds to a second cell surface receptor devoid of guanylyl cyclase activity (NPR-C), which is thought to control the local concentrations of ANP available for binding to GC-A. A third receptor, GC-B demonstrates high affinity for C-type natriuretic factor (Potter and Hunter 2001).

The presence of particulate guanylyl cyclase in cultured human tracheal smooth muscle cells was demonstrated by the ability of ANP to significantly elevate the levels of intracellular cyclic GMP within these cells (Hamad et al. 1997). Moreover, this increase in cyclic GMP in response to ANP was 2–3 times greater than that achieved following stimulation of soluble guanylyl cyclase with sodium nitroprusside, suggesting that smooth muscle function may be modulated by endogenous release of ANP. However, there is some discrepancy in the literature concerning the relaxant efficacy of ANP in human airway tissues. ANP failed to significantly relax human isolated bronchial preparations even in the presence of the neutral endopeptidase inhibitor, thiorphan or following epithelium removal in order to prevent degradation of this peptide (Candenas et al. 1991). Unlike the lack of efficacy observed with ANP in human airways, significant relaxation of guinea-pig tracheal smooth muscle was observed in this study. In contrast, ANP exhibited weak relaxant activity, which was augmented following inhibition of neutral endopeptidase in human isolated tissues (Angus et al. 1994b). Whether methodological differences can account for this discrepancy in the literature remains to be established. Nevertheless, an important observation from both of these studies is that the concentration of ANP required to elicit relaxation of airway smooth muscle in vitro would not be achieved with the doses that have been reported to induce bronchodilation in man (see below). In other studies, ANP inhibited the proliferation of human airway smooth muscle cells in culture but whether such effects extrapolate to inhibition of smooth muscle proliferation in vivo remains to be addressed (Hamad et al. 1999).

A number of clinical studies have shown that when inhaled, ANP provides significant bronchoprotection against methacholine in asthmatic subjects (Angus et al. 1994a). Similarly, when administered intravenously at doses that did not alter systemic blood pressure, ANP afforded significant bronchoprotection against nebulized distilled water (Mcalpine et al. 1992) and histamine (Hulks et al. 1991). Since ANP is susceptible to enzymatic destruction in the airways, the bronchoprotective effect of this agent when administered by the inhaled route can be significantly improved if neutral endopeptidase activity is inhibited (Angus et al. 1995). This suggests that the synthesis of enzyme-resistant analogues or GC-A-selective agonists could lead to potentially novel bronchoprotective agents with long duration of action and of greater importance, depending on whether airway over systemic selectivity can be demonstrated. Urodilatin is another member of the natriuretic peptide family formed in the kidney and activates GC-A (Heim et al. 1989). When administered intravenously, urodilatin produced bronchodilation of similar magnitude to that observed with salbutamol, and whilst this effect was short-lived, it was not at the expense of alterations in blood pressure (Fluge et al. 1995). Hence, it may be possible to obtain selective airway activity with GC-A-activating agents, and any potentially unwanted systemic effect might be avoided by aerosol delivery.

The poor correlation between ANP's efficacy to relax airway smooth muscle in vitro and to provide bronchoprotection in vivo suggests that ANP may not directly activate airway smooth muscle in vivo (Candenas et al. 1991). The possibility that ANP might induce the release of endogenous catecholamines, which in turn mediate bronchoprotection, can be ruled out since significant bronchoprotection against histamine was observed in the absence of any change in circulating plasma catecholamines (Hulks et al. 1991). Alternatively, the possibility that ANP might interfere with afferent neuronal activity (Schultz et al. 1988) could offer one explanation for the reported bronchoprotective action of ANP against stimuli which in themselves do not directly activate airway smooth muscle, e.g. distilled water, but that would interfere with reflex bronchoconstriction.

5
Phosphodiesterase 3 Inhibitors

5.1
In Vitro Studies

Relaxation of airway smooth muscle following activation of β_2-adrenoceptors or following activation of guanylyl cyclase is dependent upon elevations in the second intracellular messengers, cyclic AMP and cyclic GMP, respectively. The biological activity of these cyclic nucleotides is terminated by a family of phosphodiesterases, and biochemical investigations have documented the presence of phosphodiesterase (PDE)1–5 in human airway smooth muscle (De Boer et al. 1992; Rabe et al. 1993; Torphy et al. 1993). The physiological consequence of inactivation of these PDEs to airway smooth muscle function has been investigated in human isolated airway preparations under different experimental conditions.

Under basal conditions, human bronchial tissues generate tone spontaneously, and reversal of this contractile response can be demonstrated following inhibition of PDE3 (SKF94120). Moreover, this inhibitor was more efficacious in relaxing human airway smooth muscle compared with PDE4 (rolipram) or PDE5 (zaprinast) selective inhibitors (Qian et al. 1993; Rabe et al. 1993). Thus, PDE3 seems to play a role in regulating baseline airway function. The spontaneous generation of tone in human isolated preparations is due to the release of leukotrienes from the tissue (Ellis and Undem 1994), and inhibition of PDE3 may have induced relaxation of airway smooth muscle indirectly by inhibiting the release and/or formation of these mediators. However, whilst the PDE3 inhibitor motapizone demonstrated greater efficacy than rolipram in suppressing contraction induced by exogenously administered leukotriene C4, neither agent significantly inhibited contraction of passively immunized tissues to allergen

(Schmidt et al. 2000). The implication of these findings is that inhibition of PDE3 mediates relaxation of spontaneously contracted human airway smooth muscle by a direct effect on this cell and not via suppression of mediator release. In relation to other spasmogens, inhibition of PDE3 inhibited contraction induced by histamine and muscarinic agonists and in general is more efficacious than inhibition of PDE4 or PDE5 alone (De Boer et al. 1992; Qian et al. 1993). However, inhibition of both PDE3 and PDE4 is clearly more effective in suppressing smooth muscle contraction of human airways (Schmidt et al. 2000). Whilst most attention has focussed on the role of PDE3 in the regulation of airway smooth muscle contraction, the anti-proliferative activity following selective inhibition of PDE has also been investigated. It appears that PDE3 plays a major role in reducing the proliferation of human airway smooth muscle cells and DNA synthesis in response to growth factor, whilst inhibition of PDE4 was without significant effect on cell proliferation (Billington et al. 1999).

5.2
In Vivo Studies

The oral administration of cilostazol (PDE3 inhibitor) caused bronchodilation and bronchoprotection against methacholine challenge in non-diseased subjects but at the expense of mild to severe headache (Fujimura et al. 1995). The potential unwanted increases in heart rate and falls in blood pressure associated with this drug class appear to be avoided when inhaled, as in the case of olprinone, which lead to significant improvements in forced expiratory volume in 1 s (FEV_1) commensurate with that observed following inhalation of salbutamol (Myou et al. 1999). Therefore, it might be possible to separate the cardiovascular and respiratory actions of this drug class when administered via the inhaled route. The PDE3 inhibitor MKS492 has been reported to attenuate the early and late asthmatic response in atopic asthmatics, albeit modestly, and suggests that PDE3 inhibition may provide bronchoprotective action but would not be sufficient to suppress inflammatory cell function (Bardin et al. 1998).

6
Potassium Channel Openers

There are four major classes of potassium channels including voltage-dependent (K_v), ATP-dependent (K_{ATP}), calcium-dependent (K_{Ca}) and inward rectifier (K_{IR}) (Janssen 2002; Pelaia et al. 2002). In general, the opening of potassium channels leads to membrane hyperpolarization and is often associated with airway smooth muscle relaxation. The possibility that the development of selective agonists for potassium channels might lead to novel bronchodilator agents for use in respiratory disease has received considerable interest.

6.1
Voltage-Dependent Channels

Small unitary conductance (1–20 pS) and delay in current activation characterize these channels, hence they are often referred to as "delayed rectifier". These channels are selectively blocked by 4-aminopyridine (Grissmer et al. 1994; Adda et al. 1996; Oonuma et al. 2002). Human airway smooth muscle expresses various members of the Kv1 gene family including Kv1.1, Kv1.2 and Kv1.5, whilst no transcripts for Kv1.6, Kv3.1 nor Kv2 or Kv4 gene families were reported. This was consistent with the documentation of Kv channels in cultured human airway smooth muscle cells using patch-clamp techniques (Adda et al. 1996), although they appear to be poorly represented in freshly dissociated human airway smooth muscle cells (Snetkov et al. 1995, 1996). The role of these channels in regulating airway tone was also investigated in human bronchial tissue. The Kv inhibitor 4-aminopyridine induced a concentration-dependent increase in basal tone (Adda et al. 1996). This effect was not observed with dendrotoxin, which is reportedly more selective for Kv1.1 and Kv1.2 channels (Grissmer et al. 1994), thereby suggesting a role for Kv1.5 in the regulation of basal tone and membrane potential (Adda et al. 1996).

6.2
Calcium-Activated Potassium Channels

There a numerous reports of the existence of K_{Ca} channels in airways smooth muscle from a number of animal species which are characterized by their sensitivity to calcium, unitary conductance of various sizes—including small, intermediate and large—and selective inhibition by various toxins (Janssen 2002; Pelaia et al. 2002). The existence of a heterogeneous population of K_{Ca} channels in freshly isolated and cultured human airway smooth muscle was confirmed by their calcium dependence, large conductance (100–250 pS) and sensitivity to charybdotoxin, iberiotoxin, tetraethylammonium (TEA) but not 4-aminopyridine. The large conductance K_{Ca} channels were exclusively present in freshly isolated and cultured human bronchial smooth muscle cells (Snetkov et al. 1995, 1996) and these channels are also found in smooth muscle cells from more peripheral airways (Snetkov and Ward 1999), demonstrating that these channels are present throughout the respiratory tract.

The role of the large K_{Ca} channels in the control of airway smooth muscle is the subject of considerable debate. The K_{Ca} blockers were either ineffective or induced modest contraction of human bronchial smooth muscle, which would seem to rule out a role for these channels in the regulation of baseline tone (Miura et al. 1992a; Adda et al. 1996; Corompt et al. 1998). In contrast, a number of studies have reported that relaxation of human airway smooth muscle in response to β-adrenoceptor agonists and NO donors is, in part, a consequence of the opening of large conductance K_{Ca} (Miura et al. 1992a; Abderrahmane et al. 1998). β_2-Adrenoceptors couple to these channels either directly or indirectly through elevations in intracellular cyclic AMP (Kume et al. 1994), whilst NO appears to directly phosphorylate these channels (Abderrahmane et al. 1998) and is thought to be responsible for mediating relaxation to these agents. However, there is some debate as to whether this mechanism underlies the relaxation observed by these stimuli in airway smooth muscle (Corompt et al. 1998; Janssen 2002). An increase in probability of opening of voltage-dependent calcium channels is an indirect consequence of inhibition of K_{Ca} channels with iberiotoxin and charybdotoxin. Calcium influx via these channels would lead to the activation of the contractile apparatus, thereby resulting in functional antagonism of relaxation. Indeed, if these channels are made inoperable in the presence of L-type calcium channel blockers, then the inhibitory action of large conductance K_{Ca} channel blockers on relaxation to β_2-adrenoceptor agonist is removed (Muller-Schweinitzer and Fozard 1997; Corompt et al. 1998). Moreover, the large conductance K_{Ca} opener, NS1619, was ineffective in mediating relaxation of human bronchial tissues (Muller-Schweinitzer and Fozard 1997).

Whilst a strategy designed to open large-conductance K_{Ca} on airway smooth muscle does not appear to be a useful target for the development of novel bronchodilator agents, the possibility of interfering with the function of airway smooth muscle indirectly by targeting nerves is worth mentioning. Parasympathetic cholinergic nerves play an important role in regulating airway tone in vivo and the role of these channels in modulating the function of these nerves has been investigated. The suppression of contractile responses following stimulation of cholinergic nerves in human tracheal tissue is observed with μ-opioid agonists, which appear to be susceptible to inhibition by charybdotoxin (Miura et al. 1992b). Thus, the opening of such channels on airway parasympathetic nerves could lead to smooth muscle relaxation indirectly by reducing parasympathetic outflow. The K_{Ca} opener, NS1619, was shown to inhibit the release of acetylcholine from parasympathetic nerves and inhibit contraction of smooth muscle following nerve depolarization (Patel et al. 1998). However, the inability to reverse this effect with charybdotoxin would cast some doubt on the selectivity of NS1619 on cholinergic nerves. In contrast to these findings, NS1619 was shown to inhibit the activation of afferent nerves and contraction of airway smooth muscle following the endogenous release of neuropeptides from afferent nerves that were sensitive to iberiotoxin (Fox et al. 1997).

6.3
K_{ATP} Channels

K_{ATP} channels are characterized by their small unitary conductance (10–30 pS) and sensitivity to inhibition by ATP. Furthermore, these channels are opened by a variety of benzopyran derivatives such as levcromakalim, BRL 55834, bimakalim and a variety of other agents including pinacidil and minoxidil, whilst inhibitors of this channel include the sulphonylureas glibenclamide and tolbutamide (Edwards and Weston 1995). There is increasing evidence of the existence of multiple forms of this channel in a variety of tissues. This is determined at the molecular level by the formation of different complexes between the K_{IR} family ($K_{IR}6.x$; α-subunit) which provides the pore domain for this channel and the sulphonylurea receptor (SUR; β-subunit) which recognizes the sulphonylureas, potassium channel openers and nucleoside-diphosphate. The pancreatic K_{ATP} channel is composed of $K_{IR}6.2$/SUR1 complex (Seino 1999).

There is currently no electrophysiological evidence that K_{ATP} channels are found in human airway smooth muscle. The only indirect evidence stems from the findings that K_{ATP} channel openers relax airway smooth muscle and this response is sensitive to inhibition by sulphonylureas (Black et al. 1990; Cortijo et al. 1992; Miura et al. 1992b). It is possible that additional mechanisms might be responsible for the relaxation observed with K_{ATP} channel openers including alterations in calcium levels, inositol 1,4,5-trisphosphate (IP_3) synthesis, phosphodiesterase 3 inhibition or inducing a state of dephosphorylation of the Kv channel (Edwards and Weston 1995; Muller-Schweinitzer and Fozard 1997). Regardless of the mechanism, it appears that these channels do not regulate basal airway tone, since the addition of sulphonylureas alone does not elicit contraction of muscle (Miura et al. 1992b; Adda et al. 1996).

Since there are various isoforms of this family of ion channels, it is possible that the functional effects observed with K_{ATP} openers is mediated by different complexes of K_{IR} and SUR. Electrophysiological recordings made from cat tracheal smooth muscle cells reveal the presence of a K_{ATP} channel that is activated by levcromakalim, blocked by glibenclamide, has unitary conductance of 40 pS but is regulated by nucleoside diphosphates (NDP) (Teramoto et al. 2000). This channel has similar electrophysiological features with K_{NDP} channels observed in other smooth muscle cells, and at the molecular level it is characterized by a complex consisting of $K_{IR}6.1$/SUR2B (Sim et al. 2002; Cui et al. 2002).

The clinical efficacy of K_{ATP} channel openers has been investigated. The oral administration of cromakalim provided significant protection against nocturnal worsening of baseline lung function (Williams et al. 1990). However, there was no significant bronchoprotection against histamine or methacholine, nor was there evidence of bronchodilation following single oral administration of levcromakalim, despite the reporting of headache (Kidney et al. 1993). Similarly, the administration of bimakalim by the inhaled route in patients with chronic asthma was ineffective as a bronchodilator at doses which did not provoke any side-effect including headache or cardiovascular effects (Faurschou et al. 1994). These studies suggest that K_{ATP} channel openers have poor bronchodilator activity per se and the dose required to promote airway smooth muscle relaxation is associated with unwanted side-effects. The implication of these experimental observations if translated into the clinic are that airway-selective K_{ATP} channel openers can be developed that reduce bronchial hyperresponsiveness without any direct bronchodilator action and therefore would be a novel class of bronchoprotective agent.

In an experimental model, a range of benzopyran K_{ATP} channel openers were found to be ineffective in providing bronchoprotection against histamine in naïve animals. However, when animals were made hyperresponsive, these agents provided significant bronchoprotection (Chapman et al. 1991; Bucheit and Hofmann 1996). A similar pharmacological profile has been observed with the benzopyran, KCO912, which selectively suppresses bronchial hyperresponsiveness but has minimal direct action on airway smooth muscle. Moreover, when administered directly into the lung, it is relatively free from cardiovascular action (Bucheit et al. 2002). This suggests the possibility of developing novel K_{ATP} channel openers that have minimal action on airway smooth muscle per se, yet inhibit the development of bronchial hyperresponsiveness and are free from cardiovascular side-effects. The mechanism by which these agents selectively inhibit bronchial hyperresponsiveness is not clear, but a neural site of action is a distinct possibility since K_{ATP} channel openers have been shown to suppress cholinergic and afferent activity in the airways (Ichinose and Barnes 1990; Miura et al. 1992a) and to suppress cough (Poggioli et al. 1999).

6.4
Inward Rectifiers

Few studies have addressed the role of these channels in the regulation of airway smooth muscle function. As the name suggests, these channels conduct inward current more readily than outward current. However, a rise in extracellular potassium will lead to an alteration in the gating characteristics of the channel, leading to potassium efflux from cells, membrane hyperpolarization and smooth muscle relaxation. K_{IR} has been documented in human airway smooth muscle from peripheral airways (Snetkov and Ward 1999) and in cultured human airway smooth muscle cells (Oonuma et al. 2002). The expression of $K_{IR}2.1$ and not $K_{IR}1.1$, 2,2, 2,3 accounted for the inward rectification in these cells (Oonuma et al. 2002).

7
Conclusion

A variety of signalling pathways exists in airway smooth muscle that when activated mediate relaxation via a non-β_2-adrenoceptor mechanism (Fig. 1). It is possible that agents that stimulate these processes could potentially offer new therapeutic strategies for providing bronchoprotection and bronchodilation in respiratory disease. Many of the current agents used to stimulate these pathways suffer from a relatively short duration of action in comparison with long-acting β_2-adrenoceptor agonists, although this is a problem that can be overcome with chemistry. For example, chemical modification to VIP has led to analogues with significantly greater potency and duration of action than native VIP and with more potency than salbutamol.

Systemic side-effects remain a potential problem, since many of the processes that mediate relaxation of airway smooth muscle are also found in the cardiovascular system. However, developing drugs for the inhaled route (e.g. PDE3 inhibitor, olprinone) may eliminate this. Moreover, a new generation of potassium

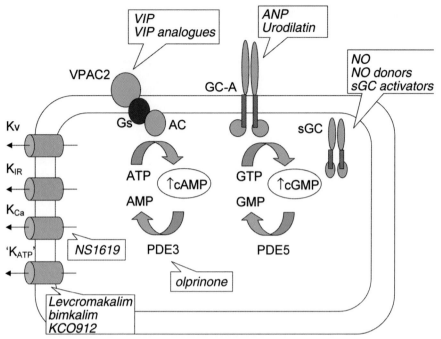

Fig. 1 Diagrammatic representation of a number of pathways that can be exploited for the development of new non-β_2-adrenoceptor bronchodilator agents. Relaxation of airway smooth muscle can be mediated following (1) activation of cyclic AMP (*cAMP*)-dependent pathways as a consequence of signalling via G protein (*Gs*)-coupled VPAC2 and adenylyl cyclase (*AC*) by VIP or VIP analogues. The intracellular levels of cAMP are controlled by phosphodiesterases (*PDE*), and drugs which inhibit PDE3 also induce relaxation of airway smooth muscle. (2) Relaxation can also be mediated following activation of cyclic GMP (*cGMP*) via soluble guanylyl cyclase (*sGC*) by NO, NO donors, sGC activators, or particulate guanylyl cyclase (*GC-A*) by ANP or urodilatin. (3) Opening of potassium channels has also been implicated in the regulation of airway smooth muscle function, although there is some debate as to the role of the different families of ion channels in the modulation of airways smooth muscle activity. To date, K_{ATP} openers have been shown to provide significant bronchoprotection in animal models of hyperresponsiveness, but interestingly, this may not be a consequence of a direct action on airway smooth muscle but on neuronal cells

channel openers appears to demonstrate airway over cardiovascular selectivity, and of particular interest is the possibility of developing agents which inhibit bronchial hyperresponsiveness that is not attributed to a direct action on airway smooth muscle (e.g. potassium channel openers). In other situations, it may be possible to stimulate multiple pathways simultaneously with a combination of drugs, in order to obtain additive effects. Thus, the co-administration of salbutamol and urodilatin provided greater bronchodilation than either agent alone (Fluge et al. 1999), whilst the addition of a NO moiety to salbutamol resulted in greater bronchodilation than the parent compound alone (Sardina et al. 2002).

There are potential avenues for the development of novel bronchodilator agents, and the challenge will be in discovering agents that are superior to long-acting β_2-adrenoceptor agonists.

Acknowledgements. The author acknowledges the financial support of the National Asthma Campaign, UK.

References

Abderrahmane A, Salvail D, Dumoulin M, Garon J, Cadieux A, Rousseau E (1998) Direct activation of K(Ca) channel in airway smooth muscle by nitric oxide: involvement of a nitrothiosylation mechanism? Am J Respir Cell Mol Biol 19:485–497

Adamou JE, Aiyar N, Van Horn S, Elshourbagy NA (1995) Cloning and functional characterization of the human vasoactive intestinal peptide (VIP)-2 receptor. Biochem Biophys Res Commun 209:385–392

Adda S, Fleischmann BK, Freedman BD, Yu M, Hay DW, Kotlikoff MI (1996) Expression and function of voltage-dependent potassium channel genes in human airway smooth muscle. J Biol Chem 271:13239–13243

Altiere RJ, Diamond L (1984) VIP as a bronchodilator. Lancet 1:162–163

Angus RM, McCallum MJ, Thomson NC (1994a) Effect of inhaled atrial natriuretic peptide on methacholine induced bronchoconstriction in asthma. Clin Exp Allergy 24:784–788

Angus RM, Millar EA, Chalmers GW, Thomson NC (1995) Effect of inhaled atrial natriuretic peptide and a neutral endopeptidase inhibitor on histamine-induced bronchoconstriction. Am J Respir Crit Care Med 151:2003–2005

Angus RM, Nally JE, McCall R, Young LC, McGrath JC, Thomson NC (1994b) Modulation of the effect of atrial natriuretic peptide in human and bovine bronchi by phosphoramidon. Clin Sci (Lond) 86:291–295

Bardin PG, Dorward MA, Lampe FC, Franke B, Holgate ST (1998) Effect of selective phosphodiesterase 3 inhibition on the early and late asthmatic responses to inhaled allergen. Br J Clin Pharmacol 45:387–391

Bernareggi M, Mitchell JA, Barnes PJ, Belvisi MG (1997) Dual action of nitric oxide on airway plasma leakage. Am J Respir Crit Care Med 155:869–874

Billington CK, Joseph SK, Swan C, Scott MG, Jobson TM, Hall IP (1999) Modulation of human airway smooth muscle proliferation by type 3 phosphodiesterase inhibition. Am J Physiol 276:L412-L419

Black JL, Armour CL, Johnson PR, Alouan LA, Barnes PJ (1990) The action of a potassium channel activator, BRL 38227 (lemakalim), on human airway smooth muscle. Am Rev Respir Dis 142:1384–1389

Blease K, Kunkel SL, Hogaboam CM (2000) Acute inhibition of nitric oxide exacerbates airway hyperresponsiveness, eosinophilia and C-C chemokine generation in a murine model of fungal asthma. Inflamm Res 49:297–304

Bolin DR, Cottrell J, Garippa R, Michalewsky J, Rinaldi N, Simko B, O'Donnell M (1995) Structure-activity studies on the vasoactive intestinal peptide pharmacophore. 1. Analogs of tyrosine. Int J Pept Protein Res 46:279–289

Buchheit KH, Hofmann A (1996) KATP channel openers reverse immune complex-induced airways hyperreactivity independently of smooth muscle relaxation. Naunyn Schmiedebergs Arch Pharmacol 354:355–361

Buchheit KH, Manley PW, Quast U, Russ U, Mazzoni L, Fozard JR (2002) KCO912: a potent and selective opener of ATP-dependent potassium (K(ATP)) channels which suppresses airways hyperreactivity at doses devoid of cardiovascular effects. Naunyn Schmiedebergs Arch Pharmacol 365:220–230

Bundgaard A, Enehjelm SD, Aggestrup S (1983) Pretreatment of exercise-induced asthma with inhaled vasoactive intestinal peptide (VIP). Eur J Respir Dis Suppl 128 (Pt 2):427–429

Busto R, Carrero I, Guijarro LG, Solano RM, Zapatero J, Noguerales F, Prieto JC (1999) Expression, pharmacological, and functional evidence for PACAP/VIP receptors in human lung. Am J Physiol 277:L42-L48

Busto R, Prieto JC, Bodega G, Zapatero J, Carrero I (2000) Immunohistochemical localization and distribution of VIP/PACAP receptors in human lung. Peptides 21:265–269

Candenas ML, Naline E, Puybasset L, Devillier P, Advenier C (1991) Effect of atrial natriuretic peptide and on atriopeptins on the human isolated bronchus. Comparison with the reactivity of the guinea-pig isolated trachea. Pulm Pharmacol 4:120–125

Carstairs JR, Barnes PJ (1986) Visualization of vasoactive intestinal peptide receptors in human and guinea pig lung. J Pharmacol Exp Ther 239:249–255

Chapman ID, Mazzoni L, Morley J (1991) Actions of SDZ PCO 400 and cromakalim on airway smooth muscle in vivo. Agents Actions Suppl 34:53–62

Corompt E, Bessard G, Lantuejoul S, Naline E, Advenier C, Devillier P (1998) Inhibitory effects of large Ca^{2+}-activated K^+ channel blockers on beta-adrenergic- and NO-donor-mediated relaxations of human and guinea-pig airway smooth muscles. Naunyn Schmiedebergs Arch Pharmacol 357:77–86

Cortijo J, Sarria B, Pedros C, Perpina M, Paris F, Morcillo E (1992) The relaxant effects of cromakalim (BRL 34915) on human isolated airway smooth muscle. Naunyn Schmiedebergs Arch Pharmacol 346:462–468

Cui Y, Tran S, Tinker A, Clapp LH (2002) The molecular composition of K(ATP) channels in human pulmonary artery smooth muscle cells and their modulation by growth. Am J Respir Cell Mol Biol 26:135–143

de Boer J, Philpott AJ, van Amsterdam RG, Shahid M, Zaagsma J, Nicholson CD (1992) Human bronchial cyclic nucleotide phosphodiesterase isoenzymes: biochemical and pharmacological analysis using selective inhibitors. Br J Pharmacol 106:1028–1034

De Sanctis GT, MacLean JA, Hamada K, Mehta S, Scott JA, Jiao A, Yandava CN, Kobzik L, Wolyniec WW, Fabian AJ, Venugopal CS, Grasemann H, Huang PL, Drazen JM (1999) Contribution of nitric oxide synthases 1, 2, and 3 to airway hyperresponsiveness and inflammation in a murine model of asthma. J Exp Med 189:1621–1630

De Sanctis GT, Mehta S, Kobzik L, Yandava C, Jiao A, Huang PL, Drazen JM (1997) Contribution of type I NOS to expired gas NO and bronchial responsiveness in mice. Am J Physiol 273:L883-L888

Dweik RA, Comhair SA, Gaston B, Thunnissen FB, Farver C, Thomassen MJ, Kavuru M, Hammel J, Abu-Soud HM, Erzurum SC (2001) NO chemical events in the human airway during the immediate and late antigen-induced asthmatic response. Proc Natl Acad Sci U S A 98:2622–2627

Edwards G, Weston AH (1995) Pharmacology of the potassium channel openers. Cardiovasc Drugs Ther 9 Suppl 2:185–193

Ellis JL (1997) Role of soluble guanylyl cyclase in the relaxations to a nitric oxide donor and to nonadrenergic nerve stimulation in guinea pig trachea and human bronchus. J Pharmacol Exp Ther 280:1215–1218

Ellis JL, Undem BJ (1992) Inhibition by L-NG-nitro-L-arginine of nonadrenergic-noncholinergic-mediated relaxations of human isolated central and peripheral airway. Am Rev Respir Dis 146:1543–1547

Ellis JL, Undem BJ (1994) Role of cysteinyl-leukotrienes and histamine in mediating intrinsic tone in isolated human bronchi. Am J Respir Crit Care Med 149:118–122

Erjefalt JS, Erjefalt I, Sundler F, Persson CG (1994) Mucosal nitric oxide may tonically suppress airways plasma exudation. Am J Respir Crit Care Med 150:227–232

Faurschou P, Mikkelsen KL, Steffensen I, Franke B (1994) The lack of bronchodilator effect and the short-term safety of cumulative single doses of an inhaled potassium channel opener (bimakalim) in adult patients with mild to moderate bronchial asthma. Pulm Pharmacol 7:293-297

Feder LS, Stelts D, Chapman RW, Manfra D, Crawley Y, Jones H, Minnicozzi M, Fernandez X, Paster T, Egan RW, Kreutner W, Kung TT (1997) Role of nitric oxide on eosinophilic lung inflammation in allergic mice. Am J Respir Cell Mol Biol 17:436-442

Ferreira HH, Bevilacqua E, Gagioti SM, De Luca IM, Zanardo RC, Teixeira CE, Sannomiya P, Antunes E, De Nucci G (1998) Nitric oxide modulates eosinophil infiltration in antigen-induced airway inflammation in rats. Eur J Pharmacol 358:253-259

Fischer A, Canning BJ, Kummer W (1996) Correlation of vasoactive intestinal peptide and nitric oxide synthase with choline acetyltransferase in the airway innervation. Ann N Y Acad Sci 805:717-722

Fischer A, Hoffmann B (1996) Nitric oxide synthase in neurons and nerve fibers of lower airways and in vagal sensory ganglia of man. Correlation with neuropeptides. Am J Respir Crit Care Med 154:209-216

Fischer A, Mundel P, Mayer B, Preissler U, Philippin B, Kummer W (1993) Nitric oxide synthase in guinea pig lower airway innervation. Neurosci Lett 149:157-160

Fluge T, Fabel H, Wagner TO, Schneider B, Forssmann WG (1995) Bronchodilating effects of natriuretic and vasorelaxant peptides compared to salbutamol in asthmatics. Regul Pept 59:357-370

Fluge T, Forssmann WG, Kunkel G, Schneider B, Mentz P, Forssmann K, Barnes PJ, Meyer M (1999) Bronchodilation using combined urodilatin—albuterol administration in asthma: a randomized, double-blind, placebo-controlled trial. Eur J Med Res 4:411-415

Fox AJ, Barnes PJ, Venkatesan P, Belvisi MG (1997) Activation of large conductance potassium channels inhibits the afferent and efferent function of airway sensory nerves in the guinea pig. J Clin Invest 99:513-519

Fujimura M, Kamio Y, Saito M, Hashimoto T, Matsuda T (1995) Bronchodilator and bronchoprotective effects of cilostazol in humans in vivo. Am J Respir Crit Care Med 151:222-225

Gaston B, Drazen JM, Jansen A, Sugarbaker DA, Loscalzo J, Richards W, Stamler JS (1994) Relaxation of human bronchial smooth muscle by S-nitrosothiols in vitro. J Pharmacol Exp Ther 268:978-984

Gaston B, Reilly J, Drazen JM, Fackler J, Ramdev P, Arnelle D, Mullins ME, Sugarbaker DJ, Chee C, Singel DJ, . (1993) Endogenous nitrogen oxides and bronchodilator S-nitrosothiols in human airways. Proc Natl Acad Sci U S A 90:10957-10961

Gaston B, Sears S, Woods J, Hunt J, Ponaman M, McMahon T, Stamler JS (1998) Bronchodilator S-nitrosothiol deficiency in asthmatic respiratory failure. Lancet 351:1317-1319

Gourlet P, Vertongen P, Vandermeers A, Vandermeers-Piret MC, Rathe J, De Neef P, Waelbroeck M, Robberecht P (1997) The long-acting vasoactive intestinal polypeptide agonist RO 25-1553 is highly selective of the VIP2 receptor subclass. Peptides 18:403-408

Grissmer S, Nguyen AN, Aiyar J, Hanson DC, Mather RJ, Gutman GA, Karmilowicz MJ, Auperin DD, Chandy KG (1994) Pharmacological characterization of five cloned voltage-gated K+ channels, types Kv1.1, 1.2, 1.3, 1.5, and 3.1, stably expressed in mammalian cell lines. Mol Pharmacol 45:1227-1234

Groneberg DA, Hartmann P, Dinh QT, Fischer A (2001a) Expression and distribution of vasoactive intestinal polypeptide receptor VPAC(2) mRNA in human airways. Lab Invest 81:749-755

Groneberg DA, Springer J, Fischer A (2001b) Vasoactive intestinal polypeptide as mediator of asthma. Pulm Pharmacol Ther 14:391-401

Hamad AM, Johnson SR, Knox AJ (1999) Antiproliferative effects of NO and ANP in cultured human airway smooth muscle. Am J Physiol 277:L910-L918

Hamad AM, Range S, Holland E, Knox AJ (1997) Regulation of cGMP by soluble and particulate guanylyl cyclases in cultured human airway smooth muscle. Am J Physiol 273:L807-L813

Hamid Q, Springall DR, Riveros-Moreno V, Chanez P, Howarth P, Redington A, Bousquet J, Godard P, Holgate S, Polak JM (1993) Induction of nitric oxide synthase in asthma. Lancet Vol 342:-8887

Harmar AJ, Arimura A, Gozes I, Journot L, Laburthe M, Pisegna JR, Rawlings SR, Robberecht P, Said SI, Sreedharan SP, Wank SA, Waschek JA (1998) International Union of Pharmacology. XVIII. Nomenclature of receptors for vasoactive intestinal peptide and pituitary adenylate cyclase-activating polypeptide. Pharmacol Rev 50:265–270

Heim JM, Kiefersauer S, Fulle HJ, Gerzer R (1989) Urodilatin and beta-ANF: binding properties and activation of particulate guanylate cyclase. Biochem Biophys Res Commun 163:37–41

Hobbs AJ (2002) Soluble guanylate cyclase: an old therapeutic target re-visited. Br J Pharmacol 136:637–640

Hogman M, Frostell CG, Hedenstrom H, Hedenstierna G (1993) Inhalation of nitric oxide modulates adult human bronchial tone. Am Rev Respir Dis 148:1474–1478

Hulks G, Mohammed AF, Jardine AG, Connell JM, Thomson NC (1991) Circulating plasma concentrations of atrial natriuretic peptide and catecholamines in response to maximal exercise in normal and asthmatic subjects. Thorax 46:824–828

Ichinose M, Barnes PJ (1990) A potassium channel activator modulates both excitatory noncholinergic and cholinergic neurotransmission in guinea pig airways. J Pharmacol Exp Ther 252:1207–1212

Iijima H, Duguet A, Eum SY, Hamid Q, Eidelman DH (2001) Nitric oxide and protein nitration are eosinophil dependent in allergen-challenged mice. Am J Respir Crit Care Med 163:1233–1240

Janssen LJ (2002) Ionic mechanisms and Ca(2+) regulation in airway smooth muscle contraction: do the data contradict dogma? Am J Physiol Lung Cell Mol Physiol 282:L1161-L1178

Kacmarek RM, Ripple R, Cockrill BA, Bloch KJ, Zapol WM, Johnson DC (1996) Inhaled nitric oxide. A bronchodilator in mild asthmatics with methacholine-induced bronchospasm. Am J Respir Crit Care Med 153:128–135

Kakuyama M, Ahluwalia A, Rodrigo J, Vallance P (1999) Cholinergic contraction is altered in nNOS knockouts. Cooperative modulation of neural bronchoconstriction by nNOS and COX. Am J Respir Crit Care Med 160:2072–2078

Kallstrom BL, Waldeck B (2001) Bronchodilating properties of the VIP receptor agonist Ro 25-1553 compared to those of formoterol on the guinea-pig isolated trachea. Eur J Pharmacol 430:335–340

Kharitonov SA, O'Connor BJ, Evans DJ, Barnes PJ (1995) Allergen-induced late asthmatic reactions are associated with elevation of exhaled nitric oxide. Am J Respir Crit Care Med 151:1894–1899

Kharitonov SA, Yates D, Robbins RA, Logan-Sinclair R, Shinebourne EA, Barnes PJ (1994) Increased nitric oxide in exhaled air of asthmatic patients. Lancet 343:133–135

Kidney JC, Fuller RW, Worsdell YM, Lavender EA, Chung KF, Barnes PJ (1993) Effect of an oral potassium channel activator, BRL 38227, on airway function and responsiveness in asthmatic patients: comparison with oral salbutamol. Thorax 48:130–133

Kinhult J, Andersson JA, Uddman R, Stjarne P, Cardell LO (2000) Pituitary adenylate cyclase-activating peptide 38 a potent endogenously produced dilator of human airways. Eur Respir J 15:243–247

Koarai A, Ichinose M, Sugiura H, Yamagata S, Hattori T, Shirato K (2000) Allergic airway hyperresponsiveness and eosinophil infiltration is reduced by a selective iNOS inhibitor, 1400 W, in mice. Pulm Pharmacol Ther 13:267–275

Kobzik L, Bredt DS, Lowenstein CJ, Drazen J, Gaston B, Sugarbaker D, Stamler JS (1993) Nitric oxide synthase in human and rat lung: immunocytochemical and histochemical localization. Am J Respir Cell Mol Biol 9:371–377

Kume H, Hall IP, Washabau RJ, Takagi K, Kotlikoff MI (1994) Beta-adrenergic agonists regulate KCa channels in airway smooth muscle by cAMP-dependent and -independent mechanisms. J Clin Invest 93:371–379

Laitinen A, Partanen M, Hervonen A, Pelto-Huikko M, Laitinen LA (1985) VIP like immunoreactive nerves in human respiratory tract. Light and electron microscopic study. Histochemistry 82:313–319

Martinez FD, Graves PE, Baldini M, Solomon S, Erickson R (1997) Association between genetic polymorphisms of the beta2-adrenoceptor and response to albuterol in children with and without a history of wheezing. J Clin Invest 100:3184–3188

McAlpine LG, Hulks G, Thomson NC (1992) Effect of atrial natriuretic peptide given by intravenous infusion on bronchoconstriction induced by ultrasonically nebulized distilled water (fog). Am Rev Respir Dis 146:912–915

Miura M, Belvisi MG, Stretton CD, Yacoub MH, Barnes PJ (1992a) Role of K+ channels in the modulation of cholinergic neural responses in guinea-pig and human airways. J Physiol 455:1–15

Miura M, Belvisi MG, Stretton CD, Yacoub MH, Barnes PJ (1992b) Role of potassium channels in bronchodilator responses in human airways. Am Rev Respir Dis 146:132–136

Moya MP, Gow AJ, McMahon TJ, Toone EJ, Cheifetz IM, Goldberg RN, Stamler JS (2001) S-nitrosothiol repletion by an inhaled gas regulates pulmonary function. Proc Natl Acad Sci U S A 98:5792–5797

Muller-Schweinitzer E, Fozard JR (1997) SCA 40: studies of the relaxant effects on cryopreserved human airway and vascular smooth muscle. Br J Pharmacol 120:1241–1248

Myou S, Fujimura M, Kamio Y, Ishiura Y, Tachibana H, Hirose T, Hashimoto T, Matsuda T (1999) Bronchodilator effect of inhaled olprinone, a phosphodiesterase 3 inhibitor, in asthmatic patients. Am J Respir Crit Care Med 160:817–820

Naline E, Bardou M, Devillier P, Molimard M, Dumas M, Chalon P, Manara L, Advenier C (2000) Inhibition by SR 59119A of isoprenaline-, forskolin- and VIP-induced relaxation of human isolated bronchi. Pulm Pharmacol Ther 13:167–174

Nicole P, Lins L, Rouyer-Fessard C, Drouot C, Fulcrand P, Thomas A, Couvineau A, Martinez J, Brasseur R, Laburthe M (2000) Identification of key residues for interaction of vasoactive intestinal peptide with human VPAC1 and VPAC2 receptors and development of a highly selective VPAC1 receptor agonist. Alanine scanning and molecular modeling of the peptide. J Biol Chem 275:24003–24012

Nohr D, Eiden LE, Weihe E (1995) Coexpression of vasoactive intestinal peptide, calcitonin gene-related peptide and substance P immunoreactivity in parasympathetic neurons of the rhesus monkey lung. Neurosci Lett 199:25–28

O'Donnell M, Garippa RJ, Rinaldi N, Selig WM, Simko B, Renzetti L, Tannu SA, Wasserman MA, Welton A, Bolin DR (1994a) Ro 25-1553: a novel, long-acting vasoactive intestinal peptide agonist. Part I: In vitro and in vivo bronchodilator studies. J Pharmacol Exp Ther 270:1282–1288

O'Donnell M, Garippa RJ, Rinaldi N, Selig WM, Tocker JE, Tannu SA, Wasserman MA, Welton A, Bolin DR (1994b) Ro 25-1553: a novel, long-acting vasoactive intestinal peptide agonist. Part II: Effect on in vitro and in vivo models of pulmonary anaphylaxis. J Pharmacol Exp Ther 270:1289–1294

Oonuma H, Iwasawa K, Iida H, Nagata T, Imuta H, Morita Y, Yamamoto K, Nagai R, Omata M, Nakajima T (2002) Inward rectifier K(+) current in human bronchial smooth muscle cells: inhibition with antisense oligonucleotides targeted to Kir2.1 mRNA. Am J Respir Cell Mol Biol 26:371–379

Palmer JB, Cuss FM, Barnes PJ (1986) VIP and PHM and their role in nonadrenergic inhibitory responses in isolated human airways. J Appl Physiol 61:1322–1328

Patel HJ, Giembycz MA, Keeling JE, Barnes PJ, Belvisi MG (1998) Inhibition of cholinergic neurotransmission in guinea pig trachea by NS1619, a putative activator of large-conductance, calcium-activated potassium channels. J Pharmacol Exp Ther 286:952–958

Pelaia G, Gallelli L, Vatrella A, Grembiale RD, Maselli R, De Sarro GB, Marsico SA (2002) Potential role of potassium channel openers in the treatment of asthma and chronic obstructive pulmonary disease. Life Sci 70:977–990

Perreault T, Gutkowska J (1995) Role of atrial natriuretic factor in lung physiology and pathology. Am J Respir Crit Care Med 151:226–242

Poggioli R, Benelli A, Arletti R, Cavazzuti E, Bertolini A (1999) Antitussive effect of K+ channel openers. Eur J Pharmacol 371:39–42

Potter LR, Hunter T (2001) Guanylyl cyclase-linked natriuretic peptide receptors: structure and regulation. J Biol Chem 276:6057–6060

Qian Y, Naline E, Karlsson JA, Raeburn D, Advenier C (1993) Effects of rolipram and siguazodan on the human isolated bronchus and their interaction with isoprenaline and sodium nitroprusside. Br J Pharmacol 109:774–778

Rabe KF, Tenor H, Dent G, Schudt C, Liebig S, Magnussen H (1993) Phosphodiesterase isozymes modulating inherent tone in human airways: identification and characterization. Am J Physiol 264:L458-L464

Ricciardolo FL, Geppetti P, Mistretta A, Nadel JA, Sapienza MA, Bellofiore S, Di Maria GU (1996) Randomised double-blind placebo-controlled study of the effect of inhibition of nitric oxide synthesis in bradykinin-induced asthma. Lancet 348:374–377

Robberecht P, Waelbroeck M, De Neef P, Camus JC, Coy DH, Christophe J (1988) Pharmacological characterization of VIP receptors in human lung membranes. Peptides 9:339–345

Said, S. I., 1998, Antiinflammatory actions of VIP in lungs and airways in Proinflammatory and Antiinflammatory Peptides: Said, S. I., Ed., Marcel Dekker, New York.

Saleh D, Ernst P, Lim S, Barnes PJ, Giaid A (1998) Increased formation of the potent oxidant peroxynitrite in the airways of asthmatic patients is associated with induction of nitric oxide synthase: effect of inhaled glucocorticoid. FASEB J 12:929–937

Sanna A, Kurtansky A, Veriter C, Stanescu D (1994) Bronchodilator effect of inhaled nitric oxide in healthy men. Am J Respir Crit Care Med 150:1702–1704

Sardina, M, Daussogne C, Del Soldato P (2002) New NO-releasing compounds for the treatment of inflammatory airway diseases. XIVth World Congress of Pharmacology [New Drugs for Respiratory Diseases, San Diego 3rd-5th July].

Schmidt DT, Ruhlmann E, Waldeck B, Branscheid D, Luts A, Sundler F, Rabe KF (2001) The effect of the vasoactive intestinal polypeptide agonist Ro 25-1553 on induced tone in isolated human airways and pulmonary artery. Naunyn Schmiedebergs Arch Pharmacol 364:314–320

Schmidt DT, Watson N, Dent G, Ruhlmann E, Branscheid D, Magnussen H, Rabe KF (2000) The effect of selective and non-selective phosphodiesterase inhibitors on allergen- and leukotriene C(4)-induced contractions in passively sensitized human airways. Br J Pharmacol 131:1607–1618

Schultz HD, Gardner DG, Deschepper CF, Coleridge HM, Coleridge JC (1988) Vagal C-fiber blockade abolishes sympathetic inhibition by atrial natriuretic factor. Am J Physiol 255:R6-13

Sears MR (2001) Short-acting beta-agonist research: a perspective. 1997. Can Respir J 8:349–355

Seino S (1999) ATP-sensitive potassium channels: a model of heteromultimeric potassium channel/receptor assemblies. Annu Rev Physiol 61:337–362

Sharara AM, Hijazi M, Tarawneh M, Ind PW (1998) Nebulized glyceryl trinitrate exerts acute bronchodilator effects in patients with acute bronchial asthma. Pulm Pharmacol Ther 11:65–70

Shaul PW, North AJ, Wu LC, Wells LB, Brannon TS, Lau KS, Michel T, Margraf LR, Star RA (1994) Endothelial nitric oxide synthase is expressed in cultured human bronchiolar epithelium. J Clin Invest 94:2231–2236

Sim JH, Yang DK, Kim YC, Park SJ, Kang TM, So I, Kim KW (2002) ATP-sensitive K(+) channels composed of Kir6.1 and SUR2B subunits in guinea pig gastric myocytes. Am J Physiol Gastrointest Liver Physiol 282:G137-G144

Snetkov VA, Hirst SJ, Twort CH, Ward JP (1995) Potassium currents in human freshly isolated bronchial smooth muscle cells. Br J Pharmacol 115:1117–1125

Snetkov VA, Hirst SJ, Ward JP (1996) Ion channels in freshly isolated and cultured human bronchial smooth muscle cells. Exp Physiol 81:791–804

Snetkov VA, Ward JP (1999) Ion currents in smooth muscle cells from human small bronchioles: presence of an inward rectifier K+ current and three types of large conductance K+ channel. Exp Physiol 84:835–8460

Sreedharan SP, Huang JX, Cheung MC, Goetzl EJ (1995) Structure, expression, and chromosomal localization of the type I human vasoactive intestinal peptide receptor gene. Proc Natl Acad Sci U S A 92:2939–2943

Tam EK, Franconi GM, Nadel JA, Caughey GH (1990) Protease inhibitors potentiate smooth muscle relaxation induced by vasoactive intestinal peptide in isolated human bronchi. Am J Respir Cell Mol Biol 2:449–452

Taylor DA, McGrath JL, O'Connor BJ, Barnes PJ (1998a) Allergen-induced early and late asthmatic responses are not affected by inhibition of endogenous nitric oxide. Am J Respir Crit Care Med 158:99–106

Taylor DA, McGrath JL, Orr LM, Barnes PJ, O'Connor BJ (1998b) Effect of endogenous nitric oxide inhibition on airway responsiveness to histamine and adenosine-5'-monophosphate in asthma. Thorax 53:483–489

Teramoto N, Nakashima T, Ito Y (2000) Properties and pharmacological modification of ATP-sensitive K(+) channels in cat tracheal myocytes. Br J Pharmacol 130:625–635

Thomassen MJ, Raychaudhuri B, Dweik RA, Farver C, Buhrow L, Malur A, Connors MJ, Drazba J, Hammel J, Erzurum SC, Kavuru MS (1999) Nitric oxide regulation of asthmatic airway inflammation with segmental allergen challenge. J Allergy Clin Immunol 104:1174–1182

Torphy TJ, Undem BJ, Cieslinski LB, Luttmann MA, Reeves ML, Hay DW (1993) Identification, characterization and functional role of phosphodiesterase isozymes in human airway smooth muscle. J Pharmacol Exp Ther 265:1213–1223

Trifilieff A, Fujitani Y, Mentz F, Dugas B, Fuentes M, Bertrand C (2000) Inducible nitric oxide synthase inhibitors suppress airway inflammation in mice through down-regulation of chemokine expression. J Immunol 165:1526–1533

Turner PJ, Maggs JR, Foreman JC (2000) Induction by inhibitors of nitric oxide synthase of hyperresponsiveness in the human nasal airway. Br J Pharmacol 131:363–369

Ward JK, Barnes PJ, Springall DR, Abelli L, Tadjkarimi S, Yacoub MH, Polak JM, Belvisi MG (1995a) Distribution of human i-NANC bronchodilator and nitric oxide-immunoreactive nerves. Am J Respir Cell Mol Biol 13:175–184

Ward JK, Barnes PJ, Tadjkarimi S, Yacoub MH, Belvisi MG (1995b) Evidence for the involvement of cGMP in neural bronchodilator responses in humal trachea. J Physiol 483 (Pt 2):525–536

Wei Y, Mojsov S (1996) Tissue specific expression of different human receptor types for pituitary adenylate cyclase activating polypeptide and vasoactive intestinal polypeptide: implications for their role in human physiology. J Neuroendocrinol 8:811–817

Williams AJ, Lee TH, Cochrane GM, Hopkirk A, Vyse T, Chiew F, Lavender E, Richards DH, Owen S, Stone P, . (1990) Attenuation of nocturnal asthma by cromakalim. Lancet 336:334–336

Xia M, Sreedharan SP, Bolin DR, Gaufo GO, Goetzl EJ (1997) Novel cyclic peptide agonist of high potency and selectivity for the type II vasoactive intestinal peptide receptor. J Pharmacol Exp Ther 281:629–633

Xiong Y, Karupiah G, Hogan SP, Foster SP, Ramsay AJ (1999) Inhibition of allergic airway inflammation in mice lacking nitric oxide synthase 2. J Immunol 162:445–452

Selective Phosphodiesterase Inhibitors in the Treatment of Respiratory Disease

N. A. Jones · D. Spina · C. P. Page

The Sackler Institute of Pulmonary Pharmacology,
Pharmacology and Therapeutics Division, GKT School of Biomedical Sciences,
King's College London, 5th Floor Hodgkin Building, Guy's Campus,
London, SE1 9RT, UK
e-mail: clive.page@kcl.ac.uk

1	Introduction	180
1.1	Classification of Phosphodiesterase Enzymes	180
2	Characteristic Features of Asthma and COPD	181
3	PDE Expression in Allergic Disease	182
3.1	Allergy	182
3.2	Asthma	183
4	Properties and Classification of PDE4	184
5	Effect of PDE Inhibition on Inflammatory Cell Function	189
5.1	Mast Cells and Basophils	189
5.2	Neutrophil	190
5.3	Eosinophil	191
5.4	T Lymphocyte	192
5.5	B Lymphocyte	194
5.6	Monocyte	195
5.7	Macrophage	196
5.8	Bronchial Epithelium	197
5.9	Vascular Endothelium	197
5.10	Vascular Smooth Muscle	199
5.11	Airway Smooth Muscle	200
6	Clinical Studies of PDE Inhibitors in Asthma and COPD	201
7	Conclusion	202
	References	202

Abstract At least 11 families of distinct phosphodiesterase (PDE) isoenzymes are known to regulate the function of many cells secondary to altering the intracellular levels of second messengers including cyclic $3',5'$-monophosphate (cyclic AMP) and cyclic $3',5'$-guanosine monophosphate (cyclic GMP). While there is a wide distribution of these enzymes throughout the body, advances in our understanding of the molecular aspects of PDEs and accurate determination of their cellular distribution has allowed development of isoenzyme-selective inhibitors as potential therapeutic agents. Cells thought to participate in the pathogenesis of inflammatory diseases, including asthma and chronic obstructive pulmonary disease (COPD), preferentially express PDE4. This finding has stimulated the search for highly selective inhibitors of these enzymes. Such drugs offer an exciting opportunity to selectively downregulate inflammatory cell function as a novel therapeutic approach in the treatment of airway disease.

Keywords Phosphodiesterase (PDE) inhibitor · Inflammation · Asthma · COPD

1
Introduction

At least 11 families of distinct phosphodiesterase (PDE) enzymes, including PDE1-7 (Torphy1998) and PDE8-11 (Soderling et al. 2000), are known to exist based upon a variety of criteria including substrate specificity, inhibitor potency, enzyme kinetics and amino acid sequence. These enzymes are distributed widely throughout the body, differentially expressed in cells and localized to different compartments within cells. The functional significance of the subcellular localization of PDEs is not completely understood, although there is a considerable body of evidence to suggest that the expression of PDE in cellular domains can tightly regulate the levels of cyclic nucleotides in the vicinity of effector proteins and are therefore implicated in the regulation of cell function (Houslay et al. 1998).

1.1
Classification of Phosphodiesterase Enzymes

PDE enzymes function by hydrolysing the phosphodiester bond of the second messenger molecules cyclic 3′,5′-adenosine monophosphate (cAMP) and cyclic 3′,5′-guanosine monophosphate (cGMP). This converts cAMP and cGMP to their inactive 5′-mononucleotides; adenosine monophosphate (AMP) and guanosine monophosphate (GMP). These products are incapable of activating specific cyclic nucleotide-dependent protein kinase cascades.

The PDE enzyme family consists of a growing number of genetically heterologous isoenzymes (Table 1).

Regarding substrate affinities, PDE4 and PDE7 are highly selective for cAMP. Although PDE3 hydrolyses cAMP and cGMP with equal affinity (K_m: 0.1–0.5 μM), its V_{max} ("velocity" of action) for cAMP is fivefold greater than for cGMP. Functionally, then, PDE3 favours cAMP. Conversely, cGMP is the preferred substrate for PDE5 and PDE6, whereas PDE1 and PDE2 hydrolyse either cyclic nucleotide. Adding further complexity to the functional role taken by various PDEs in intact tissues is the fact that several are subject to short-term allo-

Table 1 Characteristics and properties of PDE isoenzymes

Family	Specific property	K_m (μM) cAMP	K_m (μM) cGMP
1	Ca2+/calmodulin stimulated	1–30	3
2	cGMP stimulated	50	50
3	cGMP inhibited	0.2	0.3
4	cAMP specific	4	>100
5	CGMP specific	150	1
6	Photoreceptor	60	>100
7	High-affinity cAMP specific	0.2	>100
8	High-affinity cAMP specific	0.055	124
9	High-affinity cGMP specific	230	0.17
10	Unknown	0.05	3
11	Unknown	1.04	0.52

steric regulation by endogenous activators or inhibitors. For example, PDE1 is allosterically activated by Ca^{2+}/calmodulin (Sharma et al. 1986).

Each of the families of PDE isoenzymes is populated by at least one, and as many as four, distinct gene products; thus isoenzymes are further subdivided into subtypes, which have a high percentage genetic homology (70%–90%). Subtypes may also undergo further post-translational modification and this can result in a large number of splice variants. The result of this is an array of enzymes with distinct kinetic characteristics, regulatory properties and subcellular distributions, allowing specific drug design and targeting of specific isoenzymes or subtypes in affected cells or tissues. Isoenzyme-selective inhibitors are available for most PDE families. Generally, these compounds are at least 30-fold selective for the PDE against which they are directed. Most are substrate site-directed competitive inhibitors, but a few act at allosteric sites (Beavo 1995).

2
Characteristic Features of Asthma and COPD

The incidence of respiratory diseases like asthma and chronic obstructive pulmonary disease (COPD) continue to increase despite the availability of current treatment modalities, and there is therefore a need to improve our understanding of the pathophysiology of these diseases for the development of novel therapeutic agents. While the exact causes of asthma and COPD are not completely understood, it is clear that both diseases are characterized by inflammation of the airways and a decline in respiratory function. In asthma, a number of inflammatory cells are thought to contribute toward the pathogenesis of this disease, including eosinophils (Gleich 2000) and $CD4^+$ T lymphocytes (Romagnani 2000) while it is thought that $CD8^+$ lymphocytes (Kemeny et al. 1999) and neutrophils (Saetta 1999) play an important role in COPD. Another important feature of these diseases is the presence of airway wall remodelling, and there is evidence of hyperplasia/hypertrophy of airway smooth muscle, increased collagen deposition beneath the basement membrane, increased mucus production, angiogenesis and alterations in extracellular matrix in asthma (Bousquet et al. 2000). In COPD, there is evidence of mucus gland hyperplasia, increased bronchiolar smooth muscle hypertrophy, fibrosis of the small airways and in the case of emphysema, destruction of alveolar tissue (Jeffery 1998). The mainstay of treatment for asthma includes bronchodilators like β_2-adrenoceptor agonists and glucocorticosteroids, while for COPD, the muscarinic receptor antagonists, tiotropium bromide and ipratropium bromide as well as β_2-adrenoceptor agonists are used. Reducing the inflammatory response in the airways is thought to be critical in controlling many of the changes seen in these airway diseases.

3
PDE Expression in Allergic Disease

3.1
Allergy

It had long been recognised that the ability of lymphocytes to raise levels of intracellular cAMP is impaired in mild, severe atopic eczema (Parker et al. 1977; Archer et al. 1983) and atopic dermatitis (Safko et al. 1981; Grewe et al. 1982). This finding was all the more significant since, unlike asthmatics, none of the subjects were taking β-adrenoceptor agonist medication. Therefore, tachyphylaxis of β-adrenoceptors is not an issue and not a confounding factor in this disease. The increased level of cAMP in mononuclear cells from patients with atopic dermatitis was a consequence of increased cAMP PDE activity as assessed biochemically (Grewe et al. 1982). The exact splice variant responsible for this activity remains to be established; however, evidence was provided of a putative monocyte-derived PDE in atopic dermatitis that had increased cAMP PDE catalytic activity and was Ca^{2+}/calmodulin and Ro201724 sensitive (Chan et al. 1993b). However, such changes in cAMP PDE activity have not been observed in all studies in cells from patients with atopic dermatitis (Gantner et al. 1997b), and it is unclear whether differences in methodology and/or patient selection account for this discrepancy. It has been proposed that this alteration in PDE activity is responsible for the pathogenesis of this disease, since a functional consequence of increased cAMP PDE activity in atopic dermatitis includes increased IgE production by B lymphocytes, increased histamine and LTC_4 release from basophils, increased interleukin (IL)-4 release and reduced interferon (IFN)-γ and IL-10 release by mononuclear cells—physiological responses that can be inhibited by PDE4 inhibitors (Hanifin et al. 1995).

An interesting feature of the altered cAMP PDE activity in atopic dermatitis is increased susceptibility of the isoenzyme to inhibition by PDE4 inhibitors. This is reflected by increased inhibitor potency against cAMP catalytic activity (Giustina et al. 1984; Chan et al. 1993b; Hanifin et al. 1996) and increased inhibitor potency against the proliferation of mononuclear cells (Banner et al. 1995). Interestingly, the anti-proliferative effect of theophylline was not altered in atopic dermatitis. The molecular mechanism(s) responsible for the increased PDE catalytic activity and increased sensitivity to PDE4 inhibitors in atopic dermatitis remain to be established, but conformational changes in PDE4 protein (Souness et al. 1993), phosphorylation of PDE4 (O'Connell et al. 1996) and/or expression of a splice variant/novel PDE4 (Chan et al. 1993b) could explain such a phenomenon.

Few studies have investigated the effect of methylxanthines in atopic dermatitis. Mononuclear cells from atopic dermatitis have increased cAMP PDE activity that is more susceptible to PDE4 inhibitors and is reflected by increased PDE4 inhibitor potency (Giustina et al. 1984). However, cAMP PDE activity in mononuclear cells was restored to normal values in atopic dermatitis subjects who were taking theophylline to control their asthma (Giustina et al. 1984). An interpretation of this finding is that prolonged treatment with theophylline altered the activity of cAMP PDE activity to control values. This could be attributed to the anti-inflammatory effect of theophylline (see Chap. 3) resulting in a reduction in the release of inflammatory mediators and cytokines known to increase the activity and expression of PDE4. In a double-blind study, the PDE4 inhibitor Ro201724 was shown to improve psoriatic lesions (Stawiski et al. 1979) and daily topical treatment with CP80633 on one arm improved clinical scores (erythema, induration and excoriation) compared with the untreated arm (Hanifin et al. 1996) and indicated the potential use of PDE4 inhibitors in the treatment of atopic dermatitis.

3.2
Asthma

Despite our understanding of the mechanisms that can alter PDE4 activity, there is a paucity of data concerning whether there is an any alteration in the function or expression of PDEs in airway diseases like asthma. The ability of lymphocytes to raise intracellular cAMP and/or increase adenylyl cyclase activity in response to various stimuli, including isoprenaline, sodium fluoride and guanyl-5-yl-imido-biphosphate (GppNHp) was compromised following antigen challenge in atopic asthmatics (Koeter et al. 1982; Meurs et al. 1982). Similarly, the capacity of alveolar macrophages to raise intracellular cAMP in response to histamine, salbutamol, PGE_2 and 1-methyl-3-isobutylxanthine (IBMX) was reduced in asthmatics (Bachelet et al. 1991; Beusenberg et al. 1992). These studies suggest that PDE activity may be increased in inflammatory cells in asthma. However, there is no difference in PDE4 activity in alveolar macrophages (Tenor et al. 1995b) and eosinophils (Aloui et al. 1996; Tenor et al. 1996) noted between atopic and non-atopic subjects. We documented a 50% increase in total cAMP PDE activity in monocytes isolated from asthmatic subjects compared to healthy individuals (Landells et al. 2000), a finding that is supported by a previous study in atopic dermatitis subjects with a history of airway disease (Sawai et al. 1998).

While there is a considerable body of evidence of increased PDE4 inhibitor sensitivity observed in inflammatory cell populations obtained from some

(Chan et al. 1993b; Banner et al. 1995) but not all (Tenor et al. 1996; Gantner et al. 1997b) subjects with atopic dermatitis, we have failed to document a similar finding in monocytes from mild asthmatic subjects (Landells et al. 2000). This is consistent with a study examining the potency of Ro-201724 against zymosan-induced release of glucuronidase from neutrophils obtained from asthmatic subjects (Busse et al. 1981). Similarly, atopic subjects with upper and/or lower airway disease, and who did not suffer from atopic dermatitis, failed to document an increase in PDE4 inhibitor sensitivity against mononuclear cell proliferation (Essayan et al. 1994; Crocker et al. 1996). In contrast, PDE4 inhibitors attenuated tumour necrosis factor (TNF)-α and IL-10 release from mononuclear cells stimulated by mitogen to a greater extent in atopic rhinitic compared with healthy subjects (Crocker et al. 1998).

Together these studies highlight an important observation that alterations in PDE4 catalytic activity or inhibitor sensitivity is dependent upon the type of allergic disease under study, and extrapolating from atopic disease other than asthma should be made with caution. The increase in total PDE activity observed in our study was unrelated to an increase in PDE4 activity and is consistent with a lack of evidence of an alteration in the expression of mRNA for PDE4A, PDE4B and PDE4D in monocytes from mild asthmatic subjects (Landells et al. 2001b). Thus, notwithstanding the lack of evidence of an alteration in PDE4 activity in mild asthma, a rational basis for drug targeting PDE4 in the treatment of respiratory disease stems from the finding that inhibitors of PDE4 can downregulate inflammatory cell function.

4
Properties and Classification of PDE4

The selective targeting of individual PDE isoenzymes has profound implications for the treatment of disease, as recently highlighted with the introduction of the PDE5-selective inhibitor, sildenafil, for erectile dysfunction (Stief 2000). In the context of lung disease, PDE4 has been selectively targeted using chemical inhibitors on the basis of the clinical efficacy of the archetypal non-selective PDE inhibitor, theophylline which has long been used in the treatment of asthma and COPD (see Chap. 3). A number of highly potent PDE4 inhibitors have been tested in clinical trials and shown to have some therapeutic potential. However, one of the major stumbling blocks to the development of these inhibitors is the side-effect profile, including nausea and emesis, that is a characteristic feature of many of these drugs, although attempts are being made to reduce these unwanted side effects.

PDE4 is a cAMP-specific isoenzyme (K_m: 0.2–4 μM), showing very low affinity for cGMP (K_m: >1000 μM), the latter without effect on PDE4 catalytic activity. Four PDE4 subtypes (PDE4A–D) have been cloned and expressed, with additional complexity arising as a consequence of mRNA splicing resulting in isoforms with alterations in amino acid sequences within the N-terminal region (Houslay et al. 1998). In order to gain insights into the functional significance of

Table 2 Summary of the expression of mRNA for PDE4 genes in human cells

Cell type	PDE4A	PDE4B	PDE4C	PDE4D	Reference(s)
CD4 T cell	+	++ (B2)		Weak	Gantner et al. 1997b
Th1 cells	++	++	–	–	Essayan et al. 1997
Th2 cells	++	++	–	++	Essayan et al. 1997
CD8 T cell	++	++ (B2)		++	Gantner et al. 1997b
B cell	++	++ (B2)		++	Gantner et al. 1998
Monocyte	+	++ (B2)	–	Weak	Gantner et al. 1997b; Souness et al. 1996b; Manning et al. 1996; Wang et al. 1999
Eosinophil	++	++	–	++	Engels et al. 1994; Gantner et al. 1997b
Neutrophil	±	++ (B2)	–	±	Engels et al. 1994; Wang et al. 1999
Macrophage	++				Gantner et al. 1997a
Brain	++	++	±	++	Obernolte et al. 1997; Engels et al. 1994[a]
Area postrema				++	Takahashi et al. 1999[a]; Cherry et al. 1999[b]
Epithelium	+ (A5)		+ (C1)	+ (D2) + (D3)	Wright et al. 1998 Fuhrmann et al. 1999

Parentheses denote splice variant; +, presence of expression; –, absence of expression.
[a] Analysis performed in rat brain.
[b] Immunohistochemical detection using mouse brain.

PDE4, various studies have investigated the distribution of PDE4. It is clear that PDE4C is predominantly localized to the testis, skeletal muscle and human foetal lung (Obernolte et al. 1997), while PDE4A, B and D are known to be distributed in many inflammatory cells in man (Table 2).

Analysis of the amino acid sequence of PDE4 revealed a catalytic domain and two upstream conserved regions (UCRs) that is unique to this family of PDE. Using deletion analysis, studies have shown that the catalytic domain in PDE4A4B, for example, lies between amino acid sequence 332/365 to 680/772 (Jacobitz et al. 1996; Houslay et al. 1998). Similarly, PDE4B2B has a catalytic domain between amino acid sequence 152 and 528 (Lenhard et al. 1996; Rocque et al. 1997). The atomic structure of the catalytic domain of 4B2B has recently been published, showing important structural features within the binding pocket for cAMP, including the presence of two metal ions, most likely zinc and magnesium, that is important for binding the cyclic phosphate group and various other amino acids critical for cAMP binding (Xu et al. 2000).

The cDNAs for PDE4 encode for enzymes that can exist as either the long form, containing both UCR regions, and a short form, characterized by either a lack in UCR1 and intact or partially truncated UCR2 region. It is thought that the short and long forms differ with respect to enzyme activity, subcellular localization and activation by different intracellular signalling pathways. Thus, PDE4D3 catalytic activity is increased by a protein kinase A (PKA)-dependent mechanism (Alvarez et al. 1995) a consequence of phosphorylation of Ser^{54} (Sette et al. 1996). Additionally, specific sites in the UCR region are also subject to phosphorylation by mitogen-activated protein (MAP) kinase-dependent mechanisms which could have important implications for PDE4 activity

(Houslay et al. 1998). Furthermore, the N-terminal region is implicated in targeting PDE4 to specific domains within the cell by virtue of protein–protein interactions with SH3 domain-containing proteins (O'Connell et al. 1996). Regions near the carboxyl terminus are also implicated in the regulation of PDE4 function. Thus, substitution of Ser^{487} for Ala resulted in a significant attenuation of MAP kinase-dependent phosphorylation of PDE4B2B (Lenhard et al. 1996). Similarly, phosphorylation by ERK2 kinase of PDE4D3 at Ser^{579} in the carboxyl terminal region led to a significant reduction in catalytic activity (Hoffman et al. 1999). Further complexity arises with the findings that alteration in the activity of PDE4 by ERK2 kinase is also influenced by the presence of UCR regions. Thus, while phosphorylation of PDE4D3 at Ser^{579} resulted in a reduction in cAMP PDE activity, an increase in catalytic activity was observed following phosphorylation of Ser^{491} in PDE4D1, a PDE4 enzyme that lacks a UCR1 domain (MacKenzie et al. 2000). These findings suggest that different splice variants of PDE4 may be differentially regulated by intracellular signalling pathways that may have important implications in the regulation of cell function under normal physiological and pathophysiological conditions. Indeed, alterations to the N-terminal regions of these proteins has important functional consequences, as this may alter their subcellular localization (O'Connell et al. 1996), activation (Sette et al. 1994; Alvarez et al. 1995; Lenhard et al. 1996) and inhibition by PDE4 inhibitors (Alvarez et al. 1995). Moreover, the observation of alterations in PDE4 expression during cell differentiation (Gantner et al. 1997a) or following activation by cytokines, growth factors and lipid mediators (Li et al. 1992; Li et al. 1993; DiSanto et al. 1995; Lenhard et al. 1996) could have important functional consequences during an inflammatory episode.

It has long been recognized that the archetypal PDE4 inhibitor, rolipram, binds with high affinity to brain tissue compared with peripheral organs (Schneider et al. 1986), yet is at least two to three orders of magnitude less potent at inhibiting PDE4 catalytic activity in this tissue (Nemoz et al. 1989). The significance of this discrepancy was later clarified in studies expressing human recombinant PDE4 in yeast and showing that the high-affinity rolipram binding site and the PDE4 catalytic domain reside on the same gene product (Torphy et al. 1992; McLaughlin et al. 1993). Interestingly, there was little correlation between the ability of a range of compounds to displace rolipram binding from PDE4 and their ability to inhibit PDE4 catalytic activity, raising the possibility of synthesizing compounds that could selectively target these sites. The functional significance of the two domains recognized by PDE4 inhibitors was clarified further in studies examining rolipram binding and PDE4 catalytic activity in N-terminally truncated enzymes expressed in yeast COS and Sf9 cells (Jacobitz et al. 1996; Owens et al. 1997; Rocque et al. 1997). Specific regions within the N-terminal domain of PDE4A are important for determining high-affinity binding by rolipram, and the removal of this site from the protein did not abolish catalytic activity nor the ability of rolipram to inhibit PDE4 catalytic activity, suggesting that binding to the high-affinity site is not a prerequisite for inhibition of catalytic activity (Jacobitz et al. 1996; Owens et al. 1997). Similarly, expression of an N-terminal truncated PDE4B2B (152–564) resulted in a protein which lacked a high-affinity binding site for rolipram compared with PDE4B2B (81–564) (Rocque et al. 1997), suggesting that specific sequences within the N-terminal domain are necessary for the expression of high-affinity binding. However, the binding of another PDE4 inhibitor, RP 73401 to PDE4A was unaffected by the loss of this specific amino acid sequence within the N-terminal domain, but the ability of rolipram to displace RP 73401 binding was characterized by a two-site binding model (Jacobitz et al. 1996). The implication of these findings is that specific amino acid sequences outside the catalytic domain of PDE4 can alter the conformation of the protein, such that it binds rolipram with high affinity and therefore the "high-affinity" binding site represents a different conformation of the same protein (Torphy et al. 1992; Jacobitz et al. 1996; Owens et al. 1997; Rocque et al. 1997). There is biochemical evidence supporting the view that PDE4 can exist in different conformational states, as different methods employed to isolate PDE4 from cells can lead to differences in catalytic activity and inhibitor sensitivity (Kelly et al. 1996; Souness et al. 1996b). A number of intracellular processes including phosphorylation (Alvarez et al. 1995; Lenhard et al. 1996; Sette et al. 1996; Hoffman et al. 1999; MacKenzie et al. 2000) or the presence of co-factors [eg magnesium ions (Laliberte et al. 2000)] are known to alter PDE4 catalytic activity.

Pharmacological studies have been employed in order to determine structure activity relationships between different PDE4 inhibitors, and a number of functional studies have shown correlation between PDE4 inhibitor potency and the ability of PDE4 inhibitors to alter various aspects of cell function or rolipram binding. The ability of PDE4 inhibitors to induce gastric acid secretion (Barnette et al. 1995) and emesis (Duplantier et al. 1996) and inhibit formyl-methionyl-leucyl-phenylalanine (fMLP)-induced myeloperoxidase release from human neutrophils (Barnette et al. 1996), inhibition of purified solubilized PDE4 from guinea-pig eosinophils and potentiation of isoprenaline-induced cAMP accumulation from guinea-pig eosinophils (Souness et al. 1993) correlated with the ability of these inhibitors to displace high-affinity rolipram binding. In contrast, the ability of compounds to inhibit PDE4 catalytic activity correlated with the potency of these agents against lipopolysaccharide (LPS)-induced TNF-α release by human monocytes (Barnette et al. 1996; Souness et al. 1996b), fMLP-induced superoxide production by guinea-pig eosinophils (Barnette et al. 1995) and IL-2 release by murine splenocytes (Souness et al. 1997). The possibility that PDE4 may exist as different conformers has been used in an attempt to discover novel inhibitors that are selective for the "low" affinity conformer, as this subtype is suggested to be responsible for regulating cell function, while the "high" affinity conformer is linked to the side-effect profile seen with PDE4 inhibitors.

The ability of PDE4 inhibitors to activate emetic centres within the CNS may be a consequence of a peripheral action of these drugs secondary to raising intracellular levels of cAMP in gastric acid secreting cells and/or afferent neurones in the gut. Alternatively, stimulation of the area postrema, a region within the

CNS with a poorly developed blood–brain barrier, making it accessible to substances within the circulation, can lead to activation of the emetic centre within the CNS (Carpenter et al. 1988; Duplantier et al. 1996). Since emesis and gastric acid secretion correlate with the potency of PDE4 inhibitors to displace rolipram binding (high-affinity PDE) led to the suggestion that drugs with low affinity for this site may be useful in improving the side-effect profile of these drugs (Barnette et al. 1995, 1996). However, some aspects of cell function may also correlate with inhibitors that target the "high"-affinity conformer and suggest that this method may be of limited value for the future development of PDE4 inhibitors with low emetic potential (Barnette et al. 1996). It is therefore of interest that CDP840 (Hughes et al. 1996) and SB 207499 (cilomilast) (Christensen et al. 1998) demonstrate a "high"-to-"low" ratio of 5 and 1.3, respectively. In contrast, rolipram is 1–2 orders of magnitude more selective for the high-affinity binding site compared with inhibition of PDE4 catalytic activity (Torphy et al. 1992; Jacobitz et al. 1996; Hughes et al. 1996; Christensen et al. 1998). Accordingly, both compounds have low emetic potential and a reduced side-effect profile in asthma (Harbinson et al. 1997; Christensen et al. 1998; Torphy et al. 1999). Cilomilast has been shown to inhibit myeloperoxidase release from human neutrophils with an equal potency to rolipram, even though this particular cell function is modulated by the "high" PDE4 conformer (Barnette et al. 1996). This suggests that a number of additional factors may govern why these drugs demonstrate a better side-effect profile compared with other PDE inhibitors. Cilomilast is negatively charged at normal pH that may retard its ability to gain access to the area postrema, although clearly not enough to retard access across inflammatory cells (Barnette et al. 1998; Torphy et al. 1999). It is unclear whether the expression of splice variants of PDE4 in different cells also contributes to the observed correlations between cell function and PDE4 inhibitor potency or high-affinity rolipram binding because of the similarities in the expression of PDE4 subtypes in these cells (Table 2) and the lack of subtype selectivity of the PDE4 inhibitors tested in these studies.

Another approach that is being investigated is whether compounds can be synthesized which exhibit selectivity for different PDE4 subtypes in an attempt to diminish the side-effect profile and selectively target inflammatory cells. While CDP840 does not demonstrate subtype selectivity for PDE4A, B and D (Sullivan et al. 1994), Cilomilast shows a fivefold selectivity toward PDE4D compared with the other two subtypes (Manning et al. 1999; Hersperger et al. 2000). Cilomilast is considerably less emetic than rolipram and is well tolerated by subjects, although it is not free from emesis. Therefore, there is clearly a need to discover highly potent PDE4 inhibitors with an even better side-effect profile. Consequently, a number of compounds have been synthesized that demonstrate selectivity for either PDE4A/B or PDE4D with a difference of up to 55-fold (Manning et al. 1999; Hersperger et al. 2000) and structure-activity relationships have been documented. Thus, a significant correlation was found between PDE4A/B inhibitory potency and inhibition of TNF-α release from monocytes, proliferation of T lymphocytes and oxidative burst from human eosinophils. In contrast, no significant correlation between PDE4D inhibitory potency and inhibition of lymphocyte proliferation and TNF-α release from monocytes was observed, consistent with the finding of weak PDE4D expression in these cells (Manning et al. 1999) (Table 2). However, a correlation was observed against human eosinophil function and selectivity for PDE4D, consistent with the presence of PDE4D in these cells (Hersperger et al. 2000). Therefore, it may be possible to synthesize compounds that document greater subtype and cell selectivity. An important question that needs to be addressed is whether selective targeting of PDE4 subtypes will be sufficient to modulate inflammatory cell function, particularly if cells contain multiple PDE4 subtypes.

Pharmacokinetic considerations notwithstanding, there is some evidence that selective targeting of PDE4D significantly improved the ability of compounds to attenuate pulmonary eosinophil recruitment following antigen provocation in allergic rats (Hersperger et al. 2000). In contrast, mice lacking the ability to express PDE4D have impaired growth and fertility, underlying the importance of cAMP signalling in these processes (Jin et al. 1999). However, of particular interest was the lack of effect of this gene disruption on lymphocyte proliferation, IgE production, IL-4 production and eosinophil recruitment to the airways in a model of murine inflammation (Hansen et al. 2000), features which are characteristic of an allergic phenotype. This contrasts with the findings that the PDE4 inhibitor rolipram inhibited allergen-induced eosinophilia in a murine model of airway inflammation (Kung et al. 2000). The lack of effect of this gene disruption upon eosinophil recruitment suggests redundancy concerning PDE4 regulation of cAMP signalling in inflammatory cells, or alternatively, other PDE4 subtypes play a greater role in regulating allergic inflammation (Manning et al. 1999).

It remains to be established whether drug targeting of PDE4A/B offers the advantage of suppressing inflammatory cell function in vivo while exhibiting a low emetic profile, considering that PDE4D is expressed in the area postrema in rat and mouse (Cherry et al. 1999; Takahashi et al. 1999).

5
Effect of PDE Inhibition on Inflammatory Cell Function

It is readily apparent that PDEs are widely distributed throughout the body and regulate the function of many cells. Particular interest has focused on the role of PDE4 and to a lesser extent PDE3 in disease as these enzymes are found in many inflammatory cells. The following section will highlight the role of PDE isoenzymes in regulating the function of cells thought to participate in the inflammatory process.

5.1
Mast Cells and Basophils

The presence of PDE enzymes was confirmed in rat mast cells (PDE1 and PDE3–5) (Alfonso et al. 1995) and basophils from healthy human subjects (Peachell et al. 1992), using a variety of pharmacological, biochemical and molecular biochemical techniques. In human basophils, cGMP PDE activity was minimal, appearing to be that of PDE5, while cAMP PDE activity was considerably greater, comprising both PDE3 and PDE4 (Peachell et al. 1992). These observations are consistent with functional studies demonstrating inhibition of leukotriene (LT)C_4 and anti-IgE-induced histamine or IL-4 and IL-13 release from human basophils by rolipram (Kleine et al. 1992; Peachell et al. 1992; Columbo et al. 1993; Shichijo et al. 1997; Weston et al. 1997; Barnette et al. 1998), denbufylline, Ro20–1724, RP73401, nitraquazone (Weston et al. 1997) and cilomilast (Barnette et al. 1998). Certain of these compounds were found to be ineffective against IgE-induced histamine release by human lung mast cells (Weston et al. 1997), thus the nature of the PDE-regulating human lung mast cell responses remains uncertain. Although agents that induce and sustain elevations in intracellular cAMP appear to attenuate the stimulated release of mediators from both basophils and human lung mast cells, the responsiveness of human lung mast cells and basophils to selected cAMP-active agents differs markedly (Weston et al. 1998). In other studies, the PDE 4 inhibitor rolipram attenuated LTC_4 and histamine release from murine mast cells (Griswold et al. 1993) and, in combination with forskolin, inhibited anti-IgE-induced increase of intracellular calcium levels in human skin mast cells (Columbo et al. 1994).

The inhibitory effect of rolipram in basophils is potentiated by addition of the PDE3 inhibitors siguazodan (SKF95654) or cilostazol (Peachell et al. 1992; Shichijo et al. 1997), although the mixed PDE3 and 4 inhibitor zardaverine had little effect over and above the PDE4 inhibitors alone (Kleine et al. 1992). Similarly, the PDE3/4 inhibitor, benzafentrine (AH21–132) was observed to inhibit antigen-induced histamine release from human lung fragments (Nagai et al. 1995). In contrast, neither the PDE3 inhibitors siguazodan, SKF95654 or cilostazol alone (Peachell et al. 1992; Columbo et al. 1993; Shichijo et al. 1997; Weston et al. 1997) nor the PDE5 inhibitor zaprinast (M&B 22,948) (Frossard et al. 1981; Weston et al. 1997) affected histamine or cytokine release from human basophils. These inhibitors also failed to inhibit histamine release by human lung mast cells (Weston et al. 1997).

5.2
Neutrophil

A predominant PDE isoenzyme with high affinity for cAMP but insensitive to cGMP and inhibited by rolipram was documented using diethylaminoethyl-sepharose chromatography, suggesting PDE4 activity (Nielson et al. 1990; Schudt et al. 1991a,b). In addition to this, PDE4B mRNA has been described in human neutrophils (Muller et al. 1996), with PDE4B2 thought to be the predominant PDE isoform present (Wang et al. 1999). A cGMP-specific enzyme, identified as PDE5, has also been purified in human neutrophils (Prigent et al. 1990; Schudt et al. 1991b). These findings support a number of functional studies demonstrating the ability of various PDE4 inhibitors to attenuate respiratory burst (Nielson et al. 1990; Wright et al. 1990; Schudt et al. 1991b; Ferretti et al. 1994; Ottonello et al. 1996), degranulation (Busse et al. 1981; Wright et al. 1990; Barnette et al. 1996; Barnette et al. 1998), apoptosis (Yasui et al. 1997; Ottonello et al. 1998; Niwa et al. 1999), chemotaxis (Ferretti et al. 1994), leukotriene biosynthesis (Schudt et al. 1991b; Cortijo et al. 1996), chemokine release (IL-8) (Au et al. 1998) and surface expression of the $\beta 2$ integrins CD11a/CD18 and CD11b/CD18 (Derian et al. 1995) in neutrophils. In contrast, the PDE3 inhibitors amrinone, milrinone, imazodan and cilostamide had no significant effect on neutrophil superoxide anion production (Nielson et al. 1990; Wright et al. 1990) while both milrinone and bemoradan were ineffective in attenuating the expression of adhesion molecules in human neutrophils (Derian et al. 1995), milrinone has also been observed to have no inhibitory effect on human neutrophil degranulation. However, in a more recent study both amrinone and milrinone were observed to reduce superoxide, hydrogen peroxide, and hydroxyl radical levels in neutrophils, while neither were found to impair neutrophil chemotaxis or phagocytosis (Mikawa et al. 2000).

5.3
Eosinophil

The presence of mRNA for PDE4D was first documented in guinea-pig eosinophils using reverse transcription polymerase chain reaction (RT-PCR) with primers designed against specific sequences in rat PDE4 subtype DNA clones (Souness et al. 1995). Studies to elucidate the PDE profiles of human eosinophils have shown the presence of high levels of PDE4 activity (Hatzelmann et al. 1995; Tenor et al. 1996; Gantner et al. 1997b); the majority of this activity was observed in the cytosolic fraction of cells with some activity also observed in the particulate fraction (Hatzelmann et al. 1995). RT-PCR analysis of levels of PDE subtype messenger RNA expression in human eosinophils has revealed total PDE4 activity is a result of PDE4A, PDE4B and PDE4D subtype activity (Gantner et al. 1997b). Selective PDE4 inhibition in eosinophils has been shown to increase the level of intracellular cAMP (Souness et al. 1993; Dent et al. 1994; Souness et al. 1995) and attenuate superoxide anion generation (Dent et al. 1991, 1994; Hadjokas et al. 1995; Hatzelmann et al. 1995; Nicholson et al. 1995; Souness et al. 1995; Cohan et al. 1996; Ezeamuzie 2001), LTB_4-induced thromboxane release (Souness et al. 1994; Nicholson et al. 1995), and Ig- or C5a-induced secretion of cationic proteins (Souness et al. 1995; Momose et al. 1998) in both human and guinea-pig eosinophils. Moreover, PDE4 inhibitors attenuated PAF, LTB_4 and C5a-induced release of LTC_4 from eosinophils (Tenor et al. 1996), eosinophil chemotaxis in vitro (Kaneko et al. 1995; Alves et al. 1996; Cohan et al. 1996; Tenor et al. 1996; Alves et al. 1997; Santamaria et al. 1997) and PAF-induced cell-surface CD11b upregulation (Santamaria et al. 1997; Momose et al. 1998). In some studies, the efficacy of PDE4 inhibitors was significantly increased in the presence of cAMP-elevating drugs (Hadjokas et al. 1995; Hatzelmann et al. 1995; Tenor et al. 1996; Ezeamuzie 2001); and although only low levels of PDE 3 activity have been observed in eosinophil cytosolic and particulate fractions (Hatzelmann et al. 1995), co-treatment with both a PDE3 and a PDE4 inhibitor has shown increased inhibitory effects on eosinophil function (Blease et al. 1998). In one study, cAMP-elevating drugs but not rolipram inhibited eosinophil viability in culture (Hallsworth et al. 1996), while in separate studies, PDE4-selective inhibition has been shown not to inhibit C5a-induced eosinophil degranulation (Hatzelmann et al. 1995; Ezeamuzie 2001). The differing results of these studies suggest that PDE4 inhibitors alone may not be sufficient to elevate cAMP in this cell type and therefore may not inhibit all aspects of eosinophil function.

5.4
T Lymphocyte

Cyclic AMP PDE activity in the soluble and particulate fraction of enriched T lymphocytes was inhibited by Ro-201724 and the PDE3 inhibitor Cl-930 (Archer et al. 1983), and both PDE3 and PDE4 have been confirmed in membrane and cytosolic compartments of human $CD4^+$ and $CD8^+$ T lymphocytes (Tenor et al. 1995b; Giembycz 1996). On closer inspection, PDE4A, PDE4B and PDE4D were described in $CD4^+$ and $CD8^+$ human T lymphocytes (Giembycz 1996; Landells et al. 2001b). Semiquantitative RT-PCR analyses of mRNA from healthy and mild atopic subjects revealed that PDE4A and PDE4B2 were present in both $CD4^+$ and $CD8^+$ cells and that PDE4D was expressed only in $CD8^+$ cells (Gantner et al. 1997b). Increased PDE4A and PDE4B2 expression was observed in $CD4^+$ cells from atopic subjects, although this did not appear to result in significantly higher cAMP PDE activity (Gantner et al. 1997b). PDE3B has been shown to account for the PDE3 activity in lymphocytes from healthy subjects (Sheth et al. 1997), and a fragment corresponding to PDE7 has also been described (Ichimura et al. 1993; Bloom et al. 1996; Giembycz 1996).

Functional studies have shown that PDE4, and to a lesser extent PDE3 inhibitors, attenuated mitogen-, antiCD3- and allergen-induced human T lymphocyte proliferation (Averill et al. 1988; Robicsek et al. 1991; Essayan et al. 1994; Banner et al. 1995; Essayan et al. 1995; Schudt et al. 1995; Crocker et al. 1996; Banner et al. 1997; Essayan et al. 1997; Gantner et al. 1997b; Barnette et al. 1998; Banner et al. 1999). However, inhibition of lymphocyte proliferation was more pronounced if dual inhibitors or a combination of PDE3 and PDE4 inhibitors were used (Robicsek et al. 1991; Marcoz et al. 1993; Banner et al. 1995; Schudt et al. 1995; Gantner et al. 1997b). Similarly, rolipram and Ro-201724 inhibited lymphocyte proliferation and contact hypersensitivity in oxazolone-treated mice (Moodley et al. 1995). The phytohaemagglutinin (PHA)-induced or anti-CD3-induced proliferation of CD4$^+$ and CD8$^+$ T lymphocytes was inhibited in a concentration-dependent manner by rolipram but not SKF95654, consistent with the ability of rolipram to elevate intracellular cAMP in these cells (Giembycz 1996). SKF95654 increased the inhibitory potency of rolipram against CD4$^+$ and CD8$^+$ T lymphocyte proliferation, although complete inhibition was not achieved. Similarly, it has been demonstrated that PDE7 activity is increased upon activation of lymphocytes, and that this, in turn, correlates with decreased cAMP and increased proliferation (Li et al. 1999). Furthermore, when PDE7 expression is reduced by a PDE7 antisense oligonucleotide, proliferation is reduced (Li et al. 1999). Thus, it appears that PDE4 and to a lesser degree, both PDE3 and PDE7 may all play a role in regulating T lymphocyte proliferation.

Various studies have shown that elevating the level of intracellular cAMP may preferentially inhibit the synthesis and release of T helper (Th)1 cytokines. Thus, drugs which elevate intracellular level of cAMP, including forskolin and prostaglandin (PG)E$_2$ (Munoz et al. 1990; Novak et al. 1990; Betz et al. 1991; van der Pouw et al. 1992; Snijdewint et al. 1993), inhibited the production of Th1 but not Th2 cytokines, most likely via inhibition of IL-2 synthesis, reduction in $t_{1/2}$ of IL-2 mRNA and IL-2 receptor (IL2R) expression by a PKA-dependent mechanism (Averill et al. 1988; Anastassiou et al. 1992; Yoshimura et al. 1998; Kanda et al. 2001).

The production of T lymphocyte-derived cytokines is also influenced by antigen-presenting cells like monocytes. Indeed, PGE2 inhibited the release of monocyte-derived IL-12, yet augmented the release of IL-10. These cytokines are important for the proliferation of Th1 and Th2 lymphocytes, respectively (van der Pouw et al. 1995; Van der Pouw Kraan TCTM et al. 1996). In other studies, addition of exogenous PGE$_2$ to purified lymphocytes caused a marked reduction in IFN-γ release (Chan et al. 1996). Similarly, rolipram attenuated the PHA-induced or phorbol 12-myristate 13-acetate (PMA) and ionomycin-induced release of IL-2 and IFN-γ from CD4$^+$ and CD8$^+$ human T lymphocytes (Giembycz et al. 1996) and IFN-γ production by PHA-stimulated human peripheral blood mononuclear cells (Yoshimura et al. 1998). On the other hand, rolipram only inhibited T lymphocyte proliferation when the former stimulus was used, suggesting the possible involvement of other cytokines in the proliferative response (Giembycz 1996). In LPS-stimulated human peripheral blood mononuclear cells, rolipram was observed to inhibit IL-1β and TNF-α production (Yoshimura et al. 1997). In each of these studies PDE3-selective inhibitors showed little or no independent efficacy; however, they were observed to augment the efficacy of PDE4 inhibitors. Other studies have shown that rolipram significantly reduced TNF-α, and to a lesser extent, IFN-γ production in human and rat auto-reactive T lymphocytes (Sommer et al. 1995) and was only partially effective against TNF-α release from encephalitogenic T cells (Molnar et al. 1993). In general, these studies support the view that elevation of cAMP inhibits the generation of Th1-like cytokines but that PDE-mediated effects are selective.

It has now become increasingly apparent that intracellular cAMP can also regulate the expression and release of cytokines from Th2 cells. It was established in a murine Th2 cell clone that rolipram had minimal effects on anti-CD3-induced IL-4 production but enhanced IL-5 production via a PKA-dependent pathway (Schmidt et al. 1995) which is consistent with the ability of dibutyryl cAMP, in combination with PMA, to increase IL-5 mRNA expression and protein levels in a mouse thymoma line (Lee et al. 1993). The effect of cAMP on the expression of IL-5 mRNA is indirect, since there does not appear to be a cAMP response element (CRE) consensus sequence in the IL-5 promoter. Furthermore, dibutyryl cAMP inhibited the production of IL-2, IL-4 and IL-10 in these cells, confirming the ability of cAMP to regulate the expression of Th2 cytokines (Lee et al. 1993). Similarly, IBMX inhibited the synthesis of IL-2 and IL-4, yet moderately affected IFN-γ production in human T lymphocytes (Snijdewint et al. 1993) and Ro-201724 and inhibited IL-4 and IL-5 secretion in human Th2 cell lines (Crocker et al. 1996). Rolipram has also been observed to reduce IL-2, IL-4 and IL-5 production by PHA-stimulated human peripheral blood mononuclear cells (Yoshimura et al. 1998). The ability of cAMP to regulate Th2 cytokine production is not specific for T cell clones and cell lines. Rolipram inhibited ragweed (Th2)- but not tetanus toxoid (Th1)-driven proliferation of peripheral blood mononuclear cells (Essayan et al. 1994). This anti-proliferative effect of rolipram against ragweed challenge was associated with a reduction in gene expression for IL-5 and IFN-γ but not IL-4 (Essayan et al. 1995). It was initially suggested that the relative resistance to inhibition by rolipram of peripheral blood mononuclear cell proliferation to a Th1-driven stimulus may be due to the lack of PDE4B in Jurkat cells (Essayan et al. 1995), and that this may account for the inability of rolipram to effect IL-2 mRNA synthesis in these cells (Lewis et al. 1993). The differential effect of PDE inhibitors on T lymphocyte cytokine generation was also suggested to be a function of the ability of different populations of T lymphocytes to elevate cAMP (Snijdewint et al. 1993; Knudsen et al. 1995). It has since been reported that the enhanced sensitivity of Th2 cells and the relative insensitivity of Th1 to PDE inhibition is more likely to be due to differential expression of PDE4 isoforms in these cell types. Investigation by RT-PCR revealed reduced gene expression for the PDE4C isoform and a lack of gene expression for the PDE4D isoform in Th1 cells when compared to Th2 (Essayan et al. 1997).

It is clear that Th2 cell-derived cytokines can be inhibited by cAMP-elevating drugs, particularly when a physiological stimulus such as antigen is used as opposed to mitogens or anti-CD3. Another factor which may influence whether cAMP up- or downregulates the expression of Th2 cytokines is the availability of IL-2 (Hilkens et al. 1995). Finally, cAMP-elevating agents, including prostaglandin E_2, inhibited the expression of monocyte-derived IL-12, yet augmented the expression of IL-10 from monocytes, which would also be a determinant of the expression of Th1 and Th2 cytokines, particularly if antigen-presenting cells and/or antigen presenting cell-dependent stimuli are used (van der Pouw et al. 1995).

5.5
B Lymphocyte

Initially, studies of RNA from a human lymphocytic B cell line (43D-C12) revealed a cDNA that encoded a protein with 93% homology to rat PDE4B (Obernolte et al. 1993). It has since been demonstrated that cytosolic PDE4 is the predominant isoenzyme, followed by cytosolic PDE7-like activity, some PDE3 activity was also noted in the particulate fraction (Gantner et al. 1998). Molecular biology techniques were used in this study, allowing further investigation of the PDE profile of human B lymphocytes. RT-PCR revealed PDE4A, PDE4B and PDE4D to be present; in addition, small amounts of PDE3A were also detected (Gantner et al. 1998). A rise in the level of intracellular cAMP has been shown to inhibit proliferation (Kammer 1988), differentiation (Huang et al. 1995), apoptosis (Mentz et al. 1995; Baixeras et al. 1996) and promote isotype switching by IL-4 in murine and human B lymphocytes (Roper et al. 1990, 1995).

PGE2 inhibits IgE production induced by IL-4 in purified human B cells enriched with T lymphocytes (Pene et al. 1988). In contrast, the β_2-adrenoceptor agonist, salbutamol was reported to potentiate IL-4-induced IgE production in human peripheral blood mononuclear cells (Paul et al. 1993; Coqueret et al. 1995). The reason for this discrepancy remains to be established. However, the expression of IgE in B cells is regulated by a low-affinity IgE receptor (CD23) which is expressed on and released (soluble CD23) by B cells, a process that is cAMP-dependent (Paul et al. 1994). It is known that PGE2 (Pene et al. 1988) but not salbutamol (Coqueret et al. 1995) inhibits the expression of CD23 on B cells. The role of cAMP in regulating human B lymphocyte function can only be resolved with purified populations of CD40$^+$ lymphocytes.

Very few studies have investigated the effect of PDE inhibitors on B lymphocyte function. Peripheral blood mononuclear cells from individuals with atopic dermatitis have a propensity to generate IgE, which is inhibited by Ro-201724, and appear to be mediated by a direct inhibition of the cAMP PDE activity of B lymphocytes (Cooper et al. 1985). This result was reflected in a separate study that showed cAMP PDE activity to be more susceptible to inhibition by both selective PDE4 and non-selective PDE inhibitors in B lymphocyte homogenates from atopic subjects when compared to healthy subjects (Chan et al. 1993a). Rolipram and RP73401 (PDE4 inhibitor) increased intracellular cAMP levels and augmented proliferation of LPS- and IL-4 stimulated human B lymphocytes (Gantner et al. 1998). This effect was reduced by PKA inhibition, with PDDE4 activity being reduced by up to 50% in stimulated cells, thus showing stimulation of B cell proliferation to be dependent on a PDE4-mediated increase in cAMP. PDE3 inhibition was shown to have little effect in this model (Gantner et al. 1998). In a another study, rolipram and Ro-301724 were shown to be ineffective in inhibiting IL-4-induced IgE production by human B lymphocytes (Coqueret et al. 1997).

5.6
Monocyte

Many groups using various assay techniques to detect cAMP activity in cell homogenates have studied the isoenzyme profile of human monocytes. Purified human monocytes were found to contain PDE4 almost exclusively in the cytosol (Tenor et al. 1995a), consistent with the description of PDE4A, PDE4B (specifically PDE4B2) and PDE4D in these cells (Manning et al. 1996; Souness et al. 1996b; Gantner et al. 1997a). Small amounts of membrane-bound PDE3 have also been observed, and although investigated, no PDE2, PDE5 nor PDE4C expression could be described (Manning et al. 1996; Souness et al. 1996b; Gantner et al. 1997a). Functional studies demonstrated that rolipram attenuated leukotriene production (Griswold et al. 1993), cytokine secretion (Crocker et al. 1998) and arachidonic acid release (Hichami et al. 1995, 1996) from human monocytes. Furthermore, PDE4 and to a lesser extent, PDE3 inhibitors, attenuated endotoxin or LPS-induced TNF-α production in monocytes (Molnar et al. 1993; Prabhakar et al. 1994; Schudt et al. 1995; Seldon et al. 1995; Sinha et al. 1995; Verghese et al. 1995; Barnette et al. 1996; Greten et al. 1996; Souness et al. 1996a; Gantner et al. 1997a; Siegmund et al. 1997; Eigler et al. 1998; Landells et al. 2000). Similarly, the PDE4 inhibitor CP80633 inhibited the release of TNF-α induced by LPS in human monocytes (Cohan et al. 1996). The effect of PDE4 inhibitors on TNF-α production was a consequence of a reduction in TNF-α mRNA expression and protein activity(Prabhakar et al. 1994; Verghese et al. 1995; Greten et al. 1996; Siegmund et al. 1997). PDE4 inhibitors either have no effect (Prabhakar et al. 1994) or inhibited IL-1β release (Molnar et al. 1993; Verghese et al. 1995), but did not inhibit IL-1β mRNA expression (Verghese et al. 1995). As with non-selective PDE inhibition, rolipram was also observed to enhance IL-10 production, an effect that was reversed by addition of a selective PKA inhibitor (Siegmund et al. 1997; Eigler et al. 1998).

5.7
Macrophage

The PDE profile of monocyte-derived macrophages from healthy subjects has been determined; PDE4 activity was observed to be lower and PDE1 and PDE3 activities increased in comparison to monocytes (Gantner et al. 1997a). In human alveolar macrophages, large amounts of PDE1 and also PDE5 account for cGMP PDE activity, while an equivalent expression of both PDE3 and PDE4 are responsible for the cAMP PDE activities observed (Tenor et al. 1995a). PDE3 is located in both cytosolic and membrane compartments while PDE1, PDE4 and PDE5 are predominantly located in the cytosol (Schudt et al. 1995; Tenor et al. 1995a). Exposure of macrophages to inflammatory stimuli leads to a decrease in intracellular cAMP (Bachelet et al. 1991); in this way LPS-induced secretion of TNF-α by monocyte-derived macrophages was inhibited by the cAMP elevators dibutyryl cAMP, PGE$_2$ and forskolin (Gantner et al. 1997a). Similarly, 8-bromo cAMP, PGE$_2$ and cholera toxin reduced IL-1α expression and caused a downregulation of TNF-α gene expression in LPS-stimulated human macrophages (Zhong et al. 1995), while both dibutyryl cAMP and 8-bromo cAMP were observed to cause an inhibition of thromboxane B$_2$ release in alveolar macrophages (Fuller 1990; Baker et al. 1992).

Ro-201724 alone, or in combination with isoprenaline, attenuated zymosan or IgE/anti-IgE complex-induced release of TXB$_2$, LTB$_4$ and superoxide anion (Fuller et al. 1988). Similarly, rolipram, RP73401 and the dual PDE3/PDE4 inhibitor, zardaverine, inhibited LPS-induced TNF-α release from human alveolar macrophages (Schade et al. 1993; Schudt et al. 1995; Gantner et al. 1997a). In this model, motapizone (PDE3 inhibitor) alone acted as a weak inhibitor, and combination of this compound with either rolipram or RP73401 caused total inhibition of TNF-α release (Gantner et al. 1997a). Rolipram has also been shown to reduce LPS-induced TNF-α release from macrophages obtained from Lewis rats with experimental autoimmune encephalomyelitis (Molnar et al. 1993), while higher concentrations of both rolipram and zardaverine have been shown to attenuate the release of LTC$_4$ by LPS in murine resident peritoneal macrophages (Schade et al. 1993). However, fMLP-induced superoxide anion production in guinea-pig peritoneal macrophages remained unaffected by PDE4 inhibition (Turner et al. 1993).

5.8
Bronchial Epithelium

The PDE profile of bronchial epithelial cells has been identified. In an early study, PDE1–5 were isolated from airway epithelium with PDE3 predominantly localized to the membrane fraction (Rousseau et al. 1994). In more recent studies, analysis of PCR products from primary airway epithelial cell cultures revealed the presence of several PDE4 splice variants—PDE4A5, PDE4C1, PDE4D2 and PDE4D3—and also provided evidence of PDE7 expression through demonstration of PDE7 mRNA (Fuhrmann et al. 1999). Alterations in the levels of intracellular cAMP have long been recognized to regulate chloride channel activity in the epithelium. It is of interest, therefore, that airway epithelium chloride channel activity was increased in the presence of the PDE3 inhibitor milrinone, but neither rolipram, Ro-201724 nor IBMX were active (Kelley et al. 1995). This effect was mediated by a protein kinase-dependent pathway but was found to be unrelated to changes in total cAMP content, once again underlining the possibility that compartmentalization of cAMP in cells is important in regulating protein function (Kelley et al. 1995). Similarly, in functional studies, PDE inhibitors have been shown to have limited effects on bronchial epithelium. Rolipram was observed to inhibit bacteria-induced epithelial damage of bronchial mucosa (Dowling et al. 1997). However, in other studies, IBMX had no effect on basal or TNF-α-induced IL-8 release (Dent et al. 1998), and neither IBMX nor rolipram had any effect on bradykinin-induced PGE$_2$ release in human bronchial epithelial cells grown in primary culture (Dent et al. 1998).

5.9
Vascular Endothelium

Characterization of cAMP PDE revealed the presence of cAMP PDE3 and PDE4 in bovine and pig aortic endothelial cells in culture (Lugnier et al. 1990; Souness et al. 1990) and PDE2–4 in porcine pulmonary artery endothelial cells in culture (Suttorp et al. 1993). Functional studies have revealed the PDE profile of human vascular endothelial cells, which have been shown to express large amounts of PDE2, 3 and 4 (Suttorp et al. 1996). An increase in the intracellular level of cAMP within the endothelium attenuated transendothelial cell permeability (Casnocha et al. 1989; Stelzner et al. 1989). Both IBMX and pentoxifylline, inhibited thrombin- (Casnocha et al. 1989) and endotoxin-induced (Sato et al. 1991) increase in permeability of human umbilical vein and bovine pulmonary artery endothelial cell monolayers in culture, respectively. Interestingly, the effect of pentoxifylline on endothelial cell permeability was not associated with an increase in intracellular cAMP (Sato et al. 1991) and might reflect compartmentalization of cAMP within cells. Motapizone, rolipram and zardaverine significantly reduced hydrogen peroxide-induced permeability of porcine pulmonary artery endothelial cells (Suttorp et al. 1993), implicating a role for PDE3 and PDE4 in this response. Similarly, in human endothelial cell layers, adenylyl cyclase activation by either forskolin, cholera toxin or prostaglandin E1, or treatment with the PDE3 and/or PDE4 inhibitors motapizone, rolipram and zardaverine, was seen to abrogate thrombin or HlyA (*Escherichia coli* haemolysin, a membrane-perturbing bacterial endotoxin) induced hyperpermeability (Suttorp et al. 1996).

The endothelium also provides an interface for the adhesion and transmigration of inflammatory cells from the blood into sites of inflammation. The transendothelial migration of lymphocytes but not monocytes through human endothelial cells in culture was attenuated by theophylline and Ro-201724 (Lidington et al. 1996). It remains to be established whether the surface expression of adhesion proteins is inhibited, although an effect on lymphocyte mobility was observed. Similarly, R-rolipram inhibited PMA and TNF-α-stimulated guinea-pig eosinophil adhesion to human umbilical cord vein endothelial cells (HUVECs) in culture (Torphy et al. 1994).

IBMX attenuated TNF-induced expression of endothelial leukocyte adhesion molecule 1 (ELAM-1 or E-selectin), vascular cell adhesion molecule 1 (VCAM-1) but not intercellular adhesion molecule 1 (ICAM-1) in forskolin-treated human umbilical cord vein endothelial cells in culture (Pober et al. 1993). Similarly, treatment of HUVECs with selective PDE4 inhibitors has also been shown to inhibit E-selectin but not VCAM1 expression (Morandini et al. 1996). In contrast, pentoxifylline in combination with dibutyryl cAMP failed to attenuate the TNF-α-induced expression of any of these adhesion molecules (Deisher et al. 1993). Rolipram in combination with salbutamol has been shown to inhibit TNF-α-induced E-selectin expression, whilst ICAM-1 and VCAM-1 expression were not affected. In the same study, the PDE 3 inhibitor ORG 9935 had no effect on CAM expression alone, but in combination with rolipram, a synergistic inhibition of VCAM-1 and E-selectin, but not ICAM-1, expression was observed (Blease et al. 1998). In this way, a combination of both PDE3 and PDE4 inhibition appears to be more effective in reducing CAM expression than inhibition of either isoenzyme alone. Clearly, further studies are required to determine the exact role played by cAMP in expression of adhesion molecules on vascular endothelial cells.

5.10
Vascular Smooth Muscle

Cyclic nucleotide PDE activity in human, bovine and rat aorta was resolved into three peaks characterized by PDE1, PDE3 and PDE5, respectively (Lugnier et al. 1986). In later studies, PDE4 was observed in rat aorta (Yu et al. 1995) and mesenteric artery (Komas et al. 1991); and in pig aorta, PDE1 (soluble), PDE3 (soluble and particulate) and PDE4 (predominantly soluble) activity was found (Xiong et al. 1995). PDE1–5 were detected in the cytosolic fraction of human aorta (Miyahara et al. 1995), and in more recent studies, advanced molecular biology techniques on a range of vascular smooth muscle tissues have revealed more specific expression of isoenzyme subtypes. These include PDE5A1 and PDE5A2 in human aortic smooth muscle cells (Loughney et al. 1998), PDE3A and PDE3B in human blood vessel vascular smooth muscle cells (Palmer et al. 2000) and more specifically, PDE3A1 in human aortic myocytes (Choi et al. 2001). These biochemical studies are consistent with functional studies showing vasodilation of human mesenteric vessels, coronary, lung and renal arteries (Lindgren et al. 1989, 1991) and rat aorta (Yu et al. 1995) by PDE3 inhibitors, including milrinone, and vasodilation of rabbit aorta by the mixed PDE3/4 inhibitor ORG20421 (Nicholson et al. 1995). Interestingly, the ability of PDE4 and PDE5 inhibitors to induce relaxation of rat aorta is dependent on the presence of endothelium-derived nitric oxide (Lugnier et al. 1993; Yu et al. 1995). The endothelium-dependence of the relaxant response to PDE4 inhibitors was subsequently shown to be due to nitric oxide-induced elevation of cGMP which inhibited PDE3, thereby increasing the level of intracellular cAMP in vascular smooth muscle (Archer et al. 1983). A similar finding was noted for pentoxifylline and theophylline, although relaxation mediated by theophylline was endothelium-independent and has been attributed to the different affinities these drugs have for PDE3 and PDE4 (Marukawa et al. 1994). These studies highlight the crosstalk in vascular tissue between the nitric oxide/cGMP pathway and the cAMP pathway.

There is an abundance of PDE in human pulmonary artery according to the profile: PDE5=PDE3>>PDE4, while PDE1 was relatively scarce (Rabe et al. 1995). Both PDE3 and PDE5 were predominantly located in the cytosolic fraction. The biochemical data are supported by functional studies, which showed that vasodilation of human pulmonary artery by zardaverine and motapizone was greater than for rolipram (Rabe et al. 1994). Recent studies have also revealed expression of PDE2 in human pulmonary artery, more specifically PDE2A (Sadhu et al. 1999).

The role of PDE in regulating vascular smooth muscle proliferation has also been investigated. The PDE3 inhibitor cilostazol attenuated growth factor-induced [^{3}H]-thymidine incorporation into DNA and cell growth of rat aortic arterial smooth muscle cells in culture (Takahashi et al. 1992). Similarly, in a cell line derived from embryonic rat aorta that contained both PDE3 and PDE4 activity (~30% and 70%, respectively), the combined use of PDE3 and PDE4 inhibitors attenuated cell proliferation to a greater extent than either alone (Pan et al. 1994) and IBMX inhibited surgery-induced intimal thickening in organ cultures of human saphenous vein (Revel et al. 1992).

5.11
Airway Smooth Muscle

Biochemical investigations have documented PDE1–5 in dog (Torphy et al. 1990), bovine (Shahid et al. 1991), guinea-pig (Harris et al. 1989; Burns et al. 1994; Miyamoto et al. 1994) and human airway smooth muscle (de Boer et al. 1992; Rabe et al. 1993; Torphy et al. 1993), with most of the PDE activity located in the cytosol. Airway smooth muscle relaxation is observed following inhibition of PDE3 and PDE4 in canine (Silver et al. 1988; Torphy et al. 1988; Torphy et al. 1991), guinea-pig tracheal (Harris et al. 1989; Tomkinson et al. 1993; Miyamoto et al. 1994; Spina et al. 1995), and human airway preparations (de Boer et al. 1992; Cortijo et al. 1993; Qian et al. 1993; Rabe et al. 1993; Torphy et al. 1993; Fujii et al. 1998). In contrast, inhibition of PDE4 and not PDE3 correlated with smooth muscle relaxation in bovine trachea (Shahid et al. 1991).

The contribution of PDE3 and PDE4 to human airway smooth muscle relaxation has been investigated. The non-selective PDE inhibitors theophylline, pentoxifylline and IBMX, the PDE4-selective inhibitors rolipram, denbufylline and D22888, and the PDE3 inhibitor ORG9935 have all been observed to relax inherent bronchial smooth muscle tone, while the PDE5 selective inhibitor zaprinast remained ineffective (de Boer et al. 1992; Cortijo et al. 1993; Qian et al. 1993; Rabe et al. 1993; Dent et al. 1998). Similarly, the combination of PDE3 and PDE4 inhibitors, or the use of a dual PDE3/4 inhibitor resulted in significant relaxation of smooth muscle tone (de Boer et al. 1992; Rabe et al. 1993). In spontaneously contracted human bronchial preparations, relaxation by rolipram was greater than siguazodan (Qian et al. 1993) and SKF94120 was more potent than rolipram (Rabe et al. 1993). Thus, the relaxation potency of the PDE3 inhibitor ORG9935 was less when methacholine and not histamine was used as the spasmogen, which was not seen for rolipram (de Boer et al. 1992). In contrast, siguazodan was more efficacious than rolipram in spasmogen-contracted tissue (Qian et al. 1993; Torphy et al. 1993). Histamine-, acetylcholine- and methacholine-induced contraction of human bronchi were significantly attenuated by aminophylline, T440 (PDE4 inhibitor) and ORG20241 (PDE3/4 inhibitor) (Nicholson et al. 1995; Fujii et al. 1998), but although it has been demonstrated that theophylline, IBMX and zardaverine inhibit the contractile response to allergen, RP73401 (PDE4 inhibitor) and motapizone were without effect (Schmidt et al. 1997). Differences in the degree of basal tone, age and source of the tissue, variability in tissue response to relaxant agonists and methodology may account for the conflicting reports. Clearly, the greater efficacy demonstrated by mixed PDE3/4 inhibitors as relaxant agonists compared with subtype-selective enzyme inhibitors implies a role for both PDE3 and PDE4 in mediating relaxation of human airway smooth muscle (de Boer et al. 1992; Rabe et al. 1993).

The role of PDE in the regulation of airway smooth muscle proliferation has only received scant attention; nonetheless, IBMX was observed to attenuate thrombin-induced mitogenesis of human cultured airway smooth muscle cells (Tomlinson et al. 1995). In another study, the PDE3 inhibitor siguazodan and the non-selective PDE inhibitor IBMX were observed to inhibit both [3H]thymidine incorporation and the increase in cell number induced by platelet-derived growth factor in human cultured airway smooth muscle cells, while the PDE4 inhibitor rolipram had no effect (Billington et al. 1999).

6
Clinical Studies of PDE Inhibitors in Asthma and COPD

PDE inhibitors are currently being developed for the treatment of asthma and COPD although side effects including nausea and emesis have hampered the development of some examples of this class of drug into the clinic. To date, there are a limited number of clinical studies investigating the efficacy of PDE inhibitors in the treatment of asthma. Inhalation of zardaverine was shown to produce a modest bronchodilator effect in patients with asthma, although unacceptable side effects of nausea and emesis were reported in a significant number of patients (Brunnee et al. 1992), while oral administration of cilostazol (PDE3 inhibitor) caused bronchodilation and bronchoprotection against methacholine challenge in healthy subjects at the expense of mild to severe headache (Fujimura et al. 1995). AH-2132 (benzafentrine; a mixed PDE3/4 inhibitor) has also been reported to have significant bronchodilator activity in normal volunteers (Foster et al. 1992); the PDE4 inhibitor, ibudilast significantly improved baseline airways responsiveness to spasmogens by twofold after 6 months treatment (Kawasaki et al. 1992), and MKS492 (PDE3 inhibitor) has been reported to attenuate the early and late asthmatic response in atopic asthmatics (Bardin et al. 1998).

Recently, the orally active PDE4 selective inhibitors, CDP840 (Harbinson et al. 1997) and roflumilast (Nell et al. 2000) have been demonstrated to modestly attenuate the development of the late asthmatic response in mild asthmatics whilst having no effect on the acute response, with no significant side effects being reported in comparison with placebo. The ability of these novel selective PDE4 inhibitors to inhibit the late asthmatic response was not associated with bronchodilation, suggesting actions of this drug other than smooth muscle relaxation. Furthermore, the PDE4 inhibitor RP 73401, has also been shown to have no significant effect on allergen-induced bronchoconstriction in allergic asthmatic subjects (Jonker et al. 1996). These data are consistent with the suggestion that PDE3 rather than PDE4 may be the important isoenzyme regulating airway smooth muscle tone in asthmatic subjects. However, recent clinical studies with another orally active PDE4 inhibitor, cilomilast have shown that this drug can attenuate bronchoconstriction following exercise in asthmatic subjects (Nieman et al. 1998), an effect mimicked by 4 weeks of treatment with the selective PDE4 inhibitor, roflumilast (Timmer et al. 2000), although the effect of the latter drug was accompanied by a reduction in TNF-α levels. This would suggest that PDE4 inhibition can influence inflammatory cell function in vivo. The oral administration of V11294 has also been shown to reduce TNF-α levels in healthy volunteers (Landells et al. 2001a).

More recently, cilomilast administered to asthmatic subjects taking inhaled glucocorticosteroids or individuals with COPD (Torphy et al. 1999) demonstrated improvements in baseline lung function and was well tolerated with doses up to 15 mg b.i.d. The mechanism of the beneficial action observed with cilomilast is unlikely to be due to bronchodilation per se, since this drug has modest effects on airway smooth muscle function (Underwood et al. 1998). An explanation for the beneficial effect of cilomilast might include suppression of bronchial hyperresponsiveness secondary to a reduction of airway inflammation that would lead to improvements in lung function and/or reduction in afferent nerve activity and thereby reducing reflex bronchoconstriction.

7
Conclusion

Our increasing knowledge of the molecular biology of the expanding PDE family of enzymes provides exciting opportunities for the development of highly selective, even disease-specific drugs. It is already apparent that encouraging signs beginning to emerge concerning the development of novel PDE4 inhibitors will not only assist in our understanding of the role of PDE4 subtypes in the regulation of cell function, but also offer the potential to find novel treatments for respiratory diseases (Torphy et al. 2000).

References

Alfonso, A., Estevez, M., Louzao, M. C., Vieytes, M. R., and Botana, L. M. Determination of phosphodiesterase activity in rat mast cells using the fluorescent cAMP analogue anthraniloyl cAMP. Cell Signal. 7(5), 513–518. 1995

Aloui, R., Gormand, F., Prigent, A. F., PerrinFayolle, M., and Pacheco, Y. Increased respiratory burst and phosphodiesterase activity in alveolar eosinophils in chronic eosinophilic pneumonia. Eur.Respir.J. 9, 377–379. 1996

Alvarez, R., Sette, C., Yang, D., Eglen, R. M., Wilhelm, R., Shelton, E. R., and Conti, M. Activation and selective inhibition of a cyclic AMP-specific phosphodiesterase, PDE-4D3. Mol.Pharmacol. 48(4), 616–622. 1995

Alves, A. C., Pires, A. L., Lagente, V., Cordeiro, R. S., Martins, M. A., and Silva, P. M. Effect of selective phosphodiesterase inhibitors on the rat eosinophil chemotactic response in vitro. Mem.Inst.Oswaldo Cruz 92 Suppl 2, 201–204. 1997

Alves, A. C., Pires, A. L. A., Cruz, H. N., Serra, M. F., Diaz, B. L., Cordeiro, R. S. B., Lagente, V, Martins, M. A., and de Silva, P. M. R. Selective inhibition of phosphodiesterase type IV suppresses the chemotactic responsiveness of rat eosinophils in vitro. Eur.J.Pharmacol. 312, 89–96. 1996

Anastassiou, E. D., Paliogianni, F., Balow, J. P., Yamada, H., and Boumpas, D. T. Prostaglandin E2 and other cyclic AMP-elevating agents modulate IL-2 and IL-2R alpha gene expression at multiple levels. J.Immunol. 148(9), 2845–2852. 1992

Archer, C. B., Morley, J., and MacDonald, D. M. Impaired lymphocyte cyclic adenosine monophosphate responses in atopic eczema. Br.J.Dermatol. 109(5), 559–564. 1983

Au, B. T., Teixeira, M. M., Collins, P. D., and Williams, T. J. Effect of PDE4 inhibitors on zymosan-induced IL-8 release from human neutrophils: synergism with prostanoids and salbutamol. Br.J.Pharmacol. 123(6), 1260–1266. 1998

Averill, L. E., Stein, R. L., and Kammer, G. M. Control of human T-lymphocyte interleukin-2 production by a cAMP- dependent pathway. Cell Immunol. 115(1), 88–99. 1988

Bachelet, M., Vincent, D., Havet, N., Marrash, Chahla, Pradalier, A., Dry, J., and Vargaftig, B. B. Reduced responsiveness of adenylate cyclase in alveolar macrophages from patients with asthma. J.Allergy Clin.Immunol. 88(3 Pt 1), 322–328. 1991

Baixeras, E., GarciaLozano, E., and Martinez, A. C. Decrease in cAMP levels promoted by CD48-CD2 interaction correlates with inhibition of apoptosis in B cells. Scand.J.Immunol. 43, 406–412. 1996

Baker, A. J. and Fuller, R. W. Effect of cyclic adenosine monophosphate, 5'-(N- ethylcarboxyamido)-adenosine and methylxanthines on the release of thromboxane and lysosomal enzymes from human alveolar macrophages and peripheral blood monocytes in vitro. Eur.J.Pharmacol. 211(2), 157–161. 1992

Banner, K. H., Harbinson, P., Costello, J. F., and Page, C. P. Effect of PDE inhibitors on the proliferation of human peripheral blood mononuclear cells (HPBM) from mild asthmatics and normals. Am.J.Resp.Crit.Care Med. 155(4). 1997

Banner, K. H., Hoult, J. R., Taylor, M. N., Landells, L. J., and Page, C. P. Possible Contribution of Prostaglandin E2 to the antiproliferative effect of phosphodiesterase 4 inhibitors in human mononuclear cells. Biochem.Pharmacol. 58(9), 1487–1495. 1-11-1999

Banner, K. H., Roberts, N. M., and Page, C. P. Differential effect of phosphodiesterase 4 inhibitors on the proliferation of human peripheral blood mononuclear cells from normals and subjects with atopic dermatitis. Br.J.Pharmacol. 116, 3169–3174. 1995

Bardin, P. G., Dorward, M. A., Lampe, F. C., Franke, B., and Holgate, S. T. Effect of selective phosphodiesterase 3 inhibition on the early and late asthmatic responses to inhaled allergen. Br.J.Clin.Pharmacol. 45(4), 387–391. 1998

Barnette, M. S., Bartus, J. O., Burman, M., Christensen, S. B., Cieslinski, L. B., Esser, K. M., Prabhakar, U. S., Rush, J. A., and Torphy, T. J. Association of the anti-inflammatory activity of phosphodiesterase 4 (PDE4) inhibitors with either inhibition of PDE4 catalytic activity or competition for [3H]rolipram binding. Biochem.Pharmacol. 51(7), 949–956. 1996

Barnette, M. S., Christensen, S. B., Essayan, D. M., Grous, M., Prabhakar, U., Rush, J. A., Kagey, Sobotka A., and Torphy, T. J. SB 207499 (Ariflo), a potent and selective second-generation phosphodiesterase 4 inhibitor: in vitro anti-inflammatory actions. J.Pharmacol.Exp.Ther. 284(1), 420–426. 1998

Barnette, M. S., Grous, M., Cieslinski, L. B., Burman, M., Christensen, S. B., and Torphy, T. J. Inhibitors of phosphodiesterase IV (PDE IV) increase acid secretion in rabbit isolated gastric glands: correlation between function and interaction with a high-affinity rolipram binding site. J.Pharmacol.Exp.Ther. 273(3), 1396–1402. 1995

Beavo, J. A. Cyclic nucleotide phosphodiesterases: functional implications of multiple isoforms. Physiol.Rev. 75(4), 725–748. 1995

Betz, M. and Fox, B. S. Prostaglandin E2 inhibits production of Th1 lymphokines but not of Th2 lymphokines. J.Immunol. 146(1), 108–113. 1991

Beusenberg, F. D., Van Amsterdam, J. G. C., Hoogsteden, H. C., Hekking, P. R. M., Brouwers, JW, Schermers, H. P., and Bonta, I. L. Stimulation of cyclic AMP production in human alveolar macrophages induced by inflammatory mediators and beta-sympathicomimetic. European Journal of Pharmacology-Environmental Toxicology & Pharmacology Section 228(1), 57–62. 1992

Billington, C. K., Joseph, S. K., Swan, C., Scott, M. G., Jobson, T. M., and Hall, I. P. Modulation of human airway smooth muscle proliferation by type 3 phosphodiesterase inhibition. Am.J.Physiol 276(3 Pt 1), L412-L419. 1999

Blease, K., Burke-Gaffney A., and Hellewell, P. G. Modulation of cell adhesion molecule expression and function on human lung microvascular endothelial cells by inhibition of phosphodiesterases 3 and 4. Br.J.Pharmacol. 124(1), 229–237. 1998

Bloom, T. J. and Beavo, J. A. Identification and tissue-specific expression of PDE7 phosphodiesterase splice variants. Proc.Natl.Acad.Sci.U.S.A 93(24), 14188–14192. 26-11-1996

Bousquet, J., Jeffery, P. K., Busse, W. W., Johnson, M., and Vignola, A. M. Asthma—From bronchoconstriction to airways inflammation and remodeling. American Journal of Respiratory and Critical Care Medicine 161(5), 1720–1745. 2000

Brunnee, T., Engelstatter, R., Steinijans, V. W., and Kunkel, G. Bronchodilatory effect of inhaled zardaverine, a phosphodiesterase III and IV inhibitor, in patients with asthma. Eur.Respir.J. 5(8), 982–985. 1992

Burns, F., Stevens, P. A., and Pyne, N. J. The identification of apparently novel cyclic AMP and cyclic GMP phosphodiesterase activities in guinea-pig tracheal smooth muscle. Br.J.Pharmacol. 113(1), 3–4. 1994

Busse, W. W. and Anderson, C. L. The granulocyte response to the phosphodiesterase inhibitor RO 20-1724 in asthma. J.Allergy Clin.Immunol. 67(1), 70–74. 1981

Carpenter, D. O., Briggs, D. B., Knox, A. P., and Strominger, N. Excitation of area postrema neurones by transmitters, peptides and cyclic nucleotides. J.Neurophysiol. 59, 358–369. 1988

Casnocha, S. A., Eskin, S. G., Hall, E. R., and McIntire, L. V. Permeability of human endothelial monolayers: effect of vasoactive agonists and cAMP. J.Appl.Physiol. 67(5), 1997–2005. 1989

Chan, S. C. and Hanifin, J. M. Differential inhibitor effects on cyclic adenosine monophosphate- phosphodiesterase isoforms in atopic and normal leukocytes [see comments]. J.Lab.Clin.Med. 121(1), 44–51. 1993a

Chan, S. C., Henderson, W. R., Jr., Shi-Hua, L., and Hanifin, J. M. Prostaglandin E2 control of T cell cytokine production is functionally related to the reduced lymphocyte proliferation in atopic dermatitis. J.Allergy Clin.Immunol. 97, 85–94. 1996

Chan, S. C., Reifsnyder, D., Beavo, J. A., and Hanifin, J. M. Immunochemical characterization of the distinct monocyte cyclic AMP-phosphodiesterase from patients with atopic dermatitis. J.Allergy Clin.Immunol. 91(6), 1179–1188. 1993b

Cherry, J. A. and Davis, R. L. Cyclic AMP phosphodiesterases are localized in regions of the mouse brain associated with reinforcement, movement and affect. J.Comp.Neurol. 407, 287–301. 1999

Choi, Y. H., Ekholm, D., Krall, J., Ahmad, F., Degerman, E., Manganiello, V. C., and Movsesian, M. A. Identification of a novel isoform of the cyclic-nucleotide phosphodiesterase PDE3A expressed in vascular smooth-muscle myocytes. Biochem.J. 353(Pt 1), 41–50. 1-1-2001

Christensen, S. B., Guider, A., Forster, C. J., Gleason, J. G., Bender, P. E., Karpinski, J. M., DeWolf, W. E. Jr, Barnette, M. S., Underwood, D. C., Griswold, D. E., Cieslinski, L. B., Burman, M., Bochnowicz, S., Osborn, R. R., Manning, C. D., Grous, M., Hillegas, L.

M., Bartus, J. O., Ryan, M. D., Eggleston, D. S., Haltiwanger, R. C., and Torphy, T. J. 1,4-Cyclohexanecarboxylates: potent and selective inhibitors of phosphodiesterase 4 for the treatment of asthma. J.Med.Chem. 41(6), 821–835. 12-3-1998

Cohan, V. L., Showell, H. J., Fisher, D. A., Pazoles, C. J., Watson, J. W., Turner, C. R., and Cheng, J. B. In vitro pharmacology of the novel phosphodiesterase type 4 inhibitor, CP-80633. J.Pharmacol.Exp.Ther. 278, 1356–1361. 1996

Columbo, M., Botana, L. M., Horowitz, E. M., Lichtenstein, L. M., and MacGlashan, D. W., Jr. Studies of the intracellular Ca2+ levels in human adult skin mast cells activated by the ligand for the human c-kit receptor and anti-IgE. Biochem.Pharmacol. 47(12), 2137–2145. 1994

Columbo, M., Horowitz, E. M., McKenzie, White, Kagey, Sobotka, and Lichtenstein, L. M. Pharmacologic control of histamine release from human basophils induced by platelet-activating factor. Int.Arch.Allergy Immunol. 102(4), 383–390. 1993

Cooper, K. D., Kang, K., Chan, S. C., and Hanifin, J. M. Phosphodiesterase inhibition by Ro 20-1724 reduces hyper-IgE synthesis by atopic dermatitis cells in vitro. J.Invest.Dermatol. 84(6), 477–482. 1985

Coqueret, O., Boichot, E., and Lagente, V. Selective type IV phosphodiesterase inhibitors prevent IL-4-induced IgE production by human peripheral blood mononuclear cells. Clin.Exp.Allergy 27, 816–823. 1997

Coqueret, O., Dugas, B., Mencia, Huerta, J., and Braquet, P. Regulation of IgE production from human mononuclear cells by beta 2-adrenoceptor agonists [see comments]. Clin.Exp.Allergy 25(4), 304–311. 1995

Cortijo, J., Bou, J., Beleta, J., Cardelus, I, Llenas, J., Morcillo, E., and Gristwood, R. W. Investigation into the role of phosphodiesterase IV in bronchorelaxation, including studies with human bronchus. Br.J.Pharmacol. 108(2), 562–568. 1993

Cortijo, J., Villagrasa, V., Navarrete, C., Sanz, C., Berto, L., Michel, A., Bonnet, P. A., and Morcillo, E. J. Effects of SCA40 on human isolated bronchus and human polymorphonuclear leukocytes: comparison with rolipram, SKF94120 and levcromakalim. Br.J.Pharmacol. 119(1), 99–106. 1996

Crocker, I. C., Ohia, S. E., Church, M. K., and Townley, R. G. Phosphodiesterase type 4 inhibitors, but not glucocorticoids, are more potent in suppression of cytokine secretion by mononuclear cells from atopic than nonatopic donors. J.Allergy Clin.Immunol. 102, 797–804. 1998

Crocker, I. C., Townley, R. G., and Khan, M. M. Phosphodiesterase inhibitors suppress proliferation of peripheral blood mononuclear cells and interleukin-4 and −5 secretion by human T-helper type 2 cells. Immunopharmacology 31, 223–235. 1996

de Boer, J., Philpott, A. J., van Amsterdam, R. G., Shahid, M., Zaagsma, J., and Nicholson, C. D. Human bronchial cyclic nucleotide phosphodiesterase isoenzymes: biochemical and pharmacological analysis using selective inhibitors. Br.J.Pharmacol. 106(4), 1028–1034. 1992

Deisher, T. A., Garcia, I, and Harlan, J. M. Cytokine-induced adhesion molecule expression on human umbilical vein endothelial cells is not regulated by cyclic adenosine monophosphate accumulation. Life Sci. 53(4), 365–370. 1993

Dent, G., Giembycz, M. A., Evans, P. M., Rabe, K. F., and Barnes, P. J. Suppression of human eosinophil respiratory burst and cyclic AMP hydrolysis by inhibitors of type IV phosphodiesterase: interaction with the beta adrenoceptor agonist albuterol. J.Pharmacol.Exp.Ther. 271(3), 1167–1174. 1994

Dent, G., Giembycz, M. A., Rabe, K. F., and Barnes, P. J. Inhibition of eosinophil cyclic nucleotide PDE activity and opsonised zymosan-stimulated respiratory burst by 'type IV'- selective PDE inhibitors. Br.J.Pharmacol. 103(2), 1339–1346. 1991

Dent, G., White, S. R., Tenor, H., Bodtke, K., Schudt, C., Leff, A. R., Magnussen, H., and Rabe, K. F. Cyclic nucleotide phosphodiesterases in human bronchial epithelial cells: characterization of isoenzymes and functional effects of PDE inhibitors. Pulm.Pharmacol.Ther. 11, 47–56. 1998

Derian, C. K., Santulli, R. J., Rao, P. E., Solomon, H. F., and Barrett, J. A. Inhibition of chemotactic peptide-induced neutrophil adhesion to vascular endothelium by cAMP modulators. J.Immunol. 154(1), 308–317. 1995

DiSanto, M. E., Glaser, K. B., and Heaslip, R. J. Phospholipid regulation of a cyclic AMP-specific phosphodiesterase (PDE4) from U937 cells. Cell Signal. 7(8), 827–835. 1995

Dowling, R. B., Johnson, M., Cole, P. J., and Wilson, R. The effect of rolipram, a type IV phosphodiesterase inhibitor, on Pseudomonas aeruginosa infection of respiratory mucosa. Journal of Pharmacology & Experimental Therapeutics 282(3), 1565–1571. 1997

Duplantier, A. J., Biggers, M. S., Chambers, R. J., Cheng, J. B., Cooper, K., Damon, D. B., Eggler, J. F., Kraus, K. G., Marfat, A., Masamune, H., Pillar, J. S., Shirley, J. T., Umland, J. P., and Watson, J. W. Biarylcarboxylic acids and amides: inhibition of phosphodiesterase type IV versus [3H]rolipram binding activity and their relationship to emesis in the ferret. J.Med.Chem. 39, 120–125. 1996

Eigler, A., Siegmund, B., Emmerich, U., Baumann, K. H., Hartmann, G., and Endres, S. Anti-inflammatory activities of cAMP-elevating agents: enhancement of IL-10 synthesis and concurrent suppression of TNF production. J.Leukoc.Biol. 63(1), 101–107. 1998

Engels, P., Fichtel, K., and Lubbert, H. Expression and regulation of human and rat phosphodiesterase type IV isogenes. FEBS Lett. 350(2–3), 291–295. 1994

Essayan, D. M., Huang, S. K., Kagey, Sobotka, and Lichtenstein, L. M. Effects of nonselective and isozyme selective cyclic nucleotide phosphodiesterase inhibitors on antigen-induced cytokine gene expression in peripheral blood mononuclear cells. Am.J.Respir.Cell Mol.Biol. 13(6), 692–702. 1995

Essayan, D. M., Huang, S. K., Undem, B. J., Kagey, Sobotka, and Lichtenstein, L. M. Modulation of antigen- and mitogen-induced proliferative responses of peripheral blood mononuclear cells by nonselective and isozyme selective cyclic nucleotide phosphodiesterase inhibitors. J.Immunol. 153(8), 3408–3416. 1994

Essayan, D. M., Kagey-Sobotka, A., Lichtenstein, L. M., and Huang, S.-K. Differential regulation of human antigen-specific Th1 and Th2 lymphocyte responses by isozyme selective cyclic nucleotide phosphodiesterase inhibitors. Journal of Pharmacology & Experimental Therapeutics 282(1), 505–512. 1997

Ezeamuzie, C. I. Requirement of additional adenylate cyclase activation for the inhibition of human eosinophil degranulation by phosphodiesterase IV inhibitors. Eur.J.Pharmacol. 417(1–2), 11–18. 6-4-2001

Ferretti, M. E., Spisani, S., Pareschi, M. C., Buzzi, M., Cavallaro, R., Traniello, S., Reali, E., Torrini, I, Paradisi, M. P., and Zecchini, G. P. Two new formylated peptides able to activate chemotaxis and respiratory burst selectively as tools for studying human neutrophil responses. Cell Signal. 6(1), 91–101. 1994

Foster, R. W., Rakshi, K., Carpenter, J. R., and Small, R. C. Trials of the bronchodilator activity of the isoenzyme-selective phosphodiesterase inhibitor AH 21–132 in healthy volunteers during a methacholine challenge test. Br.J.Clin.Pharmacol. 34(6), 527–534. 1992

Frossard, N., Landry, Y., Pauli, G., and Ruckstuhl, M. Effects of cyclic AMP- and cyclic GMP- phosphodiesterase inhibitors on immunological release of histamine and on lung contraction. Br.J.Pharmacol. 73(4), 933–938. 1981

Fuhrmann, M., Jahn, H.-U., Seybold, J., Neurohr, C., Barnes, P. J., Hippenstiel, S., Kraemer, H. J., and Suttorp, N. Identification and function of cyclic nucleotide phosphodiesterase isoenzymes in airway epithelial cells. Am.J.Respir.Cell Mol.Biol. 20, 292–302. 1999

Fujii, K., Kohrogi, H., Iwagoe, H., Hamamoto, J., Hirata, N., Goto, E., Kawano, O., Wada, K., Yamagata, S., and Ando, M. Novel phosphodiesterase 4 inhibitor T-440 reverses and prevents human bronchial contraction induced by allergen. Journal of Pharmacology & Experimental Therapeutics 284(1), 162–169. 1998

Fujimura, M., Kamio, Y., Saito, M., Hashimoto, T., and Matsuda, T. Bronchodilator and bronchoprotective effects of cilostazol in humans in vivo. American Journal of Respiratory and Critical Care Medicine 151, 222–225. 1995

Fuller, R. W. Control of mediator release from the human alveolar macrophage: Role of cyclic AMP. European Journal of Pharmacology 183(2), 621. 1990

Fuller, R. W., O'Malley, G., Baker, A. J., and MacDermot, J. Human alveolar macrophage activation: inhibition by forskolin but not beta-adrenoceptor stimulation or phosphodiesterase inhibition. Pulm.Pharmacol. 1(2), 101–106. 1988

Gantner, F., Gotz, C., Gekeler, V., Schudt, C., Wendel, A., and Hatzelmann, A. Phosphodiesterase profile of human B lymphocytes from normal and atopic donors and the effects of PDE inhibition on B cell proliferation. Br.J.Pharmacol. 123, 1031–1038. 1998

Gantner, F., Kupferschmidt, R., Schudt, C., Wendel, A., and Hatzelmann, A. In vitro differentiation of human monocytes to macrophages: Change of PDE profile and its relationship to suppression of tumour necrosis factor-alpha release by PDE inhibitors. British Journal of Pharmacology 121(2), 221–231. 1997a

Gantner, F., Tenor, H., Gekeler, V., Schudt, C., Wendel, A., and Hatzelmann, A. Phosphodiesterase profiles of highly purified human peripheral blood leukocyte populations from normal and atopic individuals: a comparative study. J.Allergy Clin.Immunol. 100(4), 527–535. 1997b

Giembycz, M. A. Phosphodiesterase 4 and tolerance to β_2-adrenoceptor agonists in asthma. Trends Pharmacol.Sci. 17, 331–336. 1996

Giembycz, M. A., Corrigan, C. J., Seybold, J., Newton, R., and Barnes, P. J. Identification of cyclic AMP phosphodiesterases 3,4 and 7 in human $CD4^+$ and $CD8^+$ T-lymphocytes: role in regulating proliferation and the biosynthesis of interleukin-2. Br.J.Pharmacol. 118, 1945–1958. 1996

Giustina, T. A., Chan, S. C., Thiel, M. L., Baker, J. W., and Hanifin, J. M. Increased leukocyte sensitivity to phosphodiesterase inhibitors in atopic dermatitis: tachyphylaxis after theophylline therapy. J.Allergy Clin.Immunol. 74(3 Pt 1), 252–257. 1984

Gleich, G. J. Mechanisms of eosinophil-associated inflammation. J.Allergy Clin.Immunol. 105, 651–663. 2000

Greten, T. F., Sinha, B., Haslberger, C., Eigler, A., and Endres, S. Cicaprost and the type IV phosphodiesterase inhibitor, rolipram, synergize in suppression of tumor necrosis factor-alpha synthesis. Eur.J.Pharmacol. 299, 229–233. 1996

Grewe, S. R., Chan, S. C., and Hanifin, J. M. Elevated leukocyte cyclic AMP-phosphodiesterase in atopic disease: a possible mechanism for cyclic AMP-agonist hyporesponsiveness. J.Allergy Clin.Immunol. 70(6), 452–457. 1982

Griswold, D. E., Webb, E. F., Breton, J., White, J. R., Marshall, P. J., and Torphy, T. J. Effect of selective phosphodiesterase type IV inhibitor, rolipram, on fluid and cellular phases of inflammatory response. Inflammation 17(3), 333–344. 1993

Hadjokas, N. E., Crowley, J. J., Bayer, C. R., and Nielson, C. P. beta-Adrenergic regulation of the eosinophil respiratory burst as detected by lucigenin-dependent luminescence. J.Allergy Clin.Immunol. 95, 735–741. 1995

Hallsworth, M. P., Giembycz, M. A., Barnes, P. J., and Lee, T. H. Cyclic AMP-elevating agents prolong or inhibit eosinophil survival depending on prior exposure to GM-CSF. British Journal of Pharmacology 117, 79–86. 1996

Hanifin, J. M. and Chan, S. C. Monocyte phosphodiesterase abnormalities and dysregulation of lymphocyte function in atopic dermatitis. J.Invest.Dermatol. 105, 84S–88S. 1995

Hanifin, J. M., Chan, S. C., Cheng, J. B., Tofte, S. J., Henderson, W. R., Jr., Kirby, D. S., and Weiner, E. S. Type 4 phosphodiesterase inhibitors have clinical and in vitro anti-inflammatory effects in atopic dermatitis. J.Invest.Dermatol. 107, 51–56. 1996

Hansen, G., Jin, S., Umetsu, D. T., and Conti, M. Absence of muscarinic cholinergic airway responses in mice deficient in the cyclic nucleotide phosphodiesterase PDE4D. Proc Natl Acad Sci U S A 97(12), 6751–6756. 6-6-2000

Harbinson, P. L., MacLeod, D., Hawksworth, R., O'Toole, S., Sullivan, P. J., Heath, P., Kilfeather, S., Page, C. P., Costello, J., Holgate, S. T., and Lee, T. H. The effect of a novel orally active selective PDE4 isoenzyme inhibitor (CDP840) on allergen-induced responses in asthmatic subjects. Eur.Respir.J. 10(5), 1008–1014. 1997

Harris, A. L., Connell, M. J., Ferguson, E. W., Wallace, A. M., Gordon, R. J., Pagani, E. D., and Silver, P. J. Role of low K_m cyclic AMP phosphodiesterase inhibition in tracheal relaxation and bronchodilation in the guinea pig. J.Pharmacol.Exp.Ther. 251(1), 199–206. 1989

Hatzelmann, A., Tenor, H., and Schudt, C. Differential effects of non-selective and selective phosphodiesterase inhibitors on human eosinophil functions. Br.J.Pharmacol. 114(4), 821–831. 1995

Hersperger, R., Bray-French, K., Mazzoni, L., and Muller, T. Palladium-catalyzed cross-coupling reactions for the synthesis of 6, 8-disubstituted 1,7-naphthyridines: a novel class of potent and selective phosphodiesterase type 4D inhibitors. J.Med.Chem. 43(4), 675–682. 24-2-2000

Hichami, A., Boichot, E., Germain, N., Coqueret, O., and Lagente, V. Interactions between cAMP- and cGMP-dependent protein kinase inhibitors and phosphodiesterase IV inhibitors on arachidonate release from human monocytes. Life Sci. 59(16), L255-L261. 1996

Hichami, A., Boichot, E., Germain, N., Legrand, A., Moodley, I, and Lagente, V. Involvement of cyclic AMP in the effects of phosphodiesterase IV inhibitors on arachidonate release from mononuclear cells. Eur.J.Pharmacol. 291(2), 91–97. 1995

Hilkens, C. M., Vermeulen, H., van Neerven, R. J., Snijdewint, F. G., Wierenga, E. A., and Kapsenberg, M. L. Differential modulation of T helper type 1 (Th1) and T helper type 2 (Th2) cytokine secretion by prostaglandin E2 critically depends on interleukin-2. Eur.J.Immunol. 25(1), 59–63. 1995

Hoffman, R., Baillie, G. S., MacKenzie, S. J., Yarwood, S. J., and Houslay, M. D. The MAP kinase ERK2 inhibits the cyclic AMP-specific phosphodiesterase HSPDE3D3 by phosphorylating it at Ser579. EMBO J. 18, 893–903. 1999

Houslay, M. D., Sullivan, M., and Bolger, G. B. The multienzyme PDE4 cyclic adenosine monophosphate-specific phosphodiesterase family: Intracellular targeting, regulation, and selective inhibition by compounds exerting anti-inflammatory and antidepressant actions. August, J. T., Anders, M. W., Murad, F., and Coyle, J. T. Advances in Pharmacology. 225–342. 1998. London, Academic Press

Huang, R., Cioffi, J., Berg, K., London, R., Cidon, M., Maayani, S., and Mayer, L. B cell differentiation factor-induced B cell maturation: regulation via reduction in cAMP. Cell Immunol. 162(1), 49–55. 1995

Hughes, B., Howat, D., Lisle, H., Holbrook, M., Tames, T., Gozzard, N., Blease, K., Hughes, P., Kingaby, R., Warrellow, G., Alexander, R., Head, J., Boyd, E., Eaton, M., Perry, M., Wales, M., Smith, B., Owens, R., Catterall, C., Lumb, S., Russell, A., Allen, R., Merriman, M., Bloxham, D., and Higgs, G. The inhibition of antigen-induced eosinophilia and bronchoconstriction by CDP840, a novel stereo-selective inhibitor of phosphodiesterase type 4. Br.J.Pharmacol. 118, 1183–1191. 1996

Ichimura, M. and Kase, H. A new cyclic nucleotide phosphodiesterase isozyme expressed in the T-lymphocyte cell lines. Biochem.Biophys.Res.Commun. 193(3), 985–990. 1993

Jacobitz, S., McLaughlin, M. M., Livi, G. P., Burman, M., and Torphy, T. J. Mapping the functional domains of human recombinant phosphodiesterase 4A: structural requirements for catalytic activity and rolipram binding. Mol.Pharmacol. 50, 891–899. 1996

Jeffery, P. K. Structural and inflammatory changes in COPD: a comparison with asthma. Thorax 53, 129–136. 1998

Jin, S. L., Richard, F. J., Kuo, W. P., D'Ercole, A. J., and Conti, M. Impaired growth and fertility of cAMP-specific phosphodiesterase PDE4D-deficient mice. Proc Natl Acad Sci U S A 96(21), 11998–12003. 12-10-1999

Jonker, G. J., Tijhuis, G. J., and de Monchey, J. G. R. RP 73401 (a phosphodiesterase IV inhibitor) single does not prevent allergen induced bronchoconstriction during the early phase reaction in asthmatics. Eur.Respir.J. 9, 82 s. 1996

Kammer, G. M. The adenylate cyclase-cAMP-protein kinase A pathway and regulation of the immune response. Immunol.Today 9(7–8), 222–229. 1988

Kanda, N. and Watanabe, S. Gangliosides GD1b, GT1b, and GQ1b enhance IL-2 and IFN-gamma production and suppress IL-4 and IL-5 production in phytohemagglutinin-stimulated human T cells. J.Immunol. 166(1), 72–80. 1-1-2001

Kaneko, T., Alvarez, R., Ueki, I. F., and Nadel, J. A. Elevated intracellular cyclic AMP inhibits chemotaxis in human eosinophils. Cell Signal. 7(5), 527–534. 1995

Kawasaki, A., Hoshino, K., Osaki, R., Mizushima, Y., and Yano, S. Effect of ibudilast: a novel antiasthmatic agent, on airway hypersensitivity in bronchial asthma. J.Asthma 29(4), 245–252. 1992

Kelley, T. J., al Nakkash, L., and Drumm, M. L. CFTR-mediated chloride permeability is regulated by type III phosphodiesterases in airway epithelial cells. Am.J.Respir.Cell Mol.Biol. 13(6), 657–664. 1995

Kelly, J. R., Barnes, P. J., and Giembycz, M. A. Phosphodiesterase 4 in macrophages: relationship between cAMP accumulation, suppression of cAMP hydrolysis and inhibition of [^{3}H]R-(-)-rolipram binding by selective inhibitors. Biochem.J. 318, 425–436. 1996

Kemeny, D. M., Vyas, B., Vukmanovi-Stejic, M., Thomas, M. J., Nobel, A., Loh, L.-C., and O'Connor, B. J. CD8+ T cell subsets and chronic obstructive pulmonary disease. American Journal of Respiratory and Critical Care Medicine 160, S33-S37. 1999

Kleine, Tebbe, Wicht, L., Gagne, H., Friese, A., Schunack, W., Schudt, C., and Kunkel, G. Inhibition of IgE-mediated histamine release from human peripheral leukocytes by selective phosphodiesterase inhibitors. Agents Actions 36(3–4), 200–206. 1992

Knudsen, J. H., Kjaersgaard, E., and Christensen, N. J. Individual lymphocyte subset composition determines cAMP response to isoproterenol in mononuclear cell preparations from peripheral blood. Scand.J.Clin.Lab.Invest. 55, 9–14. 1995

Koeter, G. H., Meurs, H., Kauffman, H. F., and de Vries, K. The role of the adrenergic system in allergy and bronchial hyperreactivity. Eur.J.Respir.Dis.Suppl. 121, 72–78. 1982

Komas, N., Lugnier, C., Andriantsitohaina, R., and Stoclet, J. C. Characterisation of cyclic nucleotide phosphodiesterases from rat mesenteric artery. Eur.J.Pharmacol. 208(1), 85–87. 1991

Kung, T. T., Crawley, Y., Luo, B., Young, S., Kreutner, W., and Chapman, R. W. Inhibition of pulmonary eosinophilia and airway hyperresponsiveness in allergic mice by rolipram: involvement of endogenously released corticosterone and catecholamines. Br.J.Pharmacol. 130, 457–463. 2000

Laliberte, F., Han, Y., Govindaragan, A., Giroux, A., Liu, S., Bobechko, B., Lario, P., Bartlett, A., Gorseth, E., Gresser, M., and Huang, Z. Conformational difference between PDE4 apoenzyme and haloenzyme. Biochemistry 39, 6449–6458. 2000

Landells, L. J., Jensen, M. W., Spina, D., Donigi-Gale, D., Miller, A. J., Nichols, T., Smith, K., Rotshteyn, Y., Burch, R. M., Page, C., and O'Connor, B. J. Oral administration of the phosphodiesterase (PDE)4 inhibitor, V11294A inhibits ex-vivo agonist-induced cell activation. Eur Respir J 12(Suppl 28). 2001a

Landells, L. J., Spina, D., Souness, J. E., O'Connor, B. J., and Page, C. P. A biochemical and functional assessment of monocyte phosphodiesterase activity in healthy and asthmatic subjects. Pulm.Pharmacol.Ther. 13(5), 231–239. 2000

Landells, L. J., Szilagy, C. M., Jones, N. A., Banner, K. H., Allen, J. M., Doherty, A., O'Connor, B. J., Spina, D., and Page, C. P. Identification and quantification of phosphodiesterase 4 subtypes in CD4 and CD8 lymphocytes from healthy and asthmatic subjects. Br.J Pharmacol. 133(5), 722–729. 2001b

Lee, H. J., Koyano, Nakagawa, Naito, Y., Nishida, J., Arai, N., Arai, K., and Yokota, T. cAMP activates the IL-5 promoter synergistically with phorbol ester through the signaling pathway involving protein kinase A in mouse thymoma line EL-4. J.Immunol. 151(11), 6135–6142. 1993

Lenhard, J. M., Kassel, D. B., Rocque, W. J., Hamacher, L., Holmes, W. D., Patel, I, Hoffman, C., and Luther, M. Phosphorylation of a cAMP-specific phosphodiesterase (HSPDE4B2B) by mitogen-activated protein kinase. Biochem.J. 319, 751–758. 1996

Lewis, G. M., Caccese, R. G., Heaslip, R. J., and Bansbach, C. C. Effects of rolipram and CI-930 on IL-2 mRNA transcription in human Jurkat cells. Agents Actions 39 Spec No, C89–C92. 1993

Li, L., Yee, C., and Beavo, J. A. CD3- and CD28-dependent induction of PDE7 required for T cell activation. Science 283(5403), 848–849. 1999

Li, S. H., Chan, S. C., Kramer, S. M., and Hanifin, J. M. Modulation of leukocyte cyclic AMP phosphodiesterase activity by recombinant interferon-gamma: evidence for a differential effect on atopic monocytes. J.Interferon.Res. 13(3), 197–202. 1993

Li, S. H., Chan, S. C., Toshitani, A., Leung, D. Y., and Hanifin, J. M. Synergistic effects of interleukin 4 and interferon-gamma on monocyte phosphodiesterase activity. J.Invest.Dermatol. 99(1), 65–70. 1992

Lidington, E., Nohammer, C., Dominguez, M., Ferry, B., and Rose, M. L. Inhibition of the transendothelial migration of human lymphocytes but not monocytes by phosphodiesterase inhibitors. Clinical and Experimental Immunology 104, 66–71. 1996

Lindgren, S. and Andersson, K. E. Effects of selective phosphodiesterase inhibitors on isolated coronary, lung and renal arteries from man and rat. Acta Physiol Scand. 142(1), 77–82. 1991

Lindgren, S., Andersson, K. E., Belfrage, P., Degerman, E., and Manganiello, V. C. Relaxant effects of the selective phosphodiesterase inhibitors milrinone and OPC 3911 on isolated human mesenteric vessels. Pharmacol.Toxicol. 64(5), 440–445. 1989

Loughney, K., Hill, T. R., Florio, V. A., Uher, L., Rosman, G. J., Wolda, S. L., Jones, B. A., Howard, M. L., McAllister-Lucas, L. M., Sonnenburg, W. K., Francis, S. H., Corbin, J. D., Beavo, J. A., and Ferguson, K. Isolation and characterization of cDNAs encoding PDE5A, a human cGMP- binding, cGMP-specific 3',5'-cyclic nucleotide phosphodiesterase. Gene 216(1), 139–147. 17-8-1998

Lugnier, C. and Komas, N. Modulation of vascular cyclic nucleotide phosphodiesterases by cyclic GMP: role in vasodilatation. Eur.Heart J. 14 Suppl I, 141–148. 1993

Lugnier, C. and Schini, V. B. Characterization of cyclic nucleotide phosphodiesterases from cultured bovine aortic endothelial cells. Biochem.Pharmacol. 39(1), 75–84. 1990

Lugnier, C., Schoeffter, P., Le Bec, A., Strouthou, E., and Stoclet, J. C. Selective inhibition of cyclic nucleotide phosphodiesterases of human, bovine and rat aorta. Biochem.Pharmacol. 35(10), 1743–1751. 1986

MacKenzie, S. J., Baillie, G. S., McPhee, I., Bolger, G. B., and Houslay, M. D. ERK2 mitogen-activated protein kinase binding, phosphorylation, and regulation of the PDE4D cAMP-specific phosphodiesterases. J.Biol.Chem. 275, 16609–16617. 2000

Manning, C. D., Burman, M., Christensen, S. B., Cieslinski, L. B., Essayan, D. M., Grous, M., Torphy, T. J., and Barnette, M. S. Suppression of human inflammatory cell function by subtype- selective PDE4 inhibitors correlates with inhibition of PDE4A and PDE4B. British Journal of Pharmacology 128(7), 1393–1398. 1999

Manning, C. D., McLaughlin, M. M., Livi, G. P., Cieslinski, L. B., Torphy, T. J., and Barnette, M. S. Prolonged beta adrenoceptor stimulation up-regulates cAMP phosphodiesterase activity in human monocytes by increasing mRNA and protein for phosphodiesterases 4A and 4B. J.Pharmacol.Exp.Ther. 276(2), 810–818. 1996

Marcoz, P., Prigent, A. F., Lagarde, M., and Nemoz, G. Modulation of rat thymocyte proliferative response through the inhibition of different cyclic nucleotide phosphodiesterase isoforms by means of selective inhibitors and cGMP-elevating agents. Mol.Pharmacol. 44(5), 1027–1035. 1993

Marukawa, S., Hatake, K., Wakabayashi, I, and Hishida, S. Vasorelaxant effects of oxpentifylline and theophylline on rat isolated aorta. J.Pharm.Pharmacol. 46(5), 342–345. 1994

McLaughlin, M. M., Cieslinski, L. B., Burman, M., Torphy, T. J., and Livi, G. P. A low K(M), rolipram-sensitive, cAMP-specific phosphodiesterase from human brain. Cloning and expression of cDNA, biochemical characterization of recombinant protein, and tissue distribution of messenger-RNA. J.Biol.Chem. 268(9), 6470–6476. 1993

Mentz, F., Merle, Beral, Ouaaz, F., and Binet, J. L. Theophylline, a new inducer of apoptosis in B-CLL: role of cyclic nucleotides. Br.J.Haematol. 90(4), 957–959. 1995

Meurs, H., Koeter, G. H., de Vries, K., and Kauffman, H. F. The beta-adrenergic system and allergic bronchial asthma: changes in lymphocyte beta-adrenergic receptor number and adenylate cyclase activity after an allergen-induced asthmatic attack. J.Allergy Clin.Immunol. 70(4), 272–280. 1982

Mikawa, K., Akamatsu, H., Nishina, K., Shiga, M., Maekawa, N., Obara, H., and Niwa, Y. The effect of phosphodiesterase III inhibitors on human neutrophil function. Crit Care Med. 28(4), 1001–1005. 2000

Miyahara, M., Ito, M., Itoh, H., Shiraishi, T., Isaka, N., Konishi, T., and Nakano, T. Isoenzymes of cyclic nucleotide phosphodiesterase in the human aorta: characterization and the effects of E4021. Eur.J.Pharmacol. 284(1–2), 25–33. 1995

Miyamoto, K., Kurita, M., Sakai, R., Sanae, F., Wakusawa, S., and Takagi, K. Cyclic nucleotide phosphodiesterase isoenzymes in guinea-pig tracheal muscle and bronchorelaxation by alkylxanthines. Biochem.Pharmacol. 48(6), 1219–1223. 1994

Molnar, Kimber, Yonno, L., Heaslip, R., and Weichman, B. Modulation of TNF alpha and IL-1 beta from endotoxin-stimulated monocytes by selective PDE isozyme inhibitors. Agents Actions 39 Spec No, C77-C79. 1993

Momose, T., Okubo, Y., Horie, S., Suzuki, J., Isobe, M., and Sekiguchi, M. Effects of intracellular cyclic AMP modulators on human eosinophil survival, degranulation and CD11b expression. International Archives of Allergy & Immunology 117(2), 138–145. 1998

Moodley, I, Sotsios, Y., and Bertin, B. Modulation of oxazolone-induced hypersensitivity in mice by selective PDE inhibitors. Mediators of Inflammation 4, 112–116. 1995

Morandini, R., Ghanem, G., Portier-Lemarie, A., Robaye, B., Renaud, A., and Boeynaems, J. M. Action of cAMP on expression and release of adhesion molecules in human endothelial cells. Am.J.Physiol. 270, H807-H816. 1996

Muller, T., Engels, P., and Fozard, J. R. Subtypes of the type 4 cAMP phosphodiesterases: structure, regulation and selective inhibition. Trends Pharmacol.Sci. 17, 294–298. 1996

Munoz, E., Zubiaga, A. M., Merrow, M., Sauter, N. P., and Huber, B. T. Cholera toxin discriminates between T helper 1 and 2 cells in T cell receptor-mediated activation: role of cAMP in T cell proliferation. J.Exp.Med. 172(1), 95–103. 1990

Nagai, H., Takeda, H., Iwama, T., Yamaguchi, S., and Mori, H. Studies on anti-allergic action of AH 21-132, a novel isozyme- selective phosphodiesterase inhibitor in airways. Jap.J.Pharmacol. 67, 149–156. 1995

Nell, H., Louw, C., Leichtl, S., Rathgeb, F., Neuhauser, M., and Bardin, P. G. Acute anti-inflammatory effect of the novel phosphodiesterase 4 inhibitor roflumilast on allergen challenge in asthmatics after a single dose. American Journal of Respiratory and Critical Care Medicine 161, A200. 2000

Nemoz, G., Moueqqit, M., Prigent, A. F., and Pacheco, H. Isolation of similar rolipram-inhibitable cyclic-AMP-specific phosphodiesterases from rat-brain and heart. Eur.-J.Biochem. 184(3), 511–520. 1989

Nicholson, C. D., Shahid, M., Bruin, J., Barron, E., Spiers, I, de Boer, J., van Amsterdam, R. G., Zaagsma, J., Kelly, J. J., and Dent, G. Characterization of ORG 20241, a combined phosphodiesterase IV/III cyclic nucleotide phosphodiesterase inhibitor for asthma. J.Pharmacol.Exp.Ther. 274(2), 678–687. 1995

Nielson, C. P., Vestal, R. E., Sturm, R. J., and Heaslip, R. Effects of selective phosphodiesterase inhibitors on the polymorphonuclear leukocyte respiratory burst. J.Allergy Clin.Immunol. 86(5), 801–808. 1990

Nieman, R. B., Fisher, B. D., Amit, O., and Dockhorn, R. J. SB 207499 (Ariflow™), a second-generation, selective oral phosphodiesterase type 4 (PDE4) inhibitor, attenuates exercise induced bronchoconstriction in patients with asthma. American Journal of Respiratory and Critical Care Medicine 157, A413. 1998

Niwa, M., Hara, A., Kanamori, Y., Matsuno, H., Kozawa, O., Yoshimi, N., Mori, H., and Uematsu, T. Inhibition of tumor necrosis factor-alpha induced neutrophil apoptosis by cyclic AMP: involvement of caspase cascade. Eur.J.Pharmacol. 371(1), 59–67. 23-4-1999

Novak, T. J. and Rothenberg, E. V. cAMP inhibits induction of interleukin 2 but not of interleukin 4 in T cells. Proc.Natl.Acad.Sci.U.S.A. 87(23), 9353–9357. 1990

O'Connell, J. C., McCallum, J. F., McPhee, I, Wakefield, J., Houslay, E. S., Wishart, W., Bolger, G., Frame, M., and Houslay, M. D. The SH3 domain of Src tyrosyl protein kinase interacts with the N-terminal splice region of the PDE4A cAMP-specific phosphodiesterase RPDE-6 (RNPDE4A5). Biochem.J. 318, 255–262. 1996

Obernolte, R., Bhakta, S., Alvarez, R., Bach, C., Zuppan, P., Mulkins, M., Jarnagin, K., and Shelton, E. R. The cDNA of a human lymphocyte cyclic-AMP phosphodiesterase (PDE IV) reveals a multigene family. Gene 129(2), 239–247. 1993

Obernolte, R., Ratzliff, J., Baecker, P. A., Daniels, D. V., Zuppan, P., Jarnagin, K., and Shelton, E. R. Multiple splice variants of phosphodiesterase PDE4C cloned from human lung and testis. Biochemica Biophysica Acta 1353, 287–297. 1997

Ottonello, L., Gonella, R., Dapino, P., Sacchetti, C., and Dallegri, F. Prostaglandin E2 inhibits apoptosis in human neutrophilic polymorphonuclear leukocytes: role of intracellular cyclic AMP levels. Exp.Hematol. 26(9), 895–902. 1998

Ottonello, L., Morone, P., Dapino, P., and Dallegri, F. Inhibitory effect of salmeterol on the respiratory burst of adherent human neutrophils. Clinical and Experimental Immunology 106(1), 97–102. 1996

Owens, R. J., Caterall, C., Batty, D., Jappy, J., Russell, A., Smith, B., O'Connell, J., and Perry, M. J. Human phosphodiesterase 4A: characterization of full-length and truncated enzymes expressed in COS cells. Biochem.J. 326, 53–60. 1997

Palmer, D. and Maurice, D. H. Dual expression and differential regulation of phosphodiesterase 3A and phosphodiesterase 3B in human vascular smooth muscle: implications for phosphodiesterase 3 inhibition in human cardiovascular tissues. Mol.Pharmacol. 58(2), 247–252. 2000

Pan, X, Arauz, E., Krzanowski, J. J., Fitzpatrick, D. F., and Polson, J. B. Synergistic interactions between selective pharmacological inhibitors of phosphodiesterase isozyme families PDE III and PDE IV to attenuate proliferation of rat vascular smooth muscle cells. Biochem.Pharmacol. 48(4), 827–835. 1994

Parker, C. W., Kennedy, S., and Eisen, A. Z. Leukocyte and lymphocyte cyclic AMP responses in atopic eczema. J.Invest.Dermatol. 68(5), 302–306. 1977

Paul, Eugene, Kolb, J. P., Calenda, A., Gordon, J., Kikutani, H., Kishimoto, T., Mencia, Huerta, J., Braquet, P., and Dugas, B. Functional interaction between beta 2-adrenoceptor agonists and interleukin-4 in the regulation of CD23 expression and release and IgE production in human. Mol.Immunol. 30(2), 157–164. 1993

Paul, Eugene, Kolb, J. P., Damais, C., Abadie, A., Mencia, Huerta, J., Braquet, P., Bousquet, J., and Dugas, B. Beta 2-adrenoceptor agonists regulate the IL-4-induced phenotypical changes and IgE-dependent functions in normal human monocytes. J.Leukoc.-Biol. 55(3), 313–320. 1994

Peachell, P. T., Undem, B. J., Schleimer, R. P., MacGlashan, D. W., Jr., Lichtenstein, L. M., Cieslinski, L. B., and Torphy, T. J. Preliminary identification and role of phosphodiesterase isozymes in human basophils. J.Immunol. 148(8), 2503–2510. 1992

Pene, J., Rousset, F., Briere, F., Chretien, I, Bonnefoy, J. Y., Spits, H., Yokota, T., Arai, N., Arai, K., and Banchereau, J. IgE production by normal human lymphocytes is induced by interleukin 4 and suppressed by interferons gamma and alpha and prostaglandin E2. Proc.Natl.Acad.Sci.U.S.A. 85(18), 6880–6884. 1988

Pober, J. S., Slowik, M. R., De Luca, L. G., and Ritchie, A. J. Elevated cyclic AMP inhibits endothelial cell synthesis and expression of TNF-induced endothelial leukocyte adhesion molecule- 1, and vascular cell adhesion molecule-1, but not intercellular adhesion molecule-1. J.Immunol. 150(11), 5114–5123. 1993

Prabhakar, U., Lipshutz, D., Bartus, J. O., Slivjak, M. J., Smith, E. F., Lee, J. C., and Esser, K. M. Characterization of cAMP-dependent inhibition of LPS-induced TNF alpha production by rolipram, a specific phosphodiesterase IV (PDE IV) inhibitor. Int.J.Immunopharmacol. 16(10), 805–816. 1994

Prigent, A. F., Fonlupt, P., Dubois, M., Nemoz, G., Timouyasse, L., PACHECO, H., Pacheco, Y., Biot, N., and Perrin, Fayolle M. Cyclic nucleotide phosphodiesterases and methyltransferases in purified lymphocytes, monocytes and polymorphonuclear leucocytes from healthy donors and asthmatic patients. Eur.J.Clin.Invest. 20(3), 323–329. 1990

Qian, Y., Naline, E., Karlsson, J. A., Raeburn, D., and Advenier, C. Effects of rolipram and siguazodan on the human isolated bronchus and their interaction with isoprenaline and sodium nitroprusside. Br.J.Pharmacol. 109(3), 774–778. 1993

Rabe, K. F., Magnussen, H., and Dent, G. Theophylline and selective PDE inhibitors as bronchodilators and smooth muscle relaxants. Eur.Respir.J. 8(4), 637–642. 1995

Rabe, K. F., Tenor, H., Dent, G., Schudt, C., Liebig, S., and Magnussen, H. Phosphodiesterase isozymes modulating inherent tone in human airways: identification and characterization. Am.J.Physiol. 264(5 Pt 1), L458-L464. 1993

Rabe, K. F., Tenor, H., Dent, G., Schudt, C., Nakashima, M., and Magnussen, H. Identification of PDE isozymes in human pulmonary artery and effect of selective PDE inhibitors. Am.J.Physiol. 266(5 Pt 1), L536-L543. 1994

Revel, L., Colombo, S., Ferrari, F., Folco, G., Rovati, L. C., and Makovec, F. CR 2039, a new bis-(1H-tetrazol-5-yl)phenylbenzamide derivative with potential for the topical treatment of asthma. Eur.J.Pharmacol. 229(1), 45–53. 1992

Robicsek, S. A., Blanchard, D. K., Djeu, J. Y., Krzanowski, J. J., Szentivanyi, A., and Polson, J. B. Multiple high-affinity cAMP-phosphodiesterases in human T- lymphocytes. Biochem.Pharmacol. 42(4), 869–877. 1991

Rocque, W. J., Tian, G., Wiseman, J. S., Holmes, W. D., Zajac-Thompson, I., Willard, D. H., Patel, I. R., Wisely, G. B., Clay, W. C., Kadwell, S. H., Hoffman, C. R., and Luther, M. A. Human recombinant phosphodiesterase 4B2B binds (R)-rolipram at a single site with two affinities. Biochemistry 36, 14250–14261. 1997

Romagnani, S. The role of lymphocytes in allergic disease. J.Allergy Clin.Immunol. 105, 399–408. 2000

Roper, R. L., Brown, D. M., and Phipps, R. P. Prostaglandin E2 promotes B lymphocyte Ig isotype switching to IgE. J.Immunol. 154(1), 162–170. 1995

Roper, R. L., Conrad, D. H., Brown, D. M., Warner, G. L., and Phipps, R. P. Prostaglandin E2 promotes IL-4-induced IgE and IgG1 synthesis. J.Immunol. 145(8), 2644–2651. 1990

Rousseau, E., Gagnon, J., and Lugnier, C. Biochemical and pharmacological characterization of cyclic nucleotide phosphodiesterase in airway epithelium. Mol.Cell Biochem. 140(2), 171–175. 1994

Sadhu, K., Hensley, K., Florio, V. A., and Wolda, S. L. Differential expression of the cyclic GMP-stimulated phosphodiesterase PDE2A in human venous and capillary endothelial cells. J.Histochem.Cytochem. 47(7), 895–906. 1999

Saetta, M. Airway inflammation in chronic obstructive pulmonary disease. American Journal of Respiratory and Critical Care Medicine 160, S17-S20. 1999

Safko, M. J., Chan, S. C., Cooper, K. D., and Hanifin, J. M. Heterologous desensitization of leukocytes: a possible mechanism of beta adrenergic blockade in atopic dermatitis. J.Allergy Clin.Immunol. 68(3), 218–225. 1981

Santamaria, L. F., Palacios, J. M., and Beleta, J. Inhibition of eotaxin-mediated human eosinophil activation and migration by the selective cyclic nucleotide phosphodiesterase type 4 inhibitor rolipram. Br.J.Pharmacol. 121(6), 1150–1154. 1997

Sato, K., Stelzner, T. J., O'Brien, R. F., Weil, J. V., and Welsh, C. H. Pentoxifylline lessens the endotoxin-induced increase in albumin clearance across pulmonary artery endothelial monolayers with and without neutrophils. Am.J.Respir.Cell Mol.Biol. 4(3), 219–227. 1991

Sawai, T., Ikai, K., and Uehara, M. Cyclic adenosine monophosphate phosphodiesterase activity in peripheral blood mononuclear leukocytes from patients with atopic dermatitis: correlation with respiratory atopy. Br.J.Dermatol. 138, 846–848. 1998

Schade, F. U. and Schudt, C. The specific type III and IV phosphodiesterase inhibitor zardaverine suppresses formation of tumor necrosis factor by macrophages. Eur.J.Pharmacol. 230(1), 9–14. 1993

Schmidt, D., Watson, N., Morton, B. E., Dent, G., Magnussen, H., and Rabe, K. F. Effect of selective and non-selective phoshodiesterase inhibitors on allergen-induced contractions in passively sensitized human airways. Eur.Resp.J. 10(Suppl 25), 314S. 1997

Schmidt, J., Hatzelmann, A., Fleissner, S., Heimann, Weitschat, Lindstaedt, R., and Szelenyi, I. Effect of phosphodiesterase inhibition on IL-4 and IL-5 production of the murine TH2-type T cell clone D10.G4.1. Immunopharmacology 30(3), 191–198. 1995

Schneider, H. H., Schmiechen, R., Brezinski, M., and Seidler, J. Stereospecific binding of the antidepressant rolipram to brain protein structures. Eur.J.Pharmacol. 127(1-2), 105–115. 1986

Schudt, C., Tenor, H., and Hatzelmann, A. PDE isoenzymes as targets for anti-asthma drugs. Eur.Respir.J. 8(7), 1179–1183. 1995

Schudt, C., Winder, S., Eltze, M., Kilian, U., and Beume, R. Zardaverine: a cyclic AMP specific PDE III/IV inhibitor. Agents Actions Suppl. 34, 379–402. 1991a

Schudt, C., Winder, S., Forderkunz, S., Hatzelmann, A., and Ullrich, V. Influence of selective phosphodiesterase inhibitors on human neutrophil functions and levels of cAMP and Cai. Naunyn-Schmiedeberg's Arch.Pharmacol. 344(6), 682–690. 1991b

Seldon, P. M., Barnes, P. J., Meja, K., and Giembycz, M. A. Suppression of lipopolysaccharide-induced tumor necrosis factor- alpha generation from human peripheral blood monocytes by inhibitors of phosphodiesterase 4: interaction with stimulants of adenylyl cyclase. Mol.Pharmacol. 48(4), 747–757. 1995

Sette, C. and Conti, M. Phosphorylation and activation of cAMP-specific phosphodiesterase by the cAMP-dependent protein kinase. J.Biol.Chem. 271, 16526–16534. 1996

Sette, C., Vicini, E., and Conti, M. The rat PDE3/IVd phosphodiesterase gene codes for muliple proteins differentially activated by cAMP-dependent protein kinase. J.Biol.Chem. 269, 18271–18274. 1994

Shahid, M., van Amsterdam, R. G., de Boer, J., ten Berge, R. E., Nicholson, C. D., and Zaagsma, J. The presence of five cyclic nucleotide phosphodiesterase isoenzyme activities in bovine tracheal smooth muscle and the functional effects of selective inhibitors. Br.J.Pharmacol. 104(2), 471–477. 1991

Sharma, R. K. and Wang, J. H. Purification and characterization of bovine lung calmodulin- dependent cyclic nucleotide phosphodiesterase. An enzyme containing calmodulin as a subunit. J.Biol.Chem. 261(30), 14160–14166. 1986

Sheth, S. B., Chaganti, K., Bastepe, M., Ajuria, J., Brennan, K., Biradavolu, R., and Colman, R. W. Cyclic AMP phosphodiesterases in human lymphocytes. British Journal of Haematology 99(4), 784–789. 1997

Shichijo, M., Shimizu, Y., Hiramatsu, K., Inagaki, N., Tagaki, K., and Nagai, H. Cyclic AMP-elevating agents inhibit mite-antigen-induced IL-4 and IL-13 release from basophil-enriched leukocyte preparation. Int.Arch.Allergy Immunol. 114(4), 348–353. 1997

Siegmund, B., Eigler, A., Moeller, J., Greten, T. F., Hartmann, G., and Endres, S. Suppression of tumor necrosis factor-alpha production by interleukin-10 is enhanced by cAMP-elevating agents. Eur.J.Pharmacol. 321(2), 231–239. 26-2-1997

Silver, P. J., Hamel, L. T., Perrone, M. H., Bentley, R. G., Bushover, C. R., and Evans, D. B. Differential pharmacologic sensitivity of cyclic nucleotide phosphodiesterase isozymes isolated from cardiac muscle, arterial and airway smooth muscle. Eur.J.Pharmacol. 150(1-2), 85–94. 1988

Sinha, B., Semmler, J., Eisenhut, T., Eigler, A., and Endres, S. Enhanced tumor necrosis factor suppression and cyclic adenosine monophosphate accumulation by combination of phosphodiesterase inhibitors and prostanoids. Eur.J.Immunol. 25(1), 147–153. 1995

Snijdewint, F. G., Kalinski, P., Wierenga, E. A., Bos, J. D., and Kapsenberg, M. L. Prostaglandin E2 differentially modulates cytokine secretion profiles of human T helper lymphocytes. J.Immunol. 150(12), 5321–5329. 1993

Soderling, S. H. and Beavo, J. A. Regulation of cAMP and cGMP signaling: new phosphodiesterases and new functions. Curr.Opin.Cell Biol. 12, 174–179. 2000

Sommer, N., Loschmann, P. A., Northoff, G. H., Weller, M., Steinbrecher, A., Steinbach, J. P., Lichtenfels, R., Meyermann, R., Riethmuller, A., and Fontana, A. The antidepressant rolipram suppresses cytokine production and prevents autoimmune encephalomyelitis [see comments]. Nat.Med. 1(3), 244–248. 1995

Souness, J. E., Diocee, B. K., Martin, W., and Moodie, S. A. Pig aortic endothelial-cell cyclic nucleotide phosphodiesterases. Use of phosphodiesterase inhibitors to evaluate their roles in regulating cyclic nucleotide levels in intact cells. Biochem.J. 266(1), 127–132. 1990

Souness, J. E., Griffin, M., Maslen, C., Ebsworth, K, Scott, L. C., Pollock, K., Palfreyman, M. N, and Karlsson, J.-A. Evidence that cyclic AMP phosphodiesterase inhibitors suppress TNF-alpha generation from human monocytes by interacting with a 'low affinity' phosphodiesterase 4 conformer. Br.J.Pharmacol. 118, 649–658. 1996a

Souness, J. E., Griffin, M., Maslen, C., Ebsworth, K., Scott, L. C., Pollock, K., Palfreyman, M. N., and Karlsson, J. A. Evidence that cyclic AMP phosphodiesterase inhibitors suppress TNFα generation from human monocytes by interacting with a 'low-affinity' phosphodiesterase 4 conformer. Br.J.Pharmacol. 118, 649–658. 1996b

Souness, J. E., Houghton, C., Sardar, N., and Withnall, M. T. Evidence that cyclic AMP phosphodiesterase inhibitors suppress interleukin-2 release from murine splenocytes by interacting with a 'low-affinity' phosphodiesterase 4 conformer. British Journal of Pharmacology 121(4), 743–750. 1997

Souness, J. E., Maslen, C., Webber, S., Foster, M., Raeburn, D., Palfreyman, M. N., Ashton, M. J., and Karlsson, J. A. Suppression of eosinophil function by RP 73401, a potent and selective inhibitor of cyclic AMP-specific phosphodiesterase: comparison with rolipram. Br.J.Pharmacol. 115(1), 39–46. 1995

Souness, J. E. and Scott, L. C. Stereospecificity of rolipram actions on eosinophil cyclic AMP- specific phosphodiesterase. Biochem.J. 291(Pt 2), 389–395. 1993

Souness, J. E., Villamil, M. E., Scott, L. C., Tomkinson, A., Giembycz, M. A., and Raeburn, D. Possible role of cyclic AMP phosphodiesterases in the actions of ibudilast on eosinophil thromboxane generation and airways smooth muscle tone. Br.J.Pharmacol. 111(4), 1081–1088. 1994

Spina, D., Harrison, S., and Page, C. P. Regulation by phosphodiesterase isoenzymes of non-adrenergic non- cholinergic contraction in guinea-pig isolated main bronchus. Br.J.Pharmacol. 116(4), 2334–2340. 1995

Stawiski, M. A., Rusin, L. J., Burns, T. L., Weinstein, G. D., and Voorhees, J. J. Ro 20-1724: an agent that significantly improves psoriatic lesions in double-blind clinical trials. J.Invest.Dermatol. 73(4), 261–263. 1979

Stelzner, T. J., Weil, J. V., and O'Brien, R. F. Role of cyclic adenosine monophosphate in the induction of endothelial barrier properties. J.Cell Physiol. 139(1), 157–166. 1989

Stief, C. G. Phosphodiesterase inhibitors in the treatment of erectile dysfunction. Drugs Of Today 36(2-3), 93–99. 2000

Sullivan, P., Bekir, S., Jaffar, Z., Page, C., Jeffery, P., and Costello, J. Anti-inflammatory effects of low-dose oral theophylline in atopic asthma [published erratum appears in Lancet 1994 Jun 11; 343(8911):1512]. Lancet 343(8904), 1006–1008. 1994

Suttorp, N., Ehreiser, P., Hippenstiel, S., Fuhrmann, M., Krull, M., Tenor, H., and Schudt, C. Hyperpermeability of pulmonary endothelial monolayer: protective role of phosphodiesterase isoenzymes 3 and 4. Lung 174(3), 181–194. 1996

Suttorp, N., Weber, U., Welsch, T., and Schudt, C. Role of phosphodiesterases in the regulation of endothelial permeability in vitro. J.Clin.Invest. 91(4), 1421–1428. 1993

Takahashi, M., Terwilliger, R., Lane, C., Mezes, P. S., Conti, M., and Duman, R. S. Chronic antidepressant administration increases the expression of cAMP-specific phosphodiesterase 4A and 4B isoforms. J.Neurosci. 19, 610–618. 1999

Takahashi, S., Oida, K., Fujiwara, R., Maeda, H., Hayashi, S., Takai, H., Tamai, T., Nakai, T., and Miyabo, S. Effect of cilostazol, a cyclic AMP phosphodiesterase inhibitor, on the proliferation of rat aortic smooth muscle cells in culture. J.Cardiovasc.Pharmacol. 20(6), 900–906. 1992

Tenor, H., Hatzelmann, A., Church, M. K., Schudt, C., and Shute, J. K. Effects of theophylline and rolipram on leukotriene C4 (LTC4) synthesis and chemotaxis of human eosinophils from normal and atopic subjects. Br.J.Pharmacol. 118, 1727–1735. 1996

Tenor, H., Hatzelmann, A., Kupferschmidt, R., Stanciu, L., Djukanovic, R., Schudt, C., Wendel, A., Church, M. K., and Shute, J. K. Cyclic nucleotide phosphodiesterase isoenzyme activities in human alveolar macrophages. Clin.Exp.Allergy 25(7), 625–633. 1995a

Tenor, H., Staniciu, L., Schudt, C., Hatzelmann, A., Wendel, A., Djukanovic, R., Church, M. K., and Shute, J. K. Cyclic nucleotide phosphodiesterases from purified human CD4+ and CD8+ T lymphocytes. Clin.Exp.Allergy 25(7), 616–624. 1995b

Timmer, W., Leclerc, V., Birraux, G., Neuhauser, M., Hatzelmann, A., Bethke, T., and Wurst, W. Safety and efficacy of the new PDE4 inhibitor roflumilast administered to patients with excercise-induced asthma over 4 weeks. American Journal of Respiratory and Critical Care Medicine 161, A505. 2000

Tomkinson, A., Karlsson, J. A., and Raeburn, D. Comparison of the effects of selective inhibitors of phosphodiesterase types III and IV in airway smooth muscle with differing beta-adrenoceptor subtypes. Br.J.Pharmacol. 108(1), 57–61. 1993

Tomlinson, P. R., Wilson, J. W., and Stewart, A. G. Salbutamol inhibits the proliferation of human airway smooth muscle cells grown in culture: Relationship to elevated cAMP levels. Biochem.Pharmacol. 49, 1809–1819. 1995

Torphy, T. J. Phosphodiesterase isozymes: molecular targets for novel antiasthma agents. American Journal of Respiratory and Critical Care Medicine 157(2), 351–370. 1998

Torphy, T. J., Barnette, M. S., Hay, D. W., and Underwood, D. C. Phosphodiesterase IV inhibitors as therapy for eosinophil- induced lung injury in asthma. Environ.Health Perspect. 102 Suppl 10, 79–84. 1994

Torphy, T. J., Barnette, M. S., Underwood, D. C., Griswold, D. E., Christensen, S. B., Murdoch, R. D., Nieman, R. B., and Compton, C. H. Ariflow^{TM} (SB 207499), a second generation phosphodiesterase 4 inhibitor for the treatment of asthma and COPD: from concept to clinic. Pulm.Pharmacol.Therap. 12, 131–135. 1999

Torphy, T. J., Burman, M., Huang, L. B., and Tucker, S. S. Inhibition of the low km cyclic AMP phosphodiesterase in intact canine trachealis by SK&F 94836: mechanical and biochemical responses. J.Pharmacol.Exp.Ther. 246(3), 843–850. 1988

Torphy, T. J. and Cieslinski, L. B. Characterization and selective inhibition of cyclic nucleotide phosphodiesterase isozymes in canine tracheal smooth muscle. Mol.Pharmacol. 37(2), 206–214. 1990

Torphy, T. J. and Page, C. Phosphodiesterases: the journey toward therapeutics. Trends Pharmacol.Sci. 21, 157–159. 2000

Torphy, T. J., Stadel, J. M., Burman, M., Cieslinski, L. B., McLaughlin, M. M., White, J. R., and Livi, G. P. Coexpression of human cAMP-specific phosphodiesterase activity and high affinity rolipram binding in yeast. J.Biol.Chem. 267(3), 1798–1804. 1992

Torphy, T. J. and Undem, B. J. Phosphodiesterase inhibitors: new opportunities for the treatment of asthma. Thorax 46(7), 512–523. 1991

Torphy, T. J., Undem, B. J., Cieslinski, L. B., Luttmann, M. A., Reeves, M. L., and Hay, D. W. Identification, characterization and functional role of phosphodiesterase isozymes in human airway smooth muscle. J.Pharmacol.Exp.Ther. 265(3), 1213–1223. 1993

Turner, C. R., Esser, K. M., and Wheeldon, E. B. Therapeutic intervention in a rat model of ARDS: IV. Phosphodiesterase IV inhibition. Circ.Shock 39(3), 237–245. 1993

Underwood, D. C., Bochnowicz, S., Osborn, R. R., Kotzer, C. J., Luttmann, M. A., Hay, D. W. P., Gorycki, P. D., Christensen, S. B., and Torphy, T. J. Antiasthmatic activity of the second-generation phosphodiesterase 4 (PDE4) inhibitor SB 207499 (Ariflo) in the guinea pig. J.Pharmacol.Exp.Ther. 287, 988–995. 1998

Van der Pouw Kraan TCTM, Boeije, L. C. M., Snijders, A., Smeenk, R. J. T., Wijdenes, J., and Aarden, L. A. Regulation of IL-12 production by human monocytes and the influence of prostglandin E2. Annals of the New York Academy of Sciences 795(pp 147–157). 1996

van der Pouw, Kraan, Boeije, L. C., Smeenk, R. J., Wijdenes, J., and Aarden, L. A. Prostaglandin-E2 is a potent inhibitor of human interleukin 12 production. J.Exp.Med. 181(2), 775–779. 1995

van der Pouw, Kraan, van Kooten, C., Rensink, I, and Aarden, L. Interleukin (IL)-4 production by human T cells: differential regulation of IL-4 vs. IL-2 production. Eur.J.Immunol. 22(5), 1237–1241. 1992

Verghese, M. W., McConnell, R. T., Strickland, A. B., Gooding, R. C., Stimpson, S. A., Yarnall, D. P., Taylor, J. D., and Furdon, P. J. Differential regulation of human monocyte-derived TNF alpha and IL-1 beta by type IV cAMP-phosphodiesterase (cAMP-PDE) inhibitors. J.Pharmacol.Exp.Ther. 272(3), 1313–1320. 1995

Wang, P., Wu, P., Ohleth, K. M., Egan, R., and Billah, M. M. Phosphodiesterase 4B2 is the predominant phosphodiesterase species and undergoes differential regulation of gene expression in human monocytes and neutrophils. Mol.Pharmacol. 56, 170–174. 1999

Weston, M. C., Anderson, N., and Peachell, P. T. Effects of phosphodiesterase inhibitors on human lung mast cell and basophil function. Br.J.Pharmacol. 121(2), 287–295. 1997

Weston, M. C. and Peachell, P. T. Regulation of human mast cell and basophil function by cAMP. Gen.Pharmacol. 31(5), 715–719. 1998

Wright, C. D., Kuipers, P. J., Kobylarz-Singer, D., Devall, L. J., Klinkefus, B. A., and Weishaar, R. E. Differential inhibition of human neutrophil functions. Role of cyclic AMP-specific, cyclic GMP-insensitive phosphodiesterase. Biochem.Pharmacol. 40(4), 699–707. 1990

Wright, L. C., Seybold, J., Robichaud, A., Adcock, I. M., and Barnes, P. J. Phosphodiesterase expression in human epithelial cells. American Journal of Physiology—Lung Cellular & Molecular Physiology 275(4 19–4), L694-L700. 1998

Xiong, Y., Westhead, E. W., and Slakey, L. L. Role of phosphodiesterase isoenzymes in regulating intracellular cyclic AMP in adenosine-stimulated smooth muscle cells. Biochem.J. 305(Pt 2), 627–633. 1995

Xu, R. X., Hassell, A. M., Vanderwall, D., Lambert, M. H., Holmes, W. D., Luther, M. A., Rocque, W. J., Milburn, M. V., Zhao, Y., Ke, H., and Nolte, R. T. Atomic structure of PDE4: insights into phosphodiesterase mechanism a specificity. Science 288, 1822–1825. 2000

Yasui, K., Hu, B., Nakazawa, T., Agematsu, K., and Komiyama, A. Theophylline accelerates human granulocyte apoptosis not via phosphodiesterase inhibition. Journal of Clinical Investigation 100(7), 1677–1684. 1997

Yoshimura, T., Kurita, C., Nagao, T., Usami, E., Nakao, T., Watanabe, S., Kobayashi, J., Yamazaki, F., Tanaka, H., and Nagai, H. Effects of cAMP-phosphodiesterase isozyme inhibitor on cytokine production by lipopolysaccharide-stimulated human peripheral blood mononuclear cells. Gen.Pharmacol. 29(4), 633–638. 1997

Yoshimura, T., Nagao, T., Nakao, T., Watanabe, S., Usami, E., Kobayashi, J., Yamazaki, F., Tanaka, H., Inagaki, N., and Nagai, H. Modulation of Th1- and Th2-like cytokine production from mitogen- stimulated human peripheral blood mononuclear cells by phosphodiesterase inhibitors. Gen.Pharmacol. 30(2), 175–180. 1998

Yu, S. M., Cheng, Z. J., and Kuo, S. C. Antiproliferative effects of A02011-1, an adenylyl cyclase activator, in cultured vascular smooth muscle cells of rat. Br.J.Pharmacol. 114(6), 1227–1235. 1995

Zhong, W. W., Burke, P. A., Drotar, M. E., Chavali, S. R., and Forse, R. A. Effects of prostaglandin E2, cholera toxin and 8-bromo-cyclic AMP on lipopolysacchaside-induced gene expression of cytokines in human macrophages. Immunology 84(3), 446–452. 1995

Cytokine Modulators

P. J. Barnes

Department of Thoracic Medicine, National Heart and Lung Institute,
Imperial College, Dovehouse St, London, SW3 6LY, UK
e-mail: p.j.barnes@ic.ac.uk

1	Introduction	220
2	Strategies for Inhibiting Cytokines	220
3	Inhibition of Th2 Cytokines	222
3.1	Anti-IL-5	222
3.2	Anti-IL-4	224
3.3	Anti-IL-13	225
3.4	Anti-IL-9	227
3.5	Anti-IL-25	227
4	Inhibition of Proinflammatory Cytokines	228
4.1	Anti-IL-1	228
4.2	Anti-TNF	228
5	Anti-inflammatory Cytokines	229
5.1	Interleukin-10	229
5.2	Interferons	230
5.3	Interleukin-12	230
5.4	IL-18 and IL-23	231
6	Chemokine Inhibitors	231
6.1	CCR3 Inhibitors	232
6.2	CCR2 Inhibitors	233
6.3	Other CCR Inhibitors	233
6.4	CXCR Inhibitors	234
7	Other Approaches to Cytokine Inhibition	235
7.1	Corticosteroids	235
7.2	Immunomodulatory Drugs	235
7.3	Phosphodiesterase-4 Inhibitors	236
7.4	NF-κB (IKK2) Inhibitors	236
7.5	p38 MAP Kinase Inhibitors	236
8	Conclusions	237
	References	237

Abstract Multiple cytokines play a critical role in orchestrating and perpetuating inflammation in asthma and chronic obstructive pulmonary disease (COPD), and several specific cytokine and chemokine inhibitors are now in development as future therapy for these diseases. Anti-IL-5 antibody markedly reduces peripheral blood and airway eosinophils, but does not appear to be effective in symptomatic asthma. Inhibition of IL-4, despite promising early results in asthma, has been discontinued. Blocking IL-13 might be more effective. Inhibitory cytokines, such as IL-10, interferons and IL-12 are less promising, as systemic delivery produces side effects. Inhibition of TNF-α may be useful in severe asthma and for treating severe COPD with systemic features. Many chemokines are involved in the inflammatory response of asthma and COPD, and several small molecule inhibitors of chemokine receptors are in development. CCR3 antagonists (which block eosinophil chemotaxis) and CXCR2 antagonists (which block neutrophil and monocyte chemotaxis) are in clinical development for asthma and COPD, respectively. Because so many cytokines are involved in asthma, drugs that inhibit the synthesis of multiple cytokines may prove to be more useful; several such classes of drug are now in clinical development and any risk of side effects with these non-specific inhibitors may be reduced by the inhaled route.

Keywords Interleukin-5 · Interleukin-4 · Interleukin-10 · Interlekoin-12 · Tumour necrosis factor-α · Chemokines

1
Introduction

Cytokines play a critical role in the orchestration of chronic inflammation in all diseases, including asthma and chronic obstructive pulmonary disease (COPD). Multiple cytokines and chemokines have been implicated in the pathophysiology of asthma (Chung et al. 1999). There is less understanding of the inflammatory mediators involved in COPD, but as the inflammatory process is markedly different from that in asthma, it is probable that different cytokines and chemokines are involved and that therapeutic strategies are likely to differ (Barnes 2000c). There is now an intensive search for more specific therapies in asthma and for any novel therapies that may prevent the progression of airflow limitation in COPD. Inhibitors of cytokines and chemokines figure prominently in these novel therapeutic approaches (Barnes 2001a, 2002a,b) (Table 1).

2
Strategies for Inhibiting Cytokines

There are several possible approaches to inhibiting specific cytokines (Fig. 1). These range from drugs that inhibit cytokine synthesis [glucocorticoids, cyclosporin A, tacrolimus, rapamycin, mycophenolate, T helper (Th)2 selective inhibitors], humanised blocking antibodies to cytokines or their receptors, soluble receptors to mop up secreted cytokines, small molecule receptor antagonists

Table 1 Cytokine modulators for asthma and COPD

Anti-cytokines	Inhibitory cytokines
Anti-IL-5	IL-1 receptor antagonist
Anti-IL-4	IL-10
Anti-IL-13	IL-12
Anti-IL-9	Interferons (IFN-α, IFN-γ)
Anti-IL-25	IL-18
Anti-IL-1	IL-23
Anti-TNF-α	
Chemokine inhibitors	Cytokine synthesis inhibitors
CCR3 antagonists	Corticosteroids
CCR2 antagonists	Immunomodulators
CCR4 antagonists	Phosphodiesterase-4 inhibitors
CCR8 antagonists	NF-κB inhibitors (IKK2 inhibitors)
CXCR2 antagonists	p38 MAP kinase inhibitors
CXCR4 antagonists	

IKK, inhibitor of NF-κB kinase; IL, interleukin; MAP, mitogen-activated protein; NF-κB, nuclear factor-κB.

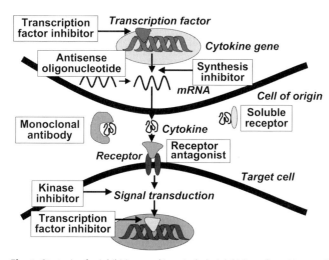

Fig. 1 Strategies for inhibiting cytokines include inhibition of cytokine synthesis, inhibition of secreted cytokines with blocking antibodies or soluble receptors, or blocking of cytokine receptors and their signal transduction pathways

or drugs that block the signal transduction pathways activated by cytokine receptors. On the other hand, there are cytokines that themselves suppress the allergic inflammatory process, and these may have therapeutic potential in asthma and COPD (Barnes 2000a).

3
Inhibition of Th2 Cytokines

Th2 lymphocytes play a key role in orchestrating the eosinophilic inflammatory response in asthma, suggesting that blocking the release or effects of these cytokines may have therapeutic potential. This has been strongly supported by studies in experimental animals, including mice with deletion of the specific Th2 cytokine genes. By contrast, Th2 cells are unlikely to play any role in COPD, and there is no evidence that Th2 cytokines are increased in the airways.

3.1
Anti-IL-5

Interleukin (IL)-5 plays an essential role in orchestrating the eosinophilic inflammation of asthma (Greenfeder et al. 2001) (Fig. 2). In IL-5 gene knock-out mice the eosinophilic response to allergen and the subsequent airway hyperresponsiveness (AHR) are markedly suppressed, and yet animals have a normal survival, validating the strategy to inhibit IL-5. This has also been achieved using blocking antibodies to IL-5. Blocking antibodies to IL-5 inhibits eosinophilic inflammation and AHR in animal models of asthma, including primates (Egan et al. 1996). This blocking effect may last for up to 3 months after a single intravenous injection of antibody in primates, making treatment of chronic asthma with such a therapy a feasible proposition. Humanised monoclonal antibodies

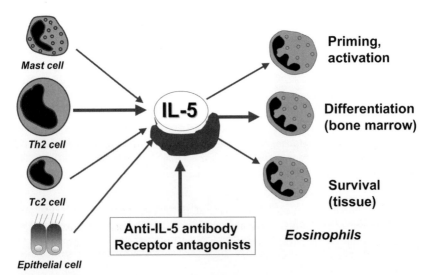

Fig. 2 Inhibition of interleukin-5. IL-5 is released predominantly from Th2 lymphocytes and other cells and its only effects are on eosinophils, resulting in differentiation in the bone marrow and priming, activation and increased survival in the airways. IL-5 may be blocked by blocking antibodies (such as mepolizumab) or theoretically by receptor antagonists

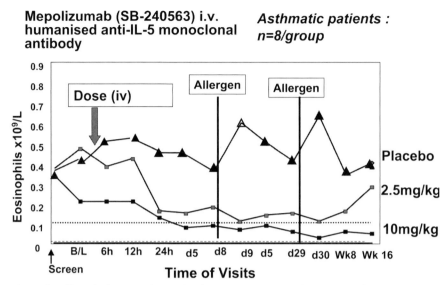

Fig. 3 The effect of a humanised monoclonal antibody against interleukin-5 (mepolizumab) on circulating eosinophils in patients with mild asthma, demonstrating a profound and very prolonged inhibitory effect. (Adapted from Leckie et al. 2000)

to IL-5 have been developed, and a single intravenous infusion of one of these antibodies (mepolizumab) markedly reduces blood eosinophils for several weeks and prevents eosinophil recruitment in to the airways after allergen challenge in patients with mild asthma (Leckie et al. 2000) (Fig. 3). However, this treatment has no significant effect on the early or late response to allergen challenge or on baseline AHR, suggesting that eosinophils may not be of critical importance for these responses in humans (Fig. 4). A clinical study in patients with moderate to severe asthma who had not been controlled on inhaled corticosteroids therapy confirmed a profound reduction in circulating eosinophils, but there was no significant improvement in either asthma symptoms or lung function (Kips et al. 2000). In both of these studies it would be expected that high doses of corticosteroids would improve these functional parameters. These surprising results question the critical role of eosinophils in asthma and indicate that other strategies aimed at inhibiting eosinophilic inflammation might not be effective. More recently, a biopsy study has demonstrated that anti-IL-5 antibody, while profoundly reducing eosinophils in the circulation (by over 95%), is less effective at reducing eosinophils in bronchial biopsies (by ~50%), which may explain why this treatment is not clinically effective (Flood-Page et al. 2003). Nevertheless, this suggests that blocking IL-5 is not likely to be a useful approach to asthma therapy.

Somewhat similar findings have previously been reported in some studies in mice where anti-IL-5 antibodies reduced eosinophilic responses to allergen, but not AHR, whereas AHR was reduced by anti-CD4 antibody which depletes help-

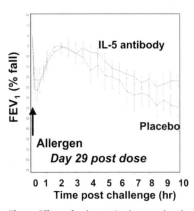 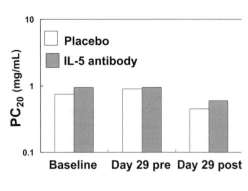

Fig. 4 Effect of a humanised monoclonal antibody against interleukin-5 (mepolizumab) on the early and late response to allergen (*left panel*) and airway hyperresponsiveness (methacholine PC_{20}) (*right panel*). (Adapted from Leckie et al. 2000)

er T cells (Hogan et al. 1998) and suggests that T cell-derived cytokines other than IL-5 must be playing a more important role in AHR.

Non-peptidic IL-5 receptor antagonists would be an alternative strategy, and there is a search for such compounds using molecular modelling of the IL-5 receptor α-chain and through large-scale throughput screening. One such molecule, YM-90709, appears to be a relatively selective inhibitor of IL-5-receptors (Morokata et al. 2002). However, the lack of clinical benefit of anti-IL-5 antibodies has made this a less attractive approach. It is possible that eosinophils are associated with more chronic aspects of asthma, such as airway remodelling, and in mice a blocking anti-IL-5 antibody prevents the increased collagen deposition in airways associated with repeated allergen exposure (Blyth et al. 2000).

3.2
Anti-IL-4

IL-4 is critical for the synthesis of IgE by B lymphocytes and is also involved in eosinophil recruitment to the airways (Steinke et al. 2001). A unique function of IL-4 is to promote differentiation of Th2 cells and it therefore acts at a proximal and critical site in the allergic response, making IL-4 an attractive target for inhibition (Fig. 5).

IL-4 blocking antibodies inhibit allergen-induced AHR, goblet cell metaplasia and pulmonary eosinophilia in a murine model (Gavett et al. 1997). Inhibition of IL-4 may therefore be effective in inhibiting allergic diseases, and soluble humanised IL-4 receptors (sIL-4r) have been tested in clinical trials. A single nebulised dose of sIL-4r prevents the fall in lung function induced by withdrawal of inhaled corticosteroids in patients with moderately severe asthma (Borish et al.

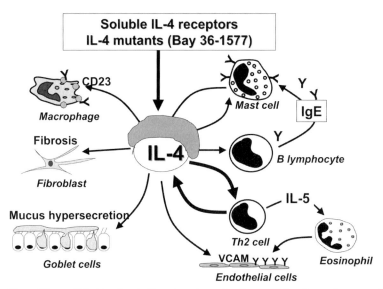

Fig. 5 Effects of blocking interleukin-4 in asthma. IL-4 has multiple effects relevant to allergic inflammation in asthma, including differentiation of Th2 cells, production of IgE from B lymphocytes, increased expression for the low-affinity receptor for IgE (FCεRII) on several inflammatory cells, increased mucus secretion and fibrosis. IL-4 may be blocked by a high-affinity soluble receptor (shuIL-4R)

1999). Subsequent studies have demonstrated that weekly nebulisation of sIL-4r improves asthma control over a 12-week period (Borish et al. 2001). Subsequent studies in patients with milder asthma proved disappointing, however, and this treatment has now been withdrawn. Another approach is blockade of IL-4 receptors with a mutated form of IL-4 (BAY 36–1677), which binds to and blocks IL-4Rα and IL-13Rα1, thus blocking both IL-4 and IL-13 (Shanafelt et al. 1998). This treatment has also been withdrawn.

IL-4 and the closely related cytokine IL-13 signal through a shared surface receptor, IL-4Rα, which activates a specific transcription factor STAT-6 (Jiang et al. 2000). Deletion of the STAT-6 gene has a similar effect to IL-4 gene knockout (Foster 1999). This has led to a search for inhibitors of STAT-6, and although peptide inhibitors that interfere with the interaction between STAT-6 and JAKs linked to IL-4Rα have been discovered, it will be difficult to deliver these intracellularly. An endogenous inhibitor of STATs, suppressor of cytokine signalling (SOCS)-1, is a potent inhibitor of IL-4 signalling pathways and offers a new therapeutic target (Jiang et al. 2000).

3.3
Anti-IL-13

There is increasing evidence that IL-13 in mice mimics many of the features of asthma, including AHR, mucus hypersecretion and airway fibrosis, indepen-

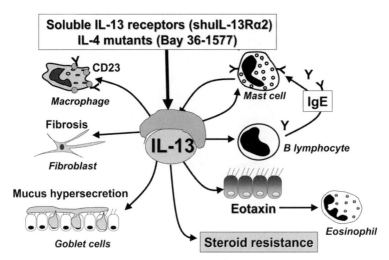

Fig. 6 Effects of blocking interleukin-13 in asthma. IL-13 has multiple effects relevant to allergic inflammation in asthma, including production of IgE from B lymphocytes, increased expression for the low-affinity receptor for IgE (FCεRII) on several inflammatory cells, increased mucus secretion and fibrosis and eotaxin release from airway epithelium. In addition IL-13 induces steroid resistance. IL-13 may be blocked by a high-affinity soluble receptor (shuIL-13R2)

dently of eosinophilic inflammation (Wills-Karp et al. 2003) (Fig. 6). It potently induces the secretion of eotaxin from airway epithelial cells and transforms airway epithelium into a secretory phenotype. Knocking out the IL-13, but not the IL-4, gene in mice prevents the development of AHR after allergen, despite a vigorous eosinophilic response (Walter et al. 2001), and the increase in AHR induced by IL-13 is only seen when the expression of STAT6 is lost in airway epithelial cells (Kuperman et al. 2002). IL-13 signals through the IL-4Rα, but may also activate different intracellular pathways via activation of IL-13Rα1 (Jiang et al. 2000), so that it may be an important target for the development of new therapies. A second specific IL-13 receptor, IL-13Rα2 exists in soluble form and has a high affinity for IL-13, thus acting as a decoy receptor for secreted IL-13. Soluble IL-13Rα2 is effective in blocking the actions of IL-13, including IgE generation, pulmonary eosinophilia and AHR in mice (Wills-Karp et al. 1998). In the murine model, IL-13Rα2 is more effective than IL-4-blocking antibodies, highlighting the potential importance of IL-13 as a mediator of allergic inflammation. Blocking IL-13 may be more important in established asthma where concentrations of IL-13 are much higher than those of IL-4. Humanised IL-13Rα2 is now in clinical development as a therapeutic approach for asthma.

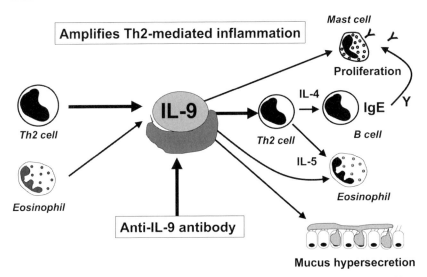

Fig. 7 Effects of blocking interleukin-9 in asthma. IL-9 has several effects, including amplification of Th2-mediated inflammatory effects on the airways. It may be inhibited by a blocking antibody

3.4
Anti-IL-9

IL-9 is a Th2 cytokines that may enhance Th2-driven inflammation and amplify mast cell mediator release and IgE production (Levitt et al. 1999) (Fig. 7). IL-9 may also enhance mucus hypersecretion (Longphre et al. 1999). IL-9 and its receptors show an increased expression in asthmatic airways (Bhathena et al. 2000; Shimbara et al. 2000). A blocking antibody to IL-9 inhibits airway inflammation and AHR in a murine model of asthma (Cheng et al. 2002). Strategies to block IL-9, including blocking humanised antibodies are now in development (Zhou et al. 2001).

3.5
Anti-IL-25

IL-25 is a newly described cytokine that stimulates the release of Th2 cytokines IL4, IL-5 and IL-13, suggesting that it may play a role in allergic inflammation (Hurst et al. 2002). It is released from mast cells via an IgE-dependent mechanism and is therefore a possible target for inhibition in the treatment of asthma (Ikeda et al. 2003).

4
Inhibition of Proinflammatory Cytokines

Pro-inflammatory cytokines, particularly IL-1β and tumour necrosis factor (TNF)-α, may amplify the inflammatory response in asthma and COPD and may be linked to disease severity. This suggests that blocking IL-1β or TNF-α may have beneficial effects, particularly in severe airway disease.

4.1
Anti-IL-1

IL-1 expression is increased in asthmatic airways (Sousa et al. 1996) and activates many inflammatory genes that are expressed in asthma. There are no small molecule inhibitors of IL-1, but a naturally occurring cytokine, IL-1 receptor antagonist (IL-1ra), binds to IL-1 receptors to block the effects of IL-1 (Arend et al. 1998). In experimental animals IL-1ra reduces AHR induced by allergen. Human recombinant IL-1ra does not appear to be effective in the treatment of asthma, however (Rosenwasser 1998). There are no published studies on the role of IL-1 in COPD.

4.2
Anti-TNF

TNF-α is expressed in asthmatic airways and may play a key role in amplifying asthmatic inflammation, through the activation of nuclear factor (NF)-κB, activator protein (AP)-1 and other transcription factors (Kips et al. 1993). TNF levels are markedly increased in induced sputum of patients with COPD (Keatings et al. 1996). Furthermore, there is evidence that COPD patients with weight loss show increased release of TNF from circulating cells and that TNF may induce apoptosis of skeletal muscle cells, resulting in the characteristic muscle wasting and cachexia seen in some patients with severe COPD (de Godoy et al. 1996).

In rheumatoid arthritis and inflammatory bowel disease, a blocking humanised monoclonal antibody to TNF-α (infliximab) and soluble TNF receptors (etanercept) have produced remarkable clinical responses, even in patients who are relatively unresponsive to steroids (Jarvis et al. 1999; Markham et al. 2000). Such TNF inhibitors are a logical approach to asthma therapy, particularly in patients with severe disease, and clinical trials are now underway. They may also be indicated in the treatment of severe COPD, particularly in patients with cachexia and clinical studies are planned.

Because of the problems associated with antibody-based therapies that have to be given by injection, there is a search for small molecule inhibitors of TNF. TNF-α-converting enzyme (TACE) is a matrix metalloproteinase (MMP)-related enzyme critical for the release of TNF from the cell surface. Small molecule TACE inhibitors are in development as oral TNF inhibitors (Barlaam et al. 1999).

5
Anti-inflammatory Cytokines

Some cytokines have anti-inflammatory effects and therefore have therapeutic potential (Barnes 2000a). While it may not be feasible or cost-effective to administer these proteins as long-term therapy, it may be possible to develop drugs in the future that increase the release of these endogenous cytokines or activate their receptors and specific signal transduction pathways.

5.1
Interleukin-10

IL-10 is a potent anti-inflammatory cytokine that inhibits the synthesis of many inflammatory proteins, including cytokines (TNF-α, GM-CSF, IL-5, chemokines) and inflammatory enzymes (iNOS) that are over-expressed in asthma (Pretolani et al. 1997) (Fig. 8). Indeed there may be a defect in IL-10 transcription and secretion from macrophages in asthma, suggesting that IL-10 might be defective in atopic diseases (Borish et al. 1996; John et al. 1998; Barnes 2001b). In sensitised animals, IL-10 is effective in suppressing the inflammatory response to allergen (Zuany-Amorim et al. 1995) and CD4$^+$ cells engineered to secrete IL-10 suppress airway inflammation in a murine model of asthma (Oh et al. 2002). Specific allergen immunotherapy results in increased production of IL-10 by Th cells and this may contribute to the beneficial effects of immunotherapy (Akdis et al. 1998).

IL-10 might also be of therapeutic value in COPD as it not only inhibits TNF and chemokines, but also certain MMPs, such as MMP-9 that may be involved in destruction of elastin in the lung parenchyma (Lacraz et al. 1995). In addi-

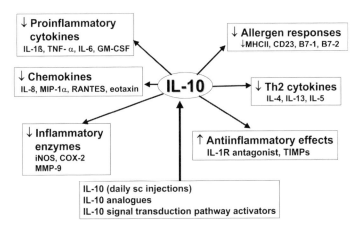

Fig. 8 Anti-inflammatory actions of interleukin-10. IL-10 has several anti-inflammatory effects and may therefore be of therapeutic value, either as the cytokine or in the future as drugs that activate the same signal transduction pathways

tion, IL-10 increases the release of the tissue inhibitors of MMPs (TIMPs), the endogenous inhibitors of MMPs.

Recombinant human IL-10 has proved to be effective in controlling inflammatory bowel disease and psoriasis, where similar cytokines are expressed, and may be given as a weekly injection (Fedorak et al. 2000). Although IL-10 is reasonably well tolerated, there are haematological side effects. In the future, drugs which activate the unique signal transduction pathways activated by the IL-10 receptor or drugs that increase endogenous production of IL-10 may be developed. In mice drugs that elevate cyclic AMP increase IL-10 production, but this does not appear to be the case in human cells (Seldon et al. 1998).

5.2
Interferons

Interferon (IFN)-γ inhibits Th2 cells and should therefore reduce atopic inflammation. In sensitised animals, nebulised IFN-γ inhibits eosinophilic inflammation induced by allergen exposure (Lack et al. 1996) and adenovirus-mediated gene transfer of IFN-γ inhibits allergic inflammation in mice (Behera et al. 2002). However, administration of IFN-γ by nebulisation to asthmatic patients did not significantly reduce eosinophilic inflammation, possibly due to the difficulty in obtaining a high enough concentration locally in the airways (Boguniewicz et al. 1995). Interestingly, allergen immunotherapy increases IFN-γ production by circulating T cells in patients with clinical benefit (Benjaponpitak et al. 1999) and increased numbers of IFN-γ-expressing cells in nasal biopsies of patients with allergic rhinitis (Durham et al. 1996). A preliminary report suggests that IFN-α may be useful in the treatment of patients with severe asthma who have reduced responsiveness to corticosteroids (Gratzl et al. 2000).

5.3
Interleukin-12

IL-12 is the endogenous regulator of Th1 cell development and determines the balance between Th1 and Th2 cells (Gately et al. 1998). IL-12 administration to rats inhibits allergen-induced inflammation (Gavett et al. 1995) and inhibits sensitisation to allergens. IL-12 releases IFN-γ, but has additional effects on T cell differentiation. IL-12 levels released from whole blood cells are lower in asthmatic patients, indicating a possible reduction in IL-12 secretion (van der Pouw Kraan et al. 1997).

Recombinant human IL-12 has been administered to humans and has several toxic effects that are diminished by slow escalation of the dose (Leonard et al. 1997). In patients with mild asthma, weekly infusions of human recombinant IL-12 in escalating doses over 4 weeks caused a progressive fall in circulating eosinophils, and a reduction in the normal rise in circulating eosinophils after allergen challenge (Bryan et al. 2000) (Fig. 9). There was a concomitant reduction in eosinophils in induced sputum. However, there was no reduction in

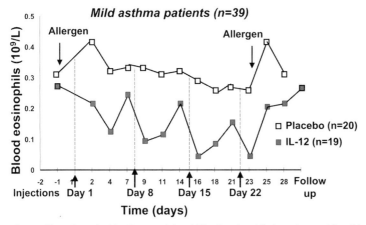

Fig. 9 Effect of interleukin-12 on peripheral blood eosinophils in patients with mild asthma. IL-12 was given in progressively increasing doses as an intravenous injection. (Adapted from Bryant et al. 2000)

either early or late response to inhaled allergen challenge or any reduction in AHR (as with anti-IL-5 therapy). Furthermore, most of the patients suffered form malaise and one out of the 12 subjects had an episode of cardiac arrhythmia. This suggests that IL-12 in not a suitable treatment for asthma. In mice, administration of an IL-12-allergen fusion protein results in the development of a specific Th1 response to the allergen, with increased production of an allergen-specific IgG2, rather than the normal Th2 response with IgE formation (Kim et al. 1997). This indicates the possibility of using local IL-12 together with specific allergens to provide a more specific immunotherapy, which might even be curative if applied early in the course of the atopic disease.

5.4
IL-18 and IL-23

IL-18 was originally described as IFN-γ-releasing factor, but has a different mechanism of action to IL-12 (Dinarello 2000). IL-12 and IL-18 appear to have a synergistic effect on inducing IFN-γ release and for inhibiting IL-4-dependent IgE production and AHR (Hofstra et al. 1998), but no clinical studies have so far been reported.

IL-23 is structurally related to IL-12 and shares some of its biological effects, so it should have a protective function in asthma (Oppmann et al. 2000). Its clinical potential has not yet been explored.

6
Chemokine Inhibitors

Many chemokines are involved in the recruitment of inflammatory cells in asthma and COPD (Lukacs 2001). Over 50 different chemokines are now recog-

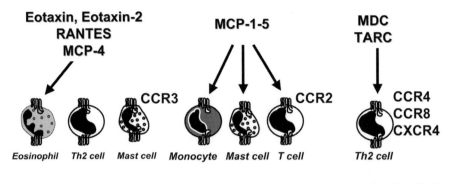

Fig. 10 Chemokine receptor antagonists in asthma. Several chemokines are likely to be involved in the pathophysiology of asthma. There are three major chemokine receptor targets in asthma: CCR3, which is most advanced in terms of small molecule inhibitor development, but also CCR2 and CCR4, for which small molecule inhibitors are now in development

nised and they activate up to 20 different surface receptors (Rossi et al. 2000). Chemokine receptors belong the seven transmembrane receptor superfamily of G protein-coupled receptors, and this makes it possible to find small molecule inhibitors, which has not yet been possible for classical cytokine receptors (Proudfoot 2002). Some chemokines appear to be selective for single chemokines, whereas others are promiscuous and mediate the effects of several related chemokines (Fig. 10). Chemokines appear to act in sequence in determining the final inflammatory response, and so inhibitors may be more or less effective depending on the kinetics of the response (Gutierrez-Ramos et al. 2000).

6.1
CCR3 Inhibitors

Several chemokines, including eotaxin, eotaxin-2, eotaxin-3, RANTES and macrophage chemoattractant protein-4 (MCP-4) activate a common receptor on eosinophils designated CCR3 (Gutierrez-Ramos et al. 1999). A neutralising antibody against eotaxin reduces eosinophil recruitment in to the lung after allergen and the associated AHR in mice (Gonzalo et al. 1996). There is increased expression of eotaxin, eotaxin-2, MCP-3, MCP-4 and CCR3 in the airways of asthmatic patients, and this is correlated with increased AHR (Ying et al. 1997, 1999). Several small molecule inhibitors of CCR3, including UCB35625, SB-297006 and SB-328437 are effective in inhibiting eosinophil recruitment in allergen models of asthma (Sabroe et al. 2000; White et al. 2000) and drugs in this class are currently undergoing clinical trials in asthma. Although it was thought that CCR3 were restricted to eosinophils, there is some evidence for their expression on Th2 cells and mast cells, so that these inhibitors may have a more widespread effect than on eosinophils alone, making them potentially more valuable in asthma treatment.

RANTES, which shows increased expression in asthmatic airways (Berkman et al. 1996) also activates CCR3, but also has effects on CCR1 and CCR5, which may play a role in T cell recruitment. Modification of the N-terminal of RANTES, met-RANTES, has a blocking effect on RANTES by inhibiting these receptors (Elsner et al. 1997).

6.2
CCR2 Inhibitors

MCP-1 activates CCR2 on monocytes and T lymphocytes. And blocking MCP-1 with neutralising antibodies reduces recruitment of both T cells and eosinophils in a murine model of ovalbumin-induced airway inflammation, with a marked reduction in AHR (Gonzalo et al. 1996). MCP-1 also recruits and activates mast cells, an effect that is mediated via CCR2 (Campbell et al. 1999). MCP-1 instilled into the airways induces marked and prolonged AHR in mice, associated with mast cell degranulation. A neutralising antibody to MCP-1 blocks the development of AHR in response to allergen (Campbell et al. 1999). MCP-1 levels are increased in bronchoalveolar lavage fluid of patients with asthma (Holgate et al. 1997). This has led to a search for small molecule inhibitors of CCR2.

CCR2 may also play an important role in COPD, as MCP-1 levels are increased in sputum and lungs of patients with COPD (de Boer et al. 2000; Traves et al. 2002). MCP-1 is a potent chemoattractant of monocytes and may therefore be involved in the recruitment of macrophages in COPD. Indeed the chemoattractant effect of induced sputum from patients with COPD is abrogated by an antibody to CCR2. Since macrophages appear to play a critical role in COPD as a source of elastases and neutrophil chemoattractants, blocking CCR2 may be a therapeutic strategy in COPD.

6.3
Other CCR Inhibitors

CCR4 and CCR8 are selectively expressed on Th2 cells and are activated by the chemokines monocyte-derived chemokine (MDC) and thymus and activation-dependent chemokine (TARC) (Lloyd et al. 2000). Inhibitors of CCR4 and CCR8 may therefore inhibit the recruitment of Th2 cells and thus persistent eosinophilic inflammation in the airways. CCR8 gene deletion does not have any effects on allergic inflammation in mice, suggesting that this receptor may not be an effective target (Chung et al. 2003). CXCR4 are also selectively expressed on Th2 cells and a small molecule inhibitor AMD3100 inhibits allergen-induced inflammation in a murine model of asthma (Lukacs et al. 2002).

CCR7 plays a role in the migration of dendritic cells to regional lymph nodes; therefore, blocking this receptor might suppress antigen presentation (Sallusto et al. 2000).

6.4
CXCR Inhibitors

CXC receptors mediate the effects of CXC chemokines, which act predominantly on neutrophils and monocytes. IL-8 levels are markedly increased in induced sputum of patients with COPD and correlate with the increased proportion of neutrophils (Keatings et al. 1996). Since IL-8 is a potent neutrophil chemoattractant, it is an attractive target for COPD therapy. Anti-IL-8 antibodies have an inhibitory effect on the chemotactic response to COPD sputum (Hill et al. 1999). IL-8 acts via two receptors: CXCR1, which is a low-affinity receptor that is specific for IL-8, and CXCR2, which has high affinity and is shared by several other CXC chemokines, including growth-related oncogene (GRO)-α, GRO-β, GRO-γ, granulocyte chemotactic protein (GCP)-2 and epithelial cell-derived neutrophil-activating peptide (ENA)-78 (Fig. 11). CXCR1 responds to high concentrations of IL-8 and is responsible for activation of neutrophils and release of superoxide anions and neutrophil elastase, whereas CXCR2 responds to low concentrations of CXC chemokines and is involved in chemotactic responses. Potent small molecule inhibitors of CXCR2, such as SB 225002, have now been developed that block the chemotactic response of neutrophils to IL-8 and GRO-α (White et al. 1998; Hay et al. 2001). This antagonist has a significant inhibitory effect on the chemotactic response to COPD sputum. Concentrations of GRO-α are also elevated in induced sputum of patients with COPD, and this mediator has a che-

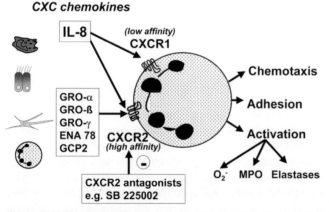

Fig. 11 CXC chemokine receptors are important for neutrophilic inflammation. Interleukin-8 (*IL-8*) signals through a low-affinity specific receptor (*CXCR1*) but also through a high-affinity receptor (*CXCR2*) shared with other CXC chemokines, such as growth-related oncogene (*GRO*)-α. Small molecule inhibitors of CXCR2 are now in development

motactic effect on neutrophils and monocytes (Traves et al. 2002). CXCR2 antagonists may therefore also reduce monocyte chemotaxis and the accumulation of macrophages in COPD patients.

Neutrophils are not a prominent feature of inflammation in patients with chronic asthma, and inflammation is dominated by eosinophils. However, there is evidence for increased neutrophils in biopsies and induced sputum of patients with severe asthma who are treated with high doses of inhaled or oral corticosteroids, and the levels of IL-8 are increased (Wenzel et al. 1997; Jatakanon et al. 1999). It is not certain whether this neutrophilic inflammation contributes to pathophysiology, but it is possible that CXCR inhibitors may have a therapeutic role in severe asthma.

7
Other Approaches to Cytokine Inhibition

Although there have been several attempts to block specific cytokines, this may not be adequate to block chronic inflammation is asthma or COPD, as so many cytokines are involved and there is considerable redundancy of effects. This has suggested that the development of drugs that have a more general effect on cytokine synthesis may be more successful. However, these drugs also affect other inflammatory processes, so their beneficial effects cannot necessarily be ascribed to inhibition of cytokine synthesis alone.

7.1
Corticosteroids

Corticosteroids are by far the most effective treatments for asthma, and part of their efficacy is due to inhibition of inflammatory cytokine expression. This is mediated via an effect on glucocorticoid receptors to reverse the acetylation of core histones that is linked to increased expression of inflammatory genes, such as those encoding cytokines and chemokines (Ito et al. 2000). However, corticosteroids are not effective in suppressing the inflammation in COPD (Barnes 2000b) and this, at least in part, may be explained by an inhibitory effect of cigarette smoking on histone deacetylation (Ito et al. 2001).

7.2
Immunomodulatory Drugs

Cyclosporin A, tacrolimus and rapamycin inhibit the transcription factor NF-AT that regulates the secretion of IL-2, IL-4, IL-5, IL-13 and GM-CSF by T lymphocytes (Rao et al. 1997). Although it has some reported beneficial steroid-sparing effects in asthma (Lock et al. 1996), the toxicity of cyclosporin A limits is usefulness, at least when given orally. More selective Th2-selective drugs may be safer for the treatment of asthma in the future. An inhibitor of Th2 cytokines, suplatast tosilate (Oda et al. 1999), is reported to provide clinical benefit in asthma (Tamaoki et al. 2000). Cytotoxic (CD8$^+$) T lymphocytes are prominent in COPD and therefore immunomodulatory drugs may also have a potential role in this disease.

7.3
Phosphodiesterase-4 Inhibitors

Phosphodiesterase (PDE)4 inhibitors inhibit the release of cytokines and chemokines from inflammatory cells via an increase in intracellular cyclic AMP (Torphy 1998). Their clinical use is limited in asthma by side effects such as nausea. In contrast to corticosteroids, PDE4 inhibitors have a potent inhibitory effect on neutrophils (Au et al. 1998), indicating that they may be useful anti-inflammatory treatments for COPD. There is preliminary evidence that a PDE4 inhibitor cilomilast improves lung function and symptoms in patients with COPD, although whether this is due to inhibition of cytokines is not yet certain (Compton et al. 2001).

7.4
NF-κB (IKK2) Inhibitors

NF-κB regulates the expression of many cytokines and chemokines involved in asthma and COPD (Barnes et al. 1997). There are several possible approaches to inhibition of NF-κB, including gene transfer of the inhibitor of NF-κB (IκB), inhibitors of IκB kinase-2 (IKK2), NF-κB-inducing kinase (NIK) and IκB ubiquitin ligase, which regulate the activity of NF-κB, and the development of drugs that inhibit the degradation of IκB (Delhase et al. 2000). One concern about this approach is that effective inhibitors of NF-κB may result in immune suppression and impair host defences, since knock-out mice which lack NF-κB proteins succumb to septicaemia. However, there are alternative pathways of NF-κB activation that might be more important in inflammatory disease (Nasuhara et al. 1999). Several small molecule inhibitors of IKK2 are now in development.

7.5
p38 MAP Kinase Inhibitors

Mitogen-activated protein (MAP) kinases play a key role in chronic inflammation and several complex enzyme cascades have now been defined. One of these, the p38 MAP kinase pathway is involved in expression of inflammatory cytokines and chemokines (Meja et al. 2000; Underwood et al. 2000). Small molecule inhibitors of p38 MAP kinase, such as SB 203580, SB 239063 and RWJ 67657, also known as cytokine synthesis anti-inflammatory drugs (CSAIDS), have now been developed and these drugs have a broad range of anti-inflammatory effects (Lee et al. 2000). There may be issues of safety, as p38 MAP kinases are involved in host defence. It is possible that the inhaled route of delivery may reduce the risk of side effects, however.

8
Conclusions

There are several specific cytokine and chemokine inhibitors now in development for the treatment of asthma and COPD. Inhibition of IL-4 with soluble IL-4 receptors in asthma showed promising early results, but this was not confirmed in larger trials and IL-13 inhibition is more promising. Anti-IL-5 antibody is very effective at inhibiting peripheral blood and airway eosinophils, but does not appear to be effective in symptomatic asthma. Inhibitory cytokines, such as IL-10, interferons and IL-12 are less promising, as systemic delivery produces side effects and it may be necessary to develop inhaled delivery systems. Inhibition of TNF-α may be useful in the treatment of severe asthma and COPD with systemic features. Many chemokines are involved in the inflammatory response of asthma and COPD and small molecule inhibitors of chemokine receptors are now in development. CCR3 and CXCR2 antagonists are now being developed for the treatment of asthma and COPD, respectively. Because so many cytokines are involved in these complex diseases, drugs that inhibit the synthesis of multiple cytokines may be more successful. Several such classes of drug are now in clinical development, including PDE4, p38 MAP kinase and IKK2 inhibitors. The risk of side effects in these non-specific inhibitors may be reduced by inhaled route of delivery.

References

Akdis CA, Blesken T, Akdis M, Wuthrich B, Blaser K (1998) Role of interleukin 10 in specific immunotherapy. J.Clin.Invest. 102:98–106
Arend WP, Malyak M, Guthridge CJ, Gabay C (1998) Interleukin-1 receptor antagonist: role in biology. Annu Rev Immunol 16:27–55
Au BT, Teixeira MM, Collins PD, Williams TJ (1998) Effect of PDE4 inhibitors on zymosan-induced IL-8 release from human neutrophils: synergism with prostanoids and salbutamol. Br.J Pharmacol. 123:1260–1266
Barlaam B, Bird TG, Lambert-Van DB, Campbell D, Foster SJ, Maciewicz R (1999) New alpha-substituted succinate-based hydroxamic acids as TNFα convertase inhibitors. J Med Chem 42:4890–4908
Barnes PJ (2000a) Endogenous inhibitory mechanisms in asthma. Am J Respir Crit Care Med 161:S176-S181
Barnes PJ (2000b) Inhaled corticosteroids are not helpful in chronic obstructive pulmonary disease. Am J Resp Crit Care Med 161:342–344
Barnes PJ (2000c) Mechanisms in COPD: differences from asthma. Chest 117:10S-14S
Barnes PJ (2001a) Cytokine modulators as novel therapies for airway disease. Eur Respir J Suppl 34:67s-77 s
Barnes PJ (2001b) IL-10: a key regulator of allergic disease. Clin Exp Allergy 31:667–669
Barnes PJ (2002a) Cytokine modulators as novel therapies for asthma. Ann Rev Pharmacol Toxicol 42:81–98
Barnes PJ (2002b) New treatments for COPD. Nature Rev Drug Disc 1:437–445
Barnes PJ, Karin M (1997) Nuclear factor-kB: a pivotal transcription factor in chronic inflammatory diseases. New Engl J Med 336:1066–1071

Behera AK, Kumar M, Lockey RF, Mohapatra SS (2002) Adenovirus-mediated interferon-g gene therapy for allergic asthma: involvement of interleukin 12 and STAT4 signaling. Hum.Gene Ther. 13:1697–1709

Benjaponpitak S, Oro A, Maguire P, Marinkovich V, DeKruyff RH, Umetsu DT (1999) The kinetics of change in cytokine production by CD4 T cells during conventional allergen immunotherapy. J.Allergy Clin.Immunol. 103:468–475

Berkman N, Krishnan VL, Gilbey T, O'Connor BJ, Barnes PJ (1996) Expression of RANTES mRNA and protein in airways of patients with mild asthma. Am J Respir Crit Care Med 15:382–389

Bhathena PR, Comhair SA, Holroyd KJ, Erzurum SC (2000) Interleukin-9 receptor expression in asthmatic airways In vivo. Lung 178:149–160

Blyth DI, Wharton TF, Pedrick MS, Savage TJ, Sanjar S (2000) Airway subepithelial fibrosis in a murine model of atopic asthma: suppression by dexamethasone or anti-interleukin-5 antibody. Am J Respir Cell Mol Biol 23:241–246

Boguniewicz M, Martin RJ, Martin D, Gibson U, Celniker A (1995) The effects of nebulized recombinant interferon-y in asthmatic airways. J Allergy Clin Immunol 95:133–135

Borish L, Aarons A, Rumbyrt J, Cvietusa P, Negri J, Wenzel S (1996) Interleukin-10 regulation in normal subjects and patients with asthma. J Allergy Clin Immunol 97:1288–1296

Borish LC, Nelson HS, Corren J, Bensch G, Busse WW, Whitmore JB, Agosti JM (2001) Efficacy of soluble IL-4 receptor for the treatment of adults with asthma. J Allergy Clin Immunol. 107:963–970

Borish LC, Nelson HS, Lanz MJ, Claussen L, Whitmore JB, Agosti JM, Garrison L (1999) Interleukin-4 Receptor in Moderate Atopic Asthma. A phase I/II randomized, placebo-controlled trial. Am J Respir.Crit Care Med 160:1816–1823

Bryan S, O'Connor BJ, Matti S, Leckie MJ, Kanabar V, Khan J, Warrington S, Renzetti L, Rames A, Bock JA, Boyce M, Hansel TT, Holgate ST, Sterk PJ, Barnes PJ (2000) Effects of recombinant human interleukin-12 on eosinophils, airway hyperreactivity and the late asthmatic response. Lancet 356:2149–2153

Campbell EM, Charo IF, Kunkel SL, Strieter RM, Boring L, Gosling J, Lukacs NW (1999) Monocyte chemoattractant protein-1 mediates cockroach allergen-induced bronchial hyperreactivity in normal but not CCR2-/- mice: the role of mast cells. J Immunol 163:2160–2167

Cheng G, Arima M, Honda K, Hirata H, Eda F, Yoshida N, Fukushima F, Ishii Y, Fukuda T (2002) Anti-interleukin-9 antibody treatment inhibits airway inflammation and hyperreactivity in mouse asthma model. Am J Respir Crit Care Med 166:409–416

Chung CD, Kuo F, Kumer J, Motani AS, Lawrence CE, Henderson WR, Jr., Venkataraman C (2003) CCR8 is not essential for the development of inflammation in a mouse model of allergic airway disease. J Immunol. 170:581–587

Chung KF, Barnes PJ (1999) Cytokines in asthma. Thorax 54:825–857

Compton CH, Gubb J, Nieman R, Edelson J, Amit O, Bakst A, Ayres JG, Creemers JP, Schultze-Werninghaus G, Brambilla C, Barnes NC (2001) Cilomilast, a selective phosphodiesterase-4 inhibitor for treatment of patients with chronic obstructive pulmonary disease: a randomised, dose-ranging study. Lancet 358:265–270

de Boer WI, Sont JK, van Schadewijk A, Stolk J, van Krieken JH, Hiemstra PS (2000) Monocyte chemoattractant protein 1, interleukin 8, and chronic airways inflammation in COPD. J Pathol 190:619–626

de Godoy I, Donahoe M, Calhoun WJ, Mancino J, Rogers RM (1996) Elevated TNF-alpha production by peripheral blood monocytes of weight- losing COPD patients. Am J Respir Crit Care Med 153:633–637

Delhase M, Li N, Karin M (2000) Kinase regulation in inflammatory response. Nature 406:367–368

Dinarello CA (2000) Interleukin-18, a proinflammatory cytokine. Eur Cytokine Netw 11:483–486
Durham SR, Ying S, Varney VA, Jacobson MR, Sudderick RM, Mackay IS, Kay AB, Hamid QA (1996) Grass pollen immunotherapy inhibits allergen-induced infiltration of CD4+ T lymphocytes and eosinophils in the nasal mucosa and increases the number of cells expressing messenger RNA for interferon-gamma. J.Allergy Clin.Immunol. 97:1356–1365
Egan RW, Umland SP, Cuss FM, Chapman RW (1996) Biology of interleukin-5 and its relevance to allergic disease. Allergy 51:71–81
Elsner J, Petering H, Hochstetter R, Kimmig D, Wells TN, Kapp A, Proudfoot AE (1997) The CC chemokine antagonist Met-RANTES inhibits eosinophil effector functions through the chemokine receptors CCR1 and CCR3. Eur.J.Immunol. 27:2892–2898
Fedorak RN, Gangl A, Elson CO, Rutgeerts P, Schreiber S, Wild G, Hanauer SB, Kilian A, Cohard M, LeBEAUT A, Feagan B (2000) Recombinant human interleukin 10 in the treatment of patients with mild to moderately active Crohn's disease. Gastroenterology 119:1473–1482
Flood-Page PT, Menzies-Gow AN, Kay AB, Robinson DS (2003) Eosinophil's role remains uncertain as anti-interleukin-5 only partially depletes numbers in asthmatic airways. Am J Respir Crit Care Med 167:199–204
Foster PS (1999) STAT6: an intracellular target for the inhibition of allergic disease. Clin.Exp.Allergy 29:12–16
Gately MK, Renzetti LM, Magram J, Stern AS, Adorini L, Gubler U, Presky DH (1998) The interleukin-12/interleukin-12-receptor system: role in normal and pathologic immune responses. Annu.Rev.Immunol. 16:495–521:495–521
Gavett SH, O'Hearn DJ, Karp CL, Patel EA, Schofield BH, Finkelman FD, Wills-Karp M (1997) Interleukin-4 receptor blockade prevents airway responses induced by antigen challenge in mice. Am.J.Physiol. 272:L253–61
Gavett SH, O'Hearn DJ, Li X, Huang SK, Finkelman FD, Wills-Karp M (1995) Interleukin 12 inhibits antigen-induced airway hyperresponsiveness, inflammation and Th2 cytokine expression in mice. J Exp Med 182:1527–1536
Gonzalo JA, Lloyd CM, Kremer L, Finger E, Martinez A, Siegelman MH, Cybulsky M, Gutierrez-Ramos JC (1996) Eosinophil recruitment to the lung in a murine model of allergic inflammation. The role of T cells, chemokines, and adhesion receptors. J Clin Invest 98:2332–2345
Gratzl S, Palca A, Schmitz M, Simon HU (2000) Treatment with IFN-α in corticosteroid-unresponsive asthma. J Allergy Clin Immunol 105:1035–1036
Greenfeder S, Umland SP, Cuss FM, Chapman RW, Egan RW (2001) The role of interleukin-5 in allergic eosinophilic disease. Respir Res 2:71–79
Gutierrez-Ramos JC, Lloyd C, Gonzalo JA (1999) Eotaxin: from an eosinophilic chemokine to a major regulator of allergic reactions. Immunol Today 20:500–504
Gutierrez-Ramos JC, Lloyd C, Kapsenberg ML, Gonzalo JA, Coyle AJ (2000) Non-redundant functional groups of chemokines operate in a coordinate manner during the inflammatory response in the lung. Immunol Rev 177:31–42
Hay DWP, Sarau HM (2001) Interleukin-8 receptor antagonists in pulmonary diseases. Curr Opin Pharmacol 1:242–247
Hill AT, Bayley D, Stockley RA (1999) The interrelationship of sputum inflammatory markers in patients with chronic bronchitis. Am J Respir Crit Care Med 160:893–898
Hofstra CL, van Ark I, Hofman G, Kool M, Nijkamp FP, Van Oosterhout AJ (1998) Prevention of Th2-like cell responses by coadministration of IL-12 and IL-18 is associated with inhibition of antigen-induced airway hyperresponsiveness, eosinophilia, and serum IgE levels. J.Immunol. 161:5054–5060
Hogan SP, Matthaei KI, Young JM, Koskinen A, Young IG, Foster PS (1998) A novel T cell-regulated mechanism modulating allergen-induced airways hyperreactivity in BALB/c mice independently of IL-4 and IL-5. J.Immunol. 161:1501–1509

Holgate ST, Bodey KS, Janezic A, Frew AJ, Kaplan AP, Teran LM (1997) Release of RANTES, MIP-1 alpha, and MCP-1 into asthmatic airways following endobronchial allergen challenge. Am J Respir Crit Care Med 156:1377–1383

Hurst SD, Muchamuel T, Gorman DM, Gilbert JM, Clifford T, Kwan S, Menon S, Seymour B, Jackson C, Kung TT, Brieland JK, Zurawski SM, Chapman RW, Zurawski G, Coffman RL (2002) New IL-17 family members promote Th1 or Th2 responses in the lung: in vivo function of the novel cytokine IL-25. J Immunol. 169:443–453

Ikeda K, Nakajima H, Suzuki K, Kagami SI, Hirose K, Suto A, Saito Y, Iwamoto I (2003) Mast cells produce interleukin-25 upon FcεRI-mediated activation. Blood:(in press).

Ito K, Barnes PJ, Adcock IM (2000) Glucocorticoid receptor recruitment of histone deacetylase 2 inhibits IL-1b-induced histone H4 acetylation on lysines 8 and 12. Mol Cell Biol 20:6891–6903

Ito K, Lim S, Caramori G, Chung KF, Barnes PJ, Adcock IM (2001) Cigarette smoking reduces histone deacetylase 2 expression, enhances cytokine expression and inhibits glucocorticoid actions in alveolar macrophages. FASEB J 15:1100–1102

Jarvis B, Faulds D (1999) Etanercept: a review of its use in rheumatoid arthritis. Drugs 57:945–966

Jatakanon A, Uasaf C, Maziak W, Lim S, Chung KF, Barnes PJ (1999) Neutrophilic inflammation in severe persistent asthma. Am J Respir Crit Care Med 160:1532–1539

Jiang H, Harris MB, Rothman P (2000) IL-4/IL-13 signaling beyond JAK/STAT. J Allergy Clin Immunol 105:1063–1070

John M, Lim S, Seybold J, Robichaud A, O'Connor B, Barnes PJ, Chung KF (1998) Inhaled corticosteroids increase IL-10 but reduce MIP-1a, GM-CSF and IFN-g release from alveolar macrophages in asthma. Am J Respir Crit Care Med 157:256–262

Keatings VM, Collins PD, Scott DM, Barnes PJ (1996) Differences in interleukin-8 and tumor necrosis factor-α in induced sputum from patients with chronic obstructive pulmonary disease or asthma. Am J Respir Crit Care Med 153:530–534

Kim TS, DeKruyff RH, Rupper R, Maecker HT, Levy S, Umetsu DT (1997) An ovalbumin-IL-12 fusion protein is more effective than ovalbumin plus free recombinant IL-12 in inducing a T helper cell type 1-dominated immune response and inhibiting antigen-specific IgE production. J.Immunol. 158:4137–4144

Kips JC, O'Connor BJ, Inman MD, Svensson K, Pauwels RA, O'Byrne PM (2000) A long-term study of the antiinflammatory effect of low-dose budesonide plus formoterol versus high-dose budesonide in asthma. Am J Respir Crit Care Med 161:996–1001

Kips JC, Tavernier JH, Joos GF, Peleman RA, Pauwels RA (1993) The potential role of tumor necrosis factor α in asthma. Clin Exp Allergy 23:247–250

Kuperman DA, Huang X, Koth LL, Chang GH, Dolganov GM, Zhu Z, Elias JA, Sheppard D, Erle DJ (2002) Direct effects of interleukin-13 on epithelial cells cause airway hyperreactivity and mucus overproduction in asthma. Nat.Med 8:885–889

Lack G, Bradley KL, Hamelmann E, Renz H, Loader J, Leung DY, Larsen G, Gelfand EW (1996) Nebulized IFN-γ inhibits the development of secondary allergic responses in mice. J.Immunol. 157:1432–1439

Lacraz S, Nicod LP, Chicheportiche R, Welgus HG, Dayer JM (1995) IL-10 inhibits metalloproteinase and stimulates TIMP-1 production in human mononuclear phagocytes. J.Clin.Invest. 96:2304–2310

Leckie MJ, ten Brincke A, Khan J, Diamant Z, O'Connor BJ, Walls CM, Mathur M, Cowley H, Chung KF, Djukanovic RJ, Hansel TT, Holgate ST, Sterk PJ, Barnes PJ (2000) Effects of an interleukin-5 blocking monoclonal antibody on eosinophils, airway hyperresponsiveness and the late asthmatic response. Lancet 356:2144–2148

Lee JC, Kumar S, Griswold DE, Underwood DC, Votta BJ, Adams JL (2000) Inhibition of p38 MAP kinase as a therapeutic strategy. Immunopharmacology 47:185–201

Leonard JP, Sherman ML, Fisher GL, Buchanan LJ, Larsen G, Atkins MB, Sosman JA, Dutcher JP, Vogelzang NJ, Ryan JL (1997) Effects of single-dose interleukin-12 exposure on interleukin-12- associated toxicity and interferon-gamma production. Blood 90:2541–2548

Levitt RC, McLane MP, MacDonald D, Ferrante V, Weiss C, Zhou T, Holroyd KJ, Nicolaides NC (1999) IL-9 pathway in asthma: new therapeutic targets for allergic inflammatory disorders. J Allergy Clin Immunol 103:S485-S491

Lloyd CM, Delaney T, Nguyen T, Tian J, Martinez A, Coyle AJ, Gutierrez-Ramos JC (2000) CC chemokine receptor (CCR)3/eotaxin is followed by CCR4/monocyte-derived chemokine in mediating pulmonary T helper lymphocyte type 2 recruitment after serial antigen challenge in vivo. J Exp Med 191:265–274

Lock SH, Kay AB, Barnes NC (1996) Double-blind, placebo-controlled study of cyclosporin A as a corticosteroid-sparing agent in corticosteroid-dependent asthma. Am J Respir Crit Care Med 153:509–514

Longphre M, Li D, Gallup M, Drori E, Ordonez CL, Redman T, Wenzel S, Bice DE, Fahy JV, Basbaum C (1999) Allergen-induced IL-9 directly stimulates mucin transcription in respiratory epithelial cells. J Clin Invest 104:1375–1382

Lukacs NW (2001) Role of chemokines in the pathogenesis of asthma. Nat.Rev.Immunol. 1:108–116

Lukacs NW, Berlin A, Schols D, Skerlj RT, Bridger GJ (2002) AMD3100, a CxCR4 antagonist, attenuates allergic lung inflammation and airway hyperreactivity. Am J Pathol. 160:1353–1360

Markham A, Lamb HM (2000) Infliximab: a review of its use in the management of rheumatoid arthritis. Drugs 59:1341–1359

Meja KK, Seldon PM, Nasuhara Y, Ito K, Barnes PJ, Lindsay MA, Giembycz MA (2000) p38 MAP kinase and MKK-1 co-operate in the generation of GM-CSF from LPS-stimulated human monocytes by an NF-κB-independent mechanism. Br J Pharmacol 131:1143–1153

Morokata T, Ida K, Yamada T (2002) Characterization of YM-90709 as a novel antagonist which inhibits the binding of interleukin-5 to interleukin-5 receptor. Int.Immunopharmacol. 2:1693–1702

Nasuhara Y, Adcock IM, Catley M, Barnes PJ, Newton R (1999) Differential IKK activation and IκBα degradation by interleukin-1β and tumor necrosis factor-α in human U937 monocytic cells: evidence for additional regulatory steps in κB-dependent transcription. J Biol Chem 274:19965–19972

Oda N, Minoguchi K, Yokoe T, Hashimoto T, Wada K, Miyamoto M, Tanaka A, Kohno Y, Adachi M (1999) Effect of suplatast tosilate (IPD-1151T) on cytokine production by allergen-specific human Th1 and Th2 cell lines. Life Sci 65:763–770

Oh JW, Seroogy CM, Meyer EH, Akbari O, Berry G, Fathman CG, DeKruyff RH, Umetsu DT (2002) CD4 T-helper cells engineered to produce IL-10 prevent allergen-induced airway hyperreactivity and inflammation. J Allergy Clin.Immunol. 110:460–468

Oppmann B, Lesley R, Blom B, Timans JC, Xu Y, Hunte B, Vega F, Yu N, Wang J, Singh K, Zonin F, Vaisberg E, Churakova T, Liu M, Gorman D, Wagner J, Zurawski S, Liu Y, Abrams JS, Moore KW, Rennick D, Waal-Malefyt R, Hannum C, Bazan JF, Kastelein RA (2000) Novel p19 protein engages IL-12p40 to form a cytokine, IL-23, with biological activities similar as well as distinct from IL-12. Immunity. 13:715–725

Pretolani M, Goldman M (1997) IL-10: a potential therapy for allergic inflammation? Immunol Today 18:277–280

Proudfoot AE (2002) Chemokine receptors: multifaceted therapeutic targets. Nat.Rev.Immunol. 2:106–115

Rao A, Luo C, Hogan PG (1997) Transcription factors of the NFAT family: regulation and function. Annu.Rev.Immunol. 15:707–47:707–747

Rosenwasser LJ (1998) Biologic activities of IL-1 and its role in human disease. J.Allergy Clin.Immunol. 102:344–350

Rossi D, Zlotnik A (2000) The biology of chemokines and their receptors. Annu Rev Immunol 18:217–242

Sabroe I, Peck MJ, Van Keulen BJ, Jorritsma A, Simmons G, Clapham PR, Williams TJ, Pease JE (2000) A small molecule antagonist of chemokine receptors CCR1 and CCR3. Potent inhibition of eosinophil function and CCR3-mediated HIV-1 entry. J Biol Chem 275:25985–25992

Sallusto F, Lanzavecchia A (2000) Understanding dendritic cell and T-lymphocyte traffic through the analysis of chemokine receptor expression. Immunol.Rev. 177:134–140

Seldon PM, Barnes PJ, Giembycz MA (1998) Interleukin-10 does not mediate the inhibitory effect of PDE4 inhibitors and other cAMP-elevating drugs on lipopolysaccharide-induced tumor necrosis factor-a generation from human peripheral blood monocytes. Cell Biochem Biophys 28:179–201

Shanafelt AB, Forte CP, Kasper JJ, Sanchez-Pescador L, Wetzel M, Gundel R, Greve JM (1998) An immune cell-selective interleukin 4 agonist. Proc Natl Acad Sci U S A 95:9454–9458

Shimbara A, Christodoulopoulos P, Soussi-Gounni A, Olivenstein R, Nakamura Y, Levitt RC, Nicolaides NC, Holroyd KJ, Tsicopoulos A, Lafitte JJ, Wallaert B, Hamid QA (2000) IL-9 and its receptor in allergic and nonallergic lung disease: increased expression in asthma. J Allergy Clin Immunol 105:108–115

Sousa AR, Lane SJ, Nakhosteen JA, Lee TH, Poston RN (1996) Expression of interleukin-1β (IL-1β) and interleukin-1 receptor antagonist (IL-1ra) on asthmatic bronchial epithelium. Am J Respir Crit Care Med 154:1061–1066

Steinke JW, Borish L (2001) Interleukin-4: its role in the pathogenesis of asthma, and targeting it for asthma treatment with interleukin-4 receptor antagonists. Respir Res 2:66–70

Tamaoki J, Kondo M, Sakai N, Aoshiba K, Tagaya E, Nakata J, Isono K, Nagai A (2000) Effect of suplatast tosilate, a Th2 cytokine inhibitor, on steroid-dependent asthma: a double-blind randomised study. Lancet 356:273–278

Torphy TJ (1998) Phosphodiesterase isoenzymes. Am J Respir Crit Care Med 157:351–370

Traves SL, Culpitt S, Russell REK, Barnes PJ, Donnelly LE (2002) Elevated levels of the chemokines GRO-a and MCP-1 in sputum samples from COPD patients. Thorax 57:590–595

Underwood DC, Osborn RR, Bochnowicz S, Webb EF, Rieman DJ, Lee JC, Romanic AM, Adams JL, Hay DW, Griswold DE (2000) SB 239063, a p38 MAPK inhibitor, reduces neutrophilia, inflammatory cytokines, MMP-9, and fibrosis in lung. Am J Physiol Lung Cell Mol Physiol 279:L895-L902

van der Pouw Kraan TC, Boeije LC, de Groot ER, Stapel SO, Snijders A, Kapsenberg ML, van der Zee JS, Aarden LA (1997) Reduced production of IL-12 and IL-12-dependent IFN-gamma release in patients with allergic asthma. J Immunol 158:5560–5565

Walter DM, McIntire JJ, Berry G, McKenzie AN, Donaldson DD, DeKruyff RH, Umetsu DT (2001) Critical role for IL-13 in the development of allergen-induced airway hyperreactivity. J Immunol 167:4668–4675

Wenzel SE, Szefler SJ, Leung DY, Sloan SI, Rex MD, Martin RJ (1997) Bronchoscopic evaluation of severe asthma. Persistent inflammation associated with high dose glucocorticoids. Am J Respir Crit Care Med 156:737–743

White JR, Lee JM, Dede K, Imburgia CS, Jurewicz AJ, Chan G, Fornwald JA, Dhanak D, Christmann LT, Darcy MG, Widdowson KL, Foley JJ, Schmidt DB, Sarau HM (2000) Identification of potent, selective non-peptide CC chemokine receptor-3 antagonist that inhibits eotaxin-, eotaxin-2-, and monocyte chemotactic protein-4-induced eosinophil migration. J Biol Chem 275:36626–36631

White JR, Lee JM, Young PR, Hertzberg RP, Jurewicz AJ, Chaikin MA, Widdowson K, Foley JJ, Martin LD, Griswold DE, Sarau HM (1998) Identification of a potent, selective non-peptide CXCR2 antagonist that inhibits interleukin-8-induced neutrophil migration. J.Biol.Chem. 273:10095–10098

Wills-Karp M, Chiaramonte M (2003) Interleukin-13 in asthma. Curr Opin Pulm Med 9:21–27

Wills-Karp M, Luyimbazi J, Xu X, Schofield B, Neben TY, Karp CL, Donaldson DD (1998) Interleukin-13: central mediator of allergic asthma. Science 282:2258–2261

Ying S, Meng Q, Zeibecoglou K, Robinson DS, Macfarlane A, Humbert M, Kay AB (1999) Eosinophil chemotactic chemokines (eotaxin, eotaxin-2, RANTES, monocyte chemoattractant protein-3 (MCP-3), and MCP-4), and C-C chemokine receptor 3 expression in bronchial biopsies from atopic and nonatopic (Intrinsic) asthmatics. J Immunol 163:6321–6329

Ying S, Robinson DS, Meng Q, Rottman J, Kennedy R, Ringler DJ, Mackay CR, Daugherty BL, Springer MS, Durham SR, Williams TJ, Kay AB (1997) Enhanced expression of eotaxin and CCR3 mRNA and protein in atopic asthma. Association with airway hyperresponsiveness and predominant co-localization of eotaxin mRNA to bronchial epithelial and endothelial cells. Eur J Immunol 27:3507–3516

Zhou Y, McLane M, Levitt RC (2001) Interleukin-9 as a therapeutic target for asthma. Respir Res 2:80–84

Zuany-Amorim C, Haile S, Leduc D, Dumarey C, Huerre M, Vargaftig BB, Pretolani M (1995) Interleukin-10 inhibits antigen-induced cellular recruitment into the airways of sensitized mice. J Clin Invest 95:2644–2651

Inhibitors of Leucocyte–Endothelial Adhesion as Potential Treatments for Respiratory Disease

R. Lever · C. P. Page

Department of Pharmacology, The School of Pharmacy,
University of London, 29–39, Brunswick Square, London, WC1N 1AX, UK
e-mail: rebecca.lever@ulsop.ac.uk

1	Inflammatory Cell Migration．	246
2	Adhesion Molecules．	247
2.1	Selectins．	247
2.2	Immunoglobulin Superfamily Adhesion Molecules．	253
2.3	Integrins．	255
3	The Potential of Adhesion Molecule Inhibitors in Inflammation．	257
4	Adhesion Molecule Inhibitors in Clinical Disease．	258
5	Discussion．	261
	References．	261

Abstract Adhesion of inflammatory cells to the vascular endothelium is an essential step in the movement of these cells out of the blood and into tissue sites and is therefore required for adequate host defence. However, the excessive recruitment of leucocytes to the lung, seen in diseases such as asthma and chronic obstructive pulmonary disease, is equally dependent upon this process. Therefore, leucocyte–endothelial adhesion presents a potential drug target for the development of new anti-inflammatory agents that would reduce the tissue damage and inflammation associated with the presence of these cells in the lung. This chapter discusses the adhesion molecules and pathways involved in adhesion of different leucocyte types to the vascular endothelium and some approaches that have been made in limiting these mechanisms.

Keywords Asthma · Chronic obstructive pulmonary disease · Adhesion molecule · Endothelium · Leucocyte · Inflammation · Selectin · Integrin · Immunoglobulin · Clinical trial

1
Inflammatory Cell Migration

An essential step in the migration of leucocytes out of blood vessels is adhesion to the vascular endothelium. Consequently, inhibition of leucocyte–endothelial adhesion presents an attractive therapeutic target as limitation of this process would therefore limit the number of leucocytes subsequently reaching the tissue and hence the damage caused by their inappropriate recruitment and activation in inflammatory disease.

Leucocyte–endothelial adhesion can be considered to occur in three overlapping stages: an initial loose tethering arrests the travel of the cell in the circulatory flow, followed by rolling of the leucocyte along the endothelium, increasing intercellular surface contact, before a firmer adhesion is mediated which requires activation of the interacting cell. Once this process is complete, the leucocyte is able to traverse the endothelium between individual cells, and travel through the basement membrane, partially under the influence of a chemotactic gradient, into the tissue. This series of events relies upon the interaction of adhesion molecules expressed on the respective cell surfaces either constitutively or in response to stimulation by cytokines or other mediators, in the event of inflammation, trauma or infection. The time scale and type of inflammatory stimulus influences which adhesion molecules are present on the vascular endothelium (Table 1).

Table 1 is not exhaustive, but serves to illustrate differences in profiles of expression of different adhesion molecules. The endothelium, in a sustained state of activation, will express both selectin and IgSF adhesion molecules.

Activated endothelial cells also synthesise and express on their surfaces inflammatory mediators and chemokines, which are able to interact with receptors on the surface of the interacting, loosely tethered leucocyte, promoting activation of the cell.

Once the leucocyte becomes activated, integrin adhesion molecules on its surface become upregulated. This process is in part a functional upregulation, as change in conformation of molecules that are already present increases avid-

Table 1 Endothelial cell adhesion molecules

Name	Constitutive expression?	Expression upregulated?	Leucocyte counterstructures	Time course of expression
Selectin adhesion molecules				
E-selectin	No	Cytokines, LPS	Carbohydrate	Medium
P-selectin	No	Thrombin, PAF, histamine, LT, etc.	Carbohydrate	Rapid
IgSF adhesion molecules				
ICAM-1	Yes	Cytokines, LPS	β_2-Integrin	Prolonged
ICAM-2	Yes	No	β_2-Integrin	Constant
VCAM-1	No	Cytokines, LPS	β_1-Integrin	Prolonged

ity for respective endothelial ligands. A numerical increase in the number of these adhesion molecules expressed on the leucocyte surface also occurs.

Conformationally altered, upregulated β_2-integrin adhesion molecules interact with endothelial immunoglobulin superfamily adhesion molecules and the leucocyte arrests on the endothelium as firm adhesion is established. At the same time, L-selectin molecules are shed from the leucocyte surface as selectin-mediated adhesion ceases to be dominant.

The adherent leucocyte then undergoes diapedesis between endothelial cells before flattening onto the basal surface of the endothelium, still interacting with adhesion molecules.

Finally, leucocytes which have crossed the endothelium dissociate from its basal surface, probably through selective downregulation of adhesion molecules. The cells then travel through to the site of tissue inflammation, under the influence of a gradient of chemotactic factors bound to the extracellular matrix.

2
Adhesion Molecules

2.1
Selectins

The selectins are a family of glycoprotein adhesion molecules comprising an epidermal growth factor-like moiety, repeating sequences mimicking those found on complement-binding proteins and an NH_2-terminal lectin domain. It is via the lectin domain that these molecules Ca^{2+}-dependently bind to carbohydrate structures on the surfaces of interacting cells. E-selectin, found on endothelial cells, L-selectin, found on leucocytes and P-selectin, found on endothelial cells and platelets show approximately 60% homology in their lectin domains (Bevilacqua et al. 1989; Johnston et al. 1989) and the genes which encode these molecules are closely located to each other on chromosome 1 (Watson et al. 1990).

All selectins interact with the tetrasaccharide sialyl Lewis x (sLe^x) on carbohydrate counterstructures (Munro et al. 1992; Alon et al. 1995; reviewed by McEver et al. 1995; Vestweber and Blanks 1999) and hence, selectin-mediated rolling is sensitive to neuraminidase, which removes sialic acid residues (Atherton and Born 1973; Spertini et al. 1991b).

Selectins are involved predominantly in the early, rolling stage of adhesion, before interacting leucocytes are activated and without which firm adhesion cannot proceed (Lawrence and Springer 1991).

E-selectin is expressed on endothelial cells as a result of de novo synthesis in response to activation by lipopolysaccharide or cytokines such as interleukin (IL)-1β or tumour necrosis factor (TNF)-α, but is not present on the surface of resting endothelial cells. Transcription is activated following endothelial cell stimulation, and surface expression peaks after 2–4 h, being over after 24 h in vitro (Bevilacqua et al. 1989). Evidence suggests that the time course in vivo is comparable (Mulligan et al. 1991).

E-selectin is known to bind to counter-structures on the surfaces of neutrophils (Luscinskas et al. 1989), eosinophils (Kyan-Aung et al. 1991), monocytes (Carlos et al. 1991) and lymphocytes (Picker et al. 1991b). Certain carcinoma cells also bind to E-selectin (Rice and Bevilacqua 1989), a factor probably involved in tumour metastasis, a process known to bear many similarities to leucocyte extravasation (Vlodavsky et al. 1992; Hulett et al. 1999).

In vitro, both neutrophils and eosinophils, under shear flow conditions, were found to accumulate and roll on E-selectin. However, high densities of E-selectin were required to elicit eosinophil accumulation, and those cells which did tether rolled at a higher velocity than neutrophils (Patel and McEver 1997), suggesting that alternative mechanisms may be more important in eosinophil adhesion.

There is much evidence, however, to suggest that this adhesion molecule is particularly important in inflammatory conditions such as bronchial asthma. E-selectin was found to be the predominant adhesion molecule involved in the adhesion of neutrophils to human lung microvascular endothelial cells (HLMVECs) that had been stimulated with TNF-α, and it was observed in the same study to contribute to the adhesion of eosinophils to these cells (Blease et al. 1998). Indeed, systemic treatment of rats with a monoclonal antibody (mAb) to E-selectin reduced by 70% the accumulation of neutrophils in both glycogen-induced peritoneal exudates and in bronchoalveolar lavage (BAL) fluid following IgG immune-complex deposition in the lung, a procedure which was shown immunohistochemically to markedly upregulate E-selectin expression in the lung microvasculature (Gundell et al. 1991; Mulligan et al. 1991). In another study in vivo, neutrophil infiltration into the lungs of sensitised cynomolgus monkeys, associated with the late-phase airway obstruction following allergen challenge, correlated directly with increased expression of E-selectin on pulmonary vascular endothelium (Gundell et al. 1991). In addition, human $CD8^+$ T cells expressing the cutaneous lymphocyte-associated antigen (CLA) were found to account for the majority of E-selectin-binding lymphocytes in a circulating population, which bound as strongly as neutrophils, whereas lymphocyte–E-selectin interactions are generally weaker (Picker et al. 1991a). In the same study, it was demonstrated that the accumulation of this subset of T cells in sites of inflamed skin correlated with dense expression of E-selectin in associated venules. Transgenic mice deficient in E-selectin, following intraperitoneal injection of *Streptococcus pneumoniae*, were found to display increased rates of mortality, with prolonged bacteraemia and morbidity than wild-type mice, effects which were also observed in P-selectin-deficient and E- plus P-selectin deficient mice (Munoz et al. 1997). However, under normal conditions, singularly E-selectin-deficient animals are no more susceptible to nascent infections than wild-type mice (Labow et al. 1994), whereas those which lack both endothelial selectins are (Bullard et al. 1996; Frenette et al. 1996).

Cross-reactivity between the selectins and carbohydrate counter-ligands is well established, notably, interactions between all three selectins and P-selectin glycoprotein ligand-1 (PSGL-1). Evidence suggests that PSGL-1 can exist in two forms, one which binds to P-selectin but not to E-selectin, and a second which contains a specific carbohydrate epitope (the cutaneous lymphocyte-associated antigen), expressed by T cells found in inflamed skin. This latter form of PSGL-1, expressed by a mouse T cell clone 4 days after antigen challenge, bound to E-selectin, but binding did not occur 8 or 12 days post-antigen. P-selectin, however, was bound on all three test days, indicating specific structural requirements of the PSGL-1 molecule for E-selectin binding (Borges et al. 1997b). Another study showed that only the first 19 amino acids of PSGL-1 are required for effective binding to E-selectin, as demonstrated by the fact that microspheres coated with different recombinant segments of PSGL-1 rolled under flow conditions on Chinese hamster ovary (CHO) cell monolayers transfected with E-selectin. However, a second region of PSGL-1 (between amino-acids 19 and 148) was found to bind E-selectin, but not P-selectin (Goetz et al. 1997). In addition, a mAb directed against P-selectin itself was shown in vitro to reduce the number of monocytes interacting with E-selectin under shear stress (Lim et al. 1998). Interestingly, neutrophils and eosinophils from patients with allergic asthma express greater levels of PSGL-1 than those from normal donors (Dang et al. 2002).

There is evidence to suggest that E-selectin can act as a signalling molecule, affecting leucocyte functions directly. For example, engagement of E-selectin by adherent neutrophils has been shown to lead to upregulation of integrin adhesion molecules on the neutrophil surface (reviewed by Crockett-Torabi 1998). Furthermore, binding of endothelial E-selectin (or P-selectin and VCAM-1, but not ICAM-1) by specific mAb or leucocyte counterreceptors, was found to raise the intracellular $[Ca^{2+}]$ of endothelial cells and induce cytoskeletal rearrangement, factors involved in optimising the endothelium to allow the transmigration of leucocytes (Lorenzon et al. 1998). E-selectin is crucially involved in the early, loose interactions between leucocytes and the endothelium which precede firm adhesion and diapedesis. The results of these studies would, however, suggest that the role of E-selectin does not end with the facilitation of cell rolling but, rather, involves the preparation of both cells for later, integrin-mediated events.

P-selectin is an adhesion molecule which is expressed by platelets and also by endothelial cells (McEver et al. 1989). This glycoprotein is stored in Weibel-Palade bodies of endothelial cells and secretory α-granules of platelets, and is rapidly externalised onto the cell surface upon cellular activation.

P-selectin is important in the interactions of platelets with vascular endothelium and with other cell types (Larsen et al. 1989; Hamburger and McEver 1990), interactions which are known to be important in inflammatory diseases such as asthma (Coyle et al. 1990; Lellouch-Tubiana et al. 1990) and inflammatory bowel disease (Collins et al. 1994). Thrombin-induced platelet–neutrophil adhesion is P-selectin dependent (de Bruijne-Admiraal et al. 1992), and is known to involve two specific sites (23–30 and 76–90) on the molecule (Chignier et al. 1994). These isolated peptides are able to inhibit thrombin-induced interactions between the two cell types. Adhesion of resting platelets to fMLP stimulated neutrophils is abolished by a mAb to CD18 but not by a mAb to P-selectin, whereas activated platelet–neutrophil interactions involve a P-selectin-dependent recognition step, followed by a tyrosine kinase-dependent signalling step, which leads to β_2-integrin-mediated adhesion of the neutrophil to the platelet (Evangelista et al. 1996). It has been demonstrated that stimulation of either neutrophils or platelets increases their heterotypic intercellular binding, a process found to utilise both P-selectin and macrophage-1 (Mac-1; CD11b/CD18) regardless of which cell type was primarily activated, supporting the involvement of transcellular activation processes (Brown et al. 1998). Similarly, tethering of monocytes to P-selectin on activated endothelium increases the generation of monocyte chemotactic protein (MCP)-1 and TNF-α by the monocytes when they are activated by platelet activating factor (PAF) (Weyrich et al. 1995), which is expressed on endothelial cells in co-operation with P-selectin (Prescott et al. 1984; Lorant et al. 1991). Moreover, adhesion of monocytes to activated platelets expressing P-selectin augments monocyte synthesis of MCP-1 and IL-8 in response to regulated upon activation, normal T cell expressed and secreted (RANTES), a chemokine released from activated platelets (Weyrich et al. 1996).

Additionally, it has been demonstrated that engagement of PSGL-1 on mouse neutrophils by recombinant P-selectin enhances their adhesion to intercellular adhesion molecule (ICAM)-1 via β_2-integrins and increases CD11b/CD18 expression, although these observations were not seen with human neutrophils (Blanks et al. 1998). Nonetheless, the results of these studies suggest a clear involvement for activated platelets in augmentation of neutrophil–endothelial interactions. Furthermore, platelets expressing P-selectin have been shown to be able to restore lymphocyte rolling on high endothelial venules (HEV) of experimental animals in which rolling has been previously inhibited (by up to 90%) by treatment with a mAb to L-selectin (Diacovo et al. 1996a), an observation that also highlights the cross-reactivity which exists between the selectins.

Whereas P-selectin expression on activated platelets is prolonged, its expression on endothelial cells is transient, following which the molecule is re-internalised. Exposure of human umbilical vein endothelial cells (HUVECs) to millimolar concentrations of H_2O_2, which is released by activated neutrophils, was seen to lead to an increased adhesiveness of the cells for neutrophils via the rapid synthesis of PAF (Lewis et al. 1988). Exposure of the cells to micromolar concentrations of H_2O_2 for longer periods was found to induce neutrophil adhesion through a mechanism which was independent of PAF synthesis and E-selectin expression, and which did not actively involve the neutrophil. This phenomenon was found to be due to P-selectin expression on endothelial cells, as confirmed by the presence of the Weibel-Palade body component, von Willebrand factor (vWF) in the culture medium and, more precisely, by the fact that mAb to P-selectin abrogated the neutrophil adhesion (Patel et al. 1991). The neutrophil is not actively involved in P-selectin mediated-adhesion to endothelium (Geng et al. 1990).

Thrombin or histamine stimulation of endothelial cells leads to an extremely rapid translocation of P-selectin to the cell surface (Hattori et al. 1989). Incubation of HUVECs with thrombin causes an increase in adhesiveness for neutrophils (Zimmerman et al. 1985) which begins 15 min following exposure to thrombin and persists for up to 120 min (Sugama et al. 1992), whereas P-selectin expression was found to peak after 5–15 min and was over within 90 min. Anti-E-selectin mAb had no effect on the thrombin induced neutrophil adhesion, whereas a mAb against P-selectin blocked the adhesion observed in the first 30 min of thrombin exposure, mAbs against ICAM-1 or CD18 being required to block the later adhesion (Sugama et al. 1992). Under flow conditions in vitro, both neutrophils and eosinophils accumulate and roll on P-selectin, although more neutrophils than eosinophils tether to a given density of P-selectin (Patel and McEver 1997). Both neutrophil and eosinophil accumulation is compromised in P-selectin-deficient mice. Eosinophil infiltration in the lungs of sensitised mice exposed to antigen is reduced in the earlier stages of the resultant inflammatory response (Gonzalo et al. 1996; Broide et al. 1998). In the study by Gonzalo and co-workers, eosinophil accumulation was reduced 3 h after allergen challenge but actually increased 7 h post-antigen, suggesting that inhibition of lung eosinophilia seen in the early phase after challenge may reflect a shift in the time-course of leucocyte trafficking rather than an actual reduction in numbers undergoing extravasation. Similar effects of P-selectin deficiency have been reported with regard to neutrophil accumulation in cerebrospinal fluid (CSF), in cytokine (TNF-α and IL-1β)-induced meningitis. P-selectin-deficient mice had less neutrophils in the CSF 4 and 6 h after intrathecal injection of the cytokines, with levels the same as in wild-type mice after 8 h. However, doubly E- plus P-selectin-deficient mice showed significant reductions in CSF neutrophils 4, 6 and 8 h after injection (Tang et al. 1996).

Rolling of human neutrophils (Sako et al. 1993; Moore et al. 1995) and monocytes (Lim et al. 1998) on P-selectin is mediated by PSGL-1. PSGL-1 is a glycoprotein expressed on leucocytes which mediates their adhesion to P-selectin expressed on activated platelets or endothelial cells, and which is also able to interact with E-selectin and L-selectin (reviewed by Moore 1998).

It has been demonstrated that only the lectin and epidermal growth factor (EGF)-like domains of P-selectin are required for binding to PSGL-1 (Mehta et al. 1997), an interaction which is dependent on a specific site within the N-terminus of PSGL-1 (Goetz et al. 1997). Indeed, it was found that a mAb to the N-terminus of PSGL-1, when administered systemically to mice, inhibited neutrophil accumulation in mouse peritoneum. This effect was augmented by concomitant administration of a mAb to P-selectin itself (Borges et al. 1997a).

When neutrophils become activated, surface ligands for P-selectin are redistributed, rendering the cells less adhesive to P-selectin (Lorant et al. 1995; Bruehl et al. 1997), a step which is probably involved in facilitation of extravasation (and which parallels the shedding of L-selectin from activated leucocytes).

L-selectin is expressed constitutively on the surface of leucocytes and is involved in their early rolling events on endothelial cells. It also plays a part in leucocyte–leucocyte interactions, in co-operation with integrin adhesion molecules (Lynam et al. 1996). Although expressed on resting leucocytes, a brief enhancement of the adhesiveness of L-selectin for its counterligands occurs when leucocytes become activated (Spertini et al. 1991a).

In a non-static, in vitro adhesion study, L-selectin was demonstrated, through specific blockade with an anti-L-selectin mAb, to mediate more than 60% of the attachment of monocytes to TNF-α-stimulated bovine aortic endothelial cells (BAEC) and, in addition, to be able to bind to unstimulated BAEC (Giuffrè et al. 1997). In vivo, the early trafficking of neutrophils to the peritoneal cavity of mice, in response to thioglycollate, was found to be inhibited by pre-administration of an L-selectin IgG chimaera, which comprises the extracellular domain of murine L-selectin linked to human IgG_1 (Watson et al. 1991). In an intravital microscopy study in rabbit mesentery, the L-selectin IgG chimaera exerted a profound inhibitory effect upon neutrophil rolling, as did an antibody directed against L-selectin itself (Ley et al. 1991). Mice deficient in L-selectin showed less neutrophil accumulation in lung microvasculature in a model of experimental bacterial pneumonia, although the numbers of emigrated cells in the lung, 6 h after bacterial instillation, did not differ from those seen in control animals (Doyle et al. 1997). Furthermore, allergic L-selectin-deficient mice did not demonstrate defective eosinophil infiltration in the lung, 3 or 6 h following inhalation of antigen (Gonzalo et al. 1996). The results of these studies illustrate that although L-selectin is involved in early leucocyte–endothelial interactions, other adhesion molecule systems mediate cell extravasation.

SLe^x is expressed on L-selectin molecules on the cell surface, which presents a potential adhesive ligand for selectin molecules on apposing endothelial cells or leucocytes (Picker et al. 1991b). In addition, PSGL-1 on the surface of leucocytes associates with L-selectin on other leucocytes and, in this way, is involved in homotypic cell interactions (reviewed by Moore 1998). In particular, it is known that neutrophils and eosinophils utilise L-selectin-PSGL-1 binding in leucocyte–leucocyte interactions, whereby the NH_2-terminus of PSGL-1 associates with the lectin domain of L-selectin (Patel and McEver 1997). The region crucial for recognition of L-selectin by PSGL-1 has now been identified as a specific tyrosine sulphation motif on the molecule (Snapp et al. 1998). As is the case for P-selectin, L-selectin can act as a signalling molecule, as induction of mRNA for IL-8 and TNF-α in neutrophils by sulphatides is dependent upon the presence of L-selectin on these cells (Laudanna et al. 1994).

A combination of anti-PSGL-1 and an anti-L-selectin mAb has been found to inhibit monocyte attachment to TNF-α-stimulated HUVECs under shear flow in vitro by more than 80% (Lim et al. 1998). However, rolling of human neutrophils and the cell line HL-60 on purified L-selectin is only partially inhibited by a mAb to PSGL-1, indicating the involvement of a distinct ligand for L-selectin on the surface of the cells (Ramos et al. 1998).

L-selectin is shed from the surface of leucocytes once they are activated (Spertini et al. 1991a) by a proteolytic mechanism (Khan et al. 1994). This may have a local regulatory role in limiting further cellular adhesion as, if soluble

L-selectin concentrations become great enough, adhesion can be partially inhibited (Schleiffenbaum et al. 1992). Indeed, it has been demonstrated that endogenous, soluble adhesion molecules in general can have an inhibitory role upon leucocyte–endothelial adhesion. One group noted that human plasma is able to inhibit neutrophil adhesive functions in vitro, an effect which is abolished by prior immunoprecipitation of soluble ICAM-1, E-selectin and P-selectin from the plasma (Ohno et al. 1997).

2.2
Immunoglobulin Superfamily Adhesion Molecules

Members of the immunoglobulin superfamily characterised to date, involved in leucocyte–endothelial adhesion, include ICAM-1, -2 and -3, vascular cell adhesion molecule-1 (VCAM-1), platelet endothelial cell adhesion molecule-1 (PECAM-1) and mucosal addressin cell adhesion molecule (MAdCAM).

ICAM-1 is a five-immunoglobulin domain structure expressed by a range of cell types (Rothlein et al. 1988), including endothelial cells, where it is constitutively expressed at a low level and significantly upregulated when the endothelium is activated by cytokines or lipopolysaccharide (LPS) (Dustin et al. 1986). This increase in expression can be detected 4–6 h after activation, is maximal after 24 h and can last for up to 48 h. Increased endothelial ICAM expression in response to cytokines occurs via a protein synthesis-dependent process (Rothlein et al. 1986). However, thrombin-stimulated HUVECs were discovered to express ICAM-1 through a process which is independent of de novo synthesis, as indicated by the absence of ICAM-1 mRNA within a relevant time period. In addition, pretreatment of the HUVECs with cycloheximide did not affect ICAM-1 expression on, nor neutrophil adhesion to, thrombin-stimulated HUVECs, whereas both parameters were blocked when TNF-α was used as the stimulus (Sugama et al. 1992). This comparatively rapid expression of ICAM-1 in response to thrombin acts to extend the adhesiveness of the endothelium to neutrophils, a factor which is initiated by P-selectin translocation.

ICAM-1 is involved in the adhesion to activated endothelium of neutrophils (Diamond et al. 1990; Kyan-Aung et al. 1991), lymphocytes (Smith et al. 1988) and eosinophils (Kyan-Aung et al. 1991; Blease et al. 1998). It is known to be a ligand for the leucocyte adhesion molecules CD11a/CD18 and CD11b/CD18, interactions which involve different extracellular domains on ICAM-1 (Diamond et al. 1990; Diamond et al. 1991).

ICAM-2 is also a ligand for CD11a/CD18. This is a two-domain IgSF adhesion molecule, homologous (35%) to the two N-terminal domains of ICAM-1, one of which is responsible for ICAM-1-CD11a/CD18 binding (Diamond et al. 1991). It has been shown that all CD11a/CD18-dependent adhesion of leucocytes to endothelial cells is mediated by either ICAM-1 or ICAM-2 (de Fougerolles et al. 1991).

ICAM-2 is constitutively expressed on the surface of endothelial cells at a much higher level than ICAM-1, but is not inducible by cytokine activation

(Staunton et al. 1989). The presence of ICAM-2 has also been demonstrated on platelets (Diacovo et al. 1994).

It is likely that ICAM-2 is predominantly concerned with normal functions, such as the recirculation of lymphocytes, due to its high constitutive expression and the lack of effect which inflammatory mediators are able to exert upon it. Conversely, ICAM-1 is probably involved largely with inflammatory and immune events, due to its low expression on resting endothelial cells and its profound upregulation by cytokines. Mice which are deficient in functional ICAM-1, expressing a variant of this structure with an incomplete extracellular domain (King et al. 1995), have been found to differ from wild-type mice in experimental allergic inflammation. Following sensitisation to ovalbumin, these animals exhibited reduced eosinophils and total cells in the airways 3 days after allergen challenge, when compared to those seen in sensitised wild-type mice (Wolyniec et al. 1998). In another study using ICAM-1 deficient mice, Broide et al. (1998) demonstrated that airway eosinophilia was also greatly reduced at earlier time points following allergen challenge (3, 12 and 24 h). Interestingly, in the first study, expression of VCAM-1 in lung microvasculature was persistent in the ICAM-1-deficient mice 3 days post-challenge (compared to only 24 h in control mice), suggesting firstly that ICAM-1 provides the predominant route for infiltrating eosinophils in the later stages of allergic airway inflammation and, secondly, that animals deficient in one adhesion molecule may compensate by increasing expression of another. Further indication that mutant mice lacking one adhesion pathway may upregulate an alternative is granted by the fact that ICAM-1-deficient mice do not exhibit reduced neutrophil accumulation in the lung in response to intratracheal instillation of LPS, whereas an anti-ICAM-1 mAb inhibits this infiltration in wild-type mice (Kumasaka et al. 1996). However, in another study, neutrophil recruitment was seen to be impaired, though not abolished, in ICAM-1-deficient mice (Sligh et al. 1993).

VCAM-1 is an IgSF adhesion molecule which is also involved in inflammation. Unlike ICAM-1, it is not expressed basally on vascular endothelium but, similarly, is synthesised in response to inflammatory mediators (Osborn et al. 1989). VCAM-1 expression is also upregulated by IL-1β, TNF-α and endotoxin. IL-4 is thought to be important in the regulation of VCAM-1 expression (Thornhill and Haskard 1990; Thornhill et al. 1990). This cytokine has been shown to act synergistically with IL-1β in the induction of lymphocyte adhesion (Masinovsky et al. 1990) and with LPS for eosinophil adhesion (Blease et al. 1998) in vitro, and with TNF-α in the recruitment of lymphocytes to skin cells in vivo (Briscoe et al. 1992), demonstrating a mechanism whereby specific leucocyte types can be recruited to tissue sites. Furthermore, the leucocyte ligand for VCAM-1, very late activation antigen (VLA)-4 (Elices et al. 1990), is an integrin expressed on monocytes, lymphocytes and eosinophils (Hemler et al. 1987; Berlin et al. 1995) but probably not on neutrophils in levels functionally important for cell adhesion. The eosinophil recruitment accompanying the late, delayed-type hypersensitivity reaction which followed chemoattractant-induced cell recruitment to mouse skin sites was found to be inhibited by an anti-α_4-integrin mAb (Teixeira and Hellewell 1998), as was eosinophil but not neutrophil adhesion to stimulated HUVECs (Walsh et al. 1991), illustrating the importance of this pathway in eosinophil accumulation and adhesion. Sensitised mice deficient in VCAM-1 have been shown to lack eosinophils in BAL fluid, as well as having reduced numbers of BAL lymphocytes, following ovalbumin challenge, effects which were also observed in ICAM-1-deficient mice (Gonzalo et al. 1996).

Of granulocytes, it is interesting to note that neutrophils, which are strongly involved early on in inflammation and host defence should be preferentially recruited via the adhesion pathway which is enabled most rapidly, the translocation of pre-formed P-selectin. Conversely, eosinophils, which are heavily implicated in the later, ongoing damage associated with diseases such as bronchial asthma, are preferentially recruited through a mechanism which relies on the much slower process of VCAM-1 upregulation.

PECAM-1 is expressed on endothelial cells, platelets, neutrophils and some mononuclear cells (reviewed by Newman 1997). It is thought to be involved in the transmigration of cells, as large numbers of this molecule are expressed constitutively at intercellular junctions of the endothelium. However, it has been suggested that the primary role of PECAM-1 may be as an inhibitory protein rather than an intercellular adhesion molecule, due the presence of cytoplasmic motifs which inhibit tyrosine kinase-mediated cellular processes (Newman 1999). Based upon this evidence, Newman proposes that PECAM-1 may have been wrongly classified as an IgSF adhesion molecule, and in fact belongs to a class of molecules now known as the Ig-ITIM (immunoreceptor tyrosine-based inhibitory motif) family. PECAM-1 is able to interact with PECAM-1 molecules on other cells (Albelda et al. 1991), a mechanism which becomes predominant as more PECAM-1 molecules are expressed on the cell surface (Sun et al. 1996), although additional, integrin-dependent interactions may be involved also. Moreover, it is known that PECAM-1 molecules are shed from the surface of activated neutrophils, along with L-selectin, as integrin adhesion molecules become upregulated (Wang et al. 1998).

MAdCAM is an adhesion molecule which is involved in the recirculation of lymphocytes. It is known to be an endothelial ligand for L-selectin which mediates lymphocyte rolling on lymph nodes (Berg et al. 1993). MAdCAM also interacts with the lymphocyte integrin $\alpha_4\beta_7$, facilitating the adhesion of these cells to Peyer's patches and is thought to be required for mucosal lymphocyte homing (Hamann et al. 1994.)

2.3
Integrins

The leucocyte β_2-integrins, lymphocyte function-related antigen (LFA-1), macrophage-1 (Mac-1), p150,95 and CD11d/CD18 are characterised by a common β-subunit (CD18), associated non-covalently with the α-subunits CD11a, CD11b, CD11c and CD11d, respectively. All three α-subunits include a cation-binding region, which is likely to be related to the cation-dependent nature of the adhesion process in general, as the engagement of these molecules with their counterligands is essential for the firm adherence of leucocytes to endothelium. The crucial importance of the β_2-integrins in inflammatory cell trafficking is illustrated by the lack of adequate host defence in patients suffering from the rare abnormality known as leucocyte adhesion deficiency (LAD)-1, where the CD18 molecule is either absent or dysfunctional (Crowley et al. 1980).

CD11a/CD18 is expressed constitutively by all leucocyte types (Smith et al. 1988). The number of molecules of CD11a/CD18 expressed does not increase in response to cytokines or chemoattractants (Carlos and Harlan 1990). However, T cell receptor cross-linkage leads to a functional upregulation of the molecule, whereby avidity for ICAM-1 and ICAM-2 is increased via a conformational alteration to the β_2-subunit (Dustin and Springer 1989; Diamond et al. 1991). This event is pivotal in T lymphocyte functions (Larson and Springer 1990).

CD11b/CD18 is expressed constitutively on the surface of granulocytes and monocytes, and is additionally present in granules of neutrophils and when the cell is activated by inflammatory mediators, additional molecules are translocated to the surface of the neutrophil. A functional upregulation also occurs and it is this latter element which appears to contribute most significantly to the increased adhesiveness of the leucocyte, as inhibition of quantitative upregulation of CD11b/CD18 does not prevent neutrophil adhesion to endothelial cells (Vedder and Harlan 1988).

Unlike CD11a/CD18, CD11b/CD18 does not bind to ICAM-2, and effects neutrophil adhesion and extravasation through binding to ICAM-1. As well as being essential for firm adhesion of neutrophils to endothelium, CD11b/CD18 is required for neutrophil adhesion to and migration across surface adherent activated platelets (Diacovo et al. 1996b) and is involved in homotypic leucocyte–leucocyte interactions and neutrophil–platelet interactions (Brown et al. 1998). The results of a recent study indicate that additional ligands for CD11b/CD18, aside from ICAM-1, are present on endothelium in that CD11b/CD18-dependent transendothelial migration of neutrophils, i.e. that which occurred in the presence of a mAb to CD11a/CD18, persisted in the presence of mAb to both ICAM-1 and ICAM-2 (Issekutz et al. 1999). Mice have been generated which are deficient in CD11b/CD18 but which express CD11a/CD18 normally, through selective disruption of CD11b. These animals were found to retain the ability to recruit neutrophils to the peritoneal cavity, a process which was found to be more dependent upon CD11a/CD18, given that 78% of the neutrophil recruitment seen in the mutant mice could be blocked by a mAb to CD11a. In the same study, neutrophils from CD11b-deficient mice did, however, display impaired adhesion, aggregation, phagocytosis and degranulation; suggesting that both molecules are necessary for a complete range of normal neutrophil functions (Lu et al. 1997).

CD11c/CD18 is constitutively expressed on the surface of granulocytes and monocytes. This molecule will support adhesion of activated endothelial cells but not resting endothelial cells to a CD11c/CD18-coated surface (Stacker and

Springer 1991). CD11d/CD18 is a further β_2-integrin produced by granulocytes. The gene which encodes the CD11d α-subunit is located very close to that for CD11c, a determinant which is thought to be involved in the fact that expression of CD11d on granulocytes is repressed during upregulation of CD11c expression, which occurs during cell differentiation (Shelley et al. 1998).

Of the β_1-integrins, VLA-4 ($\alpha_4\beta_1$) is the most extensively studied and is of particular relevance to allergic lung disease, due to its expression on eosinophils and established role in the adhesion and trafficking of this leucocyte type (Reviewed by Wardlaw 2001). This molecule is functionally regulated on eosinophils by mediators such as the chemokines RANTES and macrophage inflammatory protein (MIP)-1β (Weber et al. 1996) and, in addition to its direct involvement in eosinophil recruitment, has been found to possess a role in mast cell degranulation (Hojo et al. 1998) and also in lymphocyte recirculation (Berlin-Rufenach et al. 1999). The most important known ligand of VLA-4, of relevance to leucocyte–endothelial adhesion, is VCAM-1, although fibronectin binding is also involved in the biology of this molecule (Pulido et al. 1991). Furthermore, VLA-4-dependent leucocyte–endothelial interactions have been shown to occur, in a model of chronic inflammation, via a mechanism independent of both VCAM-1 and the established binding sequence (CS-1) of fibronectin (Johnston et al. 2000). VLA-4, previously thought not to be expressed by neutrophils, has now been demonstrated on these cells, and a possible role in their activity has been suggested (Issekutz et al. 1996; Davenpeck et al. 1998; Johnston and Kubes 1999; Burns et al. 2001). However, other pathways, including those mediated by β_2-integrins, would appear to be more heavily involved in neutrophil–endothelial adhesion (McNulty et al. 1999; Tasaka et al. 2002). By contrast, the importance of the VLA-4 adhesion pathway in eosinophil adhesion has been well established in a range of in vitro and animal studies employing blocking antibodies to this adhesion molecule, suggesting that targeting of this molecule or its ligand(s) may confer some specificity in the type of leucocyte recruitment inhibited in the setting of allergic inflammatory disease.

3
The Potential of Adhesion Molecule Inhibitors in Inflammation

Inhibition of leucocyte adhesion has long been considered an attractive target for the treatment of inflammatory disease and, as can be seen, there is a wide variety of processes within this phenomenon, inhibition of which could potentially yield clinical benefit. Obvious approaches include strategies that limit the expression or activation of adhesion molecules on endothelial cells and/or leucocytes, or inhibition of the function of these molecules once present, with synthetic inhibitors, monoclonal antibodies, or analogues of endogenous inhibitors or receptors, including soluble adhesion molecules. Differences in the adhesion pathways utilised by individual leucocyte types could allow the design of drugs that target the trafficking of specific cells, relevant to individual diseases. Alternatively, a more general approach could be taken whereby weaker, non-specific inhibition of several adhesion pathways concurrently could result in an overall attenuation of the inflammatory response. Several classes of drugs currently used in the clinic are known to affect cell adhesion molecule expression generally, to a varying extent, and these effects might be involved in their overall anti-inflammatory efficacy. These include, non-exhaustively, the glucocorticosteroids, certain second and third generation antihistamines, inhibitors of phosphodiesterase enzymes and statins. The statins present a promising area for the discovery of inhibitors of cell adhesion, as this class of drugs has been found to possess anti-inflammatory properties that may be attributed to effects on aspects of the process of cell recruitment (reviewed by Weitz-Schmidt 2002). These include inhibition of selectin and integrin expression and function, as well as chemokine and cytokine production by both endothelial cells and leucocytes. Another drug with a range of non-specific actions on cell adhesion is heparin, which inhibits the activity of an array of pro-inflammatory mediators and chemokines, as well as directly binding and inhibiting the function of L- and P-selectin, CD11b/CD18 and PECAM-1 (reviewed by Tyrrell et al. 1999; Lever and Page 2002). Drugs such as heparin or the statins, whilst in their current form possessing general anti-inflammatory properties which may depend on inhibition of adhesion molecule function but which are likely due also to indirect modulation of their expression, could be used as templates for the design of specific adhesion molecule inhibitors, based on their adhesion molecule-binding properties, an example being the statin-derived LFA-1 (CD11a/CD18) inhibitor LFA703.

The leucocyte adhesion deficiencies are rare, inherited disorders in which specific types of adhesion molecule are either absent or dysfunctional and which result in severe sepsis, morbidity and usually premature death. LAD-1 patients are unable to produce functional β_2-integrins (Crowley et al. 1980), LAD-2 patients have a deficiency in carbohydrate synthesis that ultimately prevents effective selectin function (Etzioni et al. 1993; Bevilacqua et al. 1994; Phillips et al. 1995; Karsan et al. 1998; Sturla et al. 1998) and, recently, a novel form of LAD was described in which both β_1- and β_2-integrins are dysfunctional (McDowall et al. 2003). These conditions illustrate the importance of the respective adhesion molecules in normal physiology and, in themselves, act as a caveat to over-efficient inhibition of adhesion molecule function and should be borne in mind in the design of therapeutic strategies in this area.

4
Adhesion Molecule Inhibitors in Clinical Disease

A mAb against ICAM-1 was tested in a small, open-label trial in patients with rheumatoid arthritis. Inhibitory effects on lymphocyte recirculation were noted (peripheral lymphocytosis, transient cutaneous anergy). Approximately half of the patients derived clinical benefit and side effects were short-lived, the antibody otherwise being well tolerated (Kavanaugh et al. 1994). A second trial by the same group in early rheumatoid arthritis produced similar results (Kavanaugh et al. 1996).

However, a randomised, placebo-controlled trial of the anti-ICAM-1 mAb (enlimomab) for prevention of renal allograft rejection found no significant effect of the antibody (Wramner et al. 1999). The most serious and in some cases fatal side effect was infection, although no statistically significant difference in incidence was observed between treatment and control group.

Following positive results in reducing inflammatory post-ischaemic brain damage in animal models of stroke (e.g. Bowes et al. 1993; Zhang et al. 1995), the anti-ICAM-1 mAb enlimomab was assessed, in a multicentre study, in human stroke for an effect on post-ischaemic damage, due to reperfusion injury. The treatment group was found to have a significantly worse clinical outcome than the placebo group (enlimomab acute stroke trial investigators 2001). A subsequent study, carried out to investigate this discrepancy between pre-clinical and clinical outcome, reported that in rats treated with an anti-rat-ICAM-1 mAb that was raised in mice, significant blood leucocyte activation, complement activation and upregulation of other adhesion molecules, such as E- and P-selectin, occurred acutely and, upon repeated administration, anti-mouse antibodies were produced. The authors conclude that these effects would have outweighed any useful effect of ICAM-1 inhibition and postulate that similar problems may have occurred with the mouse anti-human antibody, enlimomab, in the clinical trial (Furuya et al. 2001).

Indeed, in vitro, enlimomab was found to enhance complement-dependent activation of neutrophils in human whole blood, as measured by oxidative burst, CD11b upregulation and L-selectin shedding (Vuorte et al. 1999). These effects were not seen with an IgG_1 type Ab (enlimomab is a mouse-antihuman IgG_{2a} antibody).

Pre-clinical studies in models of myocardial ischaemia-reperfusion injury have indicated a potential role for CD11/CD18 blockade in improving clinical outcome post-infarction (e.g. Ma et al. 1991; Aversano et al. 1995). Therefore, in a randomised, double-blind multicentre trial, the anti-CD11/CD18 mAb (LeukArrest) was tested—in myocardial infarction patients undergoing percutaneous transluminal coronary angioplasty—for its effect in reducing neutrophil-mediated post-ischaemic, reperfusion injury. No significant differences between study groups were observed, although it should be noted that this trial was designed primarily as a tolerability, rather than efficacy, study (Rusnak et al. 2001). It is of interest and of clear pertinence to the present article, however, that although the antibody was found to be well tolerated in the above study, a single adverse outcome of pneumonia in the treatment group was noted in a smoker with an 8-month history of cough.

A monoclonal antibody, HuEP5C7, which reacts with the human forms of both E- and P-selectin (He et al. 1998), has recently been reported, in a primate model of stroke, to reduce neutrophil accumulation, infarct size and post-ischaemic symptoms without causing complement activation or immunosuppression (Mocco et al. 2002).

In a small clinical trial, a mouse anti-human-E-selectin mAb (CY1787) was found to be effective and well tolerated as an adjunct in the management of septic shock, although anti-mouse antibodies were produced by the majority of patients (Friedman et al. 1996).

An humanised mAb to E-selectin (CDP850) was recently assessed in a randomised, placebo-controlled multicentre trial for its effects in moderate to severe psoriasis. Although the antibody was well tolerated and E-selectin levels in plasma and on skin biopsy vascular endothelium were significantly attenuated, the latter being maintained for 4 weeks post-treatment, no reduction in dermal neutrophil or lymphocyte accumulation was observed, nor any reduction in clinical severity (Bhushan et al. 2002).

Small molecule inhibitors of E-selectin are in development. TBC1269 is based upon the structure of the ligand sialyl di-Lewis^x (Kogan et al. 1998) and has been shown to inhibit neutrophil accumulation in the lung, during acute *Mannheimia haemolytica* pneumonia, in newborn calves (Radi et al. 2002). A-205804 inhibits the expression of both E-selectin and ICAM-1, without significant effect on VCAM-1 and has been found to reduce inflammation in animal models of lung and joint inflammation (Zhu et al. 2001).

The development of inhibitors of VLA-4 has been the subject of much interest over recent years. A humanised anti-α4 mAb, HP1/2, was demonstrated to inhibit the antigen-induced late-phase response and bronchial hyperresponsiveness in allergic sheep (Abraham et al. 1994), and another humanised anti-α4 mAb (Antegren) is in phase III clinical trial for inflammatory bowel disease (reviewed by Van Assche and Rutgeerts 2002). Various peptide antagonists, based structurally on the VLA-4-binding sites of VCAM-1, fibronectin and anti-α4 antibodies, respectively, have been developed (reviewed by Lin and Castro 1998). One such inhibitor, BIO-1211, based on the CS-1 sequence of fibronectin, exhibits 200-fold selectivity for the active form of VLA-4 and shows effect against antigen-induced early and late-phase reactions and airways hyperresponsiveness in allergic sheep (Lin et al. 1999).

Novel statin derivatives are being investigated as new anti-inflammatory agents, for example, the LFA-1 (CD11a/CD18) inhibitor LFA703. This molecule lacks the HMG-CoA reductase inhibitory activity of the statins, their mode of action in reducing cholesterol synthesis, yet binds and inhibits the function of the universal leucocyte integrin LFA-1 (Weitz-Schmidt et al. 2001), whereas effects of statins on cytokine and adhesion molecule synthesis and expression are thought to depend on HMG-CoA reductase inhibition.

The majority of information available on the clinical and adverse effects of specific adhesion molecule inhibition comes from trials aimed to assess their effects in post-ischaemic inflammation, such as that following stroke or myocardial infarction. Whilst these data are clearly useful, it may not be possible to extrapolate these findings to the setting of asthma or chronic obstructive pulmonary disease (COPD), where inflammatory mechanisms may differ with respect to cell type and time course. Furthermore, whereas side effects of a relatively acute treatment aimed to improve outcome following stroke may be either absent or acceptable, adverse effects of such treatments when administered over longer periods of time, as would be required for the treatment of chronic inflammation, require both further investigation and due consideration on a risk-to-benefit basis.

5
Discussion

There is a clear rationale for the inhibition of excessive leucocyte recruitment in the airways in diseases such as asthma and COPD. However, a balance must be found that allows prevention of unwanted cell recruitment with maintenance of host defence. This is particularly important in a disease such as COPD which, whilst known to involve chronic neutrophil infiltration and neutrophil-dependent tissue damage, is also characterised by persistent respiratory infection. This balance might be achieved by targeting of specific adhesion mechanisms, for example, those involved in the eosinophil accumulation typical of asthma, whilst sparing those involved in neutrophil recruitment. However, leucocyte and endothelial adhesion molecules act in concert, with much overlap between the different pathways and, therefore, the reality is unlikely to be so clear-cut. Alternatively, a greater understanding of endogenous mechanisms in place for limitation of leucocyte–endothelial adhesion (reviewed by Perretti 1997) may allow design of drugs that inhibit cell trafficking associated with chronic inflammation whilst allowing an adequate response to trauma or infection to be mounted.

References

Abraham, W.M., Sielczak, M.W., Ahmed, A., Cortes, A., Lauredo, I.T., Kim, J., Pepinsky, B., Benjamin, C.D., Leone, D.R., Lobb, R.R. Alpha-4-integrins mediate antigen-induced late bronchial responses and prolonged airway hyperresponsiveness in sheep. J. Clin. Invest. 93, 776–787. 1994

Albelda, S.M., Muller, W.A., Buck, C.A. and Newman, P.J. Molecular and cellular properties of PECAM-1 (endoCAM/CD31): a novel vascular cell-cell adhesion molecule. J. Cell Biol. 114, 1059–1068. 1991

Alon, R., Feizi, T., Fuhlbrigge, R.C. and Springer, T.A. Glycolipid ligands for selectins support leukocyte tethering and rolling under physiologic flow conditions. J. Immunol. 154, 5356–5366. 1995

Atherton, A. and Born, G.V.R. Effects of neuraminidase and N-acetyl neuraminic acid on the adhesion of circulatin granulocytes and platelets in venules. J. Physiol. 234, 66P-67P. 1973

Aversano, T., Zhou, W., Nedelman, M., Nakada, M. And Weisman, H. A chimeric IgG4 monoclonal antibody directed against CD18 reduces infarct size in a primate model of myocardial ischemia and reperfusion. J. Am. Coll. Cardiol. 25, 781–788. 1995

Berg, E.L., McEvoy, L.M., Berlin, C., Bargatze, R.F. and Butcher, E.C. L-selectin mediated rolling on MAdCAM-1. Nature. 366, 695–698. 1993

Berlin, C., Bargatze, R.F., Campbell, J.J., von Andrian, U.H., Szabo, M.C., Hasslen, S.R., Nelson, R.D., Berg, E.L., Erlandsen, S.L. and Butcher, E.C. α4 integrins mediate lymphocyte attachment and rolling under physiologic flow. Cell. 80, 413–422. 1995

Berlin-Rufenach, C., Otto, F., Mathies, M., Westermann, J., Owen, M.J., Hamann, A. and Hogg, N. Lymphocyte migration in lymphocyte function-associated antigen (LFA)-1-deficient mice. J. Exp. Med. 189(9),1467–1478. 1999

Bevilacqua, M.P., Nelson, R.M., Mannori, G. and Cecconi, O. Endothelial-leukocyte adhesion molecules in human disease. 45, 361–378. 1994

Bevilacqua, M.P., Stengelin, S., Gimbrone, M.A., and Seed, B. Endothelial leukocyte adhesion molecule-1: an inducible receptor for neutrophils related to complement regulatory proteins and lectins. Science 243, 1160–1164. 1989

Bhushan, M., Bleiker, T.O., Ballsdon, A.E., Allen, M.H., Sopwith, M., Robinson, M.K., Clarke, C., Weller, R.P.J.B., Graham-Brown, R.A.C., Keefe, M., Barker, J.N.W.N. and Griffiths, C.E.M. Anti-E-selectin is ineffective in the treatment of psoriasis: a randomized trial. Br. J. Dermatol. 146, 824–831. 2002

Blanks, J.E., Moll, T., Eytner, R. and Vestweber, D. Stimulation of P-selectin glycoprotein ligand-1 on mouse neutrophils activates β_2-integrin mediated cell attachment to ICAM-1. Eur. J. Immunol. 28, 433–443. 1998

Blease, K., Burke-Gaffney, A.B. and Hellewell, P.G. Modulation of cell adhesion molecule expression and function on human lung microvascular endothelial cells by inhibition of phosphodiesterases 3 and 4. Br. J. Pharmacol. 124, 229–237. 1998

Blease, K., Seybold, J., Adcock, I.M., Hellewell, P.G. and Burke-Gaffney, A. Interleukin-4 and lipopolysaccharide synergize to induce vascular cell adhesion molecule-1 expression in human lung microvascular endothelial cells. Am. J. Respir. Cell Mol. Biol. 18, 620–630. 1998

Borges, E., Eytner, R., Moll, T., Steegmaier, M., Campbell, M.A., Ley, K., Mossman, H. and Vestweber, D. The P-selectin glycoprotein ligand-1 is important for recruitment of neutrophils into inflamed mouse peritoneum. Blood 90 1934–1942. 1997a

Borges, E., Pendl, G., Eytner, R., Steegmaier, M., Zollner, O. and Vestweber, D. The binding of T cell-expressed P-selectin glycoprotein ligand-1 to E- and P-selectin is differentially regulated. J. Biol. Chem. 272, 28786–28792. 1997b

Bowes, M.P., Zivin, J.A. and Rothlein, R. Monoclonal antibody to the ICAM-1 adhesion site reduces neurological damage in a rabbit cerebral embolism stroke model. Exp. Neurol. 119, 215–229. 1993

Briscoe, D.M., Cotran, R.S. and Pober, J.S. Effects of tumor necrosis factor, lipopolysaccharide, and IL-4 on the expression of vascular cell adhesion molecule-1 in vivo: correlation with CD4+ T cell infiltration. J. Immunol. 149, 2954–2960. 1992

Broide, D.H., Sullivan, S., Gifford, T. and Sriramarao, P. Inhibition of pulmonary eosinophilia in P-selectin- and ICAM-1-deficient mice. Am. J. Respir. Cell Mol. Biol. 18, 218–225. 1998

Brown, K., Henson, P.M., Maclouf, J., Moyle, M., Ely, J.A. and Worthen, G.S. Neutrophil-platelet adhesion: Relative roles of platelet P-selectin and neutrophil β_2 (CD18) integrins. Am. J. Respir. Cell Mol. Biol. 18, 100–110. 1988

Bruehl, R.E., Moore, K.L., Lorant, D.E., Borregaard, N., Zimmerman, G.A., McEver, R.P. and Bainton, D.F. Leukocyte activation induces surface redistribution of P-selectin glycoprotein ligand-1. J. Leukoc. Biol. 61, 489–499. 1997

Bullard, D.C., Kunkel, E.J., Kubo, H., Hicks, M.J., Lorenzo, I., Doyle, N.A., Doerschuk, C.M., Ley, K. and Beaudet, A.L. Infectious susceptibility and severe deficiency of leukocyte rolling and recruitment in E-selectin and P-selectin double mutant mice. J. Exp. Med. 183, 2329–2336. 1996

Burns, J.A., Issekutz, T.B., Yagita, H. and Issekutz, A.C. The alpha 4 beta 1 (very late antigen (VLA)-4, CD49d/CD29) and alpha 5 beta 1 (VLA-5, CD49e/CD29) integrins mediate beta 2 (CD11/CD18) integrin-independent neutrophil recruitment to endotoxin-induced lung inflammation. J. Immunol. 166(7), 4644–4649. 2001

Carlos, T.M. and Harlan, J.M. Membrane proteins involved in phagocyte adherence to endothelium. Immunol Rev. 114, 5. 1990

Carlos, T.M., Kovach, N., Schwartz, B., Osborn, L., Rosa, M., Newman, B., Wayner, E.L., Lobb, R. and Harlan, J.M. Human monocytes bind to two cytokine-induced adhesive ligands on cultured human endothelial cells: Endothelial leukocyte adhesion molecule-1 and vascular cell adhesion molecule-1. Blood 77, 2266–2271. 1991

Chignier, E., Sparagano, M.H., McGregor, L., Thillier, A., Pellecchia, D. and McGregor, J.L. Two sites (23–30, 76–90) on rat P-selectin mediate thrombin activated platelet-neutrophil interactions. Comp. Biochem. Physiol. A. Physiol. 109, 881–886. 1994

Collins, C.E., Cahill, M.R., Newland, A.C. and Rampton, D.S. Platelets circulate in an activated state in inflammatory bowel disease. Gastroenterology 106, 840–845. 1994

Coyle, A.J., Page, C.P., Atkinson, L., Flanagan, R. and Metzger, W.J. The requirement fore platelets in allergen-induced late asthmatic airway obstruction. Am. Rev. Resp. Dis. 142, 587–593. 1990

Crocket-Torabi, E. Selectins and mechanisms of signal transduction. J. Leukoc. Biol. 63, 1–14. 1998

Crowley, C.A., Curnutte, J.T., Rosin, R.E., Andre-Schwartz, J., Gallin, J.I., Klempner, M., Snyderman, R., Southwick, F.S., Stossel, T.P. and Babior, B.M. An inherited abnormality of neutrophil adhesion: Its genetic transmission and its association with a missing protein. N. Engl. J. Med. 302, 1163–1168. 1980

Dang, B., Wiehler, S. and Patel, K.D. Increased PSGL-1 expression on granulocytes from allergic-asthmatic subjects results in enhanced leukocyte recruitment under flow conditions. J. Leukoc. Biol. 72, 702–710. 2002

Davenpeck, K.L., Sterbinsky, S.A. and Bochner, B.S. Rat neutrophils express $\alpha 4$ and $\beta 1$ integrins and bind to vascular cell adhesion molecule-1 (VCAM-1) and mucosal addressin cell adhesion molecule-1 (MAdCAM-1). Blood 7, 2341–2346. 1998

de Bruijne-Admiraal, L.G., Modderman, P.W., von dem Borne, A.E. and Sonnenberg, A. P-selectin mediates Ca^{2+}-dependent adhesion of activated platelets to many different types of leukocytes. Blood. 80, 134–142. 1992

de Fougerolles, A.R., Stacker, S.A., Schwarting, R. and Springer, T.A. Characterization of ICAM-2 and evidence for a third counter receptor for LFA-1. J. Exp. Med. 174, 253–267. 1991

Diacovo, T.G., de Fougerolles, A.R., Bainton, D.F. and Springer, T.A. A functional integrin ligand on the surface of platelets: Intercellular adhesion molecule-2. J. Clin. Invest. 94, 1243–1251. 1994

Diacovo, T.G., Puri, K.D., Warnock, R.A., Springer, T.A. and von Andrian, U.H. Platelet-mediated lymphocyte delivery to high endothelial venules. Science, 273 252–255. 1996a

Diacovo, T.G., Roth, S.J., Buccola, J.M., Bainton, D.F. and Springer, T.A. Neutrophil rolling, arrest and transmigration across activated, surface-adherent platelets via sequential action of P-selectin and the β_2-integrin CD11b/CD18. Blood 88, 146–157. 1996b

Diamond, M.S., Staunton, D.E., de Fougerolles, A.R., Stacker, S.A., Garcia-Aguilar, J., Hibbs, M.L. and Springer, T.A. ICAM-1 (CD54): A counter-receptor for Mac-1 (CD11b/CD18) J. Cell Biol. 111, 3129–3139. 1990

Diamond, M.S., Staunton, D.E., Marlin, S.D. and Springer, T.A. Binding of the integrin Mac-1 (CD11b/CD18) to the third immunoglobulin-like domain of ICAM-1 (CD54) and its regulation by glycosylation. Cell 65, 961–971. 1991

Doyle, N.A., Bhagwan, S.D., Meek, B.B., Kutkoski, G.J., Steeber, D.A., Tedder, T.F. and Doerschuk, C.M. Neutrophil margination, sequestration, and emigration in the lungs of L-selectin-deficient mice. J. Clin. Invest. 99, 526–533. 1997

Dustin, M.L., Rothlein, R., Bhan, A.K., Dinarello, C.A. and Springer, T.A. Induction by IL-1 and interferon-gamma: tissue distribution, biochemistry, and function of a natural adherence molecule. J. Immunol. 137, 245–254. 1986

Dustin, M.L. and Springer, T.A. T-cell receptor cross-linking transiently stimulates adhesiveness through LFA-1. Nature 341, 619–624. 1989

Elices, M.J., Osborn, L., Takada, Y., Crouse, C., Luhowskyj, S., Hemler, M.E. and Lobb, R.R. VCAM-1 on activated endothelium interacts with the leukocyte integrin VLA-4 at a site distinct from the VLA-4/fibronectin binding site. Cell 60, 577–584. 1990

Enlimomab Acute Stroke Trial Investigators. Use of anti-ICAM-1 therapy in ischemic stroke: results of the Enlimomab acute stroke trial. Neurology 57(8), 1428–1434. 2001

Etzioni, A., Harlan, J.M., Pollack, S., Phillips, L.M., Gershoni-Baruch, R. and Paulson, J.C. Leukocyte adhesion deficiency (lad) II: a new adhesion defect due to absence of sialyl lewis X, the ligand for selectins. 4, 307–308. 1993

Evangelista, V., Manarini, S., Rotondo, S., Martelli, N., Polischuk, R., McGregor, J.L., de Gaetano, G. and Cerletti, C. Platelet/polymorphonuclear leukocyte interaction in dynamic conditions: Evidence of adhesion cascade and cross talk between P-selectin and the β_2 integrin CD11b/CD18. Blood 88, 1–12. 1996

Furuya, K., Takeda, H., Azhar, S., McCarron, R.M., Chen, Y., Ruetzler, C.A., Wolcott, K.M., DeGraba, T.J., Rothlein, R., Hugli, T.E., del Zoppo, G.J. and Hallenbeck, J.M. Examination of several potential mechanisms for the negative outcome in a clinical stroke trial of enlimomab, a murine anti-human intercellular adhesion molecule-1 antibody. Stroke 32, 2665–2674. 2001

Frenette, P.S., Mayadas, T.N., Rayburn, H., Hynes, R.O. and Wagner, D.D. Susceptibility to infection and altered hematopoiesis in mice deficient in both P- and E-selectins. Cell 84, 563–574. 1996

Friedman, G., Jankpwski, S., Shalhla, M., Goldman, M., Rose, R.M., Kahn, R.J. and Vincent, J. I. Administration of an antibody to E-selectin in patients with septic shock. Crit. Care Med. 24(2), 229–233. 1996

Geng, J.G., Bevilacqua, M.P., Moore, K.L., McIntyre, T.M., Prescott, S.M., Kim, J.M., Bliss, G.A., Zimmerman, G.A. and McEver, R.P. Rapid neutrophil adhesion to activated endothelium mediated by GMP-140. Nature 343, 757–760. 1990

Giuffrè, L., Cordey, A-S., Monai, N., Tardy, Y., Schapira, M. and Spertini, O. Monocyte adhesion to activated aortic endothelium: Role of L-selectin and heparan sulfate proteoglycans. J. Cell Biol. 136, 945–956. 1997

Goetz, D.J., Greif, D.M., Ding, H., Camphausen, R.T., Howes, S., Comess, K.M., Snapp, K.R., Kansas, G.S. and Luscinskas, F.W. Isolated P-selectin glycoprotein ligand-1 dynamic adhesion to P- and E-selectin. J. Cell Biol. 137, 509–519. 1997

Gonzalo, J-A., Lloyd, C.M., Kremer, L., Finger, E., Martinez-A, C., Siegelman, M.H., Cybulsky, M. and Gutlerrez-Ramos, J-C. Eosinophil recruitment to the lung in a murine model of allergic inflammation: The role of T-cells, chemokines and adhesion receptors. J. Clin. Invest. 98, 2332–2345. 1996

Gundell, R.H., Wegner, C.D., Torcellini, C.A., Clarke, C.C., Haynes, N., Rothlein, R., Smith, C.W. and Letts, L.G. Endothelial leukocyte adhesion molecule-1 mediates antigen-induced acute airway inflammation and late-phase airway obstruction in monkeys. J. Clin. Invest. 88, 1407–1411. 1991

Hamann, A., Andrew, D.P., Jablonski-Westrich, D., Holzmann, B. and Butcher, E.C. Role of $\alpha 4$-integrins in lymphocyte homing to mucosal tissues in vivo. J. Immunol. 152, 3282–3292. 1994

Hamburger, S.A. and McEver, R.P. GMP-140 mediates adhesion of stimulated platelets to neutrophils. Blood 75, 550–554. 1990

Hattori, R., Hamilton, K.K., Fugate, R.D., McEver, R.P. and Sims, P.J. Stimulated secretion of von Willebrand factor is accompanied by rapid redistribution to the cell surface of the intracellular granule membrane protein GMP-140. J. Biol. Chem. 264, 7768–7771. 1989

He, X.Y., Xu, Z., Melrose, J., Mullowney, A., Vasquez, M., Queen, C., Vexler, V., Klingbeil, C., Co, M.S., Berg, E.L. Humanization and pharmacokinetics of a monoclonal antibody with specificity for both E- and P-selectin. J. Immunol.160(2), 1029–1035. 1998

Hemler, M.E., Huang, C., Takada, Y., Schwarz, L., Strominger, J.L. and Clabby, M.L. Characterisation of the cell surface heterodimer VLA-4 and related peptides. J. Biol. Chem. 262, 11478–11485. 1987

Hojo, M., Maghni, K., Issekutz, T.B. and Martin, J.G. Involvement of $\alpha 4$ integrins in allergic airway responses and mast cell degranulation in vivo. Am. J. Respir. Crit. Care Med. 158, 1127–1133. 1998

Hulett, M.D., Freeman, C., Hamdorf, B.J., Baker, R.T., Harris, M.J. and Parish, C.R. Cloning of mammalian heparanase, an important enzyme in tumor invasion and metastasis. Nat. Med. 5, 803–809. 1999

Issekutz, T.B., Miyasaka, M. and Issekutz, A.C. Rat blood neutrophils express very late antigen 4 and it mediates migration to arthritic joint and dermal inflammation. J. Exp. Med. 183, 2175–2184. 1996

Issekutz, A.C., Rowter, D. and Springer, T.A. Role of ICAM-1 and ICAM-2 and alternate CD11/CD18 ligands in neutrophil transendothelial migration. J. Leukoc. Biol. 65, 117–126. 1999

Johnston, B., Chee, A., Issekutz, T.B., Ugarova, T., Fox-Robichaud, A., Hickey, M.J. and Kubes, P. $\alpha 4$ integrin-dependent leukocyte recruitment does not require VCAM-1 in a chronic model of inflammation. J. Immunol. 164, 3337–3344. 2000

Johnston, G.I., Cook, R.G. and McEver, R.P. Cloning of GMP-140, a granule membrane protein of platelets and endothelium: Sequence similarity to proteins involved in cell adhesion and inflammation. Cell 56, 1033–1044. 1989

Johnston, B. and Kubes, P. The alpha-4 integrin: An alternative pathway for neutrophil recruitment? Immunol. Today 20, 545–550. 1999

Kahn, J., Ingraham, R.H., Shirley, F., Migaki, G.I. and Kishimoto, T.K. Membrane proximal cleavage of L-selectin: identification of the cleavage site and a 6-kD transmembrane peptide fragment of L-selectin. J. Cell Biol. 125, 461–470. 1994

Karsan, A., Cornejo, C.J., Winn, R.K., Schwartz, B.R., Way, W., Lannir, N., Gershoni-Baruch, R., Etzioni, A., Ochs, H.D. and Harlan, J.M. Leukocyte adhesion deficiency type II is a generalized defect of GDP-fucose biosynthesis: endothelial cell fucosylation is not required for neutrophil rolling on human nonlymphoid endothelium. 101, 2438–2445. 1998

Kavanaugh, A.F., Davis, L.S., Jain, R.I., Nichols, L.A., Norris, S.H. and Lipsky, P.E. A phase I/II open label study of the safety and efficacy of an anti-ICAM-1 (intercellular adhesion molecule-1; CD54) monoclonal antibody in early rheumatoid arthritis. J. Rheumatol. 23(8), 1338–1344. 1996

Kavanaugh, A.F., Davis, L.S., Nichols, L.A., Norris, S.H., Rothlein, R., Scharschmidt, L.A. and Lipsky, P.E. Treatment of refractory rheumatoid arthritis with a monoclonal antibody to intercellular adhesion molecule 1. Arthritis. Rheum. 37(7), 992–999. 1994

King, P.D., Sandberg, E.T., Selvakumar, A., Fang, P., Beaudet, A.L. and Dupont, B. Novel isoforms of murine intercellular adhesion molecule-1 generated by alternative RNA splicing. J. Immunol. 154, 6080–6093. 1995

Kogan, T.P., Dupre, B., Bui, H., McAbee, K.L., Kassir, J.M., Scott, I.L., Hu, X., Vanderslice, P., Beck, P.J. and Dixon, R.A. Novel synthetic inhibitors of selectin-mediated cell adhesion: synthesis of 1,6-bis[3-(3-carboxymethylphenyl)-4-(2-alpha-D-mannopyranosyloxy)phenyl]hexane (TBC1269). J. Med. Chem. 41(7), 1099–1111. 1998

Kumasaka, T., Quinlan, W.M., Doyle, N.A., Condon, T.P., Sligh, J., Takei, F., Beaudet, A.L., Bennett, C.F. and Doerschuk, C.M. Role of intercellular adhesion molecule-1 (ICAM-1) in endotoxin-induced pneumonia evaluated using ICAM-1 antisense oligonucleotides, anti-ICAM-1 monoclonal antibodies and ICAM-1 mutant mice. J. Clin. Invest. 97, 2362–2369. 1996

Kyan-Aung, U., Haskard, D.O., Poston, R.N., Thornhill, M.H. and Lee, T.H. Endothelial leukocyte adhesion molecule-1 and intercellular adhesion molecule-1 mediate the adhesion of eosinophils to endothelial cells in vitro and are expressed by endothelium in allergic cutaneous inflammation in vivo. J. Immunol. 146, 521–528. 1991

Labow, M.A., Norton, C.R., Rumberger, J.M., Lombard-Gillooly, K.M., Shuster, D.J., Hubbard, J., Bertko, R., Knaas, P.A., Terry, R.W. and Harbison, M.L. Characterization of E-selectin-deficient mice: demostration of overlapping function of the endothelial selectins. Immunity 1, 709–720. 1994

Larsen, E., Celi, A., Gilbert, G.E., Furie, B.C., Erban, J.K., Bonfanti, R., Wagner, D.D. and Furie, B. PADGEM protein: A receptor that mediates the interactions of activated platelets with neutrophils and monocytes. Cell 59, 305–312

Larson, R.S. and Springer, T.A. Structure and function of leukocyte integrins. Immunol. Rev. 114, 181–217. 1990

Laudanna, C., Constantin, G., Baron, P., Scarpini, E., Scarlato, G., Cabrini, G., Dechecchi, C., Rossi, F., Cassatella, M.A. and Berton, G. Sulfatides trigger increases of cytosolic free calcium and enhanced expression of tumour necrosis factor-α and IL-8 mRNA in human neutrophils. Evidence for a role of L-selectin as a signalling molecule. J. Biol. Chem. 269, 4021–4026. 1994

Lawrence, M.B. and Springer, T.A. Leukocytes roll on a selectin at physiologic flow rates: Distinction from and prerequisiste for adhesion through integrins. Cell 65, 859–873. 1991

Lellouch-Tubiana, A., Lefort, J., Simon, M.T., Pfister, A. and Vargaftig, B.B. Eosinophil recruitment into guinea-pig lungs after PAF-acether and allergen administration. Modulation by prostacyclins, platelet-depletion and selective antagonists. Am. Rev. Resp. Dis. 137, 948–959. 1990

Lever, R. and Page, C.P. Novel drug development opportunities for heparin. Nat. Rev. Drug Discov. 1(2), 40–48. 2002

Lewis, M.S., Whatley, R.E., Cain, P., McIntyre, T.M., Prescott, S.M. and Zimmerman, G.A. Hydrogen peroxide stimulates the synthesis of platelet-activating factor by endothelium and induces endothelial cell-dependent neutrophil adhesion. J Clin Invest. 82(6), 2045–2055. 1988

Ley, K., Gaehtgens, P., Fennie, C., Singer, M.S., Lasky, L.A. and Rosen, S.D. Lectin-like cell adhesion molecule 1 mediates leukocyte rolling in mesenteric venules in vivo. Blood 77, 2553–2555. 1991

Lim, Y.C., Snapp, K., Kansas, G.S., Camphausen, R., Ding, H. and Luscinskas, F.W. Important contributions of P-selectin glycoprotein ligand-1-mediated secondary capture to human monocyte adhesion to P-selectin, E-selectin and TNF-α-activated endothelium under flow in vitro. J. Immunol. 161, 2501–2508. 1998

Lin, K-C. and Castro, A.C. Very late antigen 4 (VLA4) antagonists as anti-inflammatory agents. Curr. Opinion Chem. Biol. 2, 453–457. 1998

Lin, K-C., Ateeq, H.S., Hsiung, S.H., Chong, L.T., Zimmerman, C.N., Castro, A., Lee, W-C., Hammond, C.E., Kalkunte, S., Chen, L-L., Pepinsky, R.B., Leone, D.R., Sprague, A.G., Abraham, W.M., Gill, A., Lobb, R.R. and Adams, S.P. Selective, tight binding inhibitors of integrin $\alpha 4\beta 1$ that inhibit allergic airway responses. J. Med. Chem. 42, 920–934. 1999

Lorant, D.E., McEver, R.P., McIntyre, T.M., Moore, K.L., Prescott, S.M. and Zimmerman, G.A. Activation of polymorphonuclear leukocytes reduces their adhesion to P-selectin and causes redistribution of ligands for P-selectin on their surfaces. J. Clin. Invest. 96, 171–182. 1995

Lorant, D.E., Patel, K.D., McIntyre, T.M. McEver, R.P., Prescott, S.M. and Zimmerman, G.A. Coexpression of GMP-140 and PAF by endothelium stimulated by histamine or thrombin: A juxtacrine system for adhesion and activation of neutrophils. J. Cell Biol. 115, 223–234. 1991

Lorenzon, P., Vecile, E., Nardon, E., Ferrero, E., Harlan, J.M., Tedesco, F. and Dobrina, A. Endothelial cell E- and P-selectin and vascular cell adhesion molecule-1 function as signalling receptors. J. Cell Biol. 142, 1381–1391. 1998

Lu, H., Smith, C.W., Perrard, J., Bullard, D., Tang, L., Shappell, S.B., Entman, M.L., Beaudet, A.L. and Ballantyne, C.M. LFA-1 is sufficient in mediating neutrophil emigration in Mac-1-deficient mice. J. Clin. Invest. 99, 1340–1350. 1997

Luscinskas, F.W., Brock, A.F., Arnaout, M.A. and Gimbrone Jr., M.A. Endothelial leukocyte adhesion molecule-1-dependent and leukocyte (CD11/CD18)-dependent mechanisms contribute to polymorphonuclear leukocyte adhesion to cytokine-activated human vascular endothelium. J. Immunol. 142, 2257–2263. 1989

Lynam, E.B., Rogelj, S., Edwards, B.S. and Sklar, L.A. Enhanced aggregation of human neutrophils by $MnCl_2$ or DTT differentiates the roles of L-selectin and β_2-integrins. J. Leukoc. Biol. 60, 356–364. 1996

Ma, X.L., Tsao, P.S. and Lefer, A.M. Antibody to CD18 exerts endothelial and cardiac protective effects in myocardial ischemia and reperfusion. J. Clin. Invest. 88, 1237–1243. 1991

Masinovsky, B., Urdal, D. and Gallatin, W.M. IL-4 acts synergistically with IL-1β to promote lymphocyte adhesion to microvascular endothelium by induction of vascular cell adhesion molecule-1. J. Immunol. 145, 2886–2895. 1990

McDowall, A., Inwald, D., Leitinger, B., Jones, A., Liesner, R., Klein, N. and Hogg, N. A novel form of integrin dysfunction involving β1, β2, and β3 integrins. J. Clin Invest. 111, 51–60. 2003

McEver, R.P., Beckstead, J.H., Moore, K.L., Marshall-Carlson, L. and Bainton, D.F. GMP-140, a platelet α-granule membrane protein, is also synthesised by vascular endothelial cells and is located in Weibel-Palade bodies. J. Clin. Invest. 84, 92–99. 1989

McEver, R.P., Moore, K.L. and Cummings, R.D. Leukocyte trafficking mediated by selectin-carbohydrate interactions. J. Biol. Chem. 19, 11025–11028. 1995

McNulty, C.A., Symin, F.A. and Wardlaw, A.J. Characterization of the integrin and activation steps mediating human eosinophil and neutrophil adhesion to chronically inflamed airway endothelium. Am. J. Respir. Cell Mol. Biol. 20, 1251–1259. 1999

Mehta, P., Patel, K.D., Laue, T.M., Erickson, H.P. and McEver, R.P. Soluble monomeric P-selectin containing only the lectin and epidermal growth factor domains binds to P-selectin glycoprotein ligand-1 on leukocytes. Blood 90, 2381–2389. 1997

Mocco, J., Choudhri, T., Huang, J., Harfeldt, E., Efros, L., Klingbeil, C., Vexler, V., Hall, W., Zhang, Y., Mack, W., Popilskis, S., Pinsky, D.J. and Connolly, E.S. HuEP5C7 as a humanized monoclonal anti-E/P-selectin neurovascular protective strategy in a blinded placebo-contralled trial of nonhuman primate stroke. Circ. Res. 91(10), 907–914. 2002

Moore, K.L. Structure and function of P-selectin glycoprotein ligand-1. Leuk. Lymphoma. 29, 1–15. 1998

Moore, K.L., Patel, K.D., Bruehl, R.E., Fugang, L., Johnson, D.A., Lichenstein, H.S., Cummings, R.D., Bainton, D.F. and McEver, R.P. P-selectin glycoprotein ligand-1 mediates rolling of human neutrophils on P-selectin. J. Cell Biol. 128, 661–671. 1995

Mulligan, M.S., Varani, J., Dame, M.K., Lane, C.L., Smith, C.W., Anderson, D.C. and Ward, P.A. Role of endothelial-leukocyte adhesion mlecule-1 (ELAM-1) in neutrophil-mediated lung injury in rats. J. Clin. Invest. 88, 1396–1406. 1991

Munos, F.M., Hawkins, E.P., Bullard, D.C., Beaudet, A.L. and Kaplan, S.L. Host defense against systemic infection with Streptococcus pneumoniae is impaired in E-, P-, and E-/P-selectin-deficient mice. J. Clin. Invest. 100, 2099–2106. 1997

Munro, J.M., Lo, S.K., Corless, C., Robertson, M.J., Lee, N.C., Barnhill, R.L., Weinberg, D.S. and Bevilacqua, M.P. Expression of sialyl-Lewis X, an E-selectin ligand, in inflammation, immune processes, and lymphoid tissues. Am. J. Pathol. 141, 1397–1408. 1992

Newman, P.J. Switched at birth: a new family for PECAM-1. J. Clin. Invest. 103, 5–9. 1999

Newman, P.J. The biology of PECAM-1. J. Clin. Invest. 99, 3–8. 1997

Ohno, N., Ichikawa, H., Coe, L., Kvietys, P.R., Granger, D.N. and Alexander, J.S. Soluble selectins and ICAM-1 modulate neutrophil-endothelial adhesion and diapedesis in vitro. Inflammation 21, 313–324. 1997

Osborn, L., Hession, C., Tizard, R., Vassalo, C., Luhowsky, S., Chi-Rosso, G. and Lobb, R. Direct expression cloning of vascular cell adhesion molecule 1 (VCAM-1), a cytokine-induced endothelial protein that binds to lymphocytes. Cell 59, 1203–1211. 1989

Patel, K.D. and McEver, R.P. Comparison of tethering and rolling of eosinophils and neutrophils through selectins and P-selectin glycoprotein ligand-1. J. Immunol. 159, 4555–4565. 1997

Patel, K.D., Zimmerman, G.A., Prescott, S.M., McEver, R.P. and McIntyre, T.M. Oxygen radicals induce human endothelial cells to express GMP-140 and bind neutrophils. J. Cell Biol. 112, 749–759. 1991

Perretti, M. Endogenous mediators that inhibit the leukocyte-endothelium interaction. Trends Pharmacol. Sci. 18, 418–425. 1997

Phillips, M.L., Schwartz, B.R., Etzioni, A., Bayer, R., Ochs, H.D., Paulson, J.C. and Harlan, J.M. Neutrophil adhesion in leukocyte adhesion deficiency syndrome type 2. J. Clin. Invest. 96, 2898–2906. 1995

Picker, L.J., Kishimoto, T.K., Smith, C.W., Warnock, R.A. and Butcher, E.C. ELAM-1 is an adhesion molecule for skin-homing T cells. Nature 349, 796–799. 1991a

Picker, L.J., Warnock, R.A., Burns, A.R., Doerschuk, C.M., Berg, E.L. and Butcher, E.C. The neutrophil selectin LECAM-1 presents carbohydrate ligands to the vascular selectins ELAM-1 and GMP-140. Cell 66, 921–933. 1991b

Prescott, S.M., Zimmerman, G.A. and McIntyre, T.M. Human endothelial cells in culture produce platelet-activating factor (1-alkyl-2-acetyl-sn-glycero-3-phosphocholine) when stimulated with thrombin. Proc. Natl. Acad. Sci. USA. 81, 3534–3538. 1984

Pulido, R., Elices, M.J., Campanero, M.R., Osborn, L., Schiffer, S., Garcia-Pardo, A., Lobb, R., Hemler, M.E. and Sanchez-Madrid. F. Functional evidence for three distinct and independently inhibitable adhesion activities mediated by the human integrin VLA-4. Correlation with distinct alpha 4 epitopes. J. Biol. Chem. 266(16),10241–10245. 1991

Radi, Z.A., Brogden, K.A., Dixon, R.A., Gallup, J.M. and Ackerman, M.R. A selectin inhibitor decreases neutrophil infiltration during acute Mannheimia haemolytica pneumonia. Vet. Pathol. 39(6), 697–705. 2002

Ramos, C.L., Smith, M.J., Snapp, K.R., Kansas, G.S., Stickney, G.W., Ley, K. and Lawrence, M.B. Functional characterization of L-selectin ligands on human neutrophils and leukemia cell lines: Evidence for mucin like ligand activity distinct from P-selectin glycoprotein ligand-1. Blood 91, 1067–1075. 1998

Rothlein, R., Dustin, M.L., Martin, S.D. and Springer, T.A. A human intercellular adhesion molecule (ICAM-1) distinct from LFA-1. J. Immunol. 137, 1270–1274. 1986

Rothlein, R., Czajkowski, M., O'Neill, M.M., Marlin, S.D., Mainolfi, E. and Merluzzi, V.J. Induction of intercellular adhesion molecule 1 on primary and continuous cell lines by pro-inflammatory cytokines: regulation by pharmacological agents and neutralizing antibodies. J. Immunol. 141, 1665–1669. 1988

Rusnak, J.M., Kopecky, S.L., Clements, I.P., Gibbons, R.J., Holland, A.E., Peterman, H.S., Martin, J.S., Saoud, J.B., Feldman, R.L., Breisblatt, W.M., Simons, M., Gessler, C.J. and Yu, A.S. An anti-CD11/CD18 monoclonal antibody in patients with acute myocardial infarction having percutaneous transluminal coronary angioplasty (the FESTIVAL study). Am. J. Cardiol. 88, 482–487. 2001

Sako, D., Chang, X-J., Barone, K.M., Vachino, G., White, H.M., Shaw, G., Veldman, G.M., Bean, K.M., Aherne, T.J., Furie, B., Cumming, D.A. and Larsen, G.R. Expression cloning of a functional glycoprotein ligand for P-selectin. Cell 75, 1179–1186. 1993

Salmela, K., Wramner, L., Ekberg, H., Hauser, I., Bentdal, O., Lins, L.E., Isoniemi, H., Backman, L., Persson, N., Neumayer, H.H., Jorgensen, P.F., Spieker, C., Hendry, B., Nicholls, A., Kirste, G. and Hasche, G. A randomized multicenter trial of the anti-

ICAM-1 monoclonal antibody (enlimomab) for the prevention of acute rejection and delayed onset of graft function in cadaveric renal transplantation: a report of the European Anti-ICAM-1 Renal Transplant Study Group. Transplantation. 67(5), 729–736. 1999

Schleiffenbaum, B., Spertini, O. and Tedder, T.F. Soluble L-selectin is present in human plasma at high levels and retains functional activity. J. Cell Biol. 119, 229–238. 1992

Shelley, C.S., da Silva, N., Georgakis, A., Chomienne, C. and Arnaout, M.A. Mapping of the human CD11c (ITGAX) and CD11d (ITGAD) genes demonstrates that they are arranged in tandem separated by no more than 11.5 kb. Genomics 49, 334–336. 1998

Sligh, J.E., Ballantyne, C.M., Rich, S.S., Hawkins, H.K., Smith, C.W., Bradley, A. and Beadet, A.L. Inflammatory and immune responses are impaired in mice deficient in intercellular adhesion molecule-1. Proc. Natl. Acad. Sci. USA. 90, 8529–8533. 1993

Smith, C.W., Rothlein, R., Hughes, B.J., Mariscalco, M.M., Rudloff, H.E., Schmalstieg, F.C. and Anderson, D.C. Recognition of an endothelial determinant for CD18-dependent human neutrophil adherence and transendothelial migration. J. Clin. Invest. 82, 1746–1756. 1988

Snapp, K.R., Ding, H., Atkins, K., Warnke, R., Luscinskas, F.W. and Kansas, G.S. A novel P-selectin glycoprotein ligand-1 monoclonal antibody recognizes an epitope within the tyrosine sulfate motif of human PSGL-1 and blocks recognition of both P- and L-selectin. Blood 91, 154–164. 1998

Spertini, O., Kansas, G.S., Munro, J.M., Griffin, J.D. and Tedder, T.F. Regulation of leukocyte migration by activation of the leukocyte adhesion molecule-1 (LAM-1) selectin. Nature 349, 691–694. 1991a

Spertini, O., Luscinskas, F.W., Kansas, G.S., Munro, J.M., Griffin, J.D., Gimbrone Jr., M.A. and Tedder, T.F. Leukocyte adhesion molecule-1 (LAM-1, L-selectin) interacts with an inducible endothelial cell ligand to support leukocyte adhesion. J. Immunol. 147, 2565–2573. 1991b

Stacker, S.A. and Springer, T.A. Leukocyte integrin p150,95 (CD11c/CD18) functions as an adhesion molecule binding to a counter-receptor on stimulated endothelium J. Immunol. 146, 648–655. 1991

Staunton, D.E., Dustin, M.L. and Springer, T.A. Functional cloning of ICAM-2, a cell adhesion ligand for LFA-1 homologous to ICAM-1. Nature. 339, 61–64. 1989

Sturla, L., Etzioni, A., Bisso, A., Zanardi, D., de Flora, G.K., Silengo, L., de Flora, A. and Tonetti, M. Defective intracellular activity of GDP-D-mannose-4,6-dehydratase on leukocyte adhesion deficiency type II syndrome. FEBS letters 429, 274–278. 1998

Sugama,Y., Tiruppathi, C., Janakidevi, K., Andersen, T.T., Fenton (II), J.W. and Malik, A.B. Thrombin-induced expression of endothelial P-selectin and intercellular adhesion molecule-1: A mechanism for stabilising neutrophil adhesion. J. Cell Biol. 119, 935–944. 1992

Tang, T., Frenette, P.S., Hynes, R.O., Wagner, D.D. and Mayadas, T.N. Cytokine-induced meningitis is dramatically attenuated in mice deficient in endothelial selectins. J. Clin. Invest. 97, 2485–2490. 1996

Tasaka, S., Richer, S.E., Mizgerd, J.P. and Doerschuk, C.M. Very late antigen-4 in CD18-independent neutrophil emigration during acute bacterial pneumonia in mice. Am. J. Respir. Crit. Care Med. 166(1):53–60. 2002

Teixeira, M.M. and Hellewell, P.G. Contribution of endothelial selectins and alpha-4 integrins to eosinophil trafficking in allergic and non-allergic inflammatory reactions in skin. J. Immunol. 161, 2516–2523. 1998

Thornhill, M.H. and Haskard, D.O. IL-4 regulates endothelial cell activation by IL-1, tumor necrosis factor, or IFN-γ. J. Immunol. 145, 865–872. 1990

Thornhill, M.H., Kyan-Aung, U. and Haskard, D.O. IL-4 increases human endothelial cell adhesiveness for T-cells but not for neutrophils. J. Immunol. 144, 3060–3065. 1990

Tyrrell, D.J., Horne, A.P., Holme, K.R., Preuss, J.M. and Page, C.P. Heparin in inflammation: potential therapeutic applications beyond anticoagulation. Adv. Pharmacol. 46, 151–208. 1999

Van Assche, G. and Rutgeerts, P. Antiadhesion molecule therapy in inflammatory bowel disease. Inflam. Bowel Dis. 8(4), 291–300. 2002

Vedder, N.B. and Harlan, J.M. Increased surface expression of CD11b/CD18 (Mac-1) is not required for stimulated neutrophil adhesion to cultured endothelium. J. Clin. Inv. 81, 676–682. 1988

Vestweber, D. and Blanks, J.E. Mechanisms that regulate the function of the selectins and their ligands. Physiol. Rev. 79, 181–213. 1999

Vlodavsky, I., Eldor, A., Haimovitz-Friedman, A., Matzner, Y., Ishai-Michaeli, R., Lider, O., Naparstek, Y., Cohen, I.R. and Fuks, Z. Expression of heparinase by platelets and circulating cells of the immune system: Possible involvment in diapedesis and extravasation. Invasion Metastasis 12, 112–127. 1992

Vourte, J., Lindsberg, P.J., Kaste, M., Meri, S., Jansson, S-E., Rothleinh, R. and Repo, H. Anti-ICAM-1 monoclonal antibody R6.5 (Enlimomab) promotes activation of neutrophils in whole blood. J. Immunol. 162, 2353–2357. 1999

Walsh, G.M., Mermod, J-J., Hartnell, A., Kay, A.B. and Wardlaw, A.J. Human eosinophil, but not neutrophil, adherence to IL-1-stimulated human umbilical vascular endothelial cells is $\alpha_4\beta_1$ (very late antigen-4) dependent. J. Immunol. 146, 3419–3423. 1991

Wang, S-Z., Smith, P.K., Lovejoy, M., Bowden, J.J., Alpers, J.H. and Forsyth, K.D. Shedding of L-selectin and PECAM-1 and upregulation of Mac-1 and ICAM-1 on neutrophils in RSV bronchiolitis. Am. J. Physiol. 275, L983-L989. 1998

Wardlaw, A.J. Eosinophil trafficking in asthma. Clin. Med. 1(3), 214–218. 2001

Watson., S.R., Fennie, C. and Lasky, L.A. Neutrophil influx into an inflammatory site inhibited by a soluble homing receptor-IgG chimaera. Nature 349, 164–167. 1991

Watson, M.L., Kingsmore, S.F., Johnston, G.I., Siegelman, M.H., Le Beau, M.M., Lemons, R.J., Bora, M.S., Howard, T.A., Weissman, I.L., McEver, R.P. and Seldin, M.F. Genomic organization of the selectin family of leukocyte adhesion molecules on human and mouse chromosome 1. J. Exp. Med. 172, 263–272. 1990

Weber, C., Kitayama, J. and Springer, T.A. Differential regulation of beta 1 and beta 2 integrin avidity by chemoattractants in eosinophils. Proc. Natl. Acad. Sci. USA. 93(20), 10939–10944. 1996

Weitz-Schmidt, G. Statins as anti-inflammatory agents. Trends Pharmacol. Sci. 23(10), 482–486. 2002

Weitz-Schmidt, G., Welzenbach, K., Brinkmann, V., Kamata, T., Kallen, J., Bruns, C., Cottens, S., Takada, Y. and Hommel, U. Statins selectively inhibit leukocyte function antigen-1 by binding to a novel regulatory integrin site. Nat. Med. 7(6), 687–692. 2001

Weyrich, A.S., Elstad, M.R., McEver, R.P., McIntyre, T.M., Moore, K.L., Morrissey, J.H., Prescott, S.M. and Zimmerman, G.A. Activated platelets signal chemokine synthesis by human monocytes. J. Clin. Invest. 97, 1525–1534. 1996

Weyrich, A.S., McIntyre, T.M., McEver, R.P., Prescott, S.M. and Zimmerman, G.A. Monocyte tethering by P-selectin regulates monocyte chemotactic protein-1 and tumor necrosis factor-α secretion. J. Clin. Invest. 95, 2297–2303. 1995

Wolyniec, W.W., de Sanctis, G.T., Nabozny, G., Torcellini, C., Haynes, N., Joetham, A., Gelfand, E.W., Drazen, J.M. and Noonan, T.C. Reduction of antigen-induced airway hyperreactivity and eosinophilia in ICAM-1-deficient mice. Am. J. Respir. Cell Mol. Biol. 18, 777–785. 1998

Zhang, R.L., Chopp, M., Jiang, N., Tang, W.X., Prostak, J., Manning, A.M. and Anderson, D.C. Anti-intercellular adhesion molecule-1 antibody reduces ischemic cell damage after transient but not permanent middle cerebral artery occlusion in the Wistar rat. Stroke 26, 1438–1443. 1995

Zhu, G.D., Arendson, D.L., Gunawardana, I.W., Boyd, S.A., Stewart, A.O., Fry, D.G., Cool, B.L., Kifle, L., Schaefer, V., Meuth, J., Marsh, K.C., Kempf-Grote, A.J., Kilgannon, P., Gallatin, W.M. and Okasinki, G.F. Selective inhibition of ICAM-1 and E-selectin expression in human endothelial cells. 2. Aryl modifications of 4-(aryloxy)thieno[2,3-c]pyridines with fine-tuning at C-2 carbamides. J. Med. Chem. 44(21), 3469–3487. 2001

Zimmerman, G.A., McIntyre, T.M. and Prescott, S.M. Thrombin stimulates the adherence of neutrophils to human endothelial cells in vitro. J. Clin. Invest. 76, 2235–2246. 1985

New Antiallergic Drugs

L. Dziadzio · W. Neaville · W. Busse

Department of Allergy and Immunology, University of Wisconsin, Highland Avenue, Madison, WI 53792, USA
e-mail: wwb@medicine.wisc.edu

1	Anti-IgE and Anti-CD23 as Therapy for Allergic Disease	274
2	DNA Vaccines as Therapy for Allergic Disease	278
2.1	ISS-ODN Vaccination	279
2.2	Plasmid Vaccination	280
2.3	ISS-ODN Plus Protein	281
3	Conclusions	282
	References	283

Abstract Atopic diseases have been increasing in prevalence over the last few years and are becoming common causes of chronic health problems for patients of all ages. Current pharmacological treatments are targeted for symptom relief (antihistamines, decongestants, β-adrenergic agents, etc.) or for reduction of allergic inflammation (corticosteroids, leukotriene modifiers, etc.). The development of new therapies to treat allergic disease is now progressing into new areas of immune modulation such as regulating the allergic inflammatory response at the level of IgE (anti-IgE and anti-CD23) and inducing immunologic tolerance by shifting from the T helper (Th)2 to Th1 phenotype (DNA vaccines). IgE plays a central role in the allergic inflammatory process; therefore, blocking the effects of IgE should be beneficial in the treatment of allergic diseases. At this time, a recombinant humanized monoclonal anti-IgE antibody has been synthesized that does not cause IgE receptor crosslinking on mast cells leading to degranulation or immune complex deposition. Studies have been done to evaluate the effectiveness of anti-IgE in the treatment of allergic rhinitis and allergic asthma. One of the most significant endpoints of benefit was the ability to reduce the corticosteroid dose for moderate-to-severe allergic asthmatics. Another possible approach to reduce the effect of IgE in the inflammatory process would be to block CD23 (low-affinity IgE receptor, FcεRII), which plays a role in IgE synthesis and triggers the release of inflammatory cytokines. Little is known about the effectiveness of anti-CD23, but preliminary studies suggest it is safe. The use of immunostimulatory DNA to provoke a Th1 phenotype has been an impetus for the development of DNA vaccines in the treatment of allergic disease. Various combinations of plasmid DNA, immunostimulatory oligodeoxynucleotide (ISS-ODN), and proteins have been studied in murine models to evaluate the effectiveness of DNA vaccination. The success in skewing the immune response towards a Th1 phenotype in mice still needs to be evaluated in humans. The use of DNA vaccination as a treatment for allergic disease remains a viable option for the future; however, at this time further evaluation is needed including concerns regarding the possible unmasking or development of Th1 diseases (lupus, rheumatoid arthritis, inflammatory bowel disease).

Keywords Anti-IgE · Anti-CD23 · DNA vaccine · Plasmid DNA · Immunostimulatory oligodeoxynucleotide (ISS-ODN) · CpG motifs

The development of new therapies to treat allergic disease, including asthma, rhinitis, and food allergy, has progressed into new areas of immune modulation. In this chapter, we will examine three new therapeutic approaches to allergic disease. We will first examine anti-IgE, which has been shown in human trials to be particularly helpful for asthmatics through its corticosteroid-sparing effects. We will also discuss anti-CD23 and DNA vaccines, which have been examined in murine models.

1
Anti-IgE and Anti-CD23 as Therapy for Allergic Disease

Atopic diseases such as allergic rhinitis, asthma, and atopic dermatitis have had increasing prevalences in recent years, especially in western societies. In fact, allergic diseases are a most common cause of chronic health problems for patients of all ages. The cornerstone of atopic disease is the underlying allergic reaction and subsequent inflammation, which may affect a variety of target organs including the skin, eyes, nasal mucosa, and airways. Even though the mechanisms of specific target organ involvement remain unknown, much has been learned about the pathophysiology of the allergic inflammatory process in recent years. Allergic inflammation includes an immediate hypersensitivity reaction which follows allergen exposure and, in certain circumstances, a late-phase reaction peaking 6–9 h after exposure. IgE plays a central role in the onset of immediate hypersensitivity reactions and has a less well-established role in the late-phase reactions. Mast cell degranulation with release of vasoactive mediators such as histamine, tryptase, prostaglandins, and cysteinyl leukotrienes is an essential component in immediate reactions. Allergen-specific IgE is bound to mast cells via the high-affinity IgE receptor (FcεRI). Once the bound IgE is crosslinked by the specific allergen, mast cell degranulation occurs, and the inflammatory process is initiated. IgE is also bound to other cell types including activated B cells, macrophages and monocytes via the low-affinity IgE receptor (FcεRII or CD23). CD23 is thought to be involved in the allergic inflammatory process by possibly triggering the release of monocyte cytokines and also by the regulation of IgE synthesis. When considering possibilities for new antiallergic drugs, it is critical to keep the underlying inflammatory process, which is a target for therapy, in mind. Blocking IgE and its effect in inflammation is an obvious target for new medications. Current investigations are evaluating the effectiveness of anti-IgE and anti-CD23 molecules for the treatment of allergic diseases.

Considering that IgE plays an essential role in allergic inflammation and serum IgE levels have been correlated with airway hyperresponsiveness (Sunyer et al. 1995; Burrows et al. 1989), blocking the effects of IgE could potentially be useful and effective in treating atopic diseases such as allergic asthma. One approach to accomplish this goal is with the use of a monoclonal antibody to IgE. An important aspect of any new medication is safety. When IgE bound to mast cells via the FcεRI becomes crosslinked, mast cell degranulation occurs, and the immediate hypersensitivity reaction is initiated. To prevent crosslinking induced by the anti-IgE antibody, the anti-IgE molecule was designed to bind to free IgE. To accomplish this, a murine anti-IgE antibody was developed which binds the same portion of the IgE molecule that attaches to the FcεRI (Chang et al. 1990; Presta et al. 1994). Due to the nonhuman nature of the anti-IgE antibody, another safety issue included the induction of immune responses in human patients (Shawler et al. 1985). To avoid the induction of these immune responses, the antibody was humanized by grafting the antigen-binding loop of the murine anti-IgE onto human IgG1, resulting in a recombinant humanized monoclonal anti-IgE antibody (Presta et al. 1993) (Fig. 1). Considering that the anti-IgE antibody binds to free IgE, another safety issue to consider is the possibility of immune complex deposition. It has been shown that anti-IgE binds with IgE to form complexes of tetramers to hexamers with the largest complex approximating 1×10^6 Da, which are too small to fix complement or cause immune complex reactions (Fox et al. 1996).

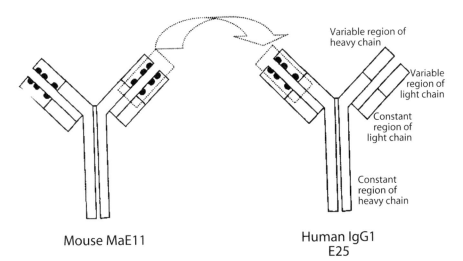

Fig. 1 Mouse monoclonal anti-IgE was humanized by exchanging most of the molecule with human IgG1 in order to reduce the risk of immune response to foreign protein. (From Lotvall et al. 1999)

Because IgE plays a central role in the allergic inflammatory process, the prospect of using an anti-IgE molecule for therapy has brought about much interest. Several studies have evaluated the effectiveness of using a recombinant humanized monoclonal antibody to IgE (rhuMAb-E25) for the treatment of allergic diseases. Currently, most studies have looked at atopic diseases of the airway such as allergic rhinitis and allergic asthma.

One study evaluated patients with ragweed-induced allergic rhinitis. Anti-IgE lowered free IgE levels in a dose-dependent fashion and did not induce antibodies to the anti-IgE molecule or cause any severe adverse reactions. In this study, there were no statistical differences in the symptom scores between the treatment and the placebo groups (Casale et al. 1997). Another study evaluated patients with birch-pollen-induced allergic rhinitis. Free IgE levels were significantly decreased, and there was no immune complex formation or other serious adverse reactions. In this study, there was, however, clinical benefit for the treatment group with significantly lower nasal symptom scores, lower numbers of rescue antihistamine use, lower proportion of any rhinitis medication use and higher scores on the quality-of-life surveys (Adelroth et al. 2000).

Studies have also evaluated the use of an anti-IgE molecule for the treatment of allergic asthma. Allergic asthmatic subjects treated with rhuMAb-E25 were compared to those receiving a placebo with endpoints of effect being the allergen PC15 [provocation concentration of allergen causing a 15% decrease in forced expiratory volume in 1 s (FEV_1)] and the methacholine PC20. For the anti-IgE treatment group, the allergen PC15 dose increased and the methacholine PC20 improved. There was no change in the methacholine PC20 in the placebo group (Boulet et al. 1997). Another study also found that allergic asthmatic subjects treated with rhuMAb-E25 required higher doses of allergen to cause an early-phase response and had a decrease in the severity of the fall of the FEV_1 in both the early and late phases compared to those patients treated with placebo (Fahy et al. 1997). A clinical trial evaluating rhuMAb-E25 (high and low dose) for the treatment of moderate to severe allergic asthma showed improvements in outcome measures such as symptom scores, use of β-agonists, peak expiratory flow rate, quality-of-life scores and dose of inhaled and oral steroids when compared to a placebo group. One of the most significant endpoints of benefit was the ability to reduce the corticosteroid dose in the anti-IgE treatment group. Oral corticosteroid use was reduced by 50% in the high-dose group, 65% in the low-dose group, and 0% in the placebo group. The inhaled corticosteroid dose was reduced by at least 50% in 51% of the high-dose group, 49% of the low-dose group, but only 38% in the placebo group (Fig. 2) (Milgrom et al. 1999).

Another possible approach to modify the effect of IgE in the allergic inflammatory process would be to block the low-affinity IgE receptor (CD23 or FcεRII). A monoclonal antibody against CD23 has been developed. Although experience with anti-CD23 is limited, murine models of asthma have shown decreased lung eosinophils and airway hyperresponsiveness following administration of anti-CD23 (Cernadas et al. 1999; Dasic et al. 1999; Haczku et al. 2000). One investigation, currently in press (Rossenwasser et al., in press), was de-

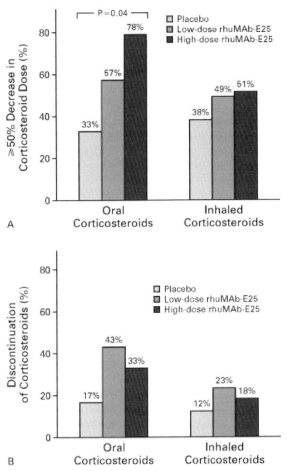

Fig. 2 Milgrom et al. (1999) found that daily oral corticosteroid dose was reduced by at least 50% in a significantly higher percentage of patients receiving anti-IgE compared to those receiving placebo (**A**). The percentage of patients in each group able to discontinue corticosteroids is presented in **B**

signed to evaluate the safety, pharmacokinetic profile and clinical activity of an anti-CD23 antibody in patients with mild to moderate persistent asthma in a single-dose, dose-escalating, placebo-controlled study. In this study, the anti-CD23 molecule was well tolerated with no serious adverse reactions noted. The C_{max} was proportional to the dose administered, and the serum half-life ranged from 2 to 10 h, depending on the dose given. There was also a dose-dependent decrease in mean IgE levels with administration of anti-CD23. There was no change in FEV_1, and there were no improvements in asthma or rhinitis symptom-free days, number of nighttime awakenings, or use of β-agonists. This study was not designed to evaluate efficacy. Nonetheless, considering the dose-dependent decrease in IgE, there may be improved clinical improvement noted with multiple doses of anti-CD23 or in the treatment of more severe allergic asthmatic subjects.

The use of anti-IgE and anti-CD23 to modulate IgE and its effects in the allergic inflammatory process is being evaluated as new antiallergic medications. Anti-IgE has been shown to be safe with no serious adverse effects and to have corticosteroid-sparing effects when used for allergic asthmatics. Work still needs to be completed to determine the appropriate positioning of this new medication in the management and treatment of atopic patients. Anti-CD23 has been studied in allergic diseases, with preliminary observations suggesting this approach is also safe. Further studies, however, are needed to evaluate the efficacy of this approach for atopic diseases. The possibility of using these new immune-modulating medications has begun to focus on the underlying allergic inflammation and is a very exciting approach which may offer new therapies not only for allergic rhinitis and asthma but also for other atopic diseases such as food allergies and stinging insect allergy.

2
DNA Vaccines as Therapy for Allergic Disease

The concept of DNA vaccines as treatment for allergic diseases has its origins both in conventional immunotherapy and the "hygiene hypothesis" of atopic diseases. Immunotherapy has been used as treatment for allergic diseases for over a century including treatment for allergic rhinitis and stinging insect allergy and less successfully for allergic asthma. Immunotherapy for food allergy, in fact, is contraindicated, as it has been associated with life-threatening anaphylactic reactions. Moreover, its success in the treatment of seasonal rhinitis has been tempered by the risk of life-threatening anaphylaxis. Because of the potential risk involved in conventional immunotherapy, coupled with the limited success of immunotherapy in allergic asthma and food allergy, therapy that may result in immunologic tolerance and a shift from the T helper (Th)2 to Th1 phenotype would be of potential importance.

In addition, the "hygiene hypothesis," although still theoretical, purports that the increasing rates of allergic disease in the developed world result, in part, from the improved sanitation and increasing immunizations to reduce childhood illnesses. In westernized countries, the incidence of atopic disorders is higher than in other areas of the world, for example, in former West Germany compared with East Germany (von Mutius et al. 1994) and Sweden compared with the Baltic states (Braback et al. 1994). It has also been shown that in subjects with a positive purified protein derivative (PPD) skin test (Sirakawa et al. 1997), as well as those who have had measles (Shaheen et al. 1996) and hepatitis A (Matricardi et al. 1997), there was a decreased incidence of atopic disease. In addition, children with frequent viral upper respiratory tract infections early in life had a reduced risk of asthma through school age (Illi et al. 2001). Other studies, however, have found contradictory results. For example, Paunio et al.

(2001) reported higher incidence of atopic disease in children who had been infected with measles, compared with their uninfected (vaccinated) counterparts.

The use of immunostimulatory DNA to provoke a Th1 phenotype has been an impetus for the development of DNA vaccines. Vaccine development has included the use of DNA plasmids that can induce both cellular and humoral immunity to viruses, parasites, bacteria, and tumors. These vaccines consist of a plasmid DNA backbone containing an antigen-encoding gene. In addition to the use of DNA vaccines in infectious disease prevention, there is also evidence for their use as treatment for allergic disease. Plasmid DNA has been shown to induce a Th1 immune response in mice, compared with those injected with alum (Raz et al. 1996). In mice treated with the plasmid, there was induction of IgG2a and increased interferon (IFN)-γ, compared with the alum-treated mice who had increased IgG1, IgE, interleukin (IL)-4 and IL-5.

In addition to plasmid DNA evoking a Th1 response, other studies have shown that immunostimulatory DNA sequences increase plasmid vaccine immunogenicity. These CpG motifs (also known as immunostimulatory oligodeoxynucleotide, ISS-ODN) consist of unmethylated CpG dinucleotides flanked by two 5' purines and two 3' pyrimidines. CpG repeats are more common in microbial DNA compared with mammalian. When found in mammalian DNA, these repeats are more frequently methylated, rendering them less immunogenic. The unmethylated CpG motifs promote the development of a Th1-like response eliciting several cytokines including IL-12, IL-18, IFN-γ (Klinman et al. 1996), IFN-α, IFN-β, IL-6 and IL-10. Costimulatory molecules are also induced by CpG motifs, as are natural killer (NK) cells and antigen-presenting cells. B cell activation, proliferation, and antibody secretion are also enhanced by these motifs (Krieg et al. 1995; Liang et al. 1996).

Various combinations of plasmid DNA, ISS-ODN, and proteins have been studied to evaluate the effectiveness of DNA vaccination. The combinations include: plasmid alone, ISS-ODN alone, and a combination of protein allergen plus ISS-ODN (either mixed together or covalently linked).

2.1
ISS-ODN Vaccination

Plasmids containing ISS-ODN, as well as naked ISS-ODN alone, have been examined as potential nonspecific therapies to induce a Th1 response. B cell production of IL-6 and IL-12 is enhanced in mice by the addition of ISS-ODN (Klinman et al. 1996). In addition, the same mice had increased CD4 T cell production of IL-6 and IFN-γ, in vivo and in vitro, along with increased NK cell production of IFN-γ. This initial study prompted further evaluation of CpG repeats on the immune response.

To define the effect of ISS-ODN on human cells, cultured peripheral blood mononuclear cells (PBMCs) from both allergic and nonallergic individuals were incubated with the immunostimulatory sequences (Bohle et al. 1999). Compared with PBMCs from nonallergic subjects, mononuclear cells from allergic patients produced more IFN-γ in the presence of ISS-ODN. NK cells appeared to be the main source of this IFN-γ. In addition, mRNA encoding IL-12 and IL-18 was increased in PBMCs treated with ISS-ODN. Also, in PBMCs from allergic subjects, IgG and IgM to allergens to which the subjects were allergic increased after treatment with ISS-ODN, while specific IgE production was unchanged, but total IgE was reduced. Other cytokines induced by ISS-ODN include tumor necrosis factor (TNF)-α and IL-16 (Sparwasser et al. 1998). Allergen-induced airway responsiveness was decreased in mice pretreated with ISS-ODN, along with decreased eosinophilia in the airway, lung parenchyma and blood (Broide et al. 1998). In this study, mice treated with ISS-ODN and mice treated with corticosteroids both had decreased IL-5 production, but only those treated with ISS-ODN had induction of IFN-γ that was allergen (ragweed) specific.

Immunostimulatory DNA sequences also potentiate the expression of costimulatory molecules, including B7-2 and CD40 on macrophages and dendritic cells (Martin-Orozco et al 1998; Sparwasser et al 1998), B7-1 on macrophages (Martin-Orozco et al. 1998; Sparwasser et al. 1998), major histocompatibility complex (MHC) I and II molecules, and IFN-γ receptor and IL-2 receptors as well.

The skewing of immunity to a Th1 phenotype by ISS-ODN both for cytokine and immunoglobulin profiles was also shown to "pre-prime" the immune system prior to allergen sensitization (Kobayashi et al. 1999). In this study, mice were immunized with ISS-ODN intradermally and then received an injection of β-galactosidase (β-gal) up to 28 days after the ISS-ODN immunization. Mice immunized intradermally with ISS-ODN for up to 2 weeks before β-gal injection had increased anti-β-gal IgG2a, IFN-γ production and cytotoxic T cell activity against β-gal. In another mouse model, an asthma-like phenotype was prevented along with decreased antigen-specific IgE and increased IFN-γ/IL-4 ratio in mice pretreated with ISS-ODN (Sur et al. 1998). The effects noted were present for 6 weeks after treatment. These data suggest that ISS-ODN can induce a Th1 phenotype prior to allergen exposure.

It appears that even without the presence of allergen, CpG motifs can induce a Th1 phenotype in multiple cell types including B cells, antigen-presenting cells (macrophages, dendritic cells), T cells, and NK cells. The expression of Th1 cytokines along with an upregulation of costimulatory molecules on these cells underscores the importance of ISS-ODN in Th1 and innate immune responses. The persistence of a Th1 response after antigen challenge in sensitized mice is encouraging as potential therapy for allergic disease.

2.2
Plasmid Vaccination

In addition to vaccination with ISS-ODN, other methods have been evaluated to modulate the immune system. Inoculation with plasmid vectors containing allergen-encoding genes of interest is another potential therapy in the DNA vaccine armamentarium. In this approach to therapy, DNA sequences specific for allergens are integrated into plasmids. This DNA is then translated into protein within cells and presented on class I MHC. Additionally, ISS-ODN may also be encoded on the plasmid, leading to further immunomodulation.

In one study of this type of therapy, the DNA sequence encoding the latex allergen, Hev b 5, was cloned into a plasmid and then injected into mice. Mice injected with the plasmid containing sense cDNA developed IgM, IgG1, and IgG2 to Hev b 5 (Slater et al. 1998). The elevated IgG responses continued beyond 100 days. Splenocytes from mice immunized with the sense cDNA proliferated in the presence of Hev b 5/maltose-binding protein (MBP). IFN-γ was induced from the same splenocytes in the presence of Hev b 5/MBP, and IL-4 secretion was diminished. In addition, Hev b 5 RNA was detected at the injection sites, lymph nodes, and spleens of mice injected with the plasmid. Control mice injected with antisense Hev b 5 cDNA did not show any of these effects.

Mice were also immunized with plasmid DNA encoding the five classes of murine T cell epitopes on Der p 1 and Der p 2 three times each, 1 week apart (Kwon and Yoo 2001). The mice were then sensitized to Der p extracts and then challenged intranasally. The vaccinated mice had significantly reduced Der p-specific IgE, compared with controls vaccinated with empty plasmid. Vaccinated mice also showed an enhanced IgG2a response, along with increased mRNA expression of IFN-γ. Histologically, immunized mice had a reduction in lung inflammation when compared with controls.

In mice injected with a plasmid containing two isoforms of the birch allergen Bet V 1 (low and high allergenicity), with CpG motifs appended to the plasmid DNA or co-injected with the plasmid, a Th1-type response was seen with an increased IgG1/IgG2a ratio, and little IgE produced (Hartl et al. 1999). The mice treated with the highly allergenic isoform, however, showed increased allergen-specific proliferation and an increased IFN-γ and decreased IL-4 production.

Roy et al. (1999) also examined the effects of a mixture of ISS-ODN plus DNA coding for protein allergen on sensitized mice. Their novel approach to peanut desensitization involved oral administration of DNA nanoparticles (plasmid DNA containing ISS-ODN and DNA encoding the peanut allergen, Ara h 2) complexed with chitosan, a natural polysaccharide. In these mice, there was a reduction in allergen-induced anaphylaxis compared with controls, along with decreased antigen-specific IgE, plasma histamine, and vascular leakage.

2.3
ISS-ODN Plus Protein

The combination of immunostimulatory DNA sequences plus allergenic protein has also been studied. ISS-ODN may be combined in a mixture with the protein, or it may be covalently linked to the protein. The molecular processes are more involved than those involved with DNA vaccines alone. The processing of protein allergen involves uptake by antigen-presenting cells and presentation on class II MHC molecules, whereas DNA is first transfected into cells, then translated into proteins that are presented on class I MHC molecules. Multiple allergens have been studied in conjunction with CpG repeats as potential targets for DNA vaccines.

In mice vaccinated with a mixture of cedar pollen protein allergens (Cry j 1 and Cry J 2) and ISS-ODN, allergen-specific IgG2a titers were increased, with a concomitant decrease in IgG1a and IgE (Kohama et al. 1999). Cytokine changes were also seen with treated mice showing increased IFN-γ secretion and decreased IL-4 release compared with control mice injected with allergens/mutant ODN.

Horner et al. (2000) found that mice vaccinated with plasmid containing β-gal cDNA protected 5/15 (33%) animals from death due to anaphylactic challenge with β-gal after sensitization to β-gal, in conjunction with alum and pertussis. In the same study, mice were also immunized with β-gal protein plus ISS-ODN. Compared with mice immunized with plasmid alone, these mice had 42% survival after anaphylactic challenge and had reduced postchallenge plasma histamine levels. In addition, Tighe et al. (2000) conjugated two antigens (*Escherichia coli* β-gal or HIV gp120) to ISS-ODN. Splenocytes from mice immunized with the β-gal conjugate were stimulated in vitro and produced significantly higher levels of IFN-γ than those immunized with other combination vaccines, including β-gal mixed with ISS-ODN, β-gal alone, and β-gal conjugated to a non-stimulatory DNA sequence. Similar results were seen in mice immunized with the gp-120-ISS-ODN conjugate. Mice immunized with the protein–DNA conjugates were also shown to have increased antigen-specific IgG2a levels.

In mice sensitized with Amb a 1, Santeliz et al. (2002) administered Amb a 1, only Amb a 1-ISS, or Amb a 1-Non-ISS intradermally. The mice treated with Amb a 1-ISS had less ragweed-induced airway hyperresponsiveness than those treated with Amb a 1 and Amb a 1-Non-ISS. However, these mice had more significant hyperresponsiveness than those sensitized and treated with phosphate-buffered solution (PBS). Interestingly, mice treated with Amb a 1-ISS had increased Th1 and Th2 immunoglobulins. Both IgG1 and IgG2a were significantly increased in this group. When stimulated with Amb a 1, splenocytes from the Amb a 1-ISS group had increased IFN-γ compared with all controls, and had increased IL-5 compared with PBS-treated mice, but not other groups. These data suggest that ISS-ODN treatment after sensitization may help modify airway hyperresponsiveness, but altogether upregulate both Th1 and Th2 arms of the immune response.

3
Conclusions

The use of DNA vaccination as treatment for allergic disease, particularly asthma and food allergy is a viable option, although further human studies are required. Figure 3 reviews the potential therapeutic modalities and results associated with them. The success in skewing the immune response towards a Th1 phenotype in mice using immunostimulatory DNA will need to be evaluated in humans. In

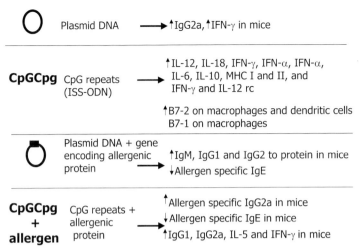

Fig. 3 The immunologic effects of multiple modes (plasmid alone, CpG repeats alone, plasmid with allergenic gene inserted, and CpG repeats with allergenic protein) are reviewed. Overall there is a shift to a Th1-type response, but some changes upregulate the Th2 arm as well

addition, concerns regarding unmasking or development of diseases considered Th1 directed (lupus, rheumatoid arthritis, inflammatory bowel disease) must be addressed.

References

Adelroth E, Rak S, Haahtela T, et al (2000) Recombinant humanized mAb-25, an anti-IgE mAb, in birch-induced seasonal allergic rhinitis. J Allergy Clin Immunol 106:253–259

Bohle B, Jahn-Schmid B, Maurer D, Kraft D, Ebnern C (1999) Oligonucleotides containing CpG motifs induce IL-12, IL-18 and IFN-γ production in cells from allergic individuals and inhibit IgE synthesis in vitro. Eur J Immunol 29:2344–2353

Boulet LP, Chapman KR, Cote J, et al (1997) Inhibitory effect of an anti-IgE antibody E25 on allergen-induced early asthmatic response. Am J Resp Crit Care Med 155:1835–1840

Braback L, Breborowicz A, Dreborg S, Knutsson A, Pieklik H, Bjorksten B (1994) Atopic sensitization and respiratory symptoms among Polish and Swedish school children. Clin Exp Allergy 24(9):826–35

Broide D, Schwarze J, Tighe H, Gifford T, Nguyen M-D, Malek S, Van Uden J, Martin-Orozco E, Gelfand, Raz E (1998) Immunostimulatory DNA sequences inhibit IL-5, eosinophilic inflammation, and airway hyperresponsiveness in mice. J Immunol 161:7054–7062

Burrows B, Martinez FD, Halonen M, Barbee RA, Cline MG (1989) Association of asthma with serum IgE levels and shin test reactivity to allergens. New Eng J Med 320:271–277

Casale TB, Bernstein IL, Busse WW et al (1997) Use of an anti-IgE humanized monoclonal antibody in ragweed-induced allergic rhinitis. J Allergy Clin Immunol 100:110–121

Cernadas M, De Sanctis GT, Krinzman SJ, Mark DA, Donovon CE, Listman JA, et al (1999) CD23 and allergic pulmonary inflammation: Potential role as an inhibitor. Am J Resp Cell Mol Biol 20:1–8

Chang TW, Davis FM, Sun NC, Sun CR, MacGlashan DW Jr (1990) Hamilton RG. Monoclonal antibodies specific for human IgE-producing B-cells, a potential therapeutic for IgE-mediated allergic diseases. Biotechnology 8:122–126

Dasic G, Juillard P, Graber P, Herren S, Angell T, Knowles R, et al (1999) Critical role of CD23 in allergen-induced bronchoconstriction in a murine model of allergic asthma. Eur J Immunol 29(9): 2957–67

Fahy JV, Fleming HE, Hofer HW, et al (1997) The effect of an anti-IgE monoclonal antibody on the early- and late-phase responses to allergen inhalation in asthmatic subjects. Am J Respir Crit Care Med 155:1828–1834

Fox JA, Hotaling TE, Struble C, Ruppel J, Bates DJ, Schoonhoff MB (1996) Tissue distribution and complex formation with IgE of an anti-IgE antibody after intravenous administration in cynomolgus monkeys. J Pharmacol Exp Ther 279:1000–1008

Haczku A, Takeda K, Hamelmann E, Loader J, Joetham A, Redai I, et al (2000) CD23 exhibits negative regulatory effects on allergic sensitization and airway hyperresponsiveness. Am J Resp Crit Care Med 161(3 pt 1):952–60

Hartl A, Kiesslich J, Weiss R, Bernhaupt A, Mostbock S, Scheiblhofer S, Flockner H, Sippl M, Ebner C, Ferreira F, Thalhamer J (1999) Isoforms of the major allergen of birch pollen induce different immune responses after genetic immunization. Int Arch Allergy Immunol 120:17–29

Horner AA, Nguyen M-D, Ronaghy A, Cinman BA, Verbeek S, Raz E (2000) DNA-based vaccination reduces the risk of lethal anaphylactic hypersensitivity in mice. J Allergy Clin Immunol 106:349–356

Illi S, von Mutius E, Lau S, Bergmann R, Niggemann B, Sommerfeld C, Wahn U, and the MAS Group (2001) Early childhood infectious diseases and the development of asthma up to school age: a birth cohort study. BMJ 322:390–395

Klinman DM, Yi A, Beaucage SL, Conover J, Krieg AM (1996) CpG motifs expressed by bacterial DNA rapidly induce lymphocytes to secrete IL-6, IL-12 and IFN-γ. Proc Natl Acad Sci 93:2879–2883

Kobayashi H, Horner AA, Takabayashi K, Nguyen M-D, Huang E, Cinman N, Raz E (1999) Immunostimulatory DNA prepriming: A novel approach for prolonged Th1-biased immunity. Cell Immunol 198:69–75

Kohama Y, Akizuki O, Hagihara K, Yamada E, Yamamoto H (1999) Immunostimulatory oligodeoxynucelotide induces Th1 immune response and inhibition of IgE antibody production to cedar pollen allergens in mice. J Allergy Clin Immunol 104:1231–1238

Krieg AM, Yi A-K, Matson S, Waldschmidt TJ, Bishop GA, Teasdale R, Koretzky GA, Klinman DM (1995) CpG motifs in bacterial DNA trigger direct B-cell activation. Nature 374:546–549

Kwon SS, Kim N, Yoo TJ (2001) The effect of vaccination with DNA encoding murine T-cell epitopes on the Der p 1 and 2 induced immunoglobulin E synthesis. Allergy 56:741–478

Lotvall J, Pullerits T (1999) Treating asthma with anti-IgE or anti-IL5. Current Pharmaceutical Design 5:757–770

Liang H, Nishioka Y, Reich CF, Pisetsky DS, Lipsky PE (1996) Activation of human B cells by phosphorothioate oligodeoxynucleotides. J Clin Invest 98:1119–1129

Martin-Orozco E, Kobayashi H, Van Uden J, Nguyen M-D, Kornbluth RS, Raz E (1999) Enhancement of antigen-presenting cell surface moleceules involved in cognate interactions by immunostimulatory DNA sequences. Int Immunol 11:1111–1118

Matricardi PM, Rosmimi F, Riondino S, Fortini M, Ferrigno L, Rapicetta M, Bonini S (1997) Exposure to foodborne and orofecal microbes versus airborne viruses in relation to atopy and allergic asthma: epidemiological study. BMJ 320:412–417

Milgrom H, Fick RB, Su JQ, et al (1999) Treatment of allergic asthma with monoclonal anti-IgE antibody. NEJM 341:1966–1973

Paunio M, Heinonen OP, Virtanen M, Leinikki P, Patja A, Peltola H (2000) Measles history and atopic diseases: a population-based cross-sectional study. JAMA 283:343–346

Presta LG, Lahr SJ, Shields RL, Porter JP, Gorman CM, Fendley BM, Jardieu PM (1993) Humanization of an antibody directed against IgE. J Immunol 151:2623–2632

Presta LG, Shields R, O'Connell LY, Lahr S, Porter J, Gorman C, Jardieu P (1994) The binding site on human IgE for its high affinity receptor. J Biol Chem 269:26368–26373

Raz E, Tighe H, Sato Y, Corr M, Dudler JA, Roman M, Swain SL, Spiegelberg HL, Carson DA (1996) Preferential induction of a Th1 immune response and inhibition of specific IgE antibody formation by plasmid DNA immunization. Proc Natl Acad Sci 93:5141–5

Rossenwasser LJ, Busse WW, Lizambri RG, et al (2003) Allergic asthma and an anti-CD23 monoclonal antibody (IDEC-152): Results of a phase I, single-dose, dose-escalating clinical trial (in press)

Roy K, Mao H-Q, Huang S-K, Leong KW (1999) Oral gene delivery with chitosan-DNA nanoparticles generates immunologic protection in a murine model of peanut allergy. Nature Med 5:387–391

Santeliz JV, Van Nest G, Traquina P, Larsen E, Wills-Karp M (2002) Amb a 1-linked CpG oligodeoxynucleotides reverse established airway hyperresponsiveness in a murine model of asthma. J Allergy Clin Immunol 109:455–62

Shaheen SO, Aaby P, Hall AJ, Barker DJP, Heyes CB, Shiell AW, Goudiaby A (1996) Measles and atopy in Guinea-Bissau. Lancet 347:1792–1796

Shawler DL, Bartholomew RM, Smith LM, Dillman RO (1985) Human immune response to multiple injections of murine monoclonal IgG. J Immunol 135:1530–1535

Slater JE, Paupore E, Zhang YT, Colberg-Poley AM (1998) The latex allergen Hev B 5 transcript is widely distributed after subcutaneous injection in BALB/c mice of its DNA vaccine. J Allergy Clin Immunol 102:469–475

Sparwasser T, Koch E-S, Vabulas RM, Heeg K, Lipford GB, Ellwart JW, Wagner H (1998) Bacterial DNA and immunostimulatory CpG oligonucleotides trigger maturation and activation of murine dendritic cells. Eur J Immunol 28:2045–2054

Sur S, Wild JS, Choudhury BK, Sur N, Alam R, Klinman DM (1998) Long term prevention of allergic lung inflammation in a mouse model of asthma by CpG oligodeoxynucleotides. J Immunol 162:6284–6293

Tighe H, Takabayashi K, Schwartz D, Marsden R., Beck L, Corbeil J, Richman DD, Eiden Jr. JJ, Speigelberg HL, Raz E (2000) Conjugation of a protein to immunostimulatory DNA results in a rapid, long-lasting and potent induction of cell-mediated and humoral immunity. Eur J Immunol 30:1939–1947

Von Mutius E, Martinez FD, Fritzch C, Nicolai T, Roell G, Thiemann HH (1994) Prevalence of asthma and atopy in two areas of West and East Germany. Am J Respir Crit Care Med 149(2 Pt 1):358–64

Pharmacogenetics, Pharmacogenomics and Gene Therapy

I. P. Hall

Division of Therapeutics, University Hospital of Nottingham, D Floor,
South Block, NG7 2UH, UK
e-mail: ian.hall@nottingham.ac.uk

1	Introduction	288
2	Pharmacogenetics and Airflow Obstruction	288
3	Genetic Variability: General Considerations	290
4	Genetic Variability in Airway Targets	291
5	Genetic Variability and Pharmacokinetics	292
6	β_2-Adrenoceptor Polymorphism	292
7	Muscarinic M_2 and M_3 Receptors	294
8	5-Lipoxygenase Activity and Cys-Leukotriene 1 Receptors	294
9	Glucocorticoid Receptor Polymorphism	295
10	Phosphodiesterase Inhibitors	295
11	Pharmacogenomics	296
12	Gene Therapy	297
13	Summary	299
	References	299

Abstract The completion of a draft of the human genome has led to a revolution in the use of genetic information to study the pathophysiology of airway disease. There are three major potential uses of genetic information. First, knowledge of the precise sequence of genes can be used to develop gene therapy approaches for the management of airway disease. Second, expression profiling approaches can be used to define novel targets for therapy. Third, genetic variation in the key targets for drugs used in the management of airway disease might account for inter-individual variability in treatment response.

Keywords Pharmacogenetics · β_2-Adrenoceptors · Expression profiling · Cystic fibrosis · α_1-Antitrypsin deficiency

1
Introduction

Whilst most patients with asthma and/or COPD will exhibit good disease control with conventional medication, there is considerable interindividual variability in response to specific treatment. In addition, a small minority of patients will have disease which responds poorly to medication. With the near completion of the working draft of the human genome, a wealth of genetic information has become available which will enable novel approaches to patient management to be considered. In this chapter, three different approaches involving the use of genetic information will be reviewed. First, the potential contribution of genetic factors to variability in treatment response in asthma and COPD will be considered. Second, the use of genetic approaches for novel target identification (pharmacogenomics) will be reviewed. Finally, an outline will be provided of the potential value of gene therapy approaches to the management of asthma and/or COPD.

2
Pharmacogenetics and Airflow Obstruction

Many factors will influence whether or not a given individual responds to a specific treatment. Whilst issues such as compliance and environmental exposure are obviously important, a major component contributing to interindividual variability in treatment response is likely to be genetic factors. In broad terms, these can be separated into factors which affect the pharmacokinetic profile of a drug (e.g. genetic variability in drug metabolising enzymes) and factors which affect the pharmacodynamic characteristics of a drug (e.g. variability in the primary drug target or in downstream signalling pathways).

There have been false expectations about the potential value of pharmacogenetic information, which has led to some disillusionment in both the industry and the academic sector. This is mainly because it is unlikely that a single genetic factor will be the major determinant of drug responsiveness for the vast majority of drugs. However, combinations of genetic factors, each with a moderate effect may well prove good predictors of treatment response. Although there are few examples at present which stand up to critical appraisal (and none in the asthma/COPD field), predictive models for determining disease response on the basis of genetic variability are being constructed in other disease areas, one example being modelling of clozapine responsiveness in the treatment of schizophrenia (Arranz et al. 2001). However, to date no large scale clinical trials have been performed to examine *prospectively* the reliability of such models, and a large amount of further information will be required regarding the functional effects of genetic variability in drug targets before such models are likely to be applicable in general.

The major hope for pharmacogenetics is that it will provide a reasonable predictive model to define treatment responses for individuals within a population

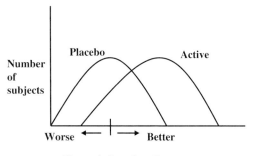

Fig. 1 Hypothetical outcome from a clinical study of a novel drug for the treatment of asthma. The figure shows responses of populations treated with placebo or active drug in a hypothetical study of a novel drug designed for the management of airway disease. Whilst no overall effect of placebo is seen, individuals on placebo may deteriorate or improve during the course of the study. In contrast, the active drug produces an improvement overall in the population. However, some individuals respond well to the active drug whereas other individuals response poorly or have no response. If a single genetic factor determined treatment response one would predict that one would obtain a bi-modal distribution of responders and non-responders in the active arm of the study. However, where multiple factors (genetic and non-genetic) contribute to inter-individual variability in treatment response, this bi-modal pattern would be lost. See text for further details

(Roses 2000, 2002). This can perhaps be best considered by examining data from a hypothetical clinical trial such as that shown in

Fig. 1. Most clinical trial data are presented with a comparison of the mean effect in the treatment group compared with the mean effect in the placebo group. Typical data obtained from a study which, for example, examines change in mean peak expiratory flow rate following treatment with an inhaled corticosteroid would be expected to be similar to that seen in Fig. 1. Here it can be seen that in the placebo group, whilst some individuals improved and some individuals deteriorated during the course of treatment, overall there is no significant effect compared with baseline. In contrast, the treatment group has a mean improvement. However, it is also clear that whilst some individuals improved markedly on treatment in the active drug arm of the study other individuals failed to respond at all or deteriorated. The hope for pharmacogenetics is that genetic information might be used to predict those individuals who are likely to fall into the high or low responder groups. Whilst this may not prove clinically important where a rapid response to treatment is observed (e.g. following administration of a short-acting β_2-agonist in asthma) it may be valuable where response to treatment takes longer to become clinically apparent, (e.g. anti-inflammatory treatment in the management of airflow obstruction).

Another possible benefit of pharmacogenetic data is the prediction of rare serious adverse drug reactions (ADRs). Again, there are currently no good examples within the respiratory field, although examples in other disease areas exist. Perhaps the best example is in the prediction of hypersensitivity reactions to abacavir, a reverse transcriptase inhibitor used in the treatment of HIV disease.

Hypersensitivity to abacavir, a serious and potentially life-threatening reaction occurred much more frequently in individuals carrying HLA B*5701 (Mallal et al. 2002), and a strong argument could be used for routine genotyping of individuals before commencing on abacavir to prevent the drug being prescribed for individuals at high risk of this ADR.

3
Genetic Variability: General Considerations

Current estimates suggest that the human genome contains around 32,000 genes and that around 1 in 1,000 bases is variable between individuals (International Human Genome Consortium 2001). By far the commonest form of variation is the single nucleotide polymorphism (SNP). Given that the majority of genes are at least 1,500 base pairs in length, most genes will contain at least one and often many more SNPs. Other forms of polymorphism/mutation occur such as deletions, insertions, and variable length di- or tri-nucleotide repeat sequences.

Given the extent of genetic variability within the human genome (currently the SNP databases contain over 2 million SNPs, and the number is expanding rapidly), simple random association approaches to pharmacogenetic studies are likely to prove problematic because of false positive and false negative results. The situation is further complicated by the existence of linkage disequilibrium. Linkage disequilibrium occurs because any given mutation must arise on a particular genetic background. Hence, SNPs which are close together will not be randomly assorted in the population, as recombination between two adjacent SNPs will be a rare event. In practise, the rate of recombination is not random across the human genome, with the result that recombination hot spots occur. This leads to the existence of regions showing tight linkage disequilibrium (haplotype blocks) and regions where linkage disequilibrium is less tight. The existence of haplotype blocks may be an advantage in that genotyping a small number of polymorphisms within a given haplotype block may provide a good genetic profile of that region, reducing the amount of genotyping required in an association study. On the other hand, the precise localisation of a causal mutation within such a region will be more difficult.

Given that SNPs arise by mutation in a single individual in a population, a reasonable question would be, How is it that a given SNP becomes common within the general population? Population genetic approaches have suggested that only very small selection advantages are required to drive a polymorphism within the population, giving rise to the concept of the "nearly neutral mutation". For humans, a selection advantage of around 1 in 10,000 has been suggested to be adequate to drive selection.

The presence of a selection advantage implies functionality for the given polymorphism.

A further way of rationalising approaches to the uses of genetic information in pharmacogenetics is to assess functionality of a given polymorphism. In principle one would predict that a coding region polymorphism which alters

Table 1 Important questions in assessing the functional consequences of polymorphic variation

Is the polymorphism in a coding region or regulatory region governing gene expression?
In coding regions, does the polymorphism alter the amino acid sequence of the protein?
In non-coding regions, does the polymorphism alter transcription factor recognition sites or splice sites?
Does the polymorphism produce a non-conservative change in amino acid sequence (e.g. introduce a charge alteration)?
Can functional effects for the different alleles be demonstrated using in vitro model systems?
Does the polymorphism show association with a biologically relevant end-point in clinical studies?

the amino acid code of a protein is likely to be more functionally important than a polymorphism which does not alter amino acid sequence. Similarly in non-coding regions polymorphisms in important regulatory regions (i.e. those which govern the level of expression for the gene of interest) might be predicted to be more important than polymorphisms in the rest of the non-coding region at that genetic locus. Some of the questions to consider when assessing potential functionality of a given polymorphism are shown in Table 1.

4
Genetic Variability in Airway Targets

Over the last 10 years a number of groups have systematically screened the coding regions of the major drug targets for airway disease to determine the extent of genetic variability within these genes. Table 2 shows the current status of this effort. From this table it is clear that, whilst the coding regions of most important primary drug targets in the airways have been screened, far less is known about the regulatory regions of the relevant genes, which in some cases have yet

Table 2 Current status of pharmacogenetic screening of major airway targets. The table shows major targets for the treatment of airway disease and the current status of genetic screening examining for polymorphisms within the key targets

Drug target	Screening		Clinical studies
	Coding	Regulatory	
β_2-Adrenoceptor	Yes	Yes	Yes
M_2 receptor	Yes	No	No
M_3 receptor	Yes	No	No
Glucocorticoid receptor α/β	Yes	No	Limited
H_1 receptor	Yes	Yes	No
Cys LT1 receptor	Limited	No	No
5-Lipoxygenase	Yes	Yes	Yes
PDE4	No	No	No

Whereas the majority of the coding regions have been screened, the regulatory regions for these genes have often not been defined or screened. Relatively few clinical studies have been performed to date (see text for further details). In addition, the downstream signalling pathways from these targets may also contain significant genetic variability, which might impinge on treatment profiles.

to be properly defined. Also, there are relatively few large-scale clinical studies which have been performed to date. A summary of the major polymorphisms which have been studied is given in the next section.

5
Genetic Variability and Pharmacokinetics

Factors affecting drug metabolism are also potentially important in determining treatment response. However, the majority of drugs used in the treatment of airway disease have wide therapeutic windows (with the exception of theophylline) and do not show marked variability in their pharmacokinetics between individuals. The best examples for pharmacokinetic effects due to genetic variability are for drugs metabolised by certain members of the cytochrome P450 pathway. Polymorphic variation in cytochrome P450 family members leads to altered metabolism of several commonly used drugs.

A good example is the role of cytochrome P450 2C9 polymorphism in the control of warfarin metabolism. Those individuals carrying either *2 or *3 alleles at this locus show reduced ability to metabolise warfarin, leading to prolongation of anticoagulant effects (Daly et al. 2002). Another example of this is the role of cytochrome P450 2D6 polymorphism in the metabolism of some antidepressants. Two other enzymes important in the metabolism of specific drugs are N-acetyltransferase (isoniazid) and thiopurine methyl transferase (TPMT) (azathioprine). Perhaps of these, the most relevant to the management of asthma is TPMT; azathioprine has occasionally been used in the management of difficult asthma to reduce prednisolone requirements. TPMT activity is a strong predictor of haematological toxicity in, for example, patients with inflammatory bowel disease with individuals who are slower acetylators being much more likely to develop toxicity (Lennard 2002) due to increased drug levels.

6
β_2-Adrenoceptor Polymorphism

The best studied example for a pharmacogenetic contribution to variability in treatment response in airway disease is that of the β_2-adrenoceptor. The β_2-adrenoceptor is an intronless gene situated on chromosome 5q31. Nine SNPs have been described within the coding region for this gene, of which four alter the amino acid sequence (Reihsaus et al. 1993). Functional data exist to suggest significant effects for the Arg-Gly16 polymorphism, the Gln-Glu27 polymorphism and the Thr-Ile 164 polymorphism. The Ile 164 form of the receptor is rare (allelic frequency 3%, i.e. 3% of individuals will be heterozygous for this polymorphism and one would predict that 50% of their β_2-adrenoceptors will carry the Ile 164 variant). In transformed cell systems the Ile 164 form of the receptor demonstrates a reduced binding affinity for catechol ligands (Green et al. 1993). In addition, because the polymorphism is in the fourth transmembrane domain of the receptor close to the binding site for the lipophilic tail of salmeterol, it is conceivable that individuals carrying this polymorphism will have altered responses to salmeterol. However, given the rarity of the polymorphism, the contribution to population variability in treatment response will be at best small. Individuals carrying two copies of the Ile 164 variant would be expected to exhibit a more extreme phenotype than heterozygous individuals. However, despite genotyping several hundred individuals at this locus, we have yet to identify a homozygous Ile 164 subject.

The two N-terminal polymorphisms at codon 16 and 27 are, in contrast, much commoner. Neither alter the ligand binding characteristics of the receptor but both have been proposed to alter receptor cycling (Green et al. 1994, 1995). Transformed cell system approaches suggest that the Gly16 form of the receptor down-regulates to a greater extent than wild type, whilst the Glu 27 form is relatively protected from down-regulation. In practice, because of linkage disequilibrium, the Gly 16 Glu 27 combination is commoner than would be predicted from chance based upon assessment of the allelic frequencies of the two SNPs, and hence the two polymorphisms frequently co-exist (Dewar et al. 1998). Interestingly the Glu 27 variant is much less common in populations of AfroCarribean descent (Candy et al. 2000).

Liggett and co-workers have put forward a dynamic model for receptor cycling which proposes that individuals carrying the Arg16 form of the receptor demonstrate an increased response to the acute effects of β_2-bronchodilator but have greater potential for down-regulation following long-term treatment. A number of clinical studies have been performed examining responses to β_2-agonists following acute or chronic treatment. Some although not all of these support the possibility that individuals with the Arg16 variant have increased acute responses but reduced longer term responses to β_2-agonists (Turki et al. 1995; Martinez et al. 1997; Hancox et al. 1998; Israel et al. 2000). No association was seen in one study with fatal or near fatal asthma (Weir et al. 1998), and in preliminary work our group has failed to see a contribution to reversibility in subjects with chronic obstructive pulmonary disease (COPD). At present, therefore, whilst there is good evidence for functional effects for these β_2-adrenoceptor polymorphisms, their overall contribution to treatment response remains to be fully determined. In addition there is considerable genetic variability in the regulatory region controlling expression of the β_2-adrenoceptor gene (Scott et al. 1999; Drysdale et al. 2000). Whilst many of the SNPs in this region are likely to be functionally silent, data exist suggesting that the −47, C-T SNP which is in a short open reading frame upstream from the gene alters downstream regulation of β_2-adrenoceptor expression. However, in one study no difference in levels of β_2-adrenoceptor expression on peripheral blood mononuclear cells was seen in individuals carrying either the −47C or T allele or haplotypes where this polymorphism was considered in conjunction with the coding region polymorphisms (Lipworth et al. 2002).

7
Muscarinic M_2 and M_3 Receptors

Both the muscarinic M_2 and M_3 receptor coding regions have been screened for polymorphic variation and been found to exhibit far less variability than the β_2-adrenoceptor. No common amino acid substitution has been reported in either the M_2 or M_3 receptors, although relatively rare SNPs which do not alter the amino acid code are present in at least the M_2 receptor (Fenech et al. 2001). The regulatory regions important for transcription of the M_2 and M_3 receptor have only recently been defined and no functional data on polymorphic variation within these regions have been presented to date. However, it remains possible that the level of expression of M_2 and/or M_3 receptors (as opposed to the ligand binding characteristics of the receptors) in the airways is in part genetically determined and this might account in part for variability in treatment response to anticholinergic agents.

8
5-Lipoxygenase Activity and Cys-Leukotriene 1 Receptors

Following the introduction of Cys-leukotriene 1 receptor antagonists such as montelukast and zafirlukast for the treatment of airway disease, it has become clear that a significant number of patients with asthma fail to respond to these agents. The level of leukotriene D4 in the airways is determined in part by the activity of the 5-lipoxygenase enzyme. Studies by Drazen's group in Boston have suggested that levels of 5-lipoxygenase activity are in part genetically determined by the number of SP1 repeats in the regulatory region of the 5-lipoxygenase gene (In et al. 1997; Silverman and Drazen 2000). In one clinical study examining changes in lung function following the administration of a 5-lipoxygenase inhibitor (ABT761), individuals with no clinical response were much more likely to have non-wild-type alleles at this locus (Drazen et al. 1999). These non-wild-type alleles were all found in complementary in vitro functional studies to be associated with lower levels of enzyme activity. Preliminary studies have examined the potential contribution of these polymorphisms to clinical responses to Cys-leukotriene 1 receptor antagonists in the treatment of asthma. In one study, the responses of individuals who were heterozygous for 5-lipoxygenase promoter non-wild-type alleles were not different from individuals carrying the wild-type alleles on both chromosomes (Fowler et al. 2002); however, the effectiveness of Cys-leukotriene-1 receptor antagonists in individuals with asthma homozygous for non-wild-type alleles at the 5-lipoxygenase locus remains to be determined. In Caucasian populations, such individuals are relatively rare (<10%), and hence the potential contribution of homozygosity for these polymorphisms to treatment response at the population level would be small. The relevance of the recently cloned Cys-leukotriene 2 receptor (Lynch et al. 1999) to the control of airway responses to leukotriene D4 and other inflammatory leukotriene mediators remains to be fully determined (Heise et al. 2000), although the majority of the treatment response seen with montelukast and zafirlukast is mediated through actions at the Cys-leukotriene 1 receptor. The potential contribution of polymorphic variation in this Cys-leukotrine 2 receptor to airway responses remains to be examined.

9
Glucocorticoid Receptor Polymorphism

Given the existence of rare individuals who appear to be resistant to the effects of glucocorticoids, considerable interest has arisen in the possibility that polymorphism within the glucocorticoid receptor may underlie variability in the effectiveness of steroid medication. The glucocorticoid receptor exists as two splice variants (GRα and GRβ) which arise due to differential splicing of the ninth exon. A number of SNPs have been described within the glucocorticoid receptor including several which cause amino acid substitutions within the receptor. The only good functional data presented to date suggest that a polymorphism in codon 363 which results in a serine substitution is a gain of function mutation with individuals carrying this substitution showing altered profiles in dexamethasone suppression tests (Huizenga et al. 1998). However, this polymorphism is rare. The possibility that individuals carrying this polymorphism may be at increased risk of glucocorticoid-associated side effects has not yet been studied in a population with airway disease. Other studies have identified rare polymorphisms which may underlie cortisol resistance in individual family members in isolated families (e.g. Malchoff et al. 1993) but have not in general identified association between commoner SNPs within the GR gene and resistance to treatment (Koper et al. 1997).

10
Phosphodiesterase Inhibitors

In contrast with the relatively simple genomic organisation of genes coding for G protein-coupled receptors, the phosphodiesterase family of genes is extremely complicated with the existence of multiple families each containing several family members of which many splice variants also exist. Because the most important phosphodiesterase isoforms for the control of cell signalling in different airway cells have not been fully defined, pharmacogenetic studies examining phosphodiesterase isoforms have not yet been performed. It is probable that members of the phosphodiesterase type 4 family may prove important, with phosphodiesterase 4B and 4D potentially being implicated as the major isoforms controlling responses in airway smooth muscle and inflammatory cells (Bolger et al. 1997; Schmidt et al. 1999). Pharmacogenetic studies are likely to be made even more difficult by the complex structures of the phosphodiesterase genes which have been examined to date. Most genes contain multiple exons/introns and span relatively large genomic regions, with the result that they are likely to contain multiple SNPs. For example, the human phosphodiesterase 4D gene covers around 1 megabase and has 17 exons (Le Jeune et al. 2002). Thus, designing informative pharmacogenetic studies on these genes will prove problematic until further data are available.

11
Pharmacogenomics

As mentioned earlier, current estimates suggest there are around 32,000 genes in the human genome (International Human Genome Sequencing Consortium 2001). The available drugs used in the treatment of human disease at present target only around 500 gene products. Thus it is likely that many novel targets exist which may be amenable to modulation by drugs, thereby providing the opportunity for the development of novel therapeutic strategies. Pharmacogenomic approaches have been used to identify potentially important gene products by taking advantage of advances in micro-array and gene chip technologies. The basic principle behind these approaches is to examine the levels of either messenger RNA (RNA expression profiling) or protein (proteomics) for large numbers of genes with the aim of identifying gene products which are up-regulated or down-regulated under conditions relevant to disease activity. The production of gene chips which include specific probes for 10,000 or more human genes means that expression profiles for the majority of genes within the human genome can easily be obtained within a short space of time. However, the wealth of data generated and the bioinformatic difficulties of handling these data have produced major problems in interpreting results obtained from experiments using expression profiling approaches (Chambers 2002). At its simplest these approaches would take an individual cell type and examine expression profiles before and after a single treatment. It is, however, critical in such experiments to observe simple rules regarding experimental design. Thus experiments should be performed with multiple replicates at appropriate time points and with appropriate controls included under carefully controlled conditions to help increase confidence in the data generated. There is no general agreement about the change in expression which is indicative of a significant change in regulation, although many investigators have used a cut off of a twofold increase or decrease in RNA expression to define targets. Proteomic approaches have used a combination of 2D gel electrophoresis coupled with mass spectrometry sampling of proteins with altered expression/migration. The difficulty here is that the technique tends to select abundant proteins rather than identify proteins of a lower abundance which may still be important, and in addition some of the proteins identified on 2D gel electrophoresis as being up- or down-regulated cannot be easily identified using mass spectrometry approaches.

The major problem with expression profiling approaches is in selecting targets which may be important. Whilst genes which previously were known to be regulated in models of airway disease will be identifiable using expression profiling approaches (an example being induction of COX2 by interleukin-1β) this is obviously not relevant in defining novel therapeutic targets. By definition, the gene products of most potential interest will be those about which relatively little is known regarding function. In order, therefore, to handle the hits obtained using expression profiling approaches appropriate strategies need to be designed. One potential way forward is to first validate hits using an alternative technology. Usually a quantitative polymerase chain reaction (PCR)-based approach (e.g. using TaqMan) is used to validate targets. In addition, one can prioritise targets depending upon their likely role in the cell. For example, a novel G protein-coupled receptor (GCPR) would be more likely to be a promising therapeutic target than a structural protein involved in cytoskeletal reorganisation.

To date relatively few studies using expression profiling approaches have been published as full peer-reviewed studies, although many are currently ongoing. One study using cultured airway smooth muscle cells identified over 40 genes up-regulated by interleukin-1β, providing a list of novel potential targets (Hakonarson et al. 2001). No proteomic data have been published to date although, again, a number of studies are currently ongoing.

One interesting approach for dealing with the large number of novel targets generated by expression profiling is to use linkage data from genome screen studies in the disease of interest to prioritise targets. There have been a number of genome screens published on different asthmatic populations which have identified a number of regions of moderate linkage, suggesting that a number of genes of moderate effect may be important in disease initiation (Daniels et al. 1996; CSGA 1997; Ober et al. 1998; Wjst et al. 1999, Lonjou et al. 2000). By prioritising genes identified using expression profiling approaches comparing asthmatic and non-asthmatic cells and limiting analyses to those genes in regions of linkage, it may be possible to narrow the number of targets to be examined.

12
Gene Therapy

Gene therapy approaches can potentially be of value in the treatment both of airway disease caused by single gene defects and airway disease in which multiple genes probably play a role. The best examples of the former category are cystic fibrosis and α_1-antitrypsin deficiency. In both these situations, where disease is predominantly related to the lack of expression of the normal protein product, transfer of the relevant wild-type gene should in principle prevent development of disease if performed at an early stage. In contrast, gene therapy approaches may potentially be of benefit in polygenic diseases such as asthma, where overexpression of a relevant transgene (e.g. an anti-inflammatory cytokine) might ameliorate the disease-causing mechanisms. Such approaches are only likely to be cost effective in small numbers of patients with severe disease unresponsive to standard medication. To date, there are no clinical data in human subjects with airway disease using this approach and most of the data published on this approach have been generated from animal models. One interesting example is in the generation of a β_2-adrenoceptor-overexpressing mouse which demonstrates increased protection against bronchoconstrictor stimuli compared with the wild-type mouse (McGraw et al. 2000).

The major airway disease which has to date been targeted by gene therapy approaches is cystic fibrosis (reviewed in Griesenbach et al. 2002). Because defective function of the cystic fibrosis transmembrane regulator (CFTR) is believed to explain at least the majority of the pathological changes seen in patients with cystic fibrosis, correction of the underlying molecular defect by gene therapy replacement might be expected to correct the physiological defect and also to potentially improve airway function. A number of studies have demonstrated reconstitution of the chloride (Cl^-) current carried by CFTR following gene transfer in cultured cell lines. Effective transfer of CFTR was also seen in animal models of cystic fibrosis which led to the approval of human gene therapy trials in patients with cystic fibrosis. These employed one of two approaches. In the first approach, modified adenovirus vectors were used as the vehicle for gene transfer. In the second approach, liposomal carriers were used to transfer CFTR to the airway epithelial cells. Both led to the successful expression of CFTR in human airway or nasal epithelial cells; however, the level of expression of the transgene was in general low and pro-inflammatory effects of the adenovirus vector were seen when administered at high doses, which led to the cessation of clinical trials using this approach in one centre. To date there are no data suggesting improvement in lung function or other clinically relevant endpoints in patients with cystic fibrosis, although inevitably subjects treated were those with moderate or relatively severe disease in whom profound structural changes had already occurred. Once substantial pathological changes have occurred in airway architecture it is unlikely that these would be reversed following successful expression of CFTR in the airways.

Despite the relative lack of success of clinical trials in cystic fibrosis, it is likely that gene therapy approaches will prove successful for the management of airway disease in the future. In conditions where an underlying defect is being corrected such as cystic fibrosis or α_1-antitrypsin deficiency, these approaches will need to be commenced at a relatively early stage in disease to prevent long-term structural changes. On the other hand, the alternative approach of over expression of anti-inflammatory mediators, for example for the management of severe asthma, could be considered where conventional therapy has failed to produce benefit. Such approaches will, however, depend upon obtaining adequate expression of the transgene in the airways without inducing inflammatory reactions. With this in mind, a major effort is ongoing to generate novel gene therapy transfer vehicles including work on new liposomal agents and the development of lentiviruses which may have less pro-inflammatory effects in the airway whilst achieving adequate levels of expression.

13
Summary

The near completion of the human genome project has lead to a wealth of genetic information available on publicly accessible databases which will revolutionise the way in which airway research is performed. The realisation that extensive inter-individual variation occurs at the genetic level may have implications for the management of patients. Pharmacogenetic information will be of use in explaining inter-individual variability in treatment response and potentially will be valuable in increasing understanding of adverse drug reactions. Expression profiling approaches will generate a wealth of data which will allow novel therapeutic targets in the airways to be identified. Ultimately these data may impinge upon the feasibility of gene transfer approaches to the management of airway disease although this is likely to be limited to small numbers of patients with disease which has failed to respond to conventional treatment. However, there is a major challenge in extending the wealth of genetic information into functional studies examining the integrated physiology of the airway.

References

Arranz MJ, Collier D, Kerwin RW. Pharmacogenetics for the individualization of psychiatric treatment. *Am J Pharmacogenomics*. 2001;1(1):3–10. Review

Bolger GB, Erdogan S, Jones RE *et al*. Characterization of five different proteins produced by alternatively spliced mRNAs from the human cAMP-specific phosphodiesterase PDE4D gene. *Biochem J* 1997; 328 (Pt 2):539–548

Candy G. Samani N. Norton G. Woodiwiss A. Radevski I. Wheatley A. Cockcroft J

Chambers RC, Gene expression profiling: good housekeeping and a clean message. *Thorax* 2002; 57: 754–6

CSGA. A genomewide search for asthma susceptibility loci in ethnically diverse populations. The Collaborative Study on the Genetics of Asthma (CSGA). *Nat Genet* 1997; 15(4):389–392

Daly AK, Day C P, Aithal GP, CYP2C9 polymorphism and warfarin dose requirements, *Br J Clin Pharmacol*. 2002; 5B:408–9

Daniels S, Bhattacharrya S, James A, Leaves NI, Young A, Hill MR et al. A genomewide search for quantitative trait loci underlying asthma. *Nature* 1996; 383 (6597):247–250

Dewar JC, Wheatley AP, Venn A, Morrison JF, Britton J, Hall IP. β_2-adrenoceptor polymorphisms are in linkage disequilibrium, but are not associated with asthma in an adult population. *Clin.Exp.Allergy* 1998; 28:442–448

Drazen JM, Yandava CN, Dube L *et al*. Pharmacogenetic association between ALOX5 promoter genotype and the response to anti-asthma treatment. *Nat.Genet*. 1999; 22:168–170

Drysdale C, McGraw D W., Stack C, Stephens J, Judson R S, Nandabalan K, Arnold K, Ruano G, Liggett S B. Complex promoter and coding region β_2-adrenergic receptor haplotypes alter receptor expression and predict *in vivo* responsiveness. *PNAS* 2000; 97:10483–10488

Fowler SJ, Hall IP, Wilson AM, Wheatley AP and Lipworth BJ, 5-Lipoxygenase polymorphism and in-vivo response to leukotrience receptor antagonists. *Eur J Clin Pharmacol*. 2002; 58:187–190

Green SA, Cole G, Jacinto M, Innis M, Liggett SB. A polymorphism of the human β_2-adrenergic receptor within the fourth transmembrane domain alters ligand binding and functional properties of the receptor. *J Biol Chem.* 1993; 268:23116-23121

Green SA, Turki J, Bejarano P, Hall IP, Liggett SB. Influence of β_2-adrenergic receptor genotypes on signal transduction in human airway smooth muscle cells. *Am.J.Respir.Cell Mol.Biol.* 1995; 13:25-33

Green SA, Turki J, Innis M, Liggett SB. Amino-terminal polymorphisms of the human β_2-adrenergic receptor impart distinct agonist-promoted regulatory properties. *Biochemistry* 1994; 33:9414-9419

Griesenbach A, Ferrari S, Geddes DM, Alton EW. Gene therapy progress and prospects: Cystic fibrosis. *Gene Ther* 2002; 20: 1344-1350

Hakonarson H, Halapi E, Whelan R, Gulcher J, Stefansson K, Grunstein MM, Association between Il-1beta/TNF-alpha-induced glucocorticoid – sensitive changes in multiple gene expression and altered responsiveness in airway smooth muscle. *American J Respiratory Cell Mol Biology.* 2001; 25:25 761-771

Hall IP. Association analysis of beta2 adrenoceptor polymorphisms with hypertension in a Black African population. *Journal of Hypertension* 2000; 18(2):167-172

Hancox RJ, Sears MR, Taylor DR. Polymorphism of the β_2-adrenoceptor and the response to long-term β_2-agonist therapy in asthma. *Eur.Respir.J.* 1998; 11:589-593

Heise CE, O'Dowd BF, Figueroa DJ et al. Characterization of the human cysteinyl leukotriene 2 receptor. *J Biol Chem* 2000; 275:30531-30536

Huizenga NA, Koper JW, de Lange P et al. A polymorphism in the glucocorticoid receptor gene may be associated with and increased sensitivity to glucocorticoids in vivo. *J Clin Endocrinol Metab* 1998; 83:144-151

In KH, Asano K, Beier D et al. Naturally occurring mutations in the human 5-lipoxygenase gene promoter that modify transcription factor binding and reporter gene transcription. *J.Clin.Invest.* 1997; 99:1130-1137

International Human Genome Sequencing Consortium. Initial sequencing and analysis of the human genome. *Nature.* 2001;409: 860 - 921

Israel E, Drazen JM, Liggett SB, Bolishey HA et al. The effect of polymorphisms of the beta (2)-adrenergic receptor on the response to regular use of albuterol in asthma. *American J of Respir Crit Care Med.* 2002; 162(1):75-80

Koper JW, Stolk RP, de Lange P et al. Lack of association between five polymorphisms in the human glucocorticoid receptor gene and glucocorticoid resistance. *Hum Genet* 1997; 99:663-668

Le Jeune IR, Shepherd M, Van Heeke G, Houslay MD, Hall IP. Cyclic AMP dependent transcriptional up regulation of phosphodiesterase 4D5 in human airway smooth muscle cells. Identification and characterisation of a novel PDE4D5 promoter. *J Biol Chem* 2002; 277:35980-9

Lennard L. TPMT in the treatment of Crohn's disease with azathioprine. *Gut* 2002; 51(2):143-6

Lipworth B, Koppelman G, Wheatley AP, Le Jeune I, Coutie W, Meurs H, Kauffman HF, Postma DS, Hall IP. β_2 adrenoceptor promoter polymorphisms: extended haplotypes and functional effects in peripheral blood mononuclear cells. *Thorax* 2002 57; 61-66

Lonjou C, Barnes k, Chen H, Cookson W, Deichmann K, Hall IP, Holloway JW, Laitinen T, Palmer LJ, Wjst M and Morton NE. A first trial of retrospective collaboration for positional cloning in complex inheritance: assay of the cytokine region on chromosome 5 by the consortium on asthma genetics (COAG). *PNAS* 2000; 97:20:10942-10947

Lynch KR, O'Neill GP, Liu Q et al . Characterization of the human cysteinyl leukotriene CysLT1 receptor. *Nature* 1999; 399:789-793

Malchoff DM, Brufsky A, Reardon G et al. A mutation of the glucocorticoid receptor in primary cortisol resistance. *J Clin Invest* 1993; 91:1918-1925

Mallal S, Nohan D, Witt C et al. Association between presence of HLAB*5701, HLA-DR7 and HLA-DQ3 and hypersensitivity to HIV-1 reverse transcriptase inhibitor abacavir. *Lancet* 2002; 359; 727–732

Martinez FD, Graves PE, Baldini M, Solomon S, Erickson R. Association between genetic polymorphisms of the β_2-adrenoceptor and response to albuterol in children with and without a history of wheezing. *J.Clin.Invest* 1997; 100; 3184–3188

McGraw DW, Forbes SL, Mak JC, Witte DP, Carrigan PE, Leikauf, Liggett SB. Transgenic overexpression of beta(2)-adrenergic receptors in airway epithelial cells decreases bronchoconstriction. *Am J Physiol Lung Cell Mol Physiol* 2000; 279:L379–89

Ober C, Cox NJ, Abney M, Di Rienzo A, Lander ES, Changyaleket B et al. Genomewide search for asthma susceptibility loci in a founder population. *Hum Mol Genet* 1998; 7(9):1393–1398

Reihsaus E, Innis M, MacIntyre N, Liggett SB. Mutations in the gene encoding for the β_2-adrenergic receptor in normal and asthmatic subjects. *Am J Respir Cell Mol.Biol.* 1993; 8:334–339

Roses AD. Pharmacogenetics and the practice of medicine. *Nature* 2000; 405:857–865

Roses AD. Genome-based pharmacogenetics and the pharmaceutical industry. *Nat Rev Drug Discov.* 2002 Jul; 1 (7): 541–9. Review

Schmidt D, Dent G, Rabe KF. Selective phosphodiesterase inhibitors for the treatment of bronchial asthma and chronic obstructive pulmonary disease. *Clin Exp Allergy* 1999; 29 Suppl 2:99–109

Scott MG, Swan C, Wheatley AP, Hall IP. Identification of novel polymorphisms within the promoter region of the human β_2-adrenergic receptor gene. *Br.J.Pharmacol.* 1999; 126:841–844

Silverman ES, Drazen JM. Genetic variations in the 5-lipoxygenase core promoter. *Am J Respir Crit Care Med* 2000; 161:S77-S80

Turki J, Pak J, Green SA, Martin RJ, Liggett SB. Genetic polymorphisms of the β_2 adrenergic receptor in nocturnal and nonnocturnal asthma. Evidence that Gly16 correlates with the nocturnal phenotype. *J.Clin.Invest* 1995; 95:1635–1641

Weir TD, Mallek N, Sandford AJ et al. β_2-Adrenergic receptor haplotypes in mild, moderate and fatal/near fatal asthma. *Am J Respir Crit Care Med* 1998; 158:787–791

Wjst M, Fisher G, Immervoll T, Jung M, Saar K, Rueschendorf F et al. A genome wide search for linkage to asthma. *Genomics* 1999; 58: 1–8

Evaluation of New Drugs for Asthma and COPD: Endpoints, Biomarkers and Clinical Trial Designs

P. J. Barnes · E. M. Erin · T. T. Hansel · S. Kharitonov · A. J. Tan · R. C. Tennant

NHLI Clinical Studies Unit, Royal Brompton Hospital, Fulham Rd.,
London, SW3 6HP, UK
e-mail: t.hansel@ic.ac.uk

1	**Background**	305
1.1	Medical Needs	305
1.2	International Guidelines: GOLD and GINA	305
1.3	Clinical Phases I to IV	305
1.4	Good Clinical Practice	306
1.5	Regulatory Guidelines	307
1.6	Ethical Issues	308
2	**Endpoints**	308
2.1	Lung Function	308
2.2	Symptom Diary Scores and Asthma Control Questionnaires	309
2.3	Health Status	309
2.4	Airway Hyperreactivity	310
2.5	Exercise Testing in COPD: 6MWT, ISWT, ESWT	310
2.6	Dynamic Hyperinflation	311
2.7	Muscle Testing	311
2.8	Blood Analysis	311
2.9	Exhaled Breath	312
2.10	Sputum Analysis	315
2.11	BAL and Biopsy	317
2.12	High-Resolution Computerised Tomography	317
2.13	Pharmacogenetics	318
2.14	Mortality	319
3	**Clinical Trial Designs**	319
3.1	Trial and Error with Herbal Remedies	319
3.2	Tolerability	319
3.3	Bronchodilators in Asthma	319
3.4	Bronchodilators in COPD	320
3.5	Bronchoprotection	320
3.6	Inhaled Allergen Challenge	321
3.7	Nasal Allergen Challenge	324
3.8	Exercise-Induced Asthma	325
3.9	Symptomatic Asthma	325
3.10	Studies with Inhaled Corticosteroids	326
3.11	Corticosteroid Add-on Studies	326

3.12 Corticosteroid Titration Studies 327
3.13 Corticosteroid Abrupt Discontinuation Studies 327
3.14 Severe Asthma 328
3.15 Natural History and Disease Modification 329
3.16 Weight Loss and Muscle Wasting in COPD 330
3.17 Exacerbations in Asthma: Prevention and Treatment 330
3.18 Exacerbations in COPD: Prevention and Treatment 331

4 Conclusion 331

References 332

Abstract The incidence of both asthma and chronic obstructive pulmonary disease (COPD) is increasing throughout the world, and acts as a major incentive for the development of new and improved drug therapy. For the large range of bronchodilator and anti-inflammatory agents in current clinical development, reliable decision-making is imperative in phase II, before entering large-scale phase III clinical studies. With anti-inflammatory therapies for asthma, many studies have been performed utilising the inhaled allergen challenge as a proof of concept study, effects on airway hyper-reactivity (AHR) can be assessed, and it is also possible to directly study limited numbers of symptomatic asthma patients. Additional clinical trial designs in asthma include studies to assess bronchodilation, bronchoprotection against a variety of inhaled constrictor agents, exercise tolerance, add-on and titration studies with inhaled and oral corticosteroids, and prevention and treatment of exacerbations. In contrast, it is a major issue for the development of new anti-inflammatory drugs for COPD that large-scale phase II studies are generally required in this disease in order to detect clinical efficacy. In COPD, clinical trial designs range from studies on lung function, symptoms and exercise performance, inflammatory biomarkers, natural history of chronic stable disease, prevention and treatment of exacerbations, and effects on cachexia and muscle function. Compared with asthma, inclusion criteria, monitoring parameters, comparator therapies and trial design are less well established for COPD. The large variety of potential clinical endpoints includes lung function, symptoms, walking tests, hyperinflation, health-related quality of life (HR-QOL), airway reactivity, and frequency and severity of exacerbations. In addition, surrogate biomarkers may be assessed in blood, exhaled breath, induced sputum, bronchial mucosal biopsy and bronchoalveolar lavage (BAL), and advanced radiographic imaging employed. Of particular utility is ex vivo whole blood stimulation to enable pharmacokinetic/pharmacodynamic modelling in establishing an optimal dosage regimen relatively early in human clinical studies. There have been considerable recent advances in the development of non-invasive biomarkers and novel clinical trial designs, as well as clarification of regulatory requirements, that will facilitate the development of new therapies for patients with asthma and COPD.

Keywords Asthma · COPD · Clinical trial designs · New drugs · Endpoints · Biomarkers

1
Background

1.1
Medical Needs

The health burden of asthma and chronic obstructive pulmonary disease (COPD) is increasing globally at an alarming rate, providing a strong impetus for the development of new therapeutics (Lopez and Murray 1998; Murray and Lopez 1997a,b; Beasley et al. 2000; Barnes 2002; Fabbri et al. 2002). Despite the availability of effective inhaled therapy for asthma, many patients are inadequately treated (Vermeire et al. 2002), while the severe asthmatics that do not respond to inhaled and oral corticosteroids have a major need for supplementary or alternative therapy. The extensive use of oral leukotriene antagonists of limited efficacy for mild persistent asthma in the USA illustrates the need for alternatives to inhaled corticosteroids. The situation is more urgent in COPD since patients lack effective therapy; bronchodilators have relatively small effects and anti-inflammatory therapy has only limited action in certain defined patients (Barnes 2000). In both asthma and COPD there is the need to provide more effective treatment for acute severe exacerbations of both asthma and COPD. A long-term goal is to develop disease-modifying therapy to alter the natural history and airway remodelling that occur in asthma, and to prevent or inhibit the progressive fall in lung function that contributes to mortality in COPD.

1.2
International Guidelines: GOLD and GINA

Inclusion criteria for patients with asthma and COPD in clinical trials can generally be based on definitions provided by international guidelines. Recently, the Global Initiative on Asthma (GINA) and the Global Initiative on Chronic Obstructive Lung Disease (GOLD) have released evidence-based guidelines that cover the definition and management of patients with asthma and COPD [National Institutes of Health (NIH) and National Heart Lung and Blood Institute (NHLBI) 2002; NIH, NHLBI, World Health Organisation (WHO) 2001].

1.3
Clinical Phases I to IV

Textbooks of pharmaceutical medicine contain useful background information on clinical trials and the development of medicines (Griffin et al. 1994), while the European Medicines Evaluation Agency (EMEA) has issued general considerations for clinical trials (EMEA CPMP 1997). The phases of drug development are illustrated in Fig. 1. Rising dose tolerability studies generally take place in first healthy volunteers and then patients with asthma or COPD. For some in-

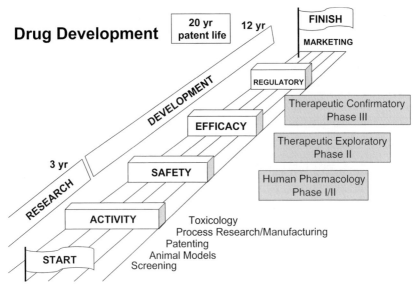

Fig. 1 A schematic diagram to illustrate the phases of preclinical and clinical drug development. For many new chemical entities (NCEs) the cost of development to registration can range from US $500 million to $1 billion. Development times can vary greatly, from 6 years to over 12 years, and should be considered in relation to a patent lifetime of 20 years. Phase I comprises rising-dose tolerability in healthy volunteers and possibly patients. Phase II involves therapeutic exploratory studies in small number of patients to demonstrate pharmacological and clinical proof of principle. Phase III involves large-scale therapeutic confirmatory studies. Phase IV involves post-registration/marketing studies to extent indications and define safety and efficacy

haled agents it may be preferable to proceed directly to patients, and tolerability studies in younger patients with asthma are generally more sensitive and have greater safety reserve then initial studies in patients with COPD. Table 1 considers some factors that can be considered in defining asthma and COPD patients.

1.4
Good Clinical Practice

Clinical studies within a drug development programme should be carried out according to good clinical practice (GCP) (Gallin 2002), although this name is imprecise, since GCP specifically relates to conduct of clinical research trials with potential new therapies for human use. A European GCP Directive (2001/20/EC) was issued in 2001, and this requires adoption into local law of member states by May 2003, and full implementation by May 2004.

Table 1 Endpoints for clinical studies in asthma and COPD

Asthma	COPD
Lung function	
Peak expiratory flow (PEF) and forced expiratory volume in 1 s (FEV$_1$)	FEV$_1$ % predicted (pre/post bronchodilator)
Reversibility and variability	FEV$_1$/forced vital capacity, FVC ratio
Airways hyperresponsiveness (AHR) to methacholine, histamine, adenosine monophosphate (AMP)	Forced inspiratory volume in 1 s, FIV$_1$ TLCO, transfer factor for carbon monoxide
Symptom diary	Symptoms and smoking rate diary
Asthma Control Questionnaire (ACQ) To include rescue short-acting β_2-agonist usage	
Health status	
Health-Related Quality of Life (HRQOL)	St. George's Respiratory Questionnaire (SGRQ)
Asthma Quality of Life Questionnaire (AQLQ)	Chronic Respiratory Disease Quest (CRDQ)
Challenge model	
Inhaled allergen challenge, leukotrienes	Cigarette, ozone, endotoxin inhalation
Exercise responses	
Exercise-induced asthma	Dynamic hyperinflation
	Six minute walk test (6MWT)
	Incremental shuttle walk test (ISWT)
	Endurance shuttle walk test (ESWT)
Systemic effects	
	Weight, fat free mass (FFM)
	Respiratory and skeletal muscle function
Laboratory tests	
Skin prick tests	Arterial blood gases
Blood	Blood
Sputum	Sputum
Exhaled breath	Exhaled breath
BAL and bronchial biopsy	BAL and bronchial biopsy
	Urine elastin/collagen degradation products
Imaging	
	High resolution computerised tomography (HRCT) scans
Natural History	
Rate of FEV$_1$ decline	Rate of FEV$_1$ decline
Airway remodelling	
Pharmacoeconomics and pharmacogenetics	
Exacerbations	
Mortality	

1.5
Regulatory Guidelines

The European Medicines Evaluation Agency (EMEA) has issued points to consider in clinical trials on drugs for the treatment of COPD (EMEA CPMP 1999), and there are also guidelines on asthma that are at draft consultative stage

(EMEA CPMP 2001). The Federal Drug Agency (FDA) has issued draft guidance for industry for metered dose inhalers (MDI) and dry powder inhalers (DPI) (US Department of Health and Human Services 1998).

1.6
Ethical Issues

There are particular ethical issues in carrying out clinical studies in patients with asthma and COPD. Patients with asthma may sometimes be required to be symptomatic in a run-in phase before being offered in a blinded manner either a potential treatment or placebo. Hence, in some instances these patients are not receiving optimal current therapy for their asthma. Investigative procedures such as bronchial biopsy and broncho-alveolar lavage (BAL) are relatively invasive and are unlikely to be required for the usual clinical management of these conditions. Inhaled allergen challenge has the potential risk of causing acute anaphylaxis, may cause considerable falls in forced expiratory volume in 1 s (FEV_1) in the first 24 h, and can affect airway reactivity for 2 weeks after exposure.

2
Endpoints

2.1
Lung Function

Peak expiratory flow (PEF) and FEV_1 are fundamental to the monitoring of patients with asthma (Crapo 1994; ATS 1995; Hughes and Pride 1999). It is important that the health professional should adequately explain, demonstrate and supervise the expiration procedure to obtain good-quality results. With the advent of hand-held portable electronic lung-function meters combined with clinical diaries, it is now possible to perform more reliable home monitoring of PEF, FEV_1 and forced vital capacity (FVC). Data can be stored for up to 1 month, and can be directly downloaded and transmitted over the internet. These electronic devices record the exact time of measurements, and can warn the patient about the need to take action in the event of an exacerbation of asthma. Whereas lung function recordings at clinic visits were previously often the primary endpoint in asthma studies, there is now an increasing tendency to use the lung function and symptom data recorded at home on electronic meters.

FEV_1 has long been established as the main outcome measure in the monitoring of patients with COPD, and it is strongly predictive of subsequent mortality from COPD (Burrows and Earle 1969; Peto et al. 1983; Anthonisen et al. 1986; Hansen et al. 1999). There are, however, limitations in the use of this measurement, since changes in FEV_1 over time are small in relation to repeatability of the measurement. Thus it can take serial measurements over several years to demonstrate a declining trend in FEV_1 in one individual. In addition, the earliest changes in smokers' lungs occur in the peripheral airways (Niewoehner et al. 1974), and may already be present whilst the FEV_1 remains normal. At present, post bronchodilator FEV_1 remains the "gold standard" to monitor progression of COPD, and to monitor the outcome of therapeutic interventions such as smoking cessation.

Several investigators have attempted to identify sensitive tests of small airways function (maximum mid-expiratory flow, MMEF) that may be used to monitor progression of smoking-related lung disease. Unfortunately it has been found that the reproducibility of the majority of these tests is low, and, in the main, abnormalities of these tests do not appear to predict the subsequent development of clinically significant airflow obstruction (Tattersall et al. 1978; Stanescu et al. 1987, 1998). Indices of the presence of emphysema include measurements reflecting air trapping (increased total lung capacity, functional residual capacity and residual volume), and loss of alveolar-capillary membrane [transfer factor, TLCO (Europe) or DLCO (USA)] (Hughes and Pride 1999). These tests may be useful in individual patient assessments, but have been shown to correlate only weakly with prognosis.

Measurement of exercise dynamic hyperinflation and exercise endurance, together with health-related quality of life and subjective measurement of dyspnoea are important additional measures of bronchodilator efficacy in COPD (O'Donnell and Webb 1993; Belman et al. 1996; O'Donnell 2000). In addition, reduction in dyspnoea following inhalation of β_2-agonists is closely correlated with the forced inspiratory volume in 1 s (FIV_1) (Taube et al. 2000).

2.2
Symptom Diary Scores and Asthma Control Questionnaires

Daytime and nocturnal asthma symptom diaries have been found to be appropriate for use as outcome measures in clinical trials of asthma therapy (Santanello et al. 1997). These symptom diaries can now be recorded on portable electronic devices that also include lung function. The Asthma Control Questionnaire (ACQ) consists of seven questions and has strong evaluative properties (Juniper et al. 1999b). Daily symptom diaries have been used to a lesser extent in studies on patients with COPD.

2.3
Health Status

Health status has become a major feature of studies in COPD (Jones and Mahler 2002). The St. George's Respiratory Questionnaire (SGRQ) is a standardised self-completed questionnaire for measuring health-related quality of life (HRQOL) in airways disease (Jones et al. 1991; Jones 2001). The final version has 76 items divided into 3 sections: "symptoms", "activity" and "impacts" and provides a total score. Scores range from 0 (perfect health) to 100 (worst possible state), with a 4-point change in score considered a worthwhile treatment effect. The SGRQ and the Chronic Respiratory Disease Questionnaire (CRDQ) have been found equivalent in a comparative study (Rutten-van Molken et al. 1999). Generic questionnaires have a place in COPD studies, but are relatively insensitive, and include the Sickness Impact Profile (SIP) and SF-36 (Engstrom et al. 2001).

The Asthma Quality of Life Questionnaire (AQLQ) has 32 questions and takes 5–10 min to complete (Juniper et al. 1992). In mild asthma, symptom diary scores have been found less sensitive than the AQLQ and SGRQ (Barley and Jones 1999; Juniper et al. 2000). More recently a standardised version, the AQLQ(S), comprising 5 generic activities instead of 5 patient-specific activities, has been validated in asthma (Juniper et al. 1999a).

2.4
Airway Hyperreactivity

The classical methods of assessment of airway hyperreactivity are to establish the provocative concentration required to cause a 20% fall (PC_{20}) in FEV_1 to inhaled histamine or methacholine. Both agents cause direct constriction of airway smooth muscle, and are said to reflect non-specific AHR. Guidelines have been published by the American Thoracic Society (ATS) for methacholine bronchoprovocation (Crapo et al. 2000), and commercial sources of methacholine produced by good manufacturing practice (GMP) are available.

Adenosine monophosphate (AMP) is a purine nucleoside that causes bronchoconstriction by indirect mediator release from human mast cells (Polosa and Holgate 1997; Polosa et al. 2002), since it is unlikely that AMP acts on smooth muscle cells in vivo (Fozard and Hannon 2000). AMP may be especially useful in detecting inflammatory changes in adult and paediatric asthma, and it has been suggested that AHR to AMP relates to allergic background while AHR to methacholine is related to decreased airway calibre (Van Den et al. 2001b; De Meer et al. 2002). AMP reactivity has been usefully employed in clinical trials with inhaled corticosteroids (Holgate et al. 2000a; Van Den et al. 2001a).

2.5
Exercise Testing in COPD: 6MWT, ISWT, ESWT

For the design of studies on exercise-induced asthma, see later in this chapter. The ATS has published recent guidelines for the 6-min walk test (6MWT), since this is an important component in the functional assessment of patients with COPD (Enright and Sherrill 1998; ATS 2002a). The 6MWT is performed indoors on a long, flat, straight corridor of at least 30 m in length, and the use of a treadmill is not recommended. In a recent review it was found that the 6MWT is easy to administer and more reflective of activities of daily living than other walk tests (Solway et al. 2001). The 12-level incremental shuttle-walking test (ISWT) provokes a symptom-limited maximal performance exercise, but it uses an audio signal from a tape cassette to guide the speed of walking of the patient back and forth between two traffic cones on a 10-m course (Singh et al. 1992; Dyer et al. 2002). The shuttle-walking test has an incremental and progressive structure to assess functional capacity, and maximum heart rates are significantly higher than for the 6MWT. The endurance shuttle walk test (ESWT) also uses a 10 m course, but has an externally controlled constant walking speed (Revill et al. 1999).

2.6
Dynamic Hyperinflation

Dynamic lung hyperinflation has been shown to improve after bronchodilator therapy in COPD (O'Donnell et al. 1999; Newton et al. 2002). Dyspnoea ratings and measurements of inspiratory capacity and endurance time are highly reproducible and responsive to change in disease status (O'Donnell et al. 1998).

2.7
Muscle Testing

An ATS/European Respiratory Society (ERS) statement has recently been made on respiratory muscle testing (ATS 2002b). Cellular adaptations in the diaphragm have been reported in COPD (Polkey et al. 1996; Levine et al. 1997), and skeletal muscle apoptosis can occur in conjunction with cachexia in COPD (Reid 2001; Agusti et al. 2002), with weakness and dysfunction of voluntary skeletal muscle (Gosselink et al. 1996; ATS 1999b; Engelen et al. 2000). Mid-thigh muscle cross-sectional area is found to be a predictor of mortality in COPD (Marquis et al. 2002).

2.8
Blood Analysis

Peripheral blood tests are generally performed in phases I and II for pharmacokinetic evaluation (PK) of the amount of drug substance and its metabolites in the blood. Before the end of phase III it is usual to have established absorption, distribution, metabolism and excretion (ADME) profiles for oral drugs, although for inhaled drugs distribution in the lung is often more relevant than systemic exposure.

In asthma it is possible to assess inflammation by measuring such parameters as peripheral blood eosinophil counts, serum eosinophil cationic protein (ECP) levels, as well as T helper (Th)2 cell activation. However, these blood parameters are not particularly useful reflections of pulmonary inflammation. Similarly, there are peripheral blood abnormalities found in COPD, including enhanced neutrophil chemotaxis (Burnett et al. 1987), oxidative burst (Noguera et al. 2001) and adhesion molecules (Noguera et al. 1998).

The utility of pharmacokinetic/pharmacodynamic (PK/PD) modelling can be illustrated by taking the example of a hypothetical chemokine antagonist, since a large range of oral compounds have been developed against the numerous chemokine receptors (Fig. 2). In relation to allergy and asthma, the chemokine C-C receptor 3 (CCR3) is an interesting target expressed on a range of cells including eosinophils and activated Th2 cells, and a range of synthetic low MW chemicals have been developed that antagonise this receptor (Erin et al. 2002). Clinical testing of CCR3 antagonists can be performed in terms of whole blood flow cytometric assays to establish the pharmacodynamic duration of chemotactic inhibitory activity on eosinophils. The GAFS (gated autofluorescence forward scatter) assay makes use of the changes seen in the forward scatter of light due to eosinophil shape change in response to chemokines (Sabroe et al. 1999; Bryan et al. 2002).

This method has been adapted to be used in whole blood samples, with the major advantage that the GAFS assay can be used for both preclinical and clinical assessment of new chemokine receptor antagonists. It can be used in laboratory screening in the preclinical work up, when dose–response curves can be determined on volunteer or macaque monkey blood to which the novel CCR3 antagonist is added ex vivo, and then the response to a specific stimulus determined (Zhang et al. 2002). The assay can then be used in clinical studies for assisting dose range finding, since serial whole blood samples taken following CCR3 antagonist ingestion can be stimulated ex vivo to give sequential pharmacodynamic capacity. Since the pharmacodynamics can last longer than the pharmacokinetics (hysteresis), this assay will be useful in monitoring the lasting effects of chemokine antagonists on leukocytes in the blood even if the drug is no longer detectable in the plasma (Fig. 2).

2.9
Exhaled Breath

A range of substances have been analysed in exhaled breath and breath condensate from adults and children with lung disease (Table 2). This includes nitric oxide (NO), ethane, isoprostanes, leukotrienes, prostaglandins, cytokines, products of lipid peroxidation, and nitrogenous derivatives. In addition, exhaled breath temperature may serve as a simple and inexpensive method for home monitoring of patients with asthma and COPD, and for assessing the effects of anti-inflammatory treatments.

There is no single test that can be used to quantify airway inflammation, and a range of different specimens and markers of airway inflammation should be considered (Kharitonov and Barnes 2000, 2001b). Peripheral blood markers are unlikely to be adequate as the most important mediator and cellular responses occur locally within airways, while induced sputum originates from more proximal rather than small airways. The "exhaled breath profile" of several markers reflecting airway inflammation and oxidative stress is a promising approach for monitoring and management of patients with asthma and COPD. However, airway inflammation is not measured directly and routinely in clinical practice.

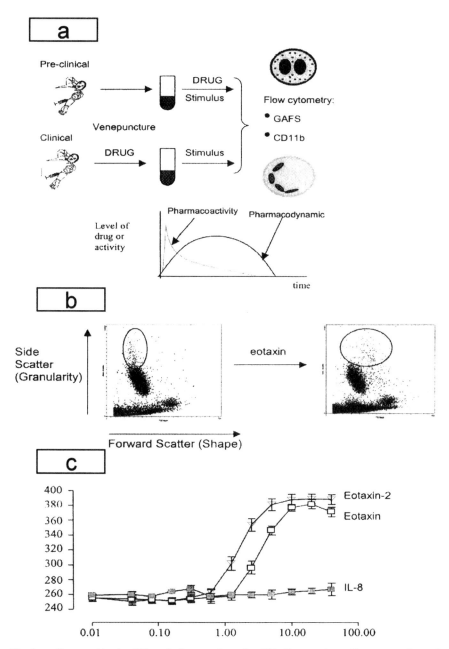

Fig. 2a–c Pharmacokinetics (PK) and pharmacodynamics (PD). The gated autofluorescence forward scatter (GAFS) method for determining eosinophil shape change. **a** Illustration of how the GAFS method can be employed to assess eosinophil shape in whole blood both pre-clinically, when novel drug is added to blood ex-vivo, and clinically, when the volunteer subject has received the therapeutic. Blood is taken at intervals to determine both the pharmacokinetics and pharmacodynamic profiles. **b** Representative flow cytometry dot plot showing the physical characteristics of defined (gated) populations of leukocytes. This illustrates the increased forward scatter of eosinophils following stimulation with eotaxin. **c** Eosinophil shape change induced by chemokines

Table 2 Exhaled breath markers

Marker	Reference(s)
Exhaled NO	
Exhaled and nasal NO	Kharitonov et al. 1994, 1996, 1997
Exhaled hydrocarbons	
Ethane	Paredi et al. 2000a,b,c
Exhaled breath condensate	
8-isoprostane	Montuschi et al. 1999, 2000a,b
Cys-LT, LTB4	Hanazawa et al. 2000a,b
IL-4, INF-γ	Shahid et al. 2002
Nitrite, nitrate, S-nitrosothiols, nitrotyrosine	Balint et al. 2001; Corradi et al. 2001
Proteins (IL-1β; TNF-α; IL-8)	Garey et al. 2000
H_2O_2	Horvath et al. 1998; Ferreira et al. 2001
Exhaled temperature	
Exhaled breath temperature	Paredi et al. 2002

Exhaled nitric oxide (NO) has been extensively studied, since levels were found to be increased in asthma (Kharitonov et al. 1994, 1997). Standardised measurements of fractional exhaled NO (FE_{NO}) provide a completely non-invasive means of monitoring of airway inflammation and anti-inflammatory treatment in asthma (ATS 1999a). It may be useful in patients using fixed combination inhalers (corticosteroids and long-acting β_2-agonist) to ensure that inflammation is controlled, as this may be difficult to assess from symptoms when a long-acting bronchodilator is taken.

Changes in serial FE_{NO} have higher predictive values for diagnosing deterioration of asthma (Jones et al. 2001) than single measurements (Kharitonov et al. 1996; Jatakanon et al. 2000). Reproducibility of FE_{NO} measurements within a single day in both adults (ICC 0.94) and children (ICC 0.94) is superior to any conventional methods of airway inflammation monitoring in asthma. This adds significantly to other advantages of FE_{NO} measurements, such as their strong association with airway inflammation (Kharitonov and Barnes 2001b), elevation even in non-symptomatic asthma patients (van Den Toorn et al. 2000), high sensitivity to steroid treatment (Kharitonov et al. 1996), and insensitivity to β_2-agonists (Kharitonov and Barnes 2001b).

Dose range finding is an important issue in development of new therapies, and there is high sensitivity and reproducibility of FE_{NO} measurements in relation to inhaled corticosteroids. Recently, a dose-dependent onset of anti-inflammatory action of inhaled corticosteroids has been demonstrated on FE_{NO} and asthma symptoms in a small number ($n=28$) of mild asthma patients who were treated with 100 or 400 μg budesonide, or placebo once daily for 3 weeks in a double-blind, placebo-controlled, parallel group study (Kharitinov et al. 2000; Kharitonov and Barnes 2001a). Long-term prospective clinical studies are necessary to prove that exhaled NO can be used to optimise doses of anti-inflammatory treatment. Breath analysis is currently a research procedure, but there is increasing evidence that it may have an important place in the diagnosis and management of lung diseases in the future (Kharitonov and Barnes 2000). This will drive the development of cheaper and more convenient analysers, which can be used in a hospital and later as personal monitoring devices for use at home by patients in clinical trials.

2.10
Sputum Analysis

The analysis of induced sputum is increasingly performed as a means of assessing airway inflammation (Gibson et al. 1989; Pin et al. 1992). Sputum induction by the inhalation of nebulised hypertonic saline has been demonstrated to be a relatively safe and effective method of enabling sputum expectoration (Fahy 1998; Vlachos-Mayer et al. 2000). Sputum processing generally consists of selecting the solid mucinous portion of sputum that is liquefied by reduction with dithiothreitol (DTT), and using the cellular portion to produce cytospins for light microscopy, while employing the supernatant for analysis of soluble proteins and mediators. In response to the expanding volume of sputum research, the European Respiratory Society has set up an international task force to standardise sputum induction and processing methodology, and their important report is now completed (European Respiratory Society Task Force 2002).

In ancient Greece phlegm was regarded by Hippocrates as one of the four humours fundamental to disease. Yet it was not until the nineteenth century that microscopy was used to demonstrate the presence of sputum eosinophilia in asthma (Gollasch 1889) that was later shown to be associated with Charcot-Leyden crystals and Curschmann's spirals (Sakula 1986). In the 1950s, sputum microscopy was used in the diagnosis and monitoring of asthmatic patients (Hansel 1953; Brown 1958). F.E. Hargeave and his team in Hamilton, Ontario (Canada) have done most in recent years to establish sputum analysis, including providing detailed methods of sputum induction, processing and microscopy (Efthimiadis et al. 1997). Nevertheless, a series of reviews and editorials reflect the considerable variations in methodology still present (O'Byrne and Inman 1996; Pavord et al. 1997; Fahy 1998; Kips et al. 1998; Kips and Pauwels 1998; Magnussen and Holz 1999; Djukanovic 2000; Holz et al. 2000; Jayaram et al. 2000; Magnussen et al. 2000).

Pretreatment with a bronchodilator, the duration of sputum induction, the output of the nebuliser, and the tonicity of saline all influence the frequency of bronchospasm during sputum induction (Schoeffel et al. 1981; Fahy 1998; Wong and Fahy 2002). However, despite bronchoprotection, it has been demonstrated—in a retrospective review of 351 inductions—that 12% of asthmatic subjects developed bronchospasm (Wong and Fahy 2002). In a study in severe asthma, sputum induction had to be stopped due to side effects in 12% of patients (de la Fuente et al. 1998). Furthermore, sputum induction has been performed in asthmatic patients with an FEV_1 of less than 60% of predicted (Vlachos-Mayer et al. 2000), while isotonic saline has been used in adults with acute severe exacerbations of asthma (Wark et al. 2001). In addition, the safety of sputum induction has been investigated in a multicentre study (Fahy et al. 2001). In these studies, subjects with bronchoconstriction induced by saline generally responded promptly to β_2-agonist treatment.

Induced sputum from asthmatics yields better quality cytospins with less squamous cell contamination than spontaneous sputum; although the leukocyte differential is similar (Pizzichini et al. 1996b). The hypertonicity of inhaled saline does not affect sputum cell composition (Bacci et al. 1996), but may cause reduced interleukin (IL)-8 levels (Belda et al. 2001). Analysis of sequential induced sputum samples has shown that the percentages of eosinophils and neutrophils are significantly higher at the beginning of the sputum induction than at the end (Holz et al. 1998a; Gershman et al. 1999; Moodley et al. 2000), these differences in sputum composition being more pronounced in healthy subjects than in asthmatic or COPD patients (Richter et al. 1999). Studies have compared two sequentially induced sputum samples performed 24 h apart in healthy and asthmatic subjects (Holz et al. 1998b; Nightingale et al. 1998), and shown a significant increase in the proportion of neutrophils with a decrease in macrophages at 24 h.

The solid mucinous portion of sputum is generally selected from saliva (Gibson et al. 1989; Pin 1992; Popov et al. 1994; Pizzichini et al. 1996a), although whole expectorate is analysed by some workers (De Gouw et al. 1996; Fahy et al. 2002). The introduction of dithiothreitol (DTT) to release the entrapped cells, and the use of cytospins to generate a monolayer of sputum cells (Fleury-Feith et al. 1987) has made the quantification and characterisation of cells more reliable (Pizzichini 1996a). DTT, while useful for obtaining cells, may interfere with immunological detection techniques (Woolhouse et al. 2002). Several reports have shown effects of DTT on the immunological detection of cell surface markers by flow cytometry (Hansel et al. 1991; Qiu and Tan 1999; Loppow et al. 2000).

An elegant methodological study of IL-5 measurement in sputum supernatants identified considerable technical problems due to DTT (Kelly et al. 2000). A range of validation experiments demonstrated that DTT did not interfere with the enzyme-linked immunosorbent assay (ELISA) and that the IL-5 molecule remained intact, but the recovery of spiked IL-5 was poor. Recently, a subsequent study by the same group suggests that proteolytic activity in the sputum sample may be interfering with the IL-5 measurements. Addition of protease inhibitors significantly increased the levels of detectable IL-5 (Kelly et al. 2001).

Differential cell counts on sputum cytospins have good interobserver consistency (Boulet et al. 1987; Popov 1994; Gelder et al. 1995; Pizzichini et al. 1996a). Similarly, the repeatability of differential cell counts and the measurement of soluble mediators in samples, obtained on different days from clinically stable patients, has been reported to be good (Boulet 1992). In a multicentre study, the reproducibility of measurements of cellular and fluid phase parameters in asthmatics were similar and without any significant centre effects (Pizzichini et al.1996a). The cell fraction and the fluid phase of induced sputum differ between asthmatics, smokers with COPD, and healthy subjects, indicating that sputum analysis can be used to assist the diagnosis of these conditions (Gibson et al. 1989; Kidney et al. 1996; Pizzichini et al. 1996a).

The relationship between the cellular content in sputum and airway tissue has been studied by comparing the cellular composition of hypertonic saline-induced sputum with bronchoalveolar lavage fluid (BAL) and bronchial mucosal biopsies. One such study compared induced sputum with bronchoscopic bronchial washing and BAL in healthy volunteers and asthmatics. It showed that sputum had a higher proportion of nonsquamous cells and higher levels of eosinophil cationic protein (ECP), albumin, and mucin-like glycoprotein. The eosinophil numbers and ECP levels in sputum correlated more closely with those in bronchial washing than in BAL. The proportion of eosinophils was higher in sputum and bronchial washing from asthmatic subjects compared with healthy volunteers (Fahy et al. 1995). In mild to moderate atopic asthmatics the percentage of eosinophils in sputum was significantly correlated with that in bronchial wash and in BAL, whilst there was a trend towards such a correlation between the number of eosinophils in sputum and the number of $EG2^+$ eosinophils in bronchial biopsies. In addition, the proportion of $CD4^+$ lymphocytes correlates between sputum and BAL (Grootendorst et al. 1997; Pizzichini et al. 1998). Another group has shown induced sputum to be relatively rich in eosinophils and neutrophils, but with less lymphocytes and macrophages compared with bronchial washings and BAL fluid (Keatings et al. 1997).

2.11
BAL and Biopsy

Studies with BAL and mucosal bronchial biopsy are important mechanistic studies for the evaluation of anti-inflammatory therapies for asthma and COPD (Robinson 1998). A European Society Task Force has issued guidelines for measurements of acellular components and standardization of BAL (Haslam et al. 1999), while the reproducibility of endobronchial biopsy has been established (Faul et al. 1999). Fluticasone has been shown to lack effects on neutrophils and $CD8^+$ T cells in biopsies from patients with COPD, but have subtle effects on mast cells (Hattotuwa et al. 2002). Fluticasone has had more convincing effects on biopsies from patients with asthma, reducing eosinophils, macrophages and T cells (Faul et al. 1998; Ward et al. 2002). There is high reproducibility of repeat measures of airway inflammation in stable atopic asthma (Faul et al. 1999).

2.12
High-Resolution Computerised Tomography

High-resolution computerised tomography (HRCT) is defined as thin-section CT (1- to 2-mm collimation scans), while spiral (helical) CT provides continuous scanning while the patient is moved through the CT gantry. HRCT is useful in studies on patients with COPD to assess the extent of airway, interstitial and vascular disease (Muller and Coxson 2002), as well as detecting bronchiectasis (O'Brien et al. 2000) and early cancerous lesions. HRCT is sufficiently sensitive to monitor longitudinal changes in the extent of emphysema (Cleverley and Muller 2000; Cosio and Snider 2001; Ferretti et al. 2001), and it has been shown that annual changes in lung density and percentage of low-attenuation area are detectable with inspiratory HRCT (Soejima et al. 2000). Inflammatory infiltration of the lung parenchyma is associated with ground glass attenuation, bronchiolitis can be visualised as parenchymal micronodules, bronchial wall thickening may occur, and emphysema is manifest by areas of decreased attenuation. Progression from parenchymal micronodules (bronchiolitis) to emphysema has been demonstrated in a proportion of smokers over 5 years (Remy-Jardin et al. 1993; Remy-Jardin et al. 2002). It appears that subtle changes in progression of lung disease related to α_1-antitrysin deficiency may be more readily detected with CT imaging than with pulmonary function testing (Dowson et al. 2001a,b), and with α_1-antitrysin augmentation therapy annual CT has proved useful in disease assessment (Dirksen et al. 1997, 1999). In addition, quantitative CT has also been used as an outcome measure in a study of all-*trans*-retinoic acid (ATRA) therapy (Mao et al. 2002). Finally, HRCT can be used to visualise the small airways and interstitium (King et al. 1999; Hansell 2001), and has been used in studies on patients with asthma (Mclean et al. 1998; Brown et al. 2001). Nevertheless, there are considerable challenges in the standardisation of CT and in the development of methodology for image processing and analysis.

2.13
Pharmacogenetics

Pharmacogenetics studies have increasing relevance to the practice of medicine (Roses 2000), and potential utility for clinical studies in patients with asthma (Palmer et al. 2002) (see previous chapter, this volume). Multiple single nucleotide polymorphisms (SNPs) in the β-adrenoceptor of smooth muscle influence responses to β-agonist (Drysdale et al. 2000), while abnormalities in the ALOX5 gene for leukotriene metabolism caused decreased responses to a 5-lipoxygenase (5LO) inhibitor (Drazen et al. 1999b); and variation in the tumour necrosis factor (TNF)-α promoter may influence response in TNF-α-directed therapy (Witte et al. 2002). In addition the ADAM33 gene, corresponding to a membrane-anchored metalloprotease (MMP) is a putative asthma susceptibility gene (Van Eerdewegh et al. 2002). Patients with COPD may have an amplification of the normal inflammatory response to cigarettes, that involves a susceptibility to an environmental factor (Sandford and Silverman 2002).

2.14
Mortality

FEV_1 has been demonstrated to correlate with mortality in COPD (Fletcher and Peto 1977). Weight loss, particularly loss of fat free mass, and a low body weight are independently unfavourable for survival of COPD patients (Schols et al. 1998). Regular use of fluticasone alone or in combination with salmeterol is associated with increased survival of COPD patients in primary care (Soriano et al. 2002). Furthermore, the regular use of low-dose inhaled corticosteroids is associated with a decreased risk of death from asthma (Suissa et al. 2000).

3
Clinical Trial Designs

3.1
Trial and Error with Herbal Remedies

Herbal remedies have been extensively used as bronchodilators; and β_2-adrenoceptor agonists, anti-cholinergics and theophyllines are all based on parent compounds from plant extracts (Bielory and Lupoli 1999). Indeed the use of herbal medications goes back to around 1000 B.C., when the ancient Chinese treated asthma with *ma huang*, an ephedrine-containing extract that has adrenoceptor bronchodilator properties. In ancient times assessment of effects of herbal extracts was based on trial and error, and many subjects were harmed before useful agents could be identified.

3.2
Tolerability

Inhaled therapies for respiratory disease may be tested in incremental ascending dose studies in asthma, when serial FEV_1 is generally the preferred endpoint, without the need for plethysmography. Many inhaled agents are well tolerated in non-asthmatic volunteers, but may be poorly tolerated in asthmatics. Novel inhaled therapies for COPD are often best studied in volunteers with mild asthma, since these subjects have responsive airways, and may be more sensitive to potential bronchoconstrictor effects.

3.3
Bronchodilators in Asthma

Asthma can be rapidly and numerically assessed on the basis of peak expiratory flow (PEF) or FEV_1. Acute bronchodilation can be studied in patients with mild asthma who have sufficient bronchoconstriction at baseline, so that there is "room to improve" FEV_1 following single dose administration (Fig. 3).

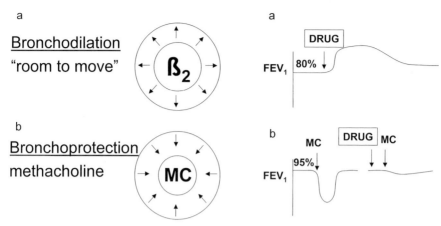

Fig. 3a, b Diagram to illustrate the study of therapies that cause bronchodilation as distinct from bronchoprotection. **a** To assess bronchodilation, the patient with asthma or COPD should have "room to move". The airway in the *upper half* of the figure is shown in cross-section and is constricted, while inhalation of a β_2-agonist (β_2) causes bronchodilation. The patient begins with an impaired 80% FEV_1 as a percentage of predicted, and inhalation of the β_2-agonist causes bronchodilation manifested as an increase in the FEV_1 to 100%. **b** For the assessment of bronchoprotection, the subject with asthma may have near normal lung function at the outset, with an FEV_1 of 95–100% predicted. Inhalation of gradually increasing doses of nebulised methacholine (*MC*) will establish the provocative concentration of methacholine to cause a 20% fall in FEV_1 (PC_{20}). Following the investigational drug, there may be protection against the bronchoconstrictor effects of methacholine so that an increased concentration of methacholine is required to cause bronchoconstriction, so there is an increased methacholine PC_{20}

3.4
Bronchodilators in COPD

The prolonged effects of the inhaled anticholinergic agent tiotropium bromide have been demonstrated in terms of bronchoprotection against methacholine-induced bronchoconstriction in asthma (O'Connor et al. 1996) and the dose–response relationship established in COPD (Maesen et al. 1995). Acute bronchodilator trials in COPD have special clinical trial requirements (Nisar et al. 1992; Rees 1998): there is the need for additional measurements of exercise dynamic hyperinflation, exercise endurance, assessment of dyspnoea and HRQOL (O'Donnell 2000).

3.5
Bronchoprotection

Bronchoprotection can be evaluated in relation to bronchoconstrictor stimuli such as histamine, methacholine (Crapo et al. 2000), adenosine monophosphate (AMP), leukotrienes, hyperventilation and exercise.

3.6
Inhaled Allergen Challenge

Inhaled allergen challenge responses have been extensively studied in relation to drugs, and effects on the early and late asthmatic reactions (EAR and LAR) as well as airway hyperreactivity provide important mechanistic insights (Fig. 4). The inhaled allergen challenge offers the opportunity to study effects on the EAR and LAR, blood and sputum eosinophils, exhaled breath nitric oxide (NO) and methacholine airway responsiveness (PC_{20}). The reproducibility of the inhaled allergen challenge is excellent, and 12 patients is adequate to reliably demonstrate 50% attenuation of the early or late asthmatic reaction with greater than 90% power (Inman et al. 1995). Recently the method of bolus as opposed to incremental allergen challenge has been validated (Arshad 2000; Taylor et al. 2000). Inhaled allergen challenge should ideally not be repeated at less than 3-week intervals, due to residual AHR (Rasmussen 1991). The use of bolus dose allergen challenge for repeated tests in the same patient is a safe and validated method to administer inhaled allergen in clinical trials with valid responses when compared to incremental dose allergen challenge (Arshad 2000; Taylor et al. 2000).

Of the anti-inflammatory drugs effective in controlling asthma, all inhibit the LAR to allergen: this includes steroids, theophyllines, leukotriene antagonists, cromones, cyclosporin A, anti-IgE (see Table 3 with references attached) (Morley 1992; Boushey and Fahy 2000). However, a number of other agents that are not effective therapy also cause some inhibition of the LAR, including furosemide, heparin, PGE2. Hence, the inhaled allergen challenge LAR is useful for position-

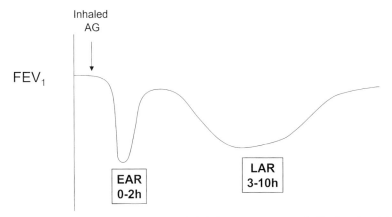

Fig. 4 The dual asthmatic response to inhaled allergen. A patient with allergic asthma inhales increasing incremental doses of allergen to which they have been demonstrated to be sensitive on epidermal skin prick testing. The allergen may be grass or tree pollen, cat or dog dander, or house dust mite. A proportion of these subjects develops a dual asthmatic response, in which there is an early asthmatic reaction (*EAR*), involving a bronchoconstriction at 0–2 h; followed by a late asthmatic reaction (*LAR*), involving bronchoconstriction at 3–10 h

Table 3 Effects of selected therapeutics on inhaled allergen challenge responses

Therapeutic Agent	Comment	Reference(s)
Inhaled corticosteroids	Fluticasone 250 μg has equivalent effects to 250 μg b.i.d. for 2 weeks on LAR	Parameswaran et al. 2000
	Fluticasone (1000 μg daily for 2 weeks) causes an 80% decrease in LAR by maximum fall in FEV_1	O'Shaughnessy et al. 1993
	Budesonide (400 μg daily for 8 days) causes 94% decrease in LAR AUC	Wood et al. 1999
Prednisone		Liu et al. 2001
Theophyllines		Sullivan et al. 1994; Pauwels 1989
Leukotriene antagonists	Montelukast: 3 oral doses: 75% inhibition of EAR and 57% inhibition of LAR	Fuller et al. 1989; Rasmussen et al. 1992; Pizzichini et al. 1998; Diamant et al. 1999; Wenzel 1999
Cromones	Nedocromil 4 mg by pressurised aerosol inhibits causes a 64% decrease in EAR and a 58.8% decrease LAR by maximum fall in FEV_1	Crimi et al. 1989; Twentyman et al. 1993; Laube et al. 1998
Salbutamol		Howarth et al. 1985; Twentyman et al. 1991, 1993
Terbutaline		Wong et al. 1994
Salmeterol		Twentyman et al. 1990; Taylor et al. 1992; Pedersen et al. 1993; Weersink et al. 1994; Dente et al. 1999; Giannini et al. 1999, 2001; Weersink et al. 1994; Calhoun et al. 2001
Ipratropium bromide		Howarth et al. 1985
Formoterol		Palmqvist et al. 1992; Wong et al. 1992
Cyclosporin A	50% decrease in LAR by AUC	Sihra et al. 1997; Khan et al. 2000
Anti-IgE	Intravenous anti-IgE inhibits the EAR by 37% and LAR by 62%	Boulet et al. 1997; Fahy et al. 1999
	Inhaled anti-IgE fails to inhibit LAR	Boulet et al. 1997
IL-12	No significant effect on LAR	Bryan et al. 2000
Anti-IL-5	No significant effect on LAR	Leckie et al. 2000
Anti-histamine		Roquet et al. 1997
Aspirin-like agents	Sodium salicylate: 23% inhibition in LAR	Sestini et al. 1999
	Indomethacin: 39% inhibition of LAR	
	Lysine acetylsalicylate: 44% inhibition of LAR	
Prostaglandin E2		Pavord et al. 1993; Gauvreau et al. 1999; Hartert et al. 2000

Table 3 (continued)

Therapeutic Agent	Comment	Reference(s)
Heparin	36% inhibition of LAR	Diamant et al. 1996
Furosemide	Inhaled furosemide: inhibition of EAR and LAR	Bianco et al. 1989
Phosphodiesterase (PDE)4 inhibitor	CDP840: 30% inhibition of LAR	Harbinson et al. 1997, Kanehiro et al. 2001
	Roflumilast: 43% inhibition of LAR	van Schalkwyk et al. 2002
Mast cell tryptase inhibitor		Krishna et al. 2001
Platelet-activating factor (PAF) antagonist	UK-74,505: no sig. effect on EAR or LAR WEB 2086: no sig. effect on EAR or LAR SR27417A: 26% inhibition of LAR	Kuitert et al. 1993 Freitag et al. 1993 Evans et al. 1997a

ing a new drug relative to established asthma therapy, but offers a relatively low hurdle due to false positives. However, we cannot identify anti-inflammatory therapeutics for asthma that are "false negatives", in which there is no effect on the LAR but clinical efficacy.

Of especial interest in relation to inhaled allergen challenge responses has been a monoclonal antibody directed against IL-5, since IL-5 has an established role as the major terminal differentiation factor during eosinopoiesis in the bone marrow (Hamelmann and Gelfand 2001; Sampson 2001). An initial study in man involved a single intravenous infusion of a humanised MoAb directed against IL-5 (SB240563) being given to mild allergic asthmatics in a parallel group, double-blind clinical trial (Leckie et al. 2000), the design of which has been criticised (O'Byrne et al. 2001; Hansel et al. 2002). There was pronounced suppression of peripheral blood eosinophil levels for 16 weeks and considerably reduced numbers of sputum eosinophils after allergen challenge. However, despite these clear effects, anti-IL-5 did not protect against the allergen-induced LAR and did not inhibit baseline or post-allergen AHR. Hence, the eosinophil did not seem to be a prerequisite for either AHR or the LAR, and this study suggests that anti-IL-5 therapy may not be clinically useful in the short term for asthma therapy. Interestingly, recent clinical studies in human allergic asthma have also found dissociation between AHR and eosinophil levels (Crimi et al. 1998; Rosi et al. 1999).

Interpretation of the anti-IL-5 effects on inhaled allergen challenge must be made with caution, because even though eosinophil numbers in blood and sputum were reduced by anti-IL-5, it has recently been reported that there are still residual eosinophils within bronchial mucosal biopsies (late abstract from Dr. P. Flood-Page et al. at American Academy for Allergy, Asthma and Immunology, AAAAI, New York 2002). This suggests that anti-IL-5 is not removing eosinophils from the airway wall, and that additional or alternative therapy will be required to ablate the eosinophil from the airways. A remaining concern is that an inhaled allergen challenge study is an artificial "model" situation, and effects on clinical symptoms and lung function in patients with symptomatic asthma were not measured. However, there are preliminary reports that MoAb versus IL-5 is not clinically effective in treating asthma of different degrees of severity (Kips et al. 2000, 2001). Hence, further clinical studies are required since anti-IL-5 could prevent or inhibit airways remodelling as well as exacerbations, although long-term studies will probably be required to assess this potential.

In the immediate future, the inhaled allergen challenge model is likely to be extensively employed in studying the pathogenesis of asthma inflammation and analysis of the relationship with bronchoconstriction and AHR. Allergen challenge is also likely to remain at the forefront for assessment of new therapeutics in initial clinical trials and to define novel and relevant targets for new drugs. However, with our increasing understanding of the allergen challenge model is recognition that this response involves complex interplay between leukocytes, numerous tissue cell types, cytokines, chemokines and inflammatory mediators.

3.7
Nasal Allergen Challenge

There is a functional and immunological relationship between the nose and bronchi, both have a ciliated respiratory epithelium, and both can be used to study mechanisms of eosinophil influx (Passalacqua et al. 2001; de Benedictis et al. 2001; Passalacqua and Canonica 2001; Vignola et al. 2001). The nose and bronchi can be challenged with allergens, chemokines and a variety of stimuli. However, it is much easier to recruit subjects with grass pollen allergic rhinitis for nasal allergen studies then it is to find subjects for an inhaled allergen challenge study. In addition, nasal challenge is less clinically stressful than inhaled challenge, and the nose can be more readily biopsied than the bronchi.

A variety of non-invasive techniques can be used to study changes in the nose following challenge. Nasal lavages permit studies on the cellular influx and release of cytokines and chemokines (Greiff et al. 1990). Following nasal allergen challenge, the kinetics of eotaxin release has been studies in nasal fluid (Greiff et al. 2001; Terada et al. 2001). Following topical nasal budesonide, nasal allergen challenge caused decreased levels of GM-CSF and IL-5 in nasal fluids collected on filter paper strips (Linden et al. 2000). Nasal challenges have also been performed with eotaxin (Hanazawa et al. 1999) and IL-8 (Douglass et al. 1994). Nasal rhinomanometry and nasal symptoms and health status can also be determined (Juniper and Guyatt 1991; Juniper et al. 2000). Nasal corticosteroids have effects on nasal late reactions (Pipkorn et al. 1987; Rak et al. 1994; Ciprandi et al. 2001), but it is difficult to characterise a well-defined late response as obtained in the bronchi (Gronborg et al. 1993).

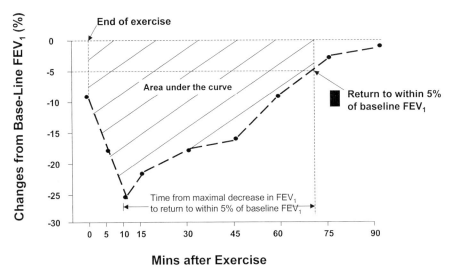

Fig. 5 Exercise challenge in asthma. After a 2-min warm up patients exercise on a treadmill for 6 min. Spirometry is performed at intervals after completion of exercise to 60 min. If after 60 min the FEV_1 has not returned to within 5% of the pre-exercise value, additional measurements can be made at 75 and 90 min after exercise. The following endpoints may be assessed: the area under the curve (AUC) for the percentage decrease in FEV_1 in the first 60 min after exercise, the maximal decrease in FEV_1 after exercise, and the time from the maximal decrease in FEV_1 to the return to within 5% of the FEV_1 value before exercise. (Adapted from Leff et al. 1998)

3.8
Exercise-Induced Asthma

For patients with asthma, the ATS has issued guidelines for exercise challenge (Crapo et al. 2000)(Fig. 5). After a 2-min warm up, patients generally exercise for a 6-min period on a treadmill while inhaling compressed dry air at room temperature through a facemask, with a nose clip in place. Exercise is performed at a level that increases the heart rate to 80%–90% of the age-predicted maximum. Spirometry is then performed at intervals up to 90 min after exercise, with the area under the curve (AUC) estimated for the first 60 min after challenge. Leukotriene antagonists have been demonstrated to be effective in exercise-induced bronchoconstriction (Manning et al. 1990; Kemp et al. 1998; Leff et al. 1998). However, H_1-receptor antagonism with loratadine does not protect against exercise-induced bronchoconstriction (Anderson and Brannan 2002; Dahlen et al. 2002).

3.9
Symptomatic Asthma

A range of recent studies have documented effects of anti-leukotrienes on patients with symptomatic mild to moderate asthma that are on inhaled β_2-agonists only (Crapo et al. 2000). These study designs provide a more clinically relevant "wild-type" population for studying potential efficacy than utilising an allergen challenge. An important study was to demonstrate that regular use of an inhaled corticosteroid (budesonide) is superior to regular use of inhaled β_2-agonist (terbutaline) (Haahtela et al. 1991), but that discontinuation of therapy is often accompanied by exacerbation of asthma (Haahtela et al. 1994). Convincing efficacy of inhaled corticosteroids can be demonstrated in groups of less than 12 symptomatic patients within 4 weeks of therapy (Jatakanon et al. 1999; Lim et al. 1999).

3.10
Studies with Inhaled Corticosteroids

A whole range of important clinical studies are still being performed on inhaled corticosteroids (ICS) themselves. These address such questions as dose–response finding (Kelly 1998; O'Byrne and Pedersen 1998; Taylor et al. 1999; Jatakanon et al 1999), the possibility of once-a-day steroids, establishing a therapeutic index (Kelly 1998; O'Byrne and Pedersen 1998) and effects on bones and growth (Efthimiou and Barnes 1998). Acute anti-inflammatory effects of inhaled budesonide can be demonstrated by a single dose of 2,500 µg (Gibson et al. 2001).

3.11
Corticosteroid Add-on Studies

Since many patients with asthma are treated with ICS, it is important to consider a range of trial designs in these subjects (Fig. 6). Theophylline and montelukast have been studied as add-on agents in asthmatics already receiving in-

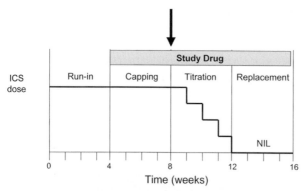

Fig. 6 The majority of patients with moderate and severe asthma are generally taking inhaled corticosteroid (*ICS*) medication. Following a run-in period when patients are generally shown to be symptomatic, the study drug can be added on in a "capping" design. Gradual titration of a dose of ICS may be performed on a weekly basis, alternatively there may be an abrupt discontinuation of ICS

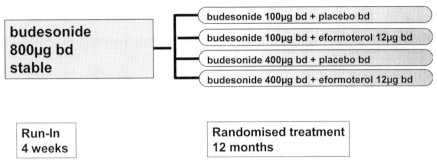

Fig. 7 The design of the FACET (Formoterol and Corticosteroid Enabling Therapy) study. The effect of inhaled formoterol and budesonide on exacerbation of asthma was studied (Pauwels et al. 1997). During the run-in period of 4 weeks, patients had to have stable asthma. In the randomisation treatment of 12 months, all patients had a reduction of budesonide, with two of the four groups receiving formoterol. *bd*, Twice daily

haled and oral corticosteroids (Evans et al. 1997b; Laviolette et al. 1999) (Fig. 7). A major clinical development in the past decade is the recognition of "add-on" therapy with long-acting β_2-agonists and moderate doses of inhaled corticosteroids (NIH and NHLBI 2002), and the availability of combination products with these agents in a single inhaler (Nelson 2001; Zetterstrom et al. 2001). The leukotriene antagonists have been demonstrated to be of lesser efficacy than long-acting β_2-agonists as add-on therapy (Drazen et al. 1999a; Laviolette et al. 1999).

3.12
Corticosteroid Titration Studies

Corticosteroid titration is a system for stepwise reduction in the dose of ICS (Gibson et al. 1992; Veen et al. 1999). Patients with asthma may need to be screened for this type of study, since only some asthmatics become immediately symptomatic on decreasing their dose of inhaled steroids. Many asthmatics will remain nonsymptomatic even on completely withdrawing their inhaled steroid, but can suddenly have an exacerbation due to infection, allergen or other triggers.

3.13
Corticosteroid Abrupt Discontinuation Studies

Soluble IL-4 receptor (Nuvance, Immunex) binds IL-4 before it can bind to the cell receptor, suppressing IgE production by B cells, and eosinophil migration into the airways. Inhaled Nuvance proved very promising in preliminary studies involving abrupt withdrawal of inhaled corticosteroids in patients with asthma

(Borish et al. 1999; Borish et al. 2001), but larger scale clinical studies in both mild and moderate asthma were discontinued in 2001 due to lack of efficacy. Discontinuation of inhaled fluticasone in COPD has been shown to cause higher risk of exacerbations (van der Valk et al. 2002).

3.14
Severe Asthma

Patients with severe asthma have disease involving more fixed airways obstruction, airways remodelling, and a tendency for neutrophilic as well as eosinophilic and IgE-mediated disease. A chimaeric monoclonal antibody against CD4 has recently been shown to be effective in chronic severe asthma (Kon et al. 1998).

Clinical studies have been carried out in patients with severe asthma to assess cyclosporin A (Lock et al. 1996) and recently with anti-IgE (Milgrom et al. 1999; Busse et al. 2001; Milgrom et al. 2001; Soler et al. 2001). Novel therapies in the future will need to be tested on patients receiving both inhaled corticosteroids and long-acting β_2-agonists (Fig. 8). This is because combined use of inhaled steroids and inhaled long-acting β_2-agonists is now established as therapy for moderate persistent asthma (NIH and NHLBI 2002).

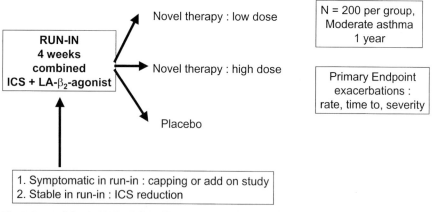

Fig. 8 Proposal for the design of a study to assess a novel therapy in patients receiving combined inhaled corticosteroids (*ICS*) and long-acting β_2-agonists (LABA). During the run-in phase patients may be either symptomatic or stable (cf. FACET). Patients with moderate persistent asthma (GINA step 3) are now recommended ICS and LABA as first-line daily controller medication [National Institutes of Health (NIH) and National Heart Lung and Blood Institute (NHLBI) 2002]. It is also possible to incorporate ICS or oral corticosteroid titration within this design

3.15
Natural History and Disease Modification

It is a major challenge to study long-term effects of drugs on the airways in the context of structural changes in the airways that comprise airways remodelling (Redington and Howarth 1997; Bousquet et al. 2000; Holgate et al. 2000b). Current asthma therapy is palliative and neither curative nor disease modifying, and only a minority of asthmatics achieve a long-lasting remission (van Essen-Zandvliet et al. 1994). There are considerable ethical and clinical trial issues in studying the influence of immunomodulatory agents in the context of allergen and peptide immunotherapy. Agents such as IL-12, CpG oligodeoxynucleotides and *Mycobacterium vaccae* (SRL 172) have potential as adjuvants, but therapy may be required in childhood for genetically susceptible individuals in the context of allergen therapy (Huang 1998; Hasko and Szabo 1999; Van Uden and Raz 1999). As with disease-modifying anti-rheumatoid therapies (DMART) it is important to demonstrate disease modification in terms of prolonged efficacy following cessation of treatment.

So far no therapeutic intervention except for smoking cessation has been shown to reduce the rate of loss of lung function in COPD (Scanlon et al. 2000). Studies on the natural history of COPD involve monitoring post-bronchodilator FEV_1 in considerable numbers of patients over a number of years. Four major studies have recently been performed on effects of inhaled corticosteroids (ICS) on the longitudinal decline in post-bronchodilator FEV_1 over 3 years in patients with mild (Pauwels et al. 1999; Vestbo et al. 1999) or moderate to severe COPD (Burge 1999; Burge et al. 2000; Lung Health Study Research Group 2000). Although effects on FEV_1 are modest, ICS are effective in preventing exacerbations in the moderate to severe group, and cause improvement in health status. The European points to consider on COPD studies suggest that studies on symptomatic relief should be for at least 6 months, while to claim that a therapy prevents disease progression will require prolonged studies (EMEA CPMP 1999). They also suggest that one design to assess effects on disease progression is to randomise patients to continue or stop treatment after a long period of treatment, and then examine the rapidity and extent at which benefit is lost. Design strategies for longitudinal spirometry have recently been elegantly reviewed (Wang et al. 2000).

Oral Ariflo is a second generation phosphodiesterase (PDE) type 4 inhibitor that selectively inhibits the low-affinity rolipram binding conformer of PDE4 (PDE4L) (Giembycz 2002), in an effort to minimise potential for nausea and vomiting (Torphy et al. 1999). In a 6-week study in 424 patients with COPD, Ariflo at 15 mg b.i.d. resulted in a maximum mean difference in trough clinic FEV_1 compared to placebo of 160 ml, representing an 11% improvement. This study included current and ex-smokers, with patients not currently receiving inhaled corticosteroids. It is encouraging that effects on FEV_1 could be demonstrated over a 6-week period in a group of approximately 100 subjects with

COPD. Large-scale studies will be required to assess whether Ariflo may be able to affect the natural history of COPD.

Roflumilast has been described as a third-generation PDE4 inhibitor that is non-selective for PDE4 isoenzymes B and D. Roflumilast has been demonstrated to be efficacious in causing 43% inhibition of the LAR following inhaled allergen challenge (van Schalkwyk et al. 2002), has achieved comparable efficacy to beclomethasone in asthma (Albrecht et al. 2002), has effects in exercise-induced asthma (Timmer et al. 2002), and has been shown to be effective on FEV_1 in COPD (Bredenbroker et al. 2002).

3.16
Weight Loss and Muscle Wasting in COPD

There is increasing evidence for a role for TNF-α in asthma (Thomas 2001), and a soluble TNF receptor construct (Nuvance, Immunex) and a monoclonal antibody (Remicade, Centocor) are licensed for use in severe rheumatoid arthritis and Crohn's disease (Feldman et al. 1998). These agents are attractive candidates as add-ons for severe asthma, as well as the treatment of cachexia associated with COPD. Should efficacy be demonstrated in proof-of-concept studies, synthetic low MW chemicals could be used to target TNF-α. Emiel Wouters and colleagues have performed extensive studies on subjects with weight loss and nutritional depletion in patients with COPD (Schols et al. 1993, 1995, 1998; Wouters 2000). Indeed it would be possible to study relatively small numbers of well-defined patients with extra-pulmonary systemic nutritional depletion in clinical trials.

3.17
Exacerbations in Asthma: Prevention and Treatment

At present we do not have a universally accepted definition of an exacerbation of asthma (Tattersfield et al. 1999). The FACET study with inhaled formoterol is an important example of a large-scale 12-month study that prospectively studied exacerbations of asthma (Pauwels et al. 1997; Kips and Pauwels 2001). A humanised anti-IgE monoclonal antibody (Xolair, Novartis-Genentech) is the first biotechnology product to be licensed for treatment of severe asthma (Chang 2000). Subcutaneous injections of recombinant human anti-IgE (rhuM-Ab-E25) have now completed a range of studies in moderate to severe asthma that have noted efficacy in terms of prevention of exacerbations and a decreased need for oral and inhaled steroids (Milgrom et al. 1999; Soler et al. 2001; Busse et al. 2001; Milgrom et al. 2001). Anti-IgE causes a reduction in the early and late phases following allergen challenge and decreased eosinophil counts in the sputum (Fahy et al. 1999).

Treatment of emergency asthma is a distinct clinical need with a requirement for more rapidly acting therapy, since hospitalisation generally occurs for a period of at least 4 days (McFadden and Hejal 1995). Children with acute severe asthma should be treated with oral prednisone and not inhaled corticosteroids (Schuh et al. 2000). Therapies that target TNF have the potential to cause rapid onset of anti-inflammatory effects in emergency asthma. It is important to stress that asthmatics may be vulnerable to exacerbations even when asthma is brought under control (Reddel et al. 1999).

3.18
Exacerbations in COPD: Prevention and Treatment

There have been efforts to provide a consensus definition for COPD exacerbations (Rodriguez-Roisin 2000), as well as evidence-based guidelines to manage acute exacerbations of COPD (Bach et al. 2001; Snow et al. 2001). An interesting design to assess the effects of therapy in *preventing* exacerbations was to employ inhaled fluticasone propionate for 6 months over the winter in COPD patients with a history of at least one winter exacerbation per year for the past 3 years (Paggiaro et al. 1998). In addition, inhaled corticosteroids have been demonstrated to prevent exacerbations in a 3-year study on patients with moderate to severe COPD (Burge et al. 2000). With regard to *treatment* of exacerbations, there has been documented moderate improvement in clinical outcome when using oral prednisolone or nebulised budesonide (Niewoehner et al. 1999; Maltais et al. 2002). However, a major disappointment was that a clinical study on treatment of exacerbations of COPD with recombinant human DNAse (dornase alfa, Genentech) had to be stopped, since a trend for increased mortality was noted on interim analysis (Hudson 1996).

4
Conclusion

Based on better understanding of the genetics and molecular pathophysiology that underlies asthma and COPD, novel rational biotechnology therapies are currently being tested in clinical trials (Hansel and Barnes 2001). Initial studies in man with protein therapeutics that target inflammation are sometimes being used as a "proof of concept", and a successful study would justify a research programme to identify rational low-molecular-weight synthetic follow-up oral compounds. Most of the new drugs for asthma and COPD inhibit components of inflammatory responses, and in the future there are real possibilities for the development of preventative and even curative treatments. We can expect clinical studies to involve assessment of genotypes and phenotypes to identify potential responders to specific therapies, as well as more sensitive monitoring of lung function, imaging methods and biomarkers of inflammation.

References

Agusti AG et al (2002) Skeletal muscle apoptosis and weight loss in chronic obstructive pulmonary disease. Am J Respir Crit Care Med 166 (4):485–489

Albrecht A et al (2002) Comparison of Roflumilast, a new orally active, selective phosphodiesterase 4 Inhibitor, with beclomethasone dipropionate in asthma control. Eur Resp J 20:304S

Anderson SD, Brannan JD (2002) Exercise-induced asthma: Is there still a case for histamine? J Allergy Clin Immunol 109 (5 Pt 1):771–773

Anthonisen NR, Wright EC, Hodgkin JE (1986) Prognosis in chronic obstructive pulmonary disease. Am Rev Respir Dis 133:14–20

Arshad SH (2000) Bronchial allergen challenge: a model for chronic allergic asthma? Clin Exp Allergy 30 (1):12–15

ATS (1995) American Thoracic Society Statement: Standardization of Spirometry. Am J Respir Crit Care Med 152:1107–1136

ATS (1999a) Recommendations for standardized procedures for the on-line and off-line measurement of exhaled lower respiratory nitric oxide and nasal nitric oxide in adults and children-1999. This official statement of the American Thoracic Society was adopted by the ATS Board of Directors, July 1999. Am J Respir Crit Care Med 160 (6):2104–2117

ATS (1999b) Supplement: American Thoracic Society/European Respiratory Society— Skeletal Muscle Dysfunction in Chronic Obstructive Pulmonary Disease. Am J of Respir Crit Care Med 159 part 2:S1-S40

ATS (2002a) ATS statement: guidelines for the six-minute walk test. Am J Respir Crit Care Med 166 (1):111–117

ATS (2002b) ATS/ERS Statement on respiratory muscle testing. Am J Respir Crit Care Med 166 (4):518–624

Bacci E et al (1996) Comparison between hypertonic and isotonic saline-induced sputum in the evaluation of airway inflammation in subjects with moderate asthma. Clin Exp Allergy 26:1395–1400

Bach PB et al (2001) Management of acute exacerbations of chronic obstructive pulmonary disease: a summary and appraisal of published evidence. Ann Intern Med 134 (7):600–620

Balint B et al (2001) Increased nitrotyrosine in exhaled breath condensate in cystic fibrosis. Eur Respir J 17 (6):1201–1207

Barley EA, Jones PW (1999) A comparison of global questions versus health status questionnaires as measures of the severity and impact of asthma. Eur Respir J 14 (3):591–596

Barnes PJ (2002) New treatments for COPD. Nature Rev Drug Disc 1 1:437–446

Barnes PJ (2000) Chronic obstructive pulmonary disease. N Engl J Med 343 (4):269–280

Beasley R et al (2000) Prevalence and etiology of asthma. J Allergy Clin Immunol 105:S466-S472

Belda J et al (2001) Sputum induction: effect of nebulizer output and inhalation time on cell counts and fluid-phase measures. Clin Exp Allergy 31 (11):1740–1744

Belman MJ, Botnick WC, Shin JW (1996) Inhaled bronchodilators reduce dynamic hyperinflation during exercise in patients with chronic obstructive pulmonary disease. Am J Respir Crit Care Med 153 (3):967–975

Bianco S et al (1989) Protective effect of inhaled furosemide on allergen-induced early and late asthmatic reactions. N Engl J Med 321:1069–1073

Bielory L, Lupoli K (1999) Herbal interventions in asthma and allergy. J Asthma 36:1–65

Borish LC et al (2001) Efficacy of soluble IL-4 receptor for the treatment of adults with asthma. J Allergy Clin Immunol. 107 (6):963–970

Borish LC et al (1999) Interleukin-4 receptor in moderate atopic asthma: a phase I/II randomized, placebo-controlled trial. Am J Resp Crit Care Med 160:1816–1823

Boulet L-P et al (1987) Comparative bronchial responses to hyperosmolar saline and methacholine in asthma. Thorax 42:953–958

Boulet LP et al (1997) Inhibitory effects of an anti-IgE antibody E25 on allergen-induced early asthmatic response. Am J Respir Crit Care Med 155 (6):1835–1840

Boushey HA, Fahy JV (2000) Targeting cytokines in asthma therapy: round one. Lancet 356 (9248):2114–2116

Bousquet J et al (2000) Asthma. From bronchoconstriction to airways inflammation and remodeling. Am J Respir Crit Care Med 161 (5):1720–1745

Bredenbroker D et al (2002) Roflumilast, a new orally active, selective phosphodiesterase 4 Inhibitor, is effective in the treatment of chronic obstructive pulmonary disease. Eur Resp J 20:374S

Brown HM (1958) Treatment of chronic asthma with prednisolone:significance of eosinophils in the sputum. Lancet ii:1245–1247

Brown RH et al (2001) High-resolution computed tomographic evaluation of airway distensibility and the effects of lung inflation on airway caliber in healthy subjects and individuals with asthma. Am J Respir Crit Care Med 163 (4):994–1001

Bryan SA et al (2002) Responses of leukocytes to chemokines in whole blood and their antagonism by novel CCR3 antagonists. Am J Respir Crit Care Med 165:1602–1609

Bryan SA et al (2000) Effects of recombinant human interleukin-12 on eosinophils, airway hyperreactivity and the late asthmatic response. Lancet 356 (2149):2153

Burge PS (1999) EUROSCOP, ISOLDE and the Copenhagen city lung study. Thorax 54 (4):287–288

Burge PS et al (2000) Randomised, double blind, placebo controlled study of fluticasone propionate in patients with moderate to severe chronic obstructive pulmonary disease: the ISOLDE trial. BMJ 320 (7245):1297–1303

Burnett D et al (1987) Neutrophils from subjects with chronic obstructive lung disease show enhanced chemotaxis and extracellular proteolysis. Lancet 2 (8567):1043–1046

Burrows B, Earle RH (1969) Course and prognosis of chronic obstructive lung disease. A prospective study of 200 patients. N Engl J Med 280 (8):397–404

Busse W et al (2001) Omalizumab, anti-IgE recombinant humanized monoclonal antibody, for the treatment of severe allergic asthma. J Allergy Clin Immunol 108 (2):184–190

Calhoun WJ, Hinton KL, Kratzenberg JJ (2001) The effect of salmeterol on markers of airway inflammation following segmental allergen challenge. Am J Respir Crit Care Med 163 (4):881–886

Chang TW (2000) The pharmacological basis of anti-IgE therapy. Nat.Biotechnol. 18 (2):157–162

Ciprandi G et al (2001) Intranasal mometasone furoate reduces late-phase inflammation after allergen challenge. Ann.Allergy Asthma Immunol. 86 (4):433–438

Cleverley JR, Muller NL (2000) Advances in radiologic assessment of chronic obstructive pulmonary disease. Clin Chest Med 21 (4):653–663

Corradi M et al (2001) Increased nitrosothiols in exhaled breath condensate in inflammatory airway diseases. Am J Respir Crit Care Med 163 (4):854–858

Cosio MG, Snider GL (2001) Chest computed tomography: is it ready for major studies of chronic obstructive pulmonary disease? Eur Respir J 17 (6):1062–1064

Crapo RO (1994) Pulmonary Function Testing. N Engl J Med 331:25–30

Crapo RO et al (2000) Guidelines for methacholine and exercise challenge testing-1999. This official statement of the American Thoracic Society was adopted by the ATS Board of Directors, July 1999. Am J Respir Crit Care Med. 161 (1):309–329

Crimi E, Brusasco V, Crimi P (1989) Effect of nedocromil sodium on the late asthmatic reaction to bronchial antigen challenge. J Allergy Clin Immunol 83:985–990

Crimi E, Brusasco V, Crimi P (1989) Effect of nedocromil sodium on the late asthmatic reaction to bronchial antigen challenge. J Allergy Clin Immunol 83 (5):985–990

Crimi E et al (1998) Dissociation between airway inflammation and airway hyperresponsiveness in allergic asthma. Am J Respir Crit Care Med 157:4–9

Dahlen B et al (2002) Influence of zafirlukast and loratadine on exercise-induced bronchoconstriction. J Allergy Clin.Immunol. 109 (5):789–793

de Benedictis FM et al (2001) Rhinitis, sinusitis and asthma: one linked airway disease. Paediatr.Respir Rev 2 (4):358–364

De Gouw HW et al (1996) Repeatability of cellular and soluble markers of inflammation in induced sputum from patients with asthma. Eur Respir J 9 (12):2441–2447

de la Fuente PT et al (1998) Safety of inducing sputum in patients with asthma of varying severity. Am J Resp Crit Care Med 157:1127–1130

De Meer G, Heederik D, Postma DS (2002) Bronchial responsiveness to adenosine 5'-monophosphate (AMP) and methacholine differ in their relationship with airway allergy and baseline FEV(1). Am J Respir Crit Care Med 165 (3):327–331

Dente FL et al (1999) Effect of a single dose of salmeterol on the increase in airway eosinophils induced by allergen challenge in asthmatic subjects. Thorax 54 (7):622–624

Diamant Z et al (1999) The effect of montelukast (MK-0476), a cysteinyl leukotriene receptor antagonist, on allergen-induced airway responses and sputum cell counts in asthma. Clin Exp Allergy 29:42–51

Diamant Z et al (1996) Effect of inhaled heparin on allergen-induced early and late asthmatic responses in patients with atopic asthma. Am J Respir Crit Care Med 153:1790–1795

Dirksen A et al (1999) A randomized clinical trial of alpha-1-antitrypsin augmentation therapy. Am J Resp Crit Care Med 160:1468–1472

Dirksen A et al (1997) Progress of emphysema in severe alpha1-antitrypsin deficiency as assessed by annual CT. Acta Radiol. 38:826–832

Djukanovic R (2000) Induced sputum-A tool with great potential but not without problems. J Allergy Clin Immunol 105 (6 Pt 1):1071–1073

Douglass JA et al (1994) Influence of interleukin-8 challenge in the nasal mucosa in atopic and nonatopic subjects. Am J Respir Crit Care Med. 150 (4):1108–1113

Dowson LJ et al (2001a) High-resolution computed tomography scanning in alpha1-antitrypsin deficiency: relationship to lung function and health status. Eur Respir J 17 (6):1097–1104

Dowson LJ, Guest PJ, Stockley RA (2001b) Longitudinal changes in physiological, radiological, and health status measurements in alpha(1)-antitrypsin deficiency and factors associated with decline. Am J Respir Crit Care Med 164 (10 Pt 1):1805–1809

Drazen JM, Israel E, O'Byrne PM (1999a) Treatment of asthma with drugs modifying the leukotriene pathway. N Engl J Med 340:197–206

Drazen JM et al (1999b) Pharmacogenetic association between ALOX5 promoter genotype and the response to anti-asthma treatment. Nat.Genet. 22 (2):168–170

Drysdale CM et al (2000) Complex promoter and coding region beta2-adrenergic receptor haplotypes alter receptor expression and predict in vivo responsiveness. PNAS 97:10483–10488

Dyer CA et al (2002) The incremental shuttle walking test in elderly people with chronic airflow limitation. Thorax 57 (1):34–38

Efthimiadis A et al (1997) Canadian Thoracic Society. Sputum Examination for Indices of Airway Inflammation: Laboratory Procedures. Lund, Sweden: Astra Draco A.B

Efthimiou J, Barnes PJ (1998) Effect of inhaled corticosteroids on bones and growth. Eur Respir J 11 (5):1167–1177

EMEA CPMP (1997) ICH Topic E 8 : General Considerations for Clinical Trials. CPMP/ICH/291/95

EMEA CPMP (1999) Points to consider on clinical investigation of medicinal products in the chronic treatment of patients with chronic obstructive pulmonary disease

(COPD). Human Medicines Evaluation Unit, European Medicines Evaluation Agency (EMEA) CPMP/EWP 562/98

EMEA CPMP (2001) Note for guidance on the clinical investigation of medicinal products in the treatment of asthma. CPMP/EWP/2922/00 draft

Engelen MP et al (2000) Skeletal muscle weakness is associated with wasting of extremity fat-free mass but not with airflow obstruction in patients with chronic obstructive pulmonary disease. Am J Clin.Nutr. 71 (3):733–738

Engstrom CP et al (2001) Health-related quality of life in COPD: why both disease-specific and generic measures should be used. Eur Respir J 18 (1):69–76

Enright PL, Sherrill DL (1998) Reference equations for the six-minute walk in healthy adults. Am J Respir Crit Care Med 158 (5 Pt 1):1384–1387

Erin EM et al (2002) Eotaxin receptor (CCR-3) antagonism in asthma and allergic disease. Current Drug Targets Inflammation & Allergy 1:201–214

European Respiratory Society Task Force (2002) Standardised methodology of sputum induction and processing. Eur Respir J 20 [Suppl 37]:1s-55 s

Evans DJ et al (1997a) Effects of a potent platelet-activating factor antagonist, SR27417A, on allergen-induced asthmatic responses. Am J Respir Crit Care Med 156 (1):11–16

Evans DJ et al (1997b) A comparison of low-dose inhaled budesonide plus theophylline and high- dose inhaled budesonide for moderate asthma. N.Engl.J Med 337 (20):1412–1418

Fabbri L et al (2002) Advances in the understanding and future therapy of COPD. Clin Exp All Rev 2:129–136

Fahy JV (1998) A safe, simple, standardized method should be used for sputum induction for research purposes [editorial; comment]. Clin Exp Allergy 28 (9):1047–1049

Fahy JV et al (2001) Safety and reproducibility of sputum induction in asthmatic subjects in a multicenter study. Am J Respir Crit Care Med 163 (6):1470–1475

Fahy JV et al (1999) Effects of aerosolized anti-IgE (E25) on airway responses to inhaled allergen in asthmatic subjects. Am J Resp Crit Care Med 160:1023–1027

Fahy JV et al (2002) Cellular and biochemical analysis of induced sputum from asthmatic and healthy individuals. Am Rev Respir Dis 147:1126–1131

Fahy JV et al (1995) Comparison of samples collected by sputum induction and bronchoscopy from asthmatic and healthy subjects. Am J Resp Crit Care Med 152:53–58

Faul JL et al (1999) The reproducibility of repeat measures of airway inflammation in stable atopic asthma. Am J Respir Crit Care Med 160 (5 Pt 1):1457–1461

Faul JL et al (1998) Fluticasone propionate induced alterations to lung function and the immunopathology of asthma over time. Thorax 53:753–761

Feldman M et al (1998) Anti-TNF alpha therapy is useful in rheumatoid arthritis and Crohn's disease: analysis of the mechanism of action predicts utility in other diseases. Transplant Proc 30 (8):4126–4127

Ferreira IM et al (2001) Exhaled nitric oxide and hydrogen peroxide in patients with chronic obstructive pulmonary disease: effects of inhaled beclomethasone. Am J Respir Crit Care Med 164 (6):1012–1015

Ferretti GR, Bricault I, Coulomb M (2001) Virtual tools for imaging of the thorax. Eur Respir J 18 (2):381–392

Fletcher C, Peto R (1977) The natural history of chronic airflow obstruction. Br Med J 1 (6077):1645–1648

Fleury-Feith J et al (1987) The effects of cytocentrifugation on differential cell counts in samples obtained by bronchoalveolar lavage. Acta Cytol 31:606–610

Fozard JR, Hannon P (2000) Species differences in adenosine receptor-mediated bronchoconstrictor responses. Clin Exp Allergy 30:1213}{2014}{5964}

Freitag A et al (1993) Effect of a platelet activating factor antagonist, WEB 2086, on allergen induced asthmatic responses. Thorax 48:594–598

Fuller RW, Black PN, Dollery CT (1989) Effect of the oral leukotriene D4 antagonist LY171883 on inhaled and intradermal challenge with antigen and leukotriene D4 in atopic subjects. J Allergy Clin Immunol 83 (5):939–944

Gallin JI (2002) Principles and Practice of Clinical Research. Academic Press, San Diego

Garey KW et al (2000) Protein, nitrite/nitrate, and cytokine concentration in exhaled breath condensate of young smokers. Am J Respir Crit Care Med 161:A175

Gauvreau GM, Watson RM, O'Byrne PM (1999) Protective effects of inhaled PGE_2 on allergen-induced airway responses and airway inflammation. Am J Resp Crit Care Med 159:31–36

Gelder CM et al (1995) Cytokine expression in normal, atopic and asthmatic subjects using the combination of sputum induction and the polymerase chain reaction. Thorax 50 (10):1033–1037

Gershman NH et al (1999) Fractional analysis of sequential induced sputum samples during sputum induction: evidence that different lung compartments are sampled at different time points. J Allergy Clin Immunol 104 (2 Pt 1):322–328

Giannini D et al (1999) Inhaled beclomethasone dipropionate reverts tolerance to the protective effect of salmeterol on allergen challenge. Chest 115 (3):629–634

Giannini D et al (2001) Tolerance to the protective effect of salmeterol on allergen challenge can be partially restored by the withdrawal of salmeterol regular treatment. Chest 119 (6):1671–1675

Gibson PG et al (1989) Cellular characteristics of sputum from patients with asthma and chronic bronchitis. Thorax 44:693–699

Gibson PG, Saltos N, Fakes K (2001) Acute anti-inflammatory effects of inhaled budesonide in asthma: a randomized controlled trial. Am J Respir Crit Care Med. 163 (1):32–36

Gibson PG et al (1992) A research method to induce and examine a mild exacerbation of asthma by withdrawal of inhaled corticosteroid. Clin Exp Allergy 22:525–532

Giembycz MA (2002) Development status of second generation PDE4 inhibitors for asthma and COPD: the story so far. Monaldi Arch Chest Dis 57 (1):48–64

Gollasch (1889) Zur kenntniss der asthmatischen sputums. Fortschr Med 7:361–365

Gosselink R, Troosters T, Decramer M (1996) Peripheral muscle weakness contributes to exercise limitation in COPD. Am J Respir Crit Care Med 153 (3):976–980

Greiff L et al (1990) The 'nasal pool' device applies controlled concentrations of solutes on human nasal airway mucosa and samples its surface exudations/secretions. Clin Exp Allergy 20:253–259

Greiff L et al (2001) Mucosal output of eotaxin in allergic rhinitis and its attenuation by topical glucocorticosteroid treatment. Clin Exp Allergy 31 (8):1321–1327

Griffin JP et al (1994) The Textbook of Pharmaceutical Medicine (including Medicines Regulation), 2nd edn. Queen's University, Belfast

Gronborg H et al (1993) Early and late nasal symptom response to allergen challenge. The effect of pretreatment with a glucocorticosteroid spray. Allergy 48 (2):87–93

Grootendorst DC et al (1997) Comparison of inflammatory cell counts in asthma:induced sputum vs bronchoalveolar lavage and bronchial biopsies. Clin Exp Allergy 27:769–779

Haahtela T et al (1994) Effects of reducing or discontinuing inhaled budesonide in patients with mild asthma. N Engl J Med 331 (11):700–705

Haahtela T et al (1991) Comparison of a beta 2-agonist, terbutaline, with an inhaled corticosteroid, budesonide, in newly detected asthma. N Engl J Med 325 (6):388–392

Hamelmann E, Gelfand EW (2001) IL-5-induced airway eosinophilia–the key to asthma? Immunol Rev 179:182–191

Hanazawa T et al (2000a) Nitrotyrosine and cystenyl leukotrienes in breath condensates are increased after withdrawal of steroid treatment in patients with asthma. Am J Respir Crit Care Med 161:A919

Hanazawa T et al (1999) Intranasal administration of eotaxin increases nasal eosinophils and nitric oxide in patients with allergic rhinitis. J Allergy Clin Immunol 105:58–64

Hanazawa T, Kharitonov SA, Barnes PJ (2000b) Increased nitrotyrosine in exhaled breath condensate of patients with asthma. Am J Respir Crit Care Med 162 (4 Pt 1):1273–1276

Hansel TT, Barnes PJ (2001) New Drugs for Asthma, Allergy and COPD. (31)

Hansel FK (1953) Clinical Allergy. In: Mosby CV (ed) St Louis

Hansel TT et al (1991) Sputum eosinophils from asthmatics express ICAM-1 and HLA-DR. Clin Exp Immunol 86:271–277

Hansel TT et al (2002) Letter to Editor: The trials and tribulations of anti-IL5. J Allergy Clin Immunol 109 (3):575

Hansell DM (2001) Small airways diseases: detection and insights with computed tomography. Eur Respir J 17 (6):1294–1313

Hansen EF et al (1999) Reversible and irreversible airflow obstruction as predictor of overall mortality in asthma and chronic obstructive pulmonary disease. Am J Respir Crit Care Med 159 (4 Pt 1):1267–1271

Harbinson PL et al (1997) The effect of a novel orally active selective PDE4 isoenzyme inhibitor (CDP840) on allergen-induced responses in asthmatic subjects. Eur Respir J 10:1008–1014

Hartert TV et al (2000) Prostaglandin E(2) decreases allergen-stimulated release of prostaglandin D(2) in airways of subjects with asthma. Am J Respir Crit Care Med 162 (2 Pt 1):637–640

Hasko G, Szabo C (1999) IL-12 as a therapeutic target for pharmacological modulation in immune- mediated and inflammatory diseases: regulation of T helper 1/T helper 2 responses. Br.J Pharmacol 127 (6):1295–1304

Haslam PL, Baughman RP, and eds (1999) Report of European Respiratory Society (ERS) Task Force: guidelines for measurement of acellular components and recommendations for standardization of bronchoalveolar lavage (BAL). Eur Respir Rev 9 (66):25–157

Hattotuwa KL et al (2002) The effects of inhaled fluticasone on airway inflammation in chronic obstructive pulmonary disease: a double-blind, placebo-controlled biopsy study. Am J Respir Crit Care Med 165 (12):1592–1596

Holgate ST et al (2000a) Mometasone furoate antagonizes AMP-induced bronchoconstriction in patients with mild asthma. J Allergy Clin Immunol 105 (5):906–911

Holgate ST et al (2000b) Epithelial-mesenchymal interactions in the pathogenesis of asthma. J Allergy Clin Immunol 105 (2 Pt 1):193–204

Holz O et al (1998a) Changes In sputum composition during sputum induction in healthy and asthmatic subjects. Clin Exp Allergy 28:284–292

Holz O, Kips J, Magnussen H (2000) Update on sputum methodology. Eur Respir J 16 (2):355–359

Holz O et al (1998b) Changes in sputum composition between two inductions performed on consecutive days. Thorax 53 (2):83–86

Horvath I et al (1998) Combined use of exhaled hydrogen peroxide and nitric oxide in monitoring asthma. Am J Respir Crit Care Med 158 (4):1042–1046

Howarth PH et al (1985) Influence of albuterol, cromolyn sodium and ipratropium bromide on the airway and circulating mediator responses to allergen bronchial provocation in asthma. Am Rev Respir Dis 132 (5):986–992

Huang S (1998) Molecular modulation of allergic responses. J Allergy Clin Immunol 102:887–892

Hudson TJ (1996) Dornase in treatment of chronic bronchitis. Ann Pharmacother 30:674–675

Hughes JM, Pride NB (1999) Lung Function Tests: Physiological Principles and Clinical Applications. WB Saunders, London

In't Veen JC, De Gouw HW, Smits HH (1994) Repeatability of cellular and soluble markers of inflammation in induced sputum from patients with asthma. Eur Respir J 9:2441–2447

Inman MD et al (1995) Reproducibility of allergen-induced early and late asthmatic responses. J Allergy Clin Immunol 95:1191–1195

Jatakanon A et al (1999) Effect of differing doses of inhaled budesonide on markers of airway inflammation in patients with mild asthma. Thorax 54 (2):108–114

Jatakanon A, Lim S, Barnes PJ (2000) Changes in sputum eosinophils predict loss of asthma control. Am J Respir Crit Care Med 161 (1):64–72

Jayaram L et al (2000) Induced sputum cell counts: their usefulness in clinical practice. Eur Respir J 16 (1):150–158

Jones PW, Mahler DA (2002) Key outcomes in COPD: health-related quality of life. Proceedings of an expert round table held July 20–22, 2001 in Boston, Massachusetts, USA. European Respiratory Review 12 (83)

Jones PW (2001) Health status measurement in chronic obstructive pulmonary disease. Thorax 56 (11):880–887

Jones PW, Quirk FH, Baveystock CM (1991) The St George's Respiratory Questionnaire. Respir Med 85 [Suppl B]:25–31

Jones SL et al (2001) The predictive value of exhaled nitric oxide measurements in assessing changes in asthma control. Am J Respir Crit Care Med 164 (5):738–743

Juniper EF, Guyatt GH (1991) Development and testing of a new measure of health status for clinical trials in rhiniconjunctivitis. Clin Exp Allergy 21:77–83

Juniper EF et al (2000) Development and validation of the mini rhinoconjunctivitis quality of life questionnaire. Clin.Exp.Allergy 30:132–140

Juniper EF et al (1999a) Validation of a standardized version of the Asthma Quality of Life Questionnaire. Chest 115 (5):1265–1270

Juniper EF et al (1992) Evaluation of impairment of health related quality of life in asthma: development of a questionnaire for use in clinical trials. Thorax 47 (2):76–83

Juniper EF et al (2000) Measuring asthma control. Clinic questionnaire or daily diary? Am J Respir Crit Care Med 162 (4 Pt 1):1330–1334

Juniper EF et al (1999b) Development and validation of a questionnaire to measure asthma control. Eur Respir J 14 (4):902–907

Kanehiro A et al (2001) Inhibition of phosphodiesterase 4 attenuates airway hyperresponsiveness and airway inflammation in a model of secondary allergen challenge. Am J Respir Crit Care Med 163 (1):173–184

Keatings VM et al (1997) Cellular profiles in asthmatic airways: a comparison of induced sputum, bronchial washings, and bronchoalveolar lavage fluid. Thorax 52 (4):372–374

Kelly HW (1998) Establishing a therapeutic index for the inhaled corticosteroids:Part 1. Pharmacokinetic/pharmacodynamic comparison of the inhaled corticosteroids. J Allergy Clin Immunol 102:S36-S51

Kelly MM et al (2001) Increased detection of interleukin-5 in sputum by addition of protease inhibitors. Eur Respir J 18 (4):685–691

Kelly MM et al (2000) Induced sputum: validity of fluid-phase IL-5 measurement. J Allergy Clin Immunol 105 (6 Pt 1):1162–1168

Kemp JP et al (1998) Montelukast once daily inhibits exercise-induced bronchoconstriction in 6 to 14 year old children with asthma. J Pediatrics 133:424–428

Khan LN et al (2000) Attenuation of the allergen-induced late asthmatic reaction by cyclosporin A is associated with inhibition of bronchial eosinophils, interleukin-5, granulocyte macrophage colony-stimulating factor, and eotaxin. Am J Respir Crit Care Med 162 (4 Pt 1):1377–1382

Kharitinov SA et al (2000) Dose-dependent onset and duration of action of 100/400mcg budesonide on exhaled nitric oxide and related changes in other potential markers of airway inflammation in mild asthma. Am J Respir Crit Care Med 161:A186

Kharitonov S, Alving K, Barnes PJ (1997) Exhaled and nasal nitric oxide measurements: recommendations. The European Respiratory Society Task Force. Eur Respir J 10 (7):1683–1693

Kharitonov SA, Yates DH, Barnes PJ (1996) Changes in the dose of inhaled steroid affect exhaled nitric oxide levels in asthmatic patients. Eur Respir J 9:196–201

Kharitonov SA, Barnes PJ (2000) Clinical aspects of exhaled nitric oxide. Eur Respir J 16 (4):781–792

Kharitonov SA, Barnes PJ (2001a) Does exhaled nitric oxide reflect asthma control? Yes, it does! Am J Respir Crit Care Med 164 (5):727–728

Kharitonov SA, Barnes PJ (2001b) Exhaled markers of pulmonary disease. Am J Respir Crit Care Med 163 (7):1693–1722

Kharitonov SA et al (1994) Increased nitric oxide in exhaled air of asthmatic patients. Lancet 343 (8890):133–135

Kharitonov SA, Yates DH, Barnes PJ (1996) Inhaled glucocortisteroids decrease nitric oxide in exhaled air of asthmatic patients. Am J Respir Crit Care Med 153:454–457

Kidney JC et al (1996) Elevated B cells in sputum of asthmatics:clear correlation with eosinophils. Am J Respir Crit Care Med 153:540–544

King GG, Muller NL, Pare PD (1999) Evaluation of airways in obstructive pulmonary disease using high-resolution computed tomography. Am J Respir Crit Care Med 159 (3):992–1004

Kips JC, Tournoy KG, Pauwels RA (2001) New anti-asthma therapies: suppression of the effect of interleukin (IL)-4 and IL-5. Eur Respir J 17 (3):499–506

Kips JC et al (1998) Methods for sputum induction and analysis of induced sputum: a method for assessing airway inflammation in asthma. Eur Respir J 11 Suppl.26: 9 s-12 s

Kips JC et al (2000) Results of a phase 1 trial with SCH55700, a humanized anti-IL-5 antibody, in severe, persistent asthma. Am J Respir Crit Care Med 161:A505

Kips JC, Pauwels RA (1998) Noninvasive indicators of airway inflammation:induced sputum in allergic diseases. Eur Respir Rev 8:1095–1097

Kips JC, Pauwels RA (2001) Long-acting inhaled beta(2)-agonist therapy in asthma. Am J Respir Crit Care Med 164 (6):923–932

Kon OM et al (1998) Randomised, dose-ranging, placebo-controlled study of chimeric antibody to CD4 (keliximab) in chronic severe asthma. Lancet 352 (9134):1109–1113

Krishna MT et al (2001) Inhibition of mast cell tryptase by inhaled APC 366 attenuates allergen-induced late-phase airway obstruction in asthma. J Allergy Clin Immunol 107 (6):1039–1045

Kuitert LM et al (1993) Effect of the platelet-activating factor antagonist UK-74,505 on the early and late response to allergen. Am Rev Respir Dis 147:82–86

Laube BL et al (1998) The efficacy of slow versus faster inhalation of cromolyn sodium in protecting against allergen challenge in patients with asthma. J Allergy Clin Immunol 101 (4 Pt 1):475–483

Laviolette M et al (1999) Montelukast added to inhaled beclomethasone in treatment of asthma. Am J Resp Crit Care Med 160:1862–1868

Leckie MJ et al (2000) Effects of an interleukin-5 blocking monoclonal antibody on eosinophils, airway hyperresponsiveness, and the response to allergen in patients with asthma. Lancet 356:2144–2148

Leff JA et al (1998) Montelukast, a leukotriene-receptor antagonist, for the treatment of mild asthma and exercise-induced bronchoconstriction. N Engl J Med 339:147–152

Levine S et al (1997) Cellular adaptations in the diaphragm in chronic obstructive pulmonary disease. N Engl J Med 337 (25):1799–1806

Lim S et al (1999) Effect of inhaled budesonide on lung function and airway inflammation. Assessment by various inflammatory markers in mild asthma. Am J Respir Crit Care Med 159:22–30

Linden M et al (2000) Immediate effect of topical budesonide on allergen challenge-induced nasal mucosal fluid levels of granulocyte-macrophage colony-stimulating factor and interleukin-5. Am J Respir Crit Care Med 162 (5):1705–1708

Liu MC et al (2001) Effects of prednisone on the cellular responses and release of cytokines and mediators after segmental allergen challenge of asthmatic subjects. J Allergy Clin Immunol 108 (1):29–38

Lock SH, Kay AB, Barnes NC (1996) Double-blind, placebo-controlled study of cyclosporin A as a corticosteroid-sparing agent in corticosteroid-dependent asthma. Am J Respir Crit Care Med 153 (2):509–514

Lopez AD, Murray CCJL (1998) The global burden of disease, 1990–2020. Nature Medicine 4:1241–1243

Loppow D et al (2000) Flow cytometric analysis of the effect of dithiothreitol on leukocyte surface markers. Eur Respir J 16 (2):324–329

Lung Health Study Research Group (2000) Effect of inhaled triamcinolone on the decline in pulmonary function in chronic obstructive pulmonary disease. N Engl J Med 343 (26):1902–1909

Maesen FPV et al (1995) Tiotropium bromide, a new long-acting antimuscarinic bronchodilator: a pharmacodynamic study in patients with chronic obstructive disease (COPD). Eur Respir J 8:1506–1513

Magnussen H, Holz O (1999) Monitoring airway inflammation in asthma by induced sputum. Eur Respir Rev 13:5–7

Magnussen H et al (2000) Noninvasive methods to measure airway inflammation: future considerations. Eur Respir J 16 (6):1175–1179

Maltais F et al (2002) Comparison of nebulized budesonide and oral prednisolone with placebo in the treatment of acute exacerbations of chronic obstructive pulmonary disease: a randomized controlled trial. Am J Respir Crit Care Med 165 (5):698–703

Manning PJ et al (1990) Inhibition of exercise-induced bronchoconstriction by MK-571, a potent leukotriene D_4-receptor antagonist. N Engl J Med 323:1736–1739

Mao JT et al (2002) A pilot study of all-trans-retinoic acid for the treatment of human emphysema. Am J Respir Crit Care Med 165 (5):718–723

Marquis K et al (2002) Midthigh muscle cross-sectional area is a better predictor of mortality than body mass index in patients with chronic obstructive pulmonary disease. Am J Respir Crit Care Med 166 (6):809–813

McFadden ER, Hejal R (1995) Asthma. Lancet 345 (8959):1215–1220

Mclean AN et al (1998) High resolution computed tomography in asthma. Thorax 53 (4):308–314

Milgrom H et al (2001) Treatment of childhood asthma with anti-immunoglobulin E antibody (omalizumab). Pediatrics 108 (2):E36

Milgrom H et al (1999) Treatment of allergic asthma with monoclonal anti-IgE antibody. rhuMAb- E25 Study Group. N Engl J Med 341 (26):1966–1973

Montuschi P et al (2000a) Exhaled prostaglandin E2: a new biomarker of airway inflammation in COPD. Am J Respir Crit Care Med 160:A821

Montuschi P et al (2000b) Exhaled 8-isoprostane as an in vivo biomarker of lung oxidative stress in patients with COPD and healthy smokers. Am J Respir Crit Care Med 162 (3 Pt 1):1175–1177

Montuschi P et al (1999) Increased 8-isoprostane, a marker of oxidative stress, in exhaled condensate of asthma patients. Am J Resp Crit Care Med 160:216–220

Moodley YP, Krishnan V, Lalloo UG (2000) Neutrophils in induced sputum arise from central airways. Eur Respir J 15:36–40

Morley J (1992) Strategies for developing novel anti-asthma drugs. In Barnes PJ (ed) IBC Technical Services Ltd, London

Muller NL, Coxson H (2002) Chronic obstructive pulmonary disease. 4: imaging the lungs in patients with chronic obstructive pulmonary disease. Thorax 57 (11):982–985

Murray CJ, Lopez AD (1997a) Global mortality, disability, and the contribution of risk factors: Global Burden of Disease Study. Lancet 349 (9063):1436–1442

Murray CJ, Lopez AD (1997b) Mortality by cause for eight regions of the world: Global Burden of Disease Study. Lancet 349 (9061):1269–1276

National Institutes of Health (NIH), National Heart Lung and Blood Institute (NHLBI). 2002. Global Initiative for Asthma (GINA) : Global Strategy for Asthma Management and Prevention. www.ginasthma.com

National Institutes of Health (NIH), National Heart Lung and Blood Institute (NHLBI), World Health Organisation (WHO). 2001. Global Initiative for Chronic Obstructive Lung Disease (GOLD) : Global Strategy for the Diagnosis, Management, and Prevention of Chronic Obstructive Pulmonary Disease NHLBI/WHO Workshop Report. www.goldcopd.com/workshop/index.html

Nelson HS (2001) Advair: combination treatment with fluticasone propionate/salmeterol in the treatment of asthma. J Allergy Clin Immunol 107:397–416

Newton MF, O'Donnell DE, Forkert L (2002) Response of lung volumes to inhaled salbutamol in a large population of patients with severe hyperinflation. Chest 121 (4):1042–1050

Niewoehner DE et al (1999) Effect of systemic glucocorticoids on exacerbations of chronic obstructive pulmonary disease. Department of Veterans Affairs Cooperative Study Group. N Engl J Med 340 (25):1941–1947

Niewoehner DE, Kleinerman J, Rice DB (1974) Pathologic changes in the peripheral airways of young cigarette smokers. N Engl J Med 291 (15):755–758

Nightingale JA, Rogers DF, Barnes PJ (1998) Effect of repeated sputum induction on cell counts in normal volunteers. Thorax 53 (2):87–90

Nisar M et al (1992) Acute bronchodilator trials in chronic obstructive pulmonary disease. Am Rev Respir Dis 146:555–559

Noguera A et al (2001) Enhanced neutrophil response in chronic obstructive pulmonary disease. Thorax 56 (6):432–437

Noguera A et al (1998) Expression of adhesion molecules and G proteins in circulating neutrophils in chronic obstructive pulmonary disease. Am J Respir Crit Care Med 158 (5 Pt 1):1664–1668

O'Brien C et al (2000) Physiological and radiological characterisation of patients diagnosed with chronic obstructive pulmonary disease in primary care. Thorax 55 (8):635–642

O'Byrne P, Inman MD (1996) Induced sputum to assess airway inflammation in asthma. Eur Respir J 9:2435–2436

O'Byrne P, Pedersen S (1998) Measuring efficacy and safety of different inhaled corticosteroid preparations. J Allergy Clin Immunol 102:879–886

O'Byrne PM, Inman MD, Parameswaran K (2001) The trials and tribulations of IL-5, eosinophils, and allergic asthma. J Allergy Clin.Immunol. 108 (4):503–508

O'Connor BJ, Towse LJ, Barnes PJ (1996) Prolonged effect of tiotropium bromide on methacholine-induced bronchoconstriction in asthma. Am J Respir Crit Care Med 154:876–880

O'Donnell DE (2000) Assessment of bronchodilator efficacy in symptomatic COPD: is spirometry useful? Chest 117 [Suppl 2]:42S–47S

O'Donnell DE, Lam M, Webb KA (1998) Measurement of symptoms, lung hyperinflation, and endurance during exercise in chronic obstructive pulmonary disease. Am J Respir Crit Care Med 158 (5 Pt 1):1557–1565

O'Donnell DE, Lam M, Webb KA (1999) Spirometric correlates of improvement in exercise performance after anticholinergic therapy in chronic obstructive pulmonary disease. Am J Respir Crit Care Med 160 (2):542–549

O'Donnell DE, Webb KA (1993) Exertional breathlessness in patients with chronic airflow limitation. The role of lung hyperinflation. Am Rev Respir Dis 148 (5):1351–1357

O'Shaughnessy KM et al (1993) Differential effects of fluticasone propionate on allergen-evoked bronchoconstriction and increased urinary leukotriene E_4 excretion. Am Rev Respir Dis 147:1472–1476

Paggiaro PL et al (1998) Multicentre randomised in placebo-controlled trial of inhaled fluticasone propionate in patients with chronic obstructive pulmonary disease. Lancet 351:773–780

Palmer LJ et al (2002) Pharmacogenetics of asthma. Am J Respir Crit Care Med 165 (7):861–866

Palmqvist M et al (1992) Late asthmatic reaction decreased after pretreatment with salbutamol and formoterol, a new long-acting beta 2-agonist. J Allergy Clin Immunol 89 (4):844–849

Parameswaran K et al (2000) Protective effects of fluticasone on allergen-induced airway responses and sputum inflammatory markers. Can Respir J 7 (4):313–319

Paredi P, Kharitonov SA, Barnes PJ (2000a) Elevation of exhaled ethane concentration in asthma. Am J Respir Crit Care Med 162 (4 Pt 1):1450–1454

Paredi P, Kharitonov SA, Barnes PJ (2002) Faster rise of exhaled breath temperature in asthma: a novel marker of airway inflammation? Am J Respir Crit Care Med 165 (2):181–184

Paredi P et al (2000b) Exhaled ethane is elevated in cystic fibrosis and correlates with carbon monoxide levels and airway obstruction. Am J Respir Crit Care Med 161 (4 Pt 1):1247–1251

Paredi P et al (2000c) Exhaled ethane, a marker of lipid peroxidation, is elevated in chronic obstructive pulmonary disease. Am J Respir Crit Care Med 162 (2 Pt 1):369–373

Passalacqua G, Canonica GW (2001) Impact of rhinitis on airway inflammation: biological and therapeutic implications. Respir Res 2 (6):320–323

Passalacqua G, Ciprandi G, Canonica GW (2001) The nose-lung interaction in allergic rhinitis and asthma: united airways disease. Curr Opin Allergy Clin Immunol 1 (1):7–13

Pauwels RA (1989) New aspects of the therapeutic potential of theophylline in asthma. J Allergy Clin.Immunol. 83 (2 Pt 2):548–553

Pauwels RA et al (1997) Effect of inhaled formoterol and budesonide on exacerbations of asthma. N Engl J Med 337:1405–1411

Pauwels RA et al (1999) Long-term treatment with inhaled budesonide in persons with mild chronic obstructive pulmonary disease who continue smoking. European Respiratory Society Study on Chronic Obstructive Pulmonary Disease. N Engl J Med 340 (25):1948–1953

Pavord ID et al (1997) The use of induced sputum to investigate airway inflammation. Thorax 52:498–501

Pavord ID et al (1993) Effect of inhaled prostaglandin E_2 on allergen-induced asthma. Am Rev Respir Dis 148:87–90

Pedersen B et al (1993) The effect of salmeterol on the early- and late-phase reaction to bronchial allergen and postchallenge variation in bronchial reactivity, blood eosinophils, serum eosinophil cationic protein, and serum eosinophil protein X. Allergy 48 (5):377–382

Peto R et al (1983) The relevance in adults of air-flow obstruction, but not of mucus hypersecretion, to mortality from chronic lung disease. Results from 20 years of prospective observation. Am Rev Respir Dis 128 (3):491–500

Pin I et al (1992) Use of induced sputum cell counts to investigate airway inflammation in asthma. Thorax 47 (1):25–29

Pipkorn U et al (1987) Inhibition of mediator release in allergic rhinitis by pretreatment with topical glucocorticosteroids. N.Engl.J Med. 316 (24):1506–1510

Pizzichini E et al (1996a) Indices of airway inflammation in induced sputum: reproducibility and validity of cell and fluid-phase measurements. Am J Respir Crit Care Med 154:308–317

Pizzichini E et al (1998) Induced sputum, bronchoalveolar lavage and blood from mild asthmatics: Inflammatory cells, lymphocyte subsets and soluble markers compared. Eur Respir J 11 (4):-834

Pizzichini MM et al (1996b) Spontaneous and induced sputum to measure indices of airway inflammation in asthma. Am J Resp Crit Care Med 154:866–869

Polkey MI et al (1996) Diaphragm strength in chronic obstructive pulmonary disease. Am J Respir Crit Care Med 154 (5):1310–1317

Polosa R, Holgate S (1997) Adenosine bronchoprovocation:a promising marker of allergic inflammation in asthma? Acta Otolaryngol 52:919–923

Polosa R, Rorke S, Holgate ST (2002) Evolving concepts on the value of adenosine hyperresponsiveness in asthma and chronic obstructive pulmonary disease. Thorax 57 (7):649–654

Popov T et al (1994) The evaluation of a cell dispersion method of sputum examination. Clin Exp Allergy 24:778–783

Qiu D, Tan WC (1999) Dithiothreitol has a dose-response effect on cell surface antigen expression. J Allergy Clin Immunol 103 (5 Pt 1):873–876

Rak S et al (1994) Influence of prolonged treatment with topical corticosteroid (fluticasone propionate) on early and late phase nasal responses and cellular infiltration in the nasal mucosa after allergen challenge. Clin Exp Allergy 24 (10):930–939

Rasmussen JB (1991) Late airway response increases at repeat allergen challenge. Allergy 46 (6):419–426

Rasmussen JB et al (1992) Leukotriene D4 receptor blockade inhibits the immediate and late bronchoconstrictor responses to inhaled antigen in patients with asthma. J Allergy Clin Immunol 90 (2):193–201

Reddel H et al (1999) Differences between asthma exacerbations and poor asthma control [published erratum appears in Lancet 1999 Feb 27;353(9154):758]. Lancet 353 (9150):364–369

Redington AE, Howarth PH (1997) Airway wall remodelling in asthma. Thorax 52:310–312

Rees PJ (1998) Bronchodilators in the therapy of chronic obstructive pulmonary disease. Eur Respir Mon 7:135–149

Reid MB (2001) COPD as a muscle disease. Am J Respir Crit Care Med. 164 (7):1101–1102

Remy-Jardin M et al (2002) Longitudinal Follow-up Study of Smoker's Lung with Thin-Section CT in Correlation with Pulmonary Function Tests. Radiology 222 (1):261–270

Remy-Jardin M et al (1993) Morphologic effects of cigarette smoking on airways and pulmonary parenchyma in healthy adult volunteers: CT evaluation and correlation with pulmonary function tests. Radiology 186 (1):107–115

Revill SM et al (1999) The endurance shuttle walk: a new field test for the assessment of endurance capacity in chronic obstructive pulmonary disease . Thorax 54 (3):213–222

Richter K et al (1999) Sequentially induced sputum in patients with asthma or chronic obstructive pulmonary disease. Eur Respir J 14 (3):697–701

Robinson D (1998) Bronchoalveolar lavage as a tool for studying airway inflammation in asthma. Eur Respir Rev 8:1072–1074

Rodriguez-Roisin R (2000) Toward a consensus definition for COPD exacerbations. Chest 117 [5 Suppl 2]:398S-401S

Roquet A et al (1997) Combined antagonism of leukotrienes and histamine produces predominant inhibition of allergen-induced early and late phase airway obstruction in asthmatics. Am J Respir Crit Care Med 155 (6):1856–1863

Roses AD (2000) Pharmacogenetics and the practice of medicine. Nature 405 (6788):857–865
Rosi E et al (1999) Sputum analysis, bronchial hyperresponsiveness, and airway function in asthma: results of a factor analysis. J Allergy Clin Immunol 103:232–237
Rutten-van Molken MP, Roos B, van Noord JA (1999) An empirical comparison of the St. George's respiratory questionnaire (SGRQ) and the chronic respiratory disease questionnaire (CRQ) in a clinical trial setting. Thorax 54 (11):995–1003
Sabroe I et al (1999) Differential regulation of eosinophil chemokine signaling via CCR3 and non-CCR3 pathways. J Immunol 162:2946–2955
Sakula A (1986) Charcot Leyden crystals and Curschmann spirals in asthmatic sputum. Thorax 41:503–507
Sampson AP (2001) IL-5 priming of eosinophil function in asthma. Clin Exp Allergy 31 (4):513–517
Sandford AJ and Silverman EK (2002) Chronic obstructive pulmonary disease. 1: Susceptibility factors for COPD the genotype-environment interaction. Thorax 57 (8):736–741
Santanello NC et al (1997) Measurement characteristics of two asthma symptom diary scales for use in clinical trials. Eur Respir J 10 (3):646–651
Scanlon PD et al (2000) Smoking Cessation and Lung Function in Mild-to-Moderate Chronic Obstructive Pulmonary Disease. The lung health study. Am J Respir Crit Care Med 161 (2 Pt 1):381–390
Schoeffel RE, Anderson SD, Altouyan REC (1981) Bronchial hyperreactivity in response to inhalation of ultrasonically nebulized solutions of distilled water and saline. Br Med J 63:459–471
Schols AM et al (1998) Weight loss is a reversible factor in the prognosis of chronic obstructive pulmonary disease. Am J Respir Crit Care Med 157 (6 Pt 1):1791–1797
Schols AM et al (1993) Prevalence and characteristics of nutritional depletion in patients with stable COPD eligible for pulmonary rehabilitation. Am Rev Respir Dis 147 (5):1151–1156
Schols AM et al (1995) Physiologic effects of nutritional support and anabolic steroids in patients with chronic obstructive pulmonary disease. A placebo- controlled randomized trial. Am J Respir Crit Care Med 152 (4 Pt 1):1268–1274
Schuh S et al (2000) A comparison of inhaled fluticasone and oral prednisone for children with severe acute asthma. N.Engl J Med 343 (10):689–694
Sestini P et al (1999) Different effects of inhaled aspirin-like drugs on allergen-induced early and late asthmatic responses. Am J Respir Crit Care Med 159 (4 Pt 1):1228–1233
Shahid SK et al (2002) Increased interleukin-4 and decreased interferon-gamma in exhaled breath condensate of children with asthma. Am J Respir Crit Care Med 165 (9):1290–1293
Sihra BS et al (1997) Effect of cyclosporin A on the allergen-induced late asthmatic reaction. Thorax 52:447–452
Singh SJ et al (1992) Development of a shuttle walking test of disability in patients with chronic airways obstruction. Thorax 47 (12):1019–1024
Snow V, Lascher S, Mottur-Pilson C (2001) Evidence base for management of acute exacerbations of chronic obstructive pulmonary disease. Ann.Intern.Med. 134 (7):595–599
Soejima K et al (2000) Longitudinal follow-up study of smoking-induced lung density changes by high-resolution computed tomography. Am J Respir Crit Care Med 161 (4 Pt 1):1264–1273
Soler M et al (2001) The anti-IgE antibody omalizumab reduces exacerbations and steroid requirement in allergic asthmatics. Eur Respir J 18 (2):254–261
Solway S et al (2001) A qualitative systematic overview of the measurement properties of functional walk tests used in the cardiorespiratory domain. Chest 119 (1):256–270

Soriano JB et al (2002) Survival in COPD patients after regular use of fluticasone propionate and salmeterol in general practice. Eur Respir J 20 (4):819–825

Spanavello A, Sharara AM, Migliori GB (1996) Reproducibility of induced sputum in asthmatics and normals. J Respir Crit Care Med 153:A289

Spanevello A et al (1997) Induced sputum to assess airway inflammation: a study of reproducibility. Clin Exp Allergy 27 (10):1138–1144

Stanescu D et al (1998) Identification of smokers susceptible to development of chronic airflow limitation: a 13-year follow-up. Chest 114 (2):416–425

Stanescu DC et al (1987) "Sensitive tests" are poor predictors of the decline in forced expiratory volume in one second in middle-aged smokers. Am Rev Respir Dis 135:585–590

Suissa S et al (2000) Low-dose inhaled corticosteroids and the prevention of death from asthma. N Engl J Med 343 (5):332–336

Sullivan P et al (1994) Anti-inflammatory effects of low-dose oral theophylline in atopic asthma. Lancet 343 (8904):1006–1008

Tattersall SF et al (1978) The use of tests of peripheral lung function for predicting future disability from airflow obstruction in middle-aged smokers. Am Rev Respir Dis 118:1035–1050

Tattersfield AE et al (1999) Exacerbations of asthma: a descriptive study of 425 severe exacerbations. The FACET International Study Group. Am J Respir Crit Care Med 160 (2):594–599

Taube C et al (2000) Factor analysis of changes in dyspnea and lung function parameters after bronchodilation in chronic obstructive pulmonary disease. Am J Respir Crit Care Med 162 (1):216–220

Taylor DA, Harris JG, O'Connor BJ (2000) Comparison of incremental and bolus dose inhaled allergen challenge in asthmatic patients. Clin Exp Allergy 30 (1):56–63

Taylor DA et al (1999) A dose-dependent effect of the novel inhaled corticosteroid ciclesonide on airway responsiveness to adenosine-5'-monophosphate in asthmatic patients. Am J Respir Crit Care Med 160 (1):237–243

Taylor IK et al (1992) A comparative study in atopic subjects with asthma of the effects of salmeterol and salbutamol on allergen-induced bronchoconstriction, increase in airway reactivity, and increase in urinary leukotriene E4 excretion. J Allergy Clin Immunol 89 (2):575–583

Terada N et al (2001) The kinetics of allergen-induced eotaxin level in nasal lavage fluid: its key role in eosinophil recruitment in nasal mucosa. Am.J Respir Crit Care Med. 164 (4):575–579

Thomas PS (2001) Tumour necrosis factor-alpha: the role of this multifunctional cytokine in asthma. Immunol.Cell Biol. 79 (2):132–140

Timmer W et al (2002) The new phosphodiesterase 4 inhibitor roflumilast is efficacious in exercise-induced asthma and leads to suppression of LPS-stimulated TNF-alpha ex vivo. J Clin Pharmacol 42 (3):297–303

Torphy TJ et al (1999) ArifloTM (SB 207499), a second generation phosphodiesterase 4 inhibitor for the treatment of asthma and COPD: from concept to clinic. Pulm Pharmacol Ther 12 (2):131–135

Twentyman OP et al (1990) Protection against allergen-induced asthma by salmeterol. Lancet 336 (8727):1338–1342

Twentyman OP, Finnerty JP, Holgate ST (1991) The inhibitory effect of nebulized albuterol on the early and late asthmatic reactions and increase in airway responsiveness provoked by inhaled allergen in asthma. Am Rev.Respir Dis. 144 (4):782–787

Twentyman OP, Sams VR, Holgate ST (1993) Albuterol and nedocromil sodium affect airway and leukocyte responses to allergen. Am Rev.Respir Dis. 147 (6 Pt 1):1425–1430

US Department of Health and Human Services, Food and Drug Adminstration Center for Drug Evaluation and Research CDER (1998) Draft Guidance for Industry: Metered

Dose Inhaler (MDI) and Dry Powder Inhaler (DPI) Drug Products. Chemistry, Manufacturing, and Controls Documentation

van Den Toorn LM et al (2000) Adolescents in clinical remission of atopic asthma have elevated exhaled nitric oxide levels and bronchial hyperresponsiveness. Am J Respir Crit Care Med 162 (3 Pt 1):953–957

Van Den BM et al (2001a) Corticosteroid-induced improvement in the PC20 of adenosine monophosphate is more closely associated with reduction in airway inflammation than improvement in the PC20 of methacholine. Am J Respir Crit Care Med. 164 (7):1127–1132

Van Den BM et al (2001b) PC(20) adenosine 5'-monophosphate is more closely associated with airway inflammation in asthma than PC(20) methacholine. Am J Respir Crit Care Med 163 (7):1546–1550

van der Valk P et al (2002) Effect of discontinuation of inhaled corticosteroids in patients with chronic obstructive pulmonary disease: the COPE study. Am J Respir Crit Care Med 166 (10):1358–1363

Van Eerdewegh P et al (2002) Associatio of the ADAM33 gene with sthma and bronchial hyperresponsiveness. Nature advance online publication (doi:10.1038/naure00878)

van Essen-Zandvliet EE et al (1994) Remission of childhood asthma after long-term treatment with an inhaled corticosteroid (budesonide): Can it be achieved? Eur Respir J 7:63–68

van Schalkwyk EM et al (2002) Dose-dependent inhibitory effect of roflumilast, a new, orally active, selective phosphodiesterase 4 inhibitor, on allergen-induced early and late asthmatic reaction. Eur Resp J 20:110S

Van Uden J, Raz E (1999) Immunostimulatory DNA and applications to allergic disease. J Allergy Clin Immunol 104 (5):902–910

Veen JC et al (1999) Lung function and sputum characteristics of patients with severe asthma during and induced exacerbation by double-blind steroid withdrawal. Am J Resp Crit Care Med 160:93–99

Vermeire PA et al (2002) Asthma control and differences in management practices across seven European countries. Respir Med 96 (3):142–149

Vestbo J et al (1999) Long-term effect of inhaled budesonide in mild and moderate chronic obstructive pulmonary disease: a randomised controlled trial. Lancet 353 (9167):1819–1823

Vignola AM et al (2001) Allergic inflammation of the upper and lower airways: a continuum of disease? Eur Respir Rev 11 (81):152–156

Vlachos-Mayer H et al (2000) Success and safety of sputum induction in the clinical setting. Eur Respir J 16 (5):997–1000

Wang ML, Gunel E, Petsonk EL (2000) Design strategies for longitudinal spirometry studies: study duration and measurement frequency. Am J Respir Crit Care Med 162 (6):2134–2138

Ward C et al (2002) Airway inflammation, basement membrane thickening and bronchial hyperresponsiveness in asthma. Thorax 57 (4):309–316

Wark PA et al (2001) Safety of sputum induction with isotonic saline in adults with acute severe asthma. Clin Exp Allergy 31 (11):1745–1753

Weersink EJM et al (1994) Partial inhibitor of the early and late asthmatic response by a single dose of salmeterol. Am J Respir Crit Care Med 150:1262–1267

Weersink EJ et al (1994) Partial inhibition of the early and late asthmatic response by a single dose of salmeterol. Am J Respir Crit Care Med 150 (5 Pt 1):1262–1267

Wenzel SE (1999) Inflammation, leukotrienes and the pathogenesis of the late asthmatic response. Clin Exp Allergy 29:1–3

Witte JS et al (2002) Relation between tumour necrosis factor polymorphism TNFalpha-308 and risk of asthma. Eur J Hum Genet 10 (1):82–85

Wong BJ et al (1992) Formoterol compared with beclomethasone and placebo on allergen-induced asthmatic responses. Am Rev Respir Dis 146 (5 Pt 1):1156–1160

Wong CS et al (1994) Effect of regular terbutaline and budesonide on bronchial reactivity to allergen challenge. Am J Respir Crit Care Med 150:1268–1273

Wong HH, Fahy JV (2002) Safety of one method of sputum induction in asthma subjects. Am J Resp Crit Care Med 156:299–303

Wood LJ et al (1999) An inhaled corticosteroid, budesonide, reduces baseline but not allergen-induced increases in bone marrow inflammatory cell progenitors in asthmatic subjects. Am J Respir Crit Care Med 159 (5 Pt 1):1457–1463

Woolhouse IS, Bayley DL, Stockley RA (2002) Effect of sputum processing with dithiothreitol on the detection of inflammatory mediators in chronic bronchitis and bronchiectasis. Thorax 57 (8):667–671

Wouters EF (2000) Nutrition and metabolism in COPD. Chest 117 [5 Suppl 1]:274S-280S

Zetterstrom O et al (2001) Improved asthma control with budesonide/formoterol in a single inhaler, compared with budesonide alone. Eur Respir J 18 (2):262–268

Zhang L et al (2002) Functional expression and characterization of macaque C-C chemokine receptor 3 (CCR3) and generation of potent antagonistic anti-macaque CCR3 monoclonal antibodies. J Biol Chem 277 (37):33799–33810

Novel Anti-inflammatory Therapies

P. J. Barnes · C. P. Page

Department of Thoracic Medicine, National Heart and Lung Institute,
Imperial College, Dovehouse St, London, SW3 6LY, UK
e-mail: p.j.barnes@ic.ac.uk

1	Introduction	350
2	**Mediator Antagonists**	350
2.1	Leukotriene Inhibitors	351
2.2	Prostaglandin Inhibitors	351
2.3	Endothelin Antagonists	352
2.4	Antioxidants	352
2.5	Purine Receptor Modulators	353
2.6	Nitric Oxide	354
3	**Protease Inhibitors**	355
3.1	Tryptase Inhibitors	355
3.2	Elastase Inhibitors	355
4	**Transcription Factor Inhibitors**	356
4.1	NF-κB Inhibitors	356
4.2	NF-AT/Calcineurin Inhibitors	357
4.3	GATA-3 Inhibitors	357
5	**Signal Transduction Inhibitors**	357
5.1	p38 MAP Kinase Inhibitors	358
5.2	JNK Inhibitors	358
5.3	Syk Kinase Inhibitors	358
5.4	Lyn Kinase Inhibitors	359
5.5	EGF Receptor Kinase	359
5.6	Phosphoinositide 3-Kinase Inhibitors	359
6	**Immunomodulators**	360
6.1	New Immunomodulators	360
6.2	Th2 Cell Inhibitors	360
6.3	Co-stimulation Inhibitors	361
7	**Preventive Strategies in Asthma**	362
7.1	Specific Allergen Vaccination (Immunotherapy)	362
7.2	Peptide Immunotherapy	362
7.3	Vaccination	362
7.4	CgP Oligonucleotides	363
8	**Remodelling Agents**	364
	References	364

Abstract Both asthma and chronic obstructive pulmonary disease (COPD) involve chronic inflammation of the airways and many new therapies are directed to suppressing or preventing the inflammatory response. The nature of the inflammatory process differs between the two diseases, so that different anti-inflammatory drugs may be needed. However, broad-spectrum anti-inflammatory therapies may be effective in both diseases. This chapter considers some of the new drugs in development that have not been included in previous chapters and discusses inhibitors of specific inflammatory mediators and proteases and inhibitors of transcription factors and signal transduction pathways (kinase inhibitors). Inhibitors of the allergic process that are relevant to asthma therapies and drugs that may reverse the tissue destruction of emphysema that are relevant to COPD are also discussed.

Keywords Anti-inflammatory · Inflammatory mediator · Cytokine · Chemokine

1
Introduction

Inflammation is a key feature of asthma and chronic obstructive pulmonary disease (COPD) and therefore a major target of chronic therapy. The importance of anti-inflammatory therapy in asthma has been underlined by the high efficacy of inhaled corticosteroids in almost all patients (see chapter "Corticosteroids", this volume). By contrast, corticosteroids have little or no anti-inflammatory action in COPD, highlighting the fundamental differences between these inflammatory diseases and reflecting marked differences in the components of inflammation (Barnes 2000). There are still concerns about systemic effects of inhaled corticosteroids when high doses are needed and there are some patients when even high does of oral or inhaled corticosteroid do not adequately control the disease. This and the lack of response to corticosteroids in COPD have prompted the search for new classes of anti-inflammatory drugs. Some of these approaches have been covered in preceding chapters, including mediator antagonists (Thomson), selective phosphodiesterase inhibitors (Jones et al.), cytokine modulators (Barnes), adhesion molecule inhibitors (Lever and Page) and anti-allergic drugs (Dziadzio et al.).

2
Mediator Antagonists

Many different inflammatory mediators have been implicated in asthma and COPD, and several specific receptor antagonists and synthesis inhibitors have been developed which have proved to be useful in working out the contribution of each mediator (Barnes et al. 1998b; Barnes 2003). As over 100 mediators probably contribute to the pathophysiology of airway diseases and many mediators have similar effects, it is unlikely that a single antagonist could have a major clinical effect, compared with non-specific agents such as β-agonists and corticosteroids. Histamine and leukotriene receptor antagonists are discussed in the earlier chapter by Thomson, but there are several other inflammatory pathways that have been considered for asthma and COPD.

2.1
Leukotriene Inhibitors

Cysteinyl-leukotriene (Cys-LT) receptor antagonists have been discussed in detail by Thomson (this volume). Leukotriene B_4, like Cys-LTs, is generated via $5'$-lipoxygenase but has neutrophil chemotactic activity mediated via distinct BLT_1-receptors. The concentration of LTB_4 is increased in exhaled breath condensate of patients with asthma, particularly in severe asthma (Csoma et al. 2002; Montuschi et al. 2002) and in exhaled breath and sputum of patients with COPD (Hill et al. 2000; Montuschi et al. 2000b). Two subtypes of receptor for LTB_4 have been described: BLT_1 receptors are mainly expressed on granulocytes and monocytes, whereas BLT_2 receptors are expressed on T lymphocytes (Yokomizo et al. 2000). Several BLT_1-antagonists have been developed, including SC-53228, CP-105,696, SB 201146 and BIIL284 (Silbaugh et al. 2000). One of these drugs, LY29311, failed to inhibit allergen-induced responses in asthmatic patients (Evans et al. 1996), but might have a role in more severe asthma and during exacerbations where neutrophil recruitment is evident. In patients with COPD, LY293111 inhibits the neutrophil chemotactic activity of sputum from COPD patients, indicating the potential clinical value of such drugs (Crooks et al. 2000). LTB_4 is synthesised by $5'$-lipoxygenase (5-LO), of which there are several inhibitors, although there have been problems in clinical development of drugs in this class because of side effects.

2.2
Prostaglandin Inhibitors

Prostaglandins have potent effects on airway function and there is increased expression of the inducible form of cyclo-oxygenase (COX)-2 in asthmatic airways (Taha et al. 2000), but inhibition of their synthesis with COX inhibitors, such as aspirin or ibuprofen, does not have any effect in most patients with asthma. Some patients have aspirin-sensitive asthma, which is more common in some ethnic groups, such as eastern Europeans and Japanese (Hirai et al. 2001). It is associated with increased expression of LTC_4 synthase, resulting in increased formation of Cys-LTs, possibly because of genetic polymorphisms (Cowburn et al. 1998). Recent evidence suggests that COX-2 inhibitors are safe in patients with aspirin-sensitive asthma, indicating that it is inhibition of COX-1 which somehow induces asthma symptoms (Martin-Garcia et al. 2002; Woessner et al. 2002). There is no evidence that COX-2 inhibitors are actually beneficial in asthma or in COPD, however.

Prostaglandin (PG)D_2 is a bronchoconstrictor prostaglandin produced predominantly by mast cells. Deletion of the PGD_2 receptors in mice significantly inhibits inflammatory responses to allergen and inhibits airway hyperresponsiveness (AHR), suggesting that this mediator may be important in asthma (Matsuoka et al. 2000). Recently it has also been discovered that PGD_2 activates a novel chemoattractant receptor termed chemoattractant receptor of T helper (Th)2 cells (CRTH2), which is expressed on Th2 cells, eosinophils and basophils and mediates chemotaxis of these cell types and may provide a link between mast cell activation and allergic inflammation (Hirai et al. 2001). There is a search for inhibitors of CRTH2, but the fact that blocking the production of PGD_2 with COX inhibitors is not beneficial in asthma, makes this approach unpromising.

2.3
Endothelin Antagonists

Endothelins are potent peptide mediators that are vasoconstrictors and bronchoconstrictors (Hay et al. 1996; Goldie et al. 1999). Endothelin-1 levels are increased in the sputum of patients with asthma; these levels are modulated by allergen exposure and steroid treatment (Chalmers et al. 1997; Redington et al. 1997). Endothelin-1 is also increased in sputum of patients with COPD, particularly during exacerbations (Roland et al. 2001). Endothelins induce airway smooth muscle cell proliferation and promote a pro-fibrotic phenotype and may therefore play a role in the chronic inflammation and airway remodelling in asthma. Endothelin-1 is released by hypoxia and may be involved in the secondary pulmonary hypertension in patients with COPD (Giaid et al. 1993).

Several relatively potent antagonists of endothelin receptors have now been developed (Benigni et al. 1999). Both ET_A and ET_B receptors may be involved in bronchoconstriction, vasoconstriction and structural changes, so that non-selective antagonists are preferred.

2.4
Antioxidants

As in all inflammatory diseases, there is increased oxidative stress in asthma and COPD, as activated inflammatory cells, such as macrophages, eosinophils and neutrophils produce reactive oxygen species. In COPD, cigarette smoking imposes an additional oxidative stress. Evidence for increased oxidative stress in asthma and COPD is provided by the increased concentrations of 8-isoprostane (a product of oxidized arachidonic acid) in exhaled breath condensates (Montuschi et al. 1999; Montuschi et al. 2000a) and increased ethane (a product of oxidative lipid peroxidation) in exhaled breath (Paredi et al. 2000a,b). Increased oxidative stress is related to disease severity and may amplify the inflammatory response and reduce responsiveness to corticosteroids, particularly in severe disease and during exacerbations and in COPD. Oxidative stress impairs responsiveness to corticosteroids by inhibition of histone deacetylase activity (probably through the formation of peroxynitrite) (Ito et al. 2001), which is the major mechanism whereby corticosteroids switch off inflammatory genes (see Barnes' chapter "Corticosteroids"). Antioxidants are therefore a logical approach in asthma therapy, particularly in severe asthma and in COPD. However, existing antioxidants are weak and are not able to neutralise the high level of oxidative stress in the airways, so that more potent antioxidants are needed in the future (Cuzzocrea et al. 2001).

2.5
Purine Receptor Modulators

It is now clear that a characteristic feature of asthma is bronchial hyperresponsiveness to a wide range of stimuli and that in particular, subjects with asthma will respond to inhaled adenosine, a substance that consistently fails to produce an airways response in non-asthmatics. As such, this feature may be phenotypic for asthma and therefore an important pathway to regulate (Spina et al. 2002). Adenosine is able to activate at least four receptor types that have been pharmacologically classified and cloned, which are referred to as A_1, A_2, A_{2b} and A_3. Whilst the receptors responsible for the exaggerated responses to adenosine in asthma have yet to be fully elucidated, the only investigation using lung tissue from subjects with asthma reports that even in vitro, airway tissues from asthmatics respond to adenosine, whereas tissues from subjects who do not have asthma do not respond to adenosine and that this effect of adenosine in asthmatic tissue is very clearly an A_1-mediated response by the use of well established pharmacological agents (Bjork et al. 1992). These data on asthmatic tissue are also supported by work in allergic rabbits which exhibit many changes that are similar to allergic asthmatics (Herd et al. 1996) in that in this species there is an exaggerated response to inhaled adenosine and selective A_1 receptor agonists such as N^6-cyclopentyladenosine (CPA)-mediated via activation of A_1 receptors (Ali et al. 1994; El-Hashim et al. 1996), a phenomenon that can also be seen in airway tissue taken from allergic, but not normal rabbits (Ali et al. 1994). Furthermore, work with an A_1 anti-sense oligonucleotide has reported that the allergen-induced changes can be abrogated in this allergic rabbit model, further supporting the role of A_1 receptors (Nyce et al. 1997). Interestingly, bamifylline, a drug that has been used orally to treat asthma and COPD in Europe for many years (Catena et al. 1998), is an effective A_1 receptor antagonist that, unlike theophylline, has no phosphodiesterase (PDE)-inhibiting activity (Abbracchio and Cattebini 1987). Analogues of bamifylline such as L-971 are currently in development for the treatment of asthma as novel orally active A_1 receptor antagonists (Obiefuna et al. 2003).

Whilst A_{2b} has been reported to be an important receptor for activation of mast cells (Foekistov et al. 1995), similar data have not been reported on other airway inflammatory cells for subjects with asthma (Ezeamuzie et al. 1999; Landells et al. 2000). The effects of adenosine are markedly potentiated by prior exposure to allergen (Hannon et al. 2001). A_{2B} receptor antagonists may therefore be of value in inhibiting mast cell activation in asthma, although it has been difficult to identify selective compounds (Fozard et al. 2002b). On the other hand, adenosine has an inhibitory effect on neutrophils, eosinophils and airway nerves mediated via adenosine A_{2A} receptors (Yukawa et al. 1989; Morimoto et al. 1993). This has led to the development of A_{2A} agonists and several selective drugs are in development, such as CGS 21680, which inhibits allergic inflammation in rats (Fozard et al. 2002a). Some groups have also implicated A_3 largely because of the observation that the A_3 receptor is expressed on eosinophils. However, recent clinical work with a monoclonal antibody to interleukin (IL)-5 (Leckie et al. 2000) and rh-IL-12 (Bryant et al. 2000) have questioned the role of the eosinophil in asthma, bringing the role of A_3 antagonists in to question as therapeutic agents for the treatment of asthma.

Adenosine 5'-triphosphate (ATP) may also be a mediator of asthma acting through P_2 receptors. ATP enhances the release of mediators from sensitised human mast cells via P_{2Y2} receptors, which are also expressed on eosinophils (Mohanty et al. 2001), suggesting that P_{2Y2} antagonists may be beneficial (Schulman et al. 1999).

2.6
Nitric Oxide

Nitric oxide (NO) is produced by several cells in the airway by NO synthases (Gaston et al. 1994; Barnes et al. 1995). Although the cellular source of NO within the lung is not known, inferences based on mathematical models suggest that it is the large airways which are the source of NO (Silkoff et al. 2000). Current data indicate that the level of NO in the exhaled air of patients with asthma is higher than the level of NO in the exhaled air of normal subjects (Kharitonov et al. 2001) and this is likely to be derived from inducible NO synthase (iNOS), which shows increased expression in asthmatic patients, particularly in airway epithelia cells and infiltrating inflammatory cells (Saleh et al. 1998; Guo et al. 2000). The combination of increased oxidative stress and NO derived from iNOS leads to the formation of the potent radical peroxynitrite that may result in nitration of proteins in the airways (Saleh et al. 1998). Peroxynitrite may nitrate certain histone deacetylases, thus impairing responsiveness to corticosteroids. This suggests that an inhibitor of iNOS might be useful in the treatment of asthma, particularly in patients with severe disease in whom it may restore steroid responsiveness and make asthma control easier to achieve. Exhaled NO is not elevated in COPD to the same extent as in asthma (Maziak et al. 1998), probably as NO gas is converted to peroxynitrite in the presence of oxidative stress and there is evidence for increased iNOS expression and nitrotyrosine formation in alveolar macrophages from patients with COPD (Ichinose et al. 2000).

Several potent and long-lasting iNOS inhibitors are now in development. One of these, L-NIL, given orally markedly reduces the levels of exhaled NO for several days, indicating that it may have the required pharmacological properties (Hansel et al. 2003).

3
Protease Inhibitors

3.1
Tryptase Inhibitors

Mast cell tryptase has several effects on airways, including increasing responsiveness of airway smooth muscle to constrictors, increasing plasma exudation, potentiating eosinophil recruitment and stimulating fibroblast and airway smooth muscle proliferation (He et al. 1997). Some of these effects are mediated by activation of the proteinase-activated receptor, PAR2, which is widely expressed in the airways of asthmatic patients (Knight et al. 2001). A tryptase inhibitor, APC366, is effective in a sheep model of allergen-induced asthma (Clark et al. 1995), but was only poorly effective in asthmatic patients (Krishna et al. 2001). More potent tryptase inhibitors and PAR2 antagonists are in development (Slusarchyk et al. 2002).

3.2
Elastase Inhibitors

There is compelling evidence for an imbalance between proteases that digest elastin (and other structural proteins) and antiproteases that protect against this. This suggests that either inhibiting these proteolytic enzymes or increasing endogenous antiproteases may be beneficial and theoretically should prevent the progression of airflow obstruction in COPD. Considerable progress has been made in identifying the enzymes involved in elastolytic activity in emphysema and in characterising the endogenous antiproteases that counteract this activity (Shapiro et al. 1999; Stockley 1999).

One approach is to give endogenous antiproteases [α_1-antitrypsin, secretory leukoprotease inhibitor, elafin, tissue inhibitors of matrix metalloproteinase (MMP)], either in recombinant form or by viral vector gene delivery. These approaches are unlikely to be cost effective, as large amounts of protein have to be delivered and gene therapy is unlikely to provide sufficient protein.

A more promising approach is to develop small molecule inhibitors of proteinases, particularly those that have elastolytic activity. Small molecule inhibitors, such as ONO-5046 and FR901277, have been developed which have high potency (Kawabata et al. 1991; Fujie et al. 1999). These drugs inhibit neutrophil elastase-induced lung injury in experimental animals, whether given by inhalation or systemically and also inhibit the other serine proteases released from neutrophils cathepsin G and proteinase-3. Small molecule inhibitors of neutrophil elastase are now entering clinical trials, but there is concern that neutrophil elastase may not play a critical role in emphysema and that other proteases are more important in elastolysis. Inhibitors of elastolytic cysteine proteases, such as cathepsins K, S and L that are released from macrophages (Punturieri et al. 2000) are also in development (Leung-Toung et al. 2002).

MMPs with elastolytic activity (such as MMP-9) may also be a target for drug development, although non-selective MMP inhibitors, such as marimastat, appear to have considerable side effects. It is possible that side effects could be reduced by increasing selectivity for specific MMPs or by targeting delivery to the lung parenchyma. MMP-9 is markedly overexpressed by alveolar macrophages from patients with COPD (Russell et al. 2002), so a selective inhibitor might be useful in the treatment of emphysema. Certain sulphated polysaccharides have also been reported to inhibit the effects of elastase, including the ability to inhibit elastase-induced emphysema in a variety of species (Lungarella et al. 1986; Lafuma et al. 1990).

4
Transcription Factor Inhibitors

Transcription factors, such as nuclear factor (NF)-κB and activator protein (AP)-1, play an important role in the orchestration of inflammation in asthma and COPD (Barnes et al. 1998a; Li et al. 2002), and this has prompted a search for specific blockers of these transcription factors.

4.1
NF-κB Inhibitors

NF-κB is naturally inhibited by the inhibitory protein IκB, which is degraded after activation by specific kinases (IKK). IKK2 is the isoenzyme that is important for activation of NF-κB by inflammatory stimuli (Delhase et al. 2000). Inhibitors of IKK2 or the proteasome, the multifunctional enzyme that degrades IκB, would thus inhibit NF-κB and there is a search for such inhibitors. There are some naturally occurring inhibitors of NF-κB, such as the fungal product gliotoxin (Pahl et al. 1996), although this compound is toxic. Several small molecule inhibitors of IKK2 have now been developed and are in clinical development (Roshak et al. 2002). These drugs should be of value in asthma and COPD, as in both diseases many of the abnormally expressed inflammatory genes are regulated by NF-κB. There are concerns that inhibition of NF-κB may cause side effects such as increased susceptibility to infections, which as been observed in gene disruption studies when components of NF-κB are inhibited (Barnes et al. 1997). One concern about long-term inhibition of NF-κB is that effective inhibitors may result in immune suppression and impair host defences, since mice which lack NF-κB genes succumb to septicaemia. However, there are alternative pathways of NF-κB activation that might be more important in inflammatory disease (Nasuhara et al. 1999).

4.2
NF-AT/Calcineurin Inhibitors

Cyclosporin A, tacrolimus and pimecrolimus inhibit T lymphocyte function by inhibiting the transcription factor NF-AT (nuclear factor of activated T cells) by blocking activation of calcineurin. This results in suppression if IL-2, IL-4, IL-5, IL-13 and granulocyte-macrophage colony-stimulating factor (GM-CSF), and therefore these drugs have therapeutic potential in asthma. However, cyclosporin A is of little value in chronic asthma as the dose is probably limited by toxicity (Evans et al. 2001). These drugs have serious side effects, particularly nephrotoxicity, and this limits their usefulness in a common disease such as asthma. Inhaled formulations of cyclosporin and tacrolimus are being tested for efficacy in asthma, but it remains to be determined whether this would give a favourable therapeutic ratio, as the drugs may be absorbed into the systemic circulation from the lungs. Rapamycin (sirolimus) has a similar action to calcineurin inhibitors, but acts more distally and has a different toxicity. The role of calcineurin inhibitors in COPD has not been explored, but the increased numbers of activated $CD8^+$ and $CD4^+$ cells, particularly in severe disease, suggest that this approach may be useful.

4.3
GATA-3 Inhibitors

GATA-3 plays a key role in differentiation of Th2 cells that drive allergic inflammation (Zheng et al. 1997). Blocking GATA-3 with an antisense oligonucleotide or a dominant-negative mutant prevents the differentiation of Th2 cells and the development of eosinophilic inflammation in mice (Zhang et al. 1999; Finotto et al. 2001), but development of a small molecule inhibitor of GATA-3 may be difficult until the specific activation pathway for this transcription factor have been identified. An alternative approach is to activate the opposing transcription factor T-bet which is reduced in asthma (Finotto et al. 2002).

5
Signal Transduction Inhibitors

Many signal transduction pathways are activated in inflammatory diseases and there is a complex interaction between these pathways. Kinases play a key role in the regulation of these pathways, many of which involve sequential phosphorylation of regulatory molecules. Many approaches in drug development involve discovering selective inhibitors of kinases (Cohen 2002).

There are three major mitogen-activated protein (MAP) kinase pathways and there is increasing recognition that these pathways are involved in chronic inflammation (Karin, 1998; Johnson et al. 2002).

5.1
p38 MAP Kinase Inhibitors

There has been particular interest in the p38 MAP kinase pathway which is involved in expression of multiple inflammatory proteins that are relevant to asthma and COPD (Carter et al. 1999; Lee et al. 2000; Meja et al. 2000). p38 MAP kinase is blocked by a novel class of drugs, the cytokine suppressant anti-inflammatory drugs (CSAIDs), such as SB203580, SB 239063 and RWJ67657. These drugs inhibit the synthesis of many inflammatory cytokines, chemokines and inflammatory enzymes. Interestingly, they appear to have a preferential inhibitory effect on synthesis of Th2 compared to Th1 cytokines, indicating their potential application in the treatment of atopic diseases (Schafer et al. 1999). Furthermore, p38 MAPK inhibitors have also been shown to decrease eosinophil survival by activating apoptotic pathways (Kankaanranta et al. 1999) and appear to be involved in corticosteroid resistance (Irusen et al. 2002). SB 239063 reduces (1) neutrophil infiltration after inhaled endotoxin and (2) the concentrations of IL-6 and MMP-9 in bronchoalveolar lavage fluid of rats, indicating its potential as an anti-inflammatory agent in COPD (Underwood et al. 2000).

p38 MAP kinase inhibitors are now in phase I development. Whether this new class of anti-inflammatory drugs will be safe in long-term studies remains to be established; it is likely that such a broad-spectrum anti-inflammatory drug will have some toxicity, but inhalation may be a feasible therapeutic approach.

5.2
JNK Inhibitors

Other MAP kinases are also involved in the inflammatory process in asthma. c-Jun N-terminal kinases (JNK) may be involved in activation of the transcription factor AP-1, and small molecule inhibitors have now been developed that have anti-inflammation effects in animal models (Huang et al. 2001). Steroid resistance in asthma is associated with increased activation of JNK (Sousa et al. 1999).

5.3
Syk Kinase Inhibitors

Syk ($p72^{Syk}$) kinase is a protein tyrosine kinase that plays a pivotal role in signalling of the high-affinity IgE receptor (FcεRI) in mast cells, and in *syk*-deficient mice mast cell degranulation is inhibited, suggesting that this might be an important potential target for the development of mast cell stabilising drugs (Costello et al. 1996). Syk is also involved in antigen receptor signalling of B and T lymphocytes and in eosinophil survival in response to IL-5 and GM-CSF (Yousefi et al. 1996), so that Syk inhibitors might have several useful beneficial effects in atopic diseases. Aerosolised Syk antisense oligodeoxynucleotide inhibits allergen-induced inflammation in a rat model, indicating that this may be a target for drug development (Stenton et al. 2002).

5.4
Lyn Kinase Inhibitors

Another tyrosine kinase, Lyn, is upstream of Syk, and an inhibitor of Lyn kinase, PP1, has an inhibitory effect on inflammatory and mast cell activation (Amoui et al. 1997). Lyn is also involved in eosinophil activation and IL-5 signalling (Adachi et al. 1999; Lynch et al. 2000), and a Lyn blocking peptide inhibits eosinophilic inflammation in a murine model (Adachi et al. 1999). Since Lyn and Syk are widely distributed in the immune system, there are doubts about the long-term safety of selective inhibitors, however.

5.5
EGF Receptor Kinase

Epidermal growth factor receptors (EGFR) may play a critical role in the regulation of mucus secretion from airways in response to multiple stimuli (Takeyama et al. 1999). An orally active small molecule inhibitor of EGFR tyrosine kinase gifitinib (Iressa) has now been developed for the treatment of epidermal cancers, but may also suppress mucus secretion (Wakeling 2002). However, treatment of the underlying inflammation that drives mucus hypersecretion in asthma, via release of IL-13 and other cytokines, is more likely to be useful, since mucus hypersecretion is not an isolated problem. Similarly, in COPD it is likely that EGFR is a major mechanism of mucus hypersecretion, so that EGFR kinase inhibitors may be successful in inhibiting this aspect of disease and may be useful in patients with chronic bronchitis (Takeyama et al. 2001). However, the mucus hypersecretion in COPD is likely to be secondary top neutrophilic inflammation in the large airways and so may be better treated by drugs that inhibit the neutrophilic inflammatory response.

5.6
Phosphoinositide 3-Kinase Inhibitors

Phosphoinositide (PI)-3Ks are a family of enzymes that lead to the generation of lipid second messengers that regulate a number of cellular events. A particular isoform, PI-3Kγ, is involved in neutrophil recruitment and activation. Knock-out of the PI-3Kγ gene results in inhibition of neutrophil migration and activation, as well as impaired T lymphocyte and macrophage function (Sasaki et al. 2000). This suggests that selective PI-3Kγ inhibitors may have relevant anti-inflammatory activity in COPD.

6
Immunomodulators

T lymphocytes may play a critical role in initiating and maintaining the inflammatory process in allergic diseases via the release of cytokines that result in eosinophilic inflammation, suggesting that T cell inhibitors may be useful in controlling asthmatic inflammation. As mentioned above, the non-specific immunomodulator cyclosporin A reduces the dose of oral steroids needed to control asthma in patients with severe asthma (Alexander et al. 1992), but its efficacy is very limited (Nizankowska et al. 1995) and a meta-analysis concluded that it was ineffective in treating chronic asthma (Evans et al. 2001). Side effects, particularly nephrotoxicity, limit its clinical use. The possibility of using inhaled cyclosporin A is now being explored, since in animal studies the inhaled drug is effective in inhibiting the inflammatory response in experimental asthma (Morley 1992).

6.1
New Immunomodulators

Immunomodulators, such as tacrolimus (FK506) and sirolimus (rapamycin), appear to be more potent but are also toxic and may offer no real advantage. Topical tacrolimus is effective in atopic dermatitis and is well tolerated, suggesting that it might also be useful in asthma (Bieber 1998). Novel immunomodulators that inhibit purine or pyrimidine pathways, such as mycophenolate mofetil, leflunomide and brequinar sodium, may be less toxic and therefore of greater potential value in asthma therapy (Thompson et al. 1993). Mycophenolate inhibits IL-5 production from a peripheral blood mononuclear cell preparation from patients with asthma, indicating that it might be useful (Powell et al. 2001), but no clinical studies of this drug have been reported in asthma.

CD4$^+$ T cells have been implicated in asthma and a chimeric antibody directed against CD4$^+$ (keliximab) which reduces circulating CD4$^+$ cells appears to have some beneficial effect in asthma (Kon et al. 1998), although long-term safety of such a treatment might be a problem. Furthermore, there is increasing evidence that CD8$^+$ cells (Tc2 cells), through release of IL-5 and other cytokines, might also be involved in allergic diseases, particularly in response to infections with certain viruses (Schwarze et al. 1999).

In COPD the potential for immunomodulators has yet to be explored. CD8$^+$ cells may play a key role in the development of emphysema and are therefore a potential target for inhibition (Cosio et al. 2002).

6.2
Th2 Cell Inhibitors

One problem with these non-specific immunomodulators is that they inhibit both Th1 and Th2 cells, and therefore do not restore the imbalance between these Th1 and Th2 cells in atopy. They also inhibit suppresser T cells (Tc cells) that may modulate the inflammatory response. Selective inhibition of Th2 cells may be more effective and better tolerated, and there is now a search for such drugs. One approach is to block the chemokine receptors expressed on Th2 cells that recruit these cells to the airways, namely CCR4, CCR8 and CXCR4, as discussed in the earlier chapter, "Cytokine Modulators". As discussed above, PGD_2 activates a novel chemoattractant receptor termed chemoattractant receptor of Th2 cells (CRTH2), which is expressed on Th2 cells, eosinophils and basophils and mediates chemotaxis of these cell types (Hirai et al. 2001; Sawyer et al. 2002). There is now a search for small molecule inhibitors of CRTH2.

Suplatast tosilate (IPD-1151T), is a dimethylsulphonium compound which inhibits the release of cytokines (IL-4, IL-5) from Th2 cells without effects on IFN-γ from Th1 cells in vitro (Oda et al. 1999). In clinical studies this drug has some clinical benefit in symptomatic asthmatic patients (Tamaoki et al. 2000) and reduces markers of inflammation and airway hyperresponsiveness (Yoshida et al. 2002). However, it has a short duration of action and it has not been compared to inhaled corticosteroids. The drug is only available in Japan and its mechanism of action is uncertain.

6.3
Co-stimulation Inhibitors

Co-stimulatory molecules may play a critical role in augmenting the interaction between antigen-presenting cells and $CD4^+$ T lymphocytes (Djukanovic 2000). The interaction between B7 and CD28 may determine whether a Th2-type cell response develops, and there is some evidence that B7-2 (CD86) skews towards a Th2 response. Blocking antibodies to B7.2 inhibits the development of specific IgE, pulmonary eosinophilia and AHR in mice, whereas antibodies to B7.1 (CD80) are ineffective (Haczku et al. 1999).

A molecule on activated T cells CTL4 appears to act as an endogenous inhibitor of T cell activation and a soluble fusion protein construct CTLA4-Ig is also effective in blocking AHR in a murine model of asthma (Van Oosterhout et al. 1997), although it appears to be less effective when the allergic inflammation is severe (Deurloo et al. 2001). Anti-CD28, anti-B7.2 and CTLA4-Ig also block the proliferative response of T cells to allergen (van Neerven et al. 1998), indicating that these are potential targets for novel therapies that should be effective in all atopic diseases.

Inducible co-stimulator (ICOS) is a costimulatory molecule related to CD28 and binds to a B7-like molecule B7RP-1. ICOS appears to be important in polarising the immune response and an antibody to ICOS blocks Th2 cell development, whereas CD28 plays a role in priming T cells (Gonzalo et al. 2001). The ICOS pathway is also critical for the development of regulatory T cells that secrete IL-10 and suppress the allergic inflammatory response (Akbari et al. 2002), so that stimulating rather than blocking its action may be beneficial.

7
Preventive Strategies in Asthma

Atopy, which underlies most asthma, appears to be due to immune deviation from Th1 to Th2 cells, which may arise because of a failure to inhibit the normal Th2 preponderance at birth as a result of lack of environmental factors (infections, and endotoxins) that stimulate a Th1 response.

7.1
Specific Allergen Vaccination (Immunotherapy)

Subcutaneous injection of small amounts of purified allergen has been used for many years in the treatment of allergy, but is not very effective in asthma and has a risk of serious side effects (Barnes 1996; Creticos et al. 1996). The molecular mechanism of desensitisation is unknown, but may be related to stimulation of IL-10 release from regulatory T cells (Akdis et al. 1998). Cloning of several common allergen genes has made it possible to prepare recombinant allergens for injection, although this purity may detract from their allergenicity, as most natural allergens contain several proteins. Intramuscular injection of rats with plasmid DNA expressing house dust mite allergen results in its long-term expression and prevents the development of IgE responses to inhaled allergen (Hsu et al. 1996). This suggests that allergen gene immunisation with a DNA vaccine might be a therapeutic strategy in the future.

7.2
Peptide Immunotherapy

Small peptide fragments of allergen (epitopes) are able to block allergen-induced T cell responses without inducing anaphylaxis (Yssel et al. 1994). T cell-derived peptides from cat allergen (*fel* d1) appear to be effective in blocking allergen responses to cat dander (Norman et al. 1996), but may induce an isolated late response to allergen by direct T cell activation followed by prolonged hyporesponsiveness (Haselden et al. 2001). *Fel* d1 peptides inhibit the cutaneous response to cat allergens, but whether this will be a useful strategy in asthma is not yet certain. One problem of this approach is that there are differences between individuals in recognising T cell peptides epitopes, so this approach may not be effective in all patients.

7.3
Vaccination

A relative lack of infections may be a factor predisposing to the development of atopy in genetically predisposed individuals, leading to the concept that vaccination to induce protective Th1 responses to prevent sensitisation and thus prevent the development of atopic diseases. Bacillus Calmette-Guérin (BCG) vaccination has been associated with a reduction in atopic diseases in Japan (Shirakawa et al. 1997), but this has not been confirmed in a Swedish population (Strannegard et al. 1998). BCG inoculation in mice 14 days before allergen sensitisation reduced the formation of specific IgE in response to allergen and the eosinophilic response and AHR responses to allergen, with an increase in production of IFN-γ (Herz et al. 1998). This has prompted several clinical trials of BCG to prevent the development of atopy. In one study, BCG vaccination has been shown to improve asthma control and reduce markers of Th2 activation (Choi et al. 2002).

Similar results have been obtained in mice with a single injection of heat-killed *Mycobacterium vaccae*, another potent inducer of Th1 responses (Wang et al. 1998) and with *Listeria*. *Lactobacillus acidophilus* in yoghurt, another potential means of tipping back the balance from Th2 to Th1 cells, weakly increases IFN-γ formation in adult asthmatic patients (Wheeler et al. 1997). In a clinical study *M. vaccae* had no effect on clinical parameters of asthma, IgE or cytokine profile, however (Shirtcliffe et al. 2001). Recent work has suggested that the effect of *M. tuberculosis* is mimicked by the chaperonins Cpn 60.1 and Cpn 10 (Riffo-Vasquez et al. 2003).

7.4
CgP Oligonucleotides

Immunostimulatory DNA sequences, such as unmethylated cytosine-guanosine dinucleotide-containing oligonucleotides (CpG ODN), are also potent inducers of Th1 cytokines through stimulation of IL-12 release (Horner et al. 2001) and in mice administration of CpG ODN increases the ratio of Th1 to Th2 cells, decreases formation of specific IgE and reduces the eosinophilic response to allergen, an effect which lasts for over 6 weeks (Sur et al. 1999). CpG ODN treatment is also able to reverse established allergen-driven eosinophilic inflammation in mice (Kline et al. 2002) and reverse airway hyperresponsiveness in mice sensitised to ragweed pollen antigen (Santeliz et al. 2002). These promising animal studies encourage the possibility that CpG ODN and DNA vaccines might prevent or cure atopic diseases in the future and clinical trials are currently underway (Agrawal et al. 2002).

Although these approaches aimed at tipping the balance back in favour of Th1 responses are promising in terms of disease modification, there are concerns that a therapeutic shift might increase the chance of developing Th1-mediated diseases, such as autoimmune diseases, multiple sclerosis, inflammatory bowel disease, rheumatoid arthritis and diabetes. These concerns particularly apply to infants.

8
Remodelling Agents

Since a major mechanism of airway obstruction in COPD is due to loss of elastic recoil due to proteolytic destruction of lung parenchyma, it seems unlikely that this could be reversible by drug therapy, although it might be possible to reduce the rate of progression by preventing the inflammatory and enzymatic disease process. Retinoic acid increases the number of alveoli in developing rats and, remarkably, reverses the histological and physiological changes induced by elastase treatment of adult rats (Massaro et al. 1997; Belloni et al. 2000). Retinoic acid activates retinoic acid receptors, which act as transcription factors to regulate the expression of many genes involved in growth and differentiation. The molecular mechanisms involved and whether this can be extrapolated to humans is not yet known. Several retinoic acid receptor subtype agonists have now been developed that may have a greater selectivity for this effect and therefore a lower risk of side effects. A short-term trial of all-*trans*-retinoic acid in patients with emphysema did not show any improvement in clinical parameters is currently underway (Mao et al. 2002).

References

Abbrachio,M.P., Cattebini,F. (1987). Selective activity of bamifylline on Adenosine A_1 receptors in rat brain. Pharmacol.Res. 19:537–545
Adachi T, Stafford S, Sur S, Alam R (1999) A novel Lyn-binding peptide inhibitor blocks eosinophil differentiation, survival, and airway eosinophilic inflammation. J Immunol. 163:939–946
Agrawal S, Kandimalla ER (2002) Medicinal chemistry and therapeutic potential of CpG DNA. Trends Mol.Med 8:114–121
Ahmed, T., Garrigo, J., Danta, I. (1993). Preventing bronchoconstriction in exercise induced asthma with inhaled haprin. N.Engl.J.Med. 329:90–95
Akbari O, Freeman GJ, Meyer EH, Greenfield EA, Chang TT, Sharpe AH, Berry G, DeKruyff RH, Umetsu DT (2002) Antigen-specific regulatory T cells develop via the ICOS-ICOS-ligand pathway and inhibit allergen-induced airway hyperreactivity. Nat.Med 8:1024–1032
Akdis CA, Blesken T, Akdis M, Wuthrich B, Blaser K (1998) Role of interleukin 10 in specific immunotherapy. J.Clin.Invest. 102:98–106
Alexander AG, Barnes NC, Kay AB (1992) Trial of cyclosporin in corticosteroid-dependent chronic severe asthma. Lancet 339:324–328
Ali, S., Mustafa, S.J., Metzger, W.J. (1994). Adenosine-induced bronchoconstriction and contraction of airways smooth muscle from allergis rabbits with late-phase airway obstruction: evidence for an inducible A_1 receptor. J.Pharm.Exp.Ther. 268:1328–1334
Amoui M, Draber P, Draberova L (1997) Src family-selective tyrosine kinase inhibitor, PP1, inhibits both Fc epsilonRI- and Thy-1-mediated activation of rat basophilic leukemia cells. Eur.J.Immunol. 27:1881–1886
Barnes PJ (1996) Immunotherapy for asthma: is it worth it? New Engl J Med 334:531–532
Barnes PJ (2000) Mechanisms in COPD: differences from asthma. Chest 117:10S-14S
Barnes PJ (2003) New concepts in COPD. Ann Rev Med 54:113–129
Barnes PJ, Adcock IM (1998a) Transcription factors and asthma. Eur Respir J 12:221–234

Barnes PJ, Chung KF, Page CP (1998b) Inflammatory mediators of asthma: an update. Pharmacol Rev 50:515–596

Barnes PJ, Karin M (1997) Nuclear factor-kB: a pivotal transcription factor in chronic inflammatory diseases. New Engl J Med 336:1066–1071

Barnes PJ, Liew FY (1995) Nitric oxide and asthmatic inflammation. Immunol Today 16:128–130

Belloni PN, Garvin L, Mao CP, Bailey-Healy I, Leaffer D (2000) Effects of all-trans-retinoic acid in promoting alveolar repair. Chest 117:235S-241S

Benigni A, Remuzzi G (1999) Endothelin antagonists. Lancet 353:133–138

Bieber T (1998) Topical tacrolimus (FK 506): a new milestone in the management of atopic dermatitis. J.Allergy Clin.Immunol. 102:555–557

Bjork, T., Gustafsson, K.E., Dahlen, E.E. (1992). Isolated bronchi from asthmatics are hyperresponsive to adenosine, which apparently acts indirectly by liberation of leukotrienes and histamine. Am.Rev.Resp.Dis. 145:1087–1091

Bryan, S.A., O'Connor, B.J., Matti, S., Leckie, M.J., Kanabar, V., Khan, J., Warnington, S.S., Renzetti, L., Rames, A., Bock, J.A., Boyce, M.J., Hansel, T., Holgate, S.T., Barnes, P.J. (2000). Effects of recombinant human interleukin-12 on eosinophils, airway hyperresponsiveness and the late asthmatic response. Lancet 356, 2149–2153

Carter AB, Monick MM, Hunninghake GW (1999) Both erk and p38 kinases are necessary for cytokine gene transcription. Am.J.Respir.Cell Mol.Biol. 20:751–758

Catena,E., Gunella,G., Monci,P.P.A., Oliani,C. (1988). Evaluation of the risk/benefit ration of bamifylline in the treatment of chronic lung disease. Italian J.Dis.Chest 42:419–426

Chalmers GW, Little SA, Patel KR, Thomson NC (1997) Endothelin-1-induced bronchoconstriction in asthma. Am.J.Respir.Crit.Care Med. 156:382–388

Choi IS, Koh YI (2002) Therapeutic effects of BCG vaccination in adult asthmatic patients: a randomized, controlled trial. Ann.Allergy Asthma Immunol. 88:584–591

Clark JM, Abraham WM, Fishman CE, Forteza R, Ahmed A, Cortes A, Warne RL, Moore WR, Tanaka RD (1995) Tryptase inhibitors block allergen-induced airway and inflammatory responses in allergic sheep. Am J Respir Crit Care Med 152:2076–2083

Cohen P (2002) Protein kinases—the major drug targets of the twenty-first century? Nat.Rev Drug Discov. 1:309–315

Cosio MG, Majo J, Cosio MG (2002) Inflammation of the airways and lung parenchyma in COPD: role of T cells. Chest 121:160S-165S

Costello PS, Turner M, Walters AE, Cunningham CN, Bauer PH, Downward J, Tybulewicz VL (1996) Critical role for the tyrosine kinase Syk in signalling through the high affinity IgE receptor of mast cells. Oncogene 13:2595–2605

Cowburn AS, Sladek K, Soja J, Adamek L, Nizankowska E, Szczeklik A, Lam BK, Penrose JF, Austen FK, Holgate ST, Sampson AP (1998) Overexpression of leukotriene C4 synthase in bronchial biopsies from patients with aspirin-intolerant asthma. J.Clin.Invest. 101:834–846

Creticos PS, Reed CE, Norman PS, Khoury J, Adkinson NF, Buncher R, Busse WW, Bush RK, Gaddie J, Li JT, Richerson HB, Rosenthal RR, Solomon WR, Steinberg P, Yunginger JW (1996) Ragweed immunotherapy in adult asthma. New Engl J Med 334:501–506

Crooks SW, Bayley DL, Hill SL, Stockley RA (2000) Bronchial inflammation in acute bacterial exacerbations of chronic bronchitis: the role of leukotriene B_4. Eur Respir J 15:274–280

Csoma Z, Kharitonov SA, Balint B, Bush A, Wilson NM, Barnes PJ (2002) Increased leukotrienes in exhaled breath condensate in childhood asthma. Am J Respir Crit Care Med 166:1345–1349

Cuzzocrea S, Riley DP, Caputi AP, Salvemini D (2001) Antioxidant Therapy: A New Pharmacological Approach in Shock, Inflammation, and Ischemia/Reperfusion Injury. Pharmacol Rev 53:135–159

Delhase M, Li N, Karin M (2000) Kinase regulation in inflammatory response. Nature 406:367–368
Deurloo DT, van Esch BC, Hofstra CL, Nijkamp FP, Van Oosterhout AJ (2001) CTLA4-IgG reverses asthma manifestations in a mild but not in a more "severe" ongoing murine model. Am J Respir Cell Mol.Biol. 25:751–760
Diamant, Z et al. (1996). Effects of inhaled heparin on allergen-induced airway early and late asthmatic responses in patients with atopic asthma. Am.J.Hosp.Crit.Care Med. 153:1790–1795
Djukanovic R (2000) The role of co-stimulation in airway inflammation. Clin.Exp.Allergy 30 Suppl 1:46–50
El-Hashim, A., D'Agostino, B., Matera, M.G., Page, C.P. (1996). Characterization of adenosine receptors involved in adenosine-induced bronchoconstriction in allergic rabbits. Br.J.Pharmacol. 119:1262–1268
Evans DJ, Barnes PJ, Coulby LJ, Spaethe SM, van Alstyne EC, Pechous PA, Mitchell MI, O'Connor BJ (1996) The effect of a leukotriene B_4 antagonist LY293111 on allergen-induced responses in asthma. Thorax 51:1178–1184
Evans DJ, Cullinan P, Geddes DM (2001) Cyclosporin as an oral corticosteroid sparing agent in stable asthma (Cochrane Review). Cochrane.Database.Syst.Rev. 2:CD002993
Ezeamuzie, C.I., Phillips, E. (1999). Adenosine A_3 receptors on human eosinophils may mediate inhibition of degranulation and superoxide anion release. Br.J.Pharmacol. 127:188–194
Feoktistov I, Biaggioni I (1998) Pharmacological characterization of adenosine A_{2B} receptors: studies in human mast cells co-expressing A_{2A} and A_{2B} adenosine receptor subtypes. Biochem.Pharmacol. 55:627–633
Finotto S, De Sanctis GT, Lehr HA, Herz U, Buerke M, Schipp M, Bartsch B, Atreya R, Schmitt E, Galle PR, Renz H, Neurath MF (2001) Treatment of allergic airway inflammation and hyperresponsiveness by antisense-induced local blockade of GATA-3 expression. J Exp.Med 193:1247–1260
Finotto S, Neurath MF, Glickman JN, Qin S, Lehr HA, Green FH, Ackerman K, Haley K, Galle PR, Szabo SJ, Drazen JM, De Sanctis GT, Glimcher LH (2002) Development of spontaneous airway changes consistent with human asthma in mice lacking T-bet. Science 295:336–338
Foekistov,I., Biaggioni,I. (1995). Adenosine 2b receptors evoke interleukin-8 secretion in human mast cells. An enprophylline sensitive mechanism with implications for asthma. J.Clin.Invest. 96:1979–1986
Fozard JR, Ellis KM, Villela Dantas MF, Tigani B, Mazzoni L (2002a) Effects of CGS 21680, a selective adenosine A_{2A} receptor agonist, on allergic airways inflammation in the rat. Eur.J Pharmacol 438:183–188
Fozard JR, McCarthy C (2002b) Adenosine receptor ligands as potential therapeutics in asthma. Curr Opin Investig.Drugs 3:69–77
Fryer,A. et al. (1997). Selective o-desulphation produces non anti-coagulant heparin that retains pharmacologic activity in the lung. J.Pharmacol.Exp.Ther. 282:209–219
Fujie K, Shinguh Y, Yamazaki A, Hatanaka H, Okamoto M, Okuhara M (1999) Inhibition of elastase-induced acute inflammation and pulmonary emphysema in hamsters by a novel neutrophil elastase inhibitor FR901277. Inflamm.Res 48:160–167
Gaston B, Drazen JM, Loscalzo J, Stamler JS (1994) The biology of nitrogen oxides in the airways. Am J Respir Crit Care Med 149:538–551
Giaid A, Yanagisawa M, Langleben D, Michel RP, Levy R, Shennib M, Kimura S, Masaki T, Duguid W, Stewart DJ (1993) Expression of endotheinl-in in the lungs of patients with pulmonary hypertension. New Engl J Med 328:1732–1739
Goldie RG, Henry PJ (1999) Endothelins and asthma. Life Sci. 65:1–15
Gonzalo JA, Tian J, Delaney T, Corcoran J, Rottman JB, Lora J, Al garawi A, Kroczek R, Gutierrez-Ramos JC, Coyle AJ (2001) ICOS is critical for T helper cell-mediated lung mucosal inflammatory responses. Nat.Immunol. 2:597–604

Guo FH, Comhair SA, Zheng S, Dweik RA, Eissa NT, Thomassen MJ, Calhoun W, Erzurum SC (2000) Molecular mechanisms of increased nitric oxide (NO) in asthma: evidence for transcriptional and post-translational regulation of NO synthesis. J Immunol 164:5970–5980

Haczku A, Takeda K, Redai I, Hamelmann E, Cieslewicz G, Joetham A, Loader J, Lee JJ, Irvin C, Gelfand EW (1999) Anti-CD86 (B7.2) treatment abolishes allergic airway hyperresponsiveness in mice. Am.J.Respir.Crit.Care Med. 159:1638–1643

Hannon JP, Tigani B, Williams I, Mazzoni L, Fozard JR (2001) Mechanism of airway hyperresponsiveness to adenosine induced by allergen challenge in actively sensitized Brown Norway rats. Br J Pharmacol 132:1509–1523

Hansel TT, Kharitonov SA, Donnelly LE, Erin EM, Currie MG, Moore WM, Manning PT, Recker DP, Barnes PJ (2003) A selective inhibitor of inducible nitric oxide synthase inhibits exhaled breath nitric oxide. FASEB J

Haselden BM, Larche M, Meng Q, Shirley K, Dworski R, Kaplan AP, Bates C, Robinson DS, Ying S, Kay AB (2001) Late asthmatic reactions provoked by intradermal injection of T-cell peptide epitopes are not associated with bronchial mucosal infiltration of eosinophils or T_H2-type cells or with elevated concentrations of histamine or eicosanoids in bronchoalveolar fluid. J Allergy Clin Immunol 108:394–401

Hay DW, Henry PJ, Goldie RG (1996) Is endothelin-1 a mediator in asthma? Am.J.Respir.Crit.Care Med. 154:1594–1597

He S, Walls AF (1997) Human mast cell tryptase: a stimulus of microvascular leakage and mast cell activation. Eur.J.Pharmacol. 328:89–97

Herd,C.M., Page,C.P. (1996). The rabbit model of asthma and the late asthmatic response. Airways smooth muscle: modelling the asthmatic response in vivo. pp147–170

Herz U, Gerhold K, Gruber C, Braun A, Wahn U, Renz H, Paul K (1998) BCG infection suppresses allergic sensitization and development of increased airway reactivity in an animal model. J.Allergy Clin.Immunol. 102:867–874

Hill AT, Campbell EJ, Hill SL, Bayley DL, Stockley RA (2000) Association between airway bacterial load and markers of airway inflammation in patients with stable chronic bronchitis. Am J Med 109:288–295

Hirai H, Tanaka K, Yoshie O, Ogawa K, Kenmotsu K, Takamori Y, Ichimasa M, Sugamura K, Nakamura M, Takano S, Nagata K (2001) Prostaglandin D2 selectively induces chemotaxis in T helper type 2 cells, eosinophils, and basophils via seven-transmembrane receptor CRTH2. J Exp Med 193:255–261

Horner AA, Van Uden JH, Zubeldia JM, Broide D, Raz E (2001) DNA-based immunotherapeutics for the treatment of allergic disease. Immunol.Rev. 179:102–118

Hsu CH, Chua KY, Tao MH, Lai YL, Wu HD, Huang SK, Hsieh KH (1996) Immunoprophylaxis of allergen-induced immunoglobulin E synthesis and airway hyperresponsiveness in vivo by genetic immunization. Nat.Med. 2:540–544

Huang TJ, Adcock IM, Chung KF (2001) A novel transcription factor inhibitor, SP100030, inhibits cytokine gene expression, but not airway eosinophilia or hyperresponsiveness in sensitized and allergen-exposed rat. Br J Pharmacol 134:1029–1036

Ichinose M, Sugiura H, Yamagata S, Koarai A, Shirato K (2000) Increase in reactive nitrogen species production in chronic obstructive pulmonary disease airways. Am J Resp Crit Care Med 160:701–706

Irusen E, Matthews JG, Takahashi A, Barnes PJ, Chung KF, Adcock IM (2002) p38 Mitogen-activated protein kinase-induced glucocorticoid receptor phosphorylation reduces its activity: Role in steroid-insensitive asthma. J Allergy Clin Immunol 109:649–657

Ito K, Lim S, Caramori G, Chung KF, Barnes PJ, Adcock IM (2001) Cigarette smoking reduces histone deacetylase 2 expression, enhances cytokine expression and inhibits glucocorticoid actions in alveolar macrophages. FASEB J 15:1100–1102

Johnson GL, Lapadat R (2002) Mitogen-activated protein kinase pathways mediated by ERK, JNK, and p38 protein kinases. Science 298:1911–1912

Kankaanranta H, Giembycz MA, Barnes PJ, Lindsay DA (1999) SB203580, an inhibitor of p38 mitogen-activated protein kinase, enhances constitutive apoptosis of cytokine-deprived human eosinophils. J Pharmacol Exp Ther 290:621–628

Karin M (1998) Mitogen-activated protein kinase cascades as regulators of stress responses. Ann.N.Y.Acad.Sci. 851:139–46:139–146

Kawabata K, Suzuki M, Sugitani M, Imaki K, Toda M, Miyamoto T (1991) ONO-5046, a novel inhibitor of human neutrophil elastase. Biochem.Biophys.Res.Commun. 177:814–820

Kharitonov SA, Barnes PJ (2001) Exhaled markers of pulmonary disease. Am J Respir Crit Care Med 163:1693–1772

Kline JN, Kitagaki K, Businga TR, Jain VV (2002) Treatment of established asthma in a murine model using CpG oligodeoxynucleotides. Am J Physiol Lung Cell Mol.Physiol 283:L170–L179

Knight DA, Lim S, Scaffidi AK, Roche N, Chung KF, Stewart GA, Thompson PJ (2001) Protease-activated receptors in human airways: upregulation of PAR-2 in respiratory epithelium from patients with asthma. J Allergy Clin Immunol 108:797–803

Kon OM, Sihra BS, Compton CH, Leonard TB, Kay AB, Barnes NC (1998) Randomised dose-ranging placebo-controlled study of chimeric anribody to CD4 (keliximab) in chronic severe asthma. Lancet 352:1109–1113

Krishna MT, Chauhan A, Little L, Sampson K, Hawksworth R, Mant T, Djukanovic R, Lee T, Holgate S (2001) Inhibition of mast cell tryptase by inhaled APC 366 attenuates allergen-induced late-phase airway obstruction in asthma. J Allergy Clin Immunol. 107:1039–1045

Lafuma C, Frisdal E, Harf A, Robert L, Hornbeck W (1991). Prevention of leukocyte elastase-induced emphysema in mice by heparin fragments. Eur.Resp.J. 4:1004–1009

Landells LJ, Jensen MW, Orr LM, Spina D, O'Connor B, Page CP (2000). The role of adenosine receptors in the action of theophylline on human peripheral blood mononuclear cells from healthy and asthmatic subjects. Br.J.Pharmacol. 129:1140–1149

Leckie MJ, Ten Brincke A, Khan J, Diamant Z, O'Connor BJ, Walls CM, Mathur AK, Cowley HC, Chung KF, Djukanovic R, Hansel T, Holgate ST, Sterk P, Barnes PJ (2000). Effects of an interleukin 5 blocking monoclonal antibody on eosinophil airway hyperresponsiveness and the late asthmatic response. Lancet 356, 2144–2158

Lee JC, Kumar S, Griswold DE, Underwood DC, Votta BJ, Adams JL (2000) Inhibition of p38 MAP kinase as a therapeutic strategy. Immunopharmacology 47:185–201

Leung-Toung R, Li W, Tam TF, Karimian K (2002) Thiol-dependent enzymes and their inhibitors: a review. Curr Med Chem. 9:979–1002

Lever,R., Page,C.P. (2002). Novel drug development opportunities for heparin. Nature Reviews Drug Discovery 1:140–148

Li Q, Verma IM (2002) NF-κB regulation in the immune system. Nat.Rev.Immunol. 2:725–734

Lungarella G, Gardi C, Fonzi L, Comparini L., Share NN, Zimmerman M, Martorana PA (1986). Effect of the novel synthetic protease inhibitor furomyl saccharin on elastase induced emphysema in rabbits and hamsters. Exp.Lung Res. 11:35–47

Lynch OT, Giembycz MA, Daniels I, Barnes PJ, Lindsay MA (2000) Pleiotropic role of *lyn* kinase in leukotriene B$_4$-induced eosinophil activation. Blood 95:3541–3547

Mao JT, Goldin JG, Dermand J, Ibrahim G, Brown MS, Emerick A, McNitt-Gray MF, Gjertson DW, Estrada F, Tashkin DP, Roth MD (2002) A pilot study of all-trans-retinoic acid for the treatment of human emphysema. Am J Respir Crit Care Med 165:718–723

Martin-Garcia C, Hinojosa M, Berges P, Camacho E, Garcia-Rodriguez R, Alfaya T, Iscar A (2002) Safety of a cyclooxygenase-2 inhibitor in patients with aspirin-sensitive asthma. Chest 121:1812–1817

Massaro G, Massaro D (1997) Retinoic acid treatment abrogates elastase-induced pulmonary emphysema in rats. Nature Med 3:675–677

Matsuoka T, Hirata M, Tanaka H, Takahashi Y, Murata T, Kabashima K, Sugimoto Y, Kobayashi T, Ushikubi F, Aze Y, Eguchi N, Urade Y, Yoshida N, Kimura K, Mizoguchi A, Honda Y, Nagai H, Narumiya S (2000) Prostaglandin D_2 as a mediator of allergic asthma. Science 287:2013–2017

Maziak W, Loukides S, Culpitt S, Sullivan P, Kharitonov SA, Barnes PJ (1998) Exhaled nitric oxide in chronic obstructive pulmonary disease. Am J Respir Crit Care Med 157:998–1002

Meja KK, Seldon PM, Nasuhara Y, Ito K, Barnes PJ, Lindsay MA, Giembycz MA (2000) p38 MAP kinase and MKK-1 co-operate in the generation of GM-CSF from LPS-stimulated human monocytes by an NF-kappaB-independent mechanism. Br J Pharmacol 131:1143–1153

Mohanty JG, Raible DG, McDermott LJ, Pelleg A, Schulman ES (2001) Effects of purine and pyrimidine nucleotides on intracellular Ca^{2+} in human eosinophils: activation of purinergic P_{2Y} receptors. J Allergy Clin.Immunol. 107:849–855

Montuschi P, Barnes PJ (2002) Exhaled leukotrienes and prostaglandins in asthma. J Allergy Clin Immunol 109:615–620

Montuschi P, Ciabattoni G, Corradi M, Nightingale JA, Collins JV, Kharitonov SA, Barnes PJ (1999) Increased 8-Isoprostane, a marker of oxidative stress, in exhaled condensates of asthmatic patients. Am J Respir Crit Care Med 160:216–220

Montuschi P, Collins JV, Ciabattoni G, Lazzeri N, Corradi M, Kharitonov SA, Barnes PJ (2000a) Exhaled 8-isoprostane as an in vivo biomarker of lung oxidative stress in patients with COPD and healthy smokers. Am J Respir Crit Care Med 162:1175–1177

Montuschi P, Kharitonov SA, Carpagnano E, Culpitt S, Russell R, Collins JV, Barnes PJ (2000b) Exhaled prostaglandin E_2: a new marker of airway inflammation in COPD. Am J Respir Crit Care Med 161:A821

Morimoto H, Yamashita M, Imazumi K, Matsuda A, Ochi T, Seki N, Mizuhara H, Fujii T, Senoh H (1993). Effect of adenosine A_2 receptor agonists on the excitation of capsaicin-sensitive nerves in airway tissues. Eur.J.Pharmacol. 240:121–126

Morley J (1992) Cyclosporin A in asthma therapy: a pharmacological rationale. J Autoimmunity 5 (Suppl A):265–269

Nasuhara Y, Adcock IM, Catley M, Barnes PJ, Newton R (1999) Differential IKK activation and IκBα degradation by interleukin-1β and tumor necrosis factor-α in human U937 monocytic cells: evidence for additional regulatory steps in κB-dependent transcription. J Biol Chem 274:19965–19972

Nizankowska E, Soja J, Pinis G, Bochenek G, Stadek K, Domgala B, Pajak A, Szczeklik A (1995) Tretment of steroid-depenent bronchial asthma with cyclosporin. Eur Resp J 8:1091–1099

Norman PS, Ohman JL, Jr., Long AA, Creticos PS, Gefter MA, Shaked Z, Wood RA, Eggleston PA, Hafner KB, Rao P, Lichtenstein LM, Jones NH, Nicodemus CF (1996) Treatment of cat allergy with T-cell reactive peptides. Am.J.Respir.Crit.Care Med. 154:1623–1628

Nyce,J.W., Metzger,W.J. (1997). DNA antisense therapy for asthma in an animal model. Nature. 385:721–725

Obiefuna, P.C.M., Qin, W., Wilson, C.N., Mustafa, S.J. (2003). A novel A_1 adenosine receptor antagonist (L-97-1) reduces allergic responses to house dust mite in a rabbit model. Am.J.Hosp.Crit.Care.Med (In press)

Oda N, Minoguchi K, Yokoe T, Hashimoto T, Wada K, Miyamoto M, Tanaka A, Kohno Y, Adachi M (1999) Effect of suplatast tosilate (IPD-1151T) on cytokine production by allergen-specific human Th1 and Th2 cell lines. Life Sci 65:763–770

Pahl HL, Krauss B, Schultze-Osthoff K, Decker T, Traenckner M, Myers C, Parks T, Warring P, Muhlbacher A, Czernilofsky A-P, Baeuerle PA (1996) The immunosuppressive fungal metabolite gliotoxin specifically inhibits transcription factor NF-kB. J Exp Med 183:1829–1840

Paredi P, Kharitonov SA, Barnes PJ (2000a) Elevation of exhaled ethane concentration in asthma. Am J Respir Crit Care Med 162:1450–1454

Paredi P, Kharitonov SA, Leak D, Ward S, Cramer D, Barnes PJ (2000b) Exhaled ethane, a marker of lipid peroxidation, is elevated in chronic obstructive pulmonary disease. Am J Respir Crit Care Med 162:369–373

Powell N, Till S, Bungre J, Corrigan C (2001) The immunomodulatory drugs cyclosporin A, mycophenolate mofetil, and sirolimus (rapamycin) inhibit allergen-induced proliferation and IL-5 production by PBMCs from atopic asthmatic patients. J Allergy Clin.Immunol. 108:915–917

Preuss, J.M.H., Page, C.P. (2000). Effect of heparin on antigen-induced airway responses and leukocyte accumulation in neonatally sensitized rabbits. Br.J.Pharmacol. 129:1585–1596

Punturieri A, Filippov S, Allen E, Caras I, Murray R, Reddy V, Weiss SJ (2000) Regulation of elastinolytic cysteine proteinase activity in normal and cathepsin K-deficient human macrophages. J Exp Med 192:789–800

Redington AE, Springall DR, Ghatei MA, Madden J, Bloom SR, Frew AJ, Polak JM, Holgate ST, Howarth PH (1997) Airway endothelin levels in asthma: influence of endobronchial allergen challenge and maintenance corticosteroid therapy. Eur.Respir.J. 10:1026–1032

Riffo-Vasquez, Y., Spina, D., Page, C.P., Tormay, P., Singh, M., Henderson, B., Coates, A. (2003). The effect of mycobacterium tuberculosis chaperonis on bronchial eosinophilia and hyperresponsiveness in a murine model of allergic inflammation. Clin.-Exp. Allergy (In press)

Roland M, Bhowmik A, Sapsford RJ, Seemungal TA, Jeffries DJ, Warner TD, Wedzicha JA (2001) Sputum and plasma endothelin-1 levels in exacerbations of chronic obstructive pulmonary disease. Thorax 56:30–35

Roshak AK, Callahan JF, Blake SM (2002) Small-molecule inhibitors of NF-kappaB for the treatment of inflammatory joint disease. Curr Opin Pharmacol 2:316–321

Russell RE, Culpitt SV, DeMatos C, Donnelly L, Smith M, Wiggins J, Barnes PJ (2002) Release and activity of matrix metalloproteinase-9 and tissue inhibitor of metalloproteinase-1 by alveolar macrophages from patients with chronic obstructive pulmonary disease. Am J Respir Cell Mol Biol 26:602–609

Saleh D, Ernst P, Lim S, Barnes PJ, Giaid A (1998) Increased formation of the potent oxidant peroxynitrite in the airways of asthmatic patients is associated with induction of nitric oxide synthase: effect of inhaled glucocorticoid. FASEB J. 12:929–937

Santeliz JV, Van Nest G, Traquina P, Larsen E, Wills-Karp M (2002) Amb a 1-linked CpG oligodeoxynucleotides reverse established airway hyperresponsiveness in a murine model of asthma. J Allergy Clin.Immunol. 109:455–462

Sasaki T, Irie-Sasaki J, Jones RG, Oliveira dSA, Stanford WL, Bolon B, Wakeham A, Itie A, Bouchard D, Kozieradzki I, Joza N, Mak TW, Ohashi PS, Suzuki A, Penninger JM (2000) Function of PI3Kgamma in thymocyte development, T cell activation, and neutrophil migration. Science 287:1040–1046

Sawyer N, Cauchon E, Chateauneuf A, Cruz RP, Nicholson DW, Metters KM, O'Neill GP, Gervais FG (2002) Molecular pharmacology of the human prostaglandin D_2 receptor, CRTH2. Br.J Pharmacol 137:1163–1172

Schafer PH, Wadsworth SA, Wang L, Siekierka JJ (1999) p38α Mitogen-activated protein kinase is activated by CD28-mediated signaling and is required for IL-4 production by human $CD4^+CD45RO^+$ T Cells and Th2 Effector Cells. J.Immunol. 162:7110–7119

Schulman ES, Glaum MC, Post T, Wang Y, Raible DG, Mohanty J, Butterfield JH, Pelleg A (1999) ATP modulates anti-IgE-induced release of histamine from human lung mast cells. Am J Respir Cell Mol.Biol. 20:530–537

Schwarze J, Cieslewicz G, Joetham A, Ikemura T, Hamelmann E, Gelfand EW (1999) CD8 T cells are essential in the development of respiratory syncytial virus-induced lung eosinophilia and airway hyperresponsiveness. J.Immunol. 162:4207–4211

Seeds,E.A.M., Page,C.P. (2001). Heparin inhibits allergen-induced eosinophil infiltration into guinea pig lung via a mechanism unrelated to its anti-coagulant activity. Pulm.Pharmacol.Ther. 14:111–119

Shapiro SD, Senior RM (1999) Matrix metalloproteinases. Matrix degradation and more. Am.J.Respir.Cell Mol.Biol. 20:1100–1102

Shirakawa T, Enomoto T, Shimazu S, Hopkin JM (1997) The inverse association between tuberculin responses and atopic disorder. Science 275:77–79

Shirtcliffe PM, Easthope SE, Cheng S, Weatherall M, Tan PL, Le Gros G, Beasley R (2001) The effect of delipidated deglycolipidated (DDMV) and heat-killed Mycobacterium vaccae in asthma. Am J Respir Crit Care Med 163:1410–1414

Silbaugh SA, Stengel PW, Cockerham SL, Froelich LL, Bendele AM, Spaethe SM, Sofia MJ, Sawyer JS, Jackson WT (2000) Pharmacologic actions of the second generation leukotriene B_4 receptor antagonist LY29311: in vivo pulmonary studies. Naunyn Schmiedebergs Arch Pharmacol 361:397–404

Silkoff PE, Sylvester JT, Zamel N, Permutt S (2000) Airway nitric oxide diffusion in asthma: Role in pulmonary function and bronchial responsiveness. Am.J.Respir.Crit Care Med. 161:1218–1228

Slusarchyk WA, Bolton SA, Hartl KS, Huang MH, Jacobs G, Meng W, Ogletree ML, Pi Z, Schumacher WA, Seiler SM, Sutton JC, Treuner U, Zahler R, Zhao G, Bisacchi GS (2002) Synthesis of potent and highly selective inhibitors of human tryptase. Bioorg.Med Chem.Lett. 12:3235–3238

Sousa AR, Lane SJ, Soh C, Lee TH (1999) In vivo resistance to corticosteroids in bronchial asthma is associated with enhanced phosyphorylation of JUN N-terminal kinase and failure of prednisolone to inhibit JUN N-terminal kinase phosphorylation. J.Allergy Clin.Immunol. 104:565–574

Spina, D., Page, C.P. (2002). Asthma, a need for a rethink. Trends Pharm.Sci. 23(7): 311–315

Stenton GR, Ulanova M, Dery RE, Merani S, Kim MK, Gilchrist M, Puttagunta L, Musat-Marcu S, James D, Schreiber AD, Befus AD (2002) Inhibition of allergic inflammation in the airways using aerosolized antisense to Syk kinase. J Immunol. 169:1028–1036

Stockley RA (1999) Neutrophils and protease/antiprotease imbalance. Am J Respir Crit Care Med 160:S49-S52

Strannegard IL, Larsson LO, Wennergren G, Strannegard O (1998) Prevalence of allergy in children in relation to prior BCG vaccination and infection with atypical mycobacteria. Allergy 53:249–254

Sur S, Wild JS, Choudhury BK, Sur N, Alam R, Klinman DM (1999) Long term prevention of allergic lung inflammation in a mouse model of asthma by CpG oligodeoxynucleotides [In Process Citation]. J.Immunol. 162:6284–6293

Taha R, Olivenstein R, Utsumi T, Ernst P, Barnes PJ, Rodger IW, Giaid A (2000) Prostaglandin H synthase 2 expression in airway cells from patients with asthma and chronic obstructive pulmonary disease. Am J Respir.Crit Care Med 161:636–640

Takeyama K, Dabbagh K, Lee HM, Agusti C, Lausier JA, Ueki IF, Grattan KM, Nadel JA (1999) Epidermal growth factor system regulates mucin production in airways. Proc.Natl.Acad.Sci.U.S.A. 96:3081–3086

Takeyama K, Jung B, Shim JJ, Burgel PR, Dao-Pick T, Ueki IF, Protin U, Kroschel P, Nadel JA (2001) Activation of epidermal growth factor receptors is responsible for mucin synthesis induced by cigarette smoke. Am J Physiol Lung Cell Mol.Physiol 280:L165-L172

Tamaoki J, Kondo M, Sakai N, Aoshiba K, Tagaya E, Nakata J, Isono K, Nagai A (2000) Effect of suplatast tosilate, a Th2 cytokine inhibitor, on steroid-dependent asthma: a double-blind randomised study. Lancet 356:273–278

Thompson AG, Starzl TC (1993) New immunosuppressive drugs: mechanistic insights and potential therapeutic advances. Immunol Rev 136:71–98

Tyrell, D.J. et al. (1999). Heparin in inflammation: potential therapeutic applications beyond anti-coagulation. Adv.Pharmacol. 46:151–208

Underwood DC, Osborn RR, Bochnowicz S, Webb EF, Rieman DJ, Lee JC, Romanic AM, Adams JL, Hay DW, Griswold DE (2000) SB 239063, a p38 MAPK inhibitor, reduces neutrophilia, inflammatory cytokines, MMP-9, and fibrosis in lung. Am J Physiol Lung Cell Mol.Physiol 279:L895-L902

Vaichieri et al. (2001). Intranasal heparin reduces eosinophil recruitment after nasal allergen challenge in patients with allergic rhinitis. J.Allergy Clin.Immunol. 108:703–708

van Neerven RJ, Van de Pol MM, van der Zee JS, Stiekema FE, De Boer M, Kapsenberg ML (1998) Requirement of CD28-CD86 costimulation for allergen-specific T cell proliferation and cytokine expression [see comments]. Clin.Exp.Allergy 28:808–816

Van Oosterhout AJ, Hofstra CL, Shields R, Chan B, van Ark I, Jardieu PM, Nijkamp FP (1997) Murine CTLA4-IgG treatment inhibits airway eosinophilia and hyperresponsiveness and attenuates IgE upregulation in a murine model of allergic asthma. Am.J.Respir.Cell Mol.Biol. 17:386–392

Wakeling AE (2002) Epidermal growth factor receptor tyrosine kinase inhibitors. Curr Opin Pharmacol 2:382–387

Wang CC, Rook GA (1998) Inhibition of an established allergic response to ovalbumin in BALB/c mice by killed *Mycobacterium vaccae*. Immunol 93:307–313

Wheeler JG, Shema SJ, Bogle ML, Shirrell MA, Burks AW, Pittler A, Helm RM (1997) Immune and clinical impact of Lactobacillus acidophilus on asthma. Ann.Allergy Asthma Immunol. 79:229–233

Woessner KM, Simon RA, Stevenson DD (2002) The safety of celecoxib in patients with aspirin-sensitive asthma. Arthritis Rheum. 46:2201–2206

Yokomizo T, Kato K, Terawaki K, Izumi T, Shimizu T (2000) A second leukotriene B_4 receptor, BLT_2. A new therapeutic target in inflammation and immunological disorders. J Exp Med 192:421–432

Yoshida M, Aizawa H, Inoue H, Matsumoto K, Koto H, Komori M, Fukuyama S, Okamoto M, Hara N (2002) Effect of suplatast tosilate on airway hyperresponsiveness and inflammation in asthma patients. J Asthma 39:545–552

Yousefi S, Hoessli DC, Blaser K, Mills GB, Simon HU (1996) Requirement of Lyn and Syk tyrosine kinases for the prevention of apoptosis by cytokines in human eosinophils. J.Exp.Med. 183:1407–1414

Yssel H, Fasler S, Lamb J, de Vries JE (1994) Induction of non-responsiveness in human allergen specific type 2 helper cells. Curr Opin Immunol 6:847–852

Yukawa T, Kroegel C, Dent G, Chanez P, Ukena D, Barnes PJ (1989) Effect of theophylline and adenosine on eosinophil function. Am Rev Respir Dis 140:327–333

Zhang DH, Yang L, Cohn L, Parkyn L, Homer R, Ray P, Ray A (1999) Inhibition of allergic inflammation in a murine model of asthma by expression of a dominant-negative mutant of GATA-3. Immunity 11:473–482

Zheng W, Flavell RA (1997) The transcription factor GATA-3 is necessary and sufficient for Th2 cytokine gene expression in CD4 T cells. Cell 89:587–596

Subject Index

Acetylation 86
Acute exacerbations 46
Adhesion 246
Adrenergic agent 44
β_2-adrenoceptor 4-6, 9, 14, 292
Adrenoceptor agonists 16, 19
β_2-adrenoceptor agonist 4, 10-13, 15, 16, 18, 19, 22, 23-25
– short-acting 10
β_2-adrenoceptor-gene 20
β_2-adrenoceptor polymorphisms 21
ADRs (adverse drug reactions) 289
Adverse effects 135, 139, 143
Adverse reactions 16
Airflow obstruction 288
Airway hyperreactivity 310
Allergen 129, 137
Allergen challenge 140
Allergen-induced asthma 129
ANP (atrial natriuretic peptide) 162
Anti-CD23 274, 277, 278
Anticholinergic agents 38, 41, 48
Antihistamines 258
Anti-IgE 274-278
Anti-inflammatory 53-57, 60, 64-66, 350
Anti-inflammatory actions of antihistamines 137
α_1-antitrypsin deficiency 298
Anti-inflammatory drugs (CSAIDs) 358
Aspirin-induced asthma 130
Asthma 4-6, 9-11, 18-25, 37, 39, 42, 43, 48, 54, 58, 64, 65, 126, 180, 181, 183, 184, 201
– aspirin-induced 130
– severe 328
– symptomatic 325
– symptom diaries 309
Asthmatic phenotype 134
Asthmatics 14, 16
ATP-dependent (K_{ATP}) 165
Atrial natriuretic peptide (ANP) 162
Azelastine 138

BAL and Biopsy 317
Baseline lung function 131
Bronchial hyperresponsiveness 162
Bronchodilation 37, 40, 43, 44, 55
Bronchodilator 42, 45, 47, 54, 319

Calcium-dependent (K_{Ca}) 165
CD11a 255
CD11a/CD18 253
CD11b 255
CD11b/CD18 253
CD18 255
CD23 276
Cetirizine 138
CFTR 298
Charybdotoxin 166
Chemokines 220, 231, 232, 234-237, 258, 350
Chloride channel blockers 139

Chronic obstructive lung disease (COLD)   305
Churg-Strauss syndrome   135
Cilostazol   165
Classification   22, 24
COLD (chronic obstructive lung disease)   305
Cold-air challenge   137
Combination with bronchodilators   46
Comparison with inhaled corticosteroids   132
Comparison with inhaled steroids   141
Comparison with placebo   131, 141
Combination therapy   4, 17
Computerised tomography   317
COPD   4-6, 18, 19, 23-25, 37, 39, 41, 42, 44, 46-48, 54, 58-60, 65, 66, 126, 143, 180, 181
Corticosteroid abrupt discontinuation   327
Corticosteroid resistance   110
Corticosteroid titration studies   327
Corticosteroids   81
CpG   279, 282
- motifs   274, 280, 281
Cromones   126, 139
CSAIDs (anti-inflammatory drugs)   358
Cys-leukotriene 1 receptor   294
Cys-LT$_1$ receptor   128
Cys-LT$_2$ receptor   128
Cysteinyl leukotrienes   127
Cytochrome P450   292
Cytokines   259, 350, 358, 360, 361

Delayed rectifier   166
Delivery   12
- devices   4
DNA vaccination   274, 282
DNA vaccines   274, 278, 280, 281, 282
Dynamic hyperinflation   309, 311

EMEA (European Medicines Evaluation Agency)   305
Enlimomab   259
Eosinophils   161, 248
E-selectin   247
European Medicines Evaluation Agency (EMEA)   305

Exacerbations in asthma   330
Exacerbations in COPD   331
Exercise-induced asthma   129, 137, 325
Exhaled breath   312
Exhaled nitric oxide (NO)   314
Expression profiling   296

FEV$_1$ (Forced expiratory volume in 1 s)   308
FLAP (5-lipoxygenase or 5-lipogenase-activating protein)   128
Forced expiratory volume in 1 s (FEV$_1$)   308
Functional upregulation   256

GAFS (gated autofluorescene forward scatter)   312
GC-B   163
2D gel electrophoresis   296
Gene therapy   297
Genetic variability   291
GINA (Global Initiative on Asthma)   305
Glucocorticoid receptor   295
Glucocorticosteroids   258
Glyceryltrinitrate   160
Glycoprotein   247
Good Clinical Practice   306
Guanylyl   159

H$_1$-receptor antagonists   126
Health-related quality of life (HRQOL)   309
Heparin   258
Herbal Remedies   319
HEV (high endothelial venules)   250
Histamine receptor antagonists   136
Four subtypes of histamine receptor   136
Histone acetylation   84
Histone deacetylation   86
HIV   289
Host defence   261
HRQOL (health-related quality of life)   309
Human genome   290

Iberiotoxin   166
ICAM-1   253

ICAM-2 253
IL-4 220, 224, 225, 227, 231, 235, 237
IL-5 222-224, 227, 231, 235
– monoclonal antibody directed against IL-5 323
IL-10 220, 229, 230, 237
IL-12 230, 231, 237
IL-13 225
Immunglobulin 253
Immunmodulatory 64, 543
Immunostimulatory oligodeoxynucleotide (ISS-ODN) 274
Incremental shuttle-walking test (ISWT) 310
Inflammation 81, 82, 84-88, 93, 94, 96, 110, 111, 113, 180, 181, 189, 352, 354, 356, 361
Inflammatory 54
Inflammatory diseases 357
Inflammatory mediators 127, 350
Inflammatory process 360
Inflammatory response 352, 360, 361
Inhaled allergen challenge 321
Inhaled corticosteroids 326
INOS 158
β_1-integrins 257
β_2-integrins 255
Interleukin-10 (see IL-10)
Interleukin-12 (see IL-12)
Interleukin-4 (see IL-4)
Interleukin-5 (see IL-5)
Interleukin-13 (see IL-13)
Inward rectifier (K_{IR}) 165
Ipratropium 40-44, 47
Ischaemia-reperfusion 259
ISS-ODN (Immunostimulatory oligodeoxynucleotide) 274, 279-282
ISWT (incremental shuttle-walking test) 310

K_{ATP} channel 167
K_{Ca} channel 166
Ketotifen 138
K_{IR} 169
K_{IR} family 168

LAD-1 258
LAD-2 258

Leucocyte adhesion deficiency 256
Leucocyte-leucocyte interactions 252
Leucocytes 246
Leukotriene-receptor 126
Leukotriene-receptor antagonist 127
Leukotriene synthesis Inhibitors 127
LFA-1 255
Linkage 297
5-lipoxygenase 294
– inhibitor 126
5-lipoxygenase or 5-lipogenase-activating protein (FLAP) 128
$_L$-NAME 162
Long-acting 17
Loratadine 138
L-selectin 251
L-type calcium channel blockers 167
Lymphocyte homing 255
Lymphocytes 248

Mac-1 255
MAdCAM 253
Medical needs 305
6-min walk test (6MWT) 310
MKS492 165
Monocytes 248
Montelukast 128
Mortality 319
Muscarinic M_2 294
Muscarinic M_3 294
Muscarinic receptors 38, 39
Muscle testing 311
Muscle wasting 330

N-acetyltransferase 292
Nasal allergen challenge 324
Natural history and disease modification 329
Nedocromil sodium 140
Neutrophils 248
Nitric oxide (NO) 158
(e)NOS 158
NO (exhaled nitric oxide) 314
NOS2 158, 161
NOS3 158
NPR-A 163
NPR-C 163
NS1619 167

Obstructive pulmonary disease 4
ODQ 159
Olprinone 165
Oral corticosteroid sparing effect 143
Oxitropium 37

PDE (phosphodiesterase) 53, 60, 62, 63, 164, 180-188, 190-192, 196, 197, 199-202, 258, 295
PDE inhibition 189
PDE inhibitors 180, 194, 201
PDE3 (SKF94120) 164
Peak expiratory flow 308
PECAM-1 253
Pharmacogenomics 296
Pharmacogenetics 19, 288, 318
Pharmacokinetic/pharmacodynamic (PK/PD) 311
Pharmacokinetics 292
Phases I to IV 305
Phosphodiesterase (PDE) 53, 60, 62, 63, 164, 180-188, 190-192, 196, 197, 199-202, 258, 295
Phosphodiesterase (PDE) inhibitor 180, 194, 201
PK/PD (pharmacokinetic/pharmacodynamic) 311
Plasmid DNA 274, 281
Plasmid vaccination 280
Platelets 249
P-selectin 249
PSGL-1 249

Regulatory Guidelines 307
Ro 25-1553 157

Schizophrenia 288
Siayl Lewis x 247
Single nucleotide polymorphism (SNP) 290

Site of action of leukotriene synthesis inhibitors and receptors antagonists 127
S-nitrosothiols 160
SNP (single nucleotide polymorphism) 290
Sodium cromoglycate 139
Soluble adhesion molecules 253
Sputum analysis 315
Stable COPD 44
Statin derivatives 260
Steroid resistance 83, 99, 111

TEA (tetraethylammonium) 166
Terfenadine 138
Tetraethylammonium (TEA) 166
Theophylline 53-55, 59-63, 65, 66
Thiopurine methyl transferase 292
Tinkelman 58
Tiotropium 37, 38, 40, 41, 46, 48
TNF-α 220, 228, 229, 237
Tolerability 319
Transcription factors 84-86, 88, 114
Translocation of P-selectin 251

Vagal activity 39
Vagal afferents 38
Vascular endothelium 246
Vasoactive intestinal peptide (VIP) 154
VCAM-1 253
VIP (vasoactive intestinal peptide) 154
VLA-4 254
Voltage-dependent K 165
VPAC1 156
VPAC2 156

Weight loss 330

Zafirlukast 128
Zileuton 129